sociology

9

sociology

exploring the architecture of everyday life | readings

9

david m. newman
DePauw University

jodi o'brien
Seattle University

Los Angeles | London | New Delhi
Singapore | Washington DC

Los Angeles | London | New Delhi
Singapore | Washington DC

FOR INFORMATION:

SAGE Publications, Inc.
2455 Teller Road
Thousand Oaks, California 91320
E-mail: order@sagepub.com

SAGE Publications Ltd.
1 Oliver's Yard
55 City Road
London EC1Y 1SP
United Kingdom

SAGE Publications India Pvt. Ltd.
B 1/I 1 Mohan Cooperative Industrial Area
Mathura Road, New Delhi 110 044
India

SAGE Publications Asia-Pacific Pte. Ltd.
3 Church Street
#10-04 Samsung Hub
Singapore 049483

Acquisitions Editor: David Repetto
Editorial Assistant: Lauren Johnson
Production Editor: Laureen Gleason
Copy Editor: Erin Livingston
Typesetter: C&M Digitals (P) Ltd.
Proofreader: Ellen Howard
Cover Designer: Candice Harman
Marketing Manager: Erica DeLuca
Permissions Editor: Karen Ehrmann

Printed in the United States of America

Library of Congress Cataloging-in-Publication Data

Sociology : exploring the architecture of everyday life : readings / editors, David M. Newman, Jodi O'Brien. — 9th ed.

p. cm.
Includes bibliographical references.

ISBN 978-1-4129-8760-8 (pbk.)

1. Sociology. I. Newman, David M., 1958– II. O'Brien, Jodi.

HM586.S64 2013
301—dc23 2012031247

This book is printed on acid-free paper.

SUSTAINABLE FORESTRY INITIATIVE
Certified Chain of Custody
Promoting Sustainable Forestry
www.sfiprogram.org
SFI-01268
SFI label applies to text stock

12 13 14 15 16 10 9 8 7 6 5 4 3 2 1

Contents

Preface ix

About the Editors xi

PART I.
THE INDIVIDUAL AND SOCIETY 1

Chapter 1. Taking a New Look at a Familiar World 3

Reading 1.1. The Sociological Imagination 5
C. Wright Mills

Reading 1.2. Invitation to Sociology 10
Peter Berger

Reading 1.3. The My Lai Massacre: A Military Crime of Obedience 14
Herbert Kelman and V. Lee Hamilton

Chapter 2. Seeing and Thinking Sociologically 27

Reading 2.1. The Metropolis and Mental Life 29
Georg Simmel

Reading 2.2. Gift and Exchange 35
Zygmunt Bauman

Reading 2.3. Culture of Fear 44
Barry Glassner

PART II.
THE CONSTRUCTION OF SELF AND SOCIETY 57

Chapter 3. Building Reality: The Social Construction of Knowledge 59

Reading 3.1. Concepts, Indicators, and Reality 61
Earl Babbie

Reading 3.2. Missing Numbers 65
Joel Best

Chapter 4. Building Order: Culture and History 75

Reading 4.1. Body Ritual among the Nacirema 77
Horace Miner

Reading 4.2. The Melting Pot 81
Anne Fadiman

Reading 4.3. McDonald's in Hong Kong: Consumerism, Dietary Change, and the Rise of a Children's Culture 91
James L. Watson

Chapter 5. Building Identity: Socialization 99

Reading 5.1. Life as the Maid's Daughter: An Exploration of the Everyday Boundaries of Race, Class, and Gender 101
Mary Romero

Reading 5.2. The Making of Culture, Identity, and Ethnicity Among Asian American Youth 110
Min Zhou and Jennifer Lee

Reading 5.3. Working 'the Code': On Girls, Gender, and Inner-City Violence 118
Nikki Jones

Chapter 6. Supporting Identity: The Presentation of Self 127

Reading 6.1. The Presentation of Self in Everyday Life: Selections 129
Erving Goffman

Reading 6.2. Public Identities: Managing Race in Public Spaces 139
Karyn Lacy

Reading 6.3. The Girl Hunt: Urban Nightlife and the Performance of Masculinity as Collective Activity 152
David Grazian

Chapter 7. Building Social Relationships: Intimacy and Family 161

Reading 7.1. The Radical Idea of Marrying for Love 163
Stephanie Coontz

Reading 7.2. Gay Parenthood and the End of Paternity as We Knew It 174
Judith Stacey

Reading 7.3. Covenant Marriage: Reflexivity and Retrenchment in the Politics of Intimacy 189
Dwight Fee

Chapter 8. Constructing Difference: Social Deviance 195

Reading 8.1. Watching the Canary 197
Lani Guinier and Gerald Torres

Reading 8.2. Healing (Disorderly) Desire: Medical-Therapeutic Regulation of Sexuality 201
P. J. McGann

Reading 8.3. Patients, "Potheads," and Dying to Get High 212
Wendy Chapkis

PART III.

SOCIAL STRUCTURE, INSTITUTIONS, AND EVERYDAY LIFE 221

Chapter 9. The Structure of Society: Organizations and Social Institutions 223

Reading 9.1. These Dark Satanic Mills 225
William Greider

Reading 9.2. The Smile Factory: Work at Disneyland 235
John Van Maanen

Reading 9.3. Creating Consumers: Freaks, Geeks, and Cool Kids 245
Murry Milner

Chapter 10. The Architecture of Stratification: Social Class and Inequality 253
Reading 10.1. Making Class Invisible 255
Gregory Mantsios
Reading 10.2. The Compassion Gap in American Poverty Policy 262
Fred Block, Anna C. Korteweg, and Kerry Woodward, with Zach Schiller and Imrul Mazid
Reading 10.3. Branded With Infamy: Inscriptions of Poverty and Class in America 271
Vivyan Adair

Chapter 11. The Architecture of Inequality: Race and Ethnicity 283
Reading 11.1. Racial and Ethnic Formation 285
Michael Omi and Howard Winant
Reading 11.2. Optional Ethnicities: For Whites Only? 292
Mary C. Waters
Reading 11.3. Silent Racism: Passivity in Well-Meaning White People 299
Barbara Trepagnier

Chapter 12. The Architecture of Inequality: Sex and Gender 309
Reading 12.1. Black Women and a New Definition of Womanhood 311
Bart Landry
Reading 12.2. Still a Man's World: Men Who Do "Women's Work" 323
Christine L. Williams
Reading 12.3. New Biomedical Technologies, New Scripts, New Genders 333
Eve Shapiro

Chapter 13. Global Dynamics and Population Demographic Trends 347
Reading 13.1. Age-Segregation in Later Life: An Examination of Personal Networks 349
Peter Uhlenberg and Jenny de Jong Gierveld
Reading 13.2. Love and Gold 357
Arlie Russell Hochschild
Reading 13.3. Cyberbrides and Global Imaginaries: Mexican Women's Turn from the National to the Foreign 365
Felicity Schaeffer-Grabiel

Chapter 14. The Architects of Change: Reconstructing Society 377
Reading 14.1. Muslim American Immigrants After 9/11: The Struggle for Civil Rights 379
Pierrette Hondagneu-Sotelo
Reading 14.2. The Seattle Solidarity Network: A New Approach to Working Class Social Movements 388
Walter Winslow
Reading 14.3. "Aquí estamos y no nos vamos!" Global Capital and Immigrant Rights 400
William I. Robinson

Credits 411

Preface

One of the greatest challenges we face as teachers of sociology is getting our students to see the relevance of the course material to their own lives and to fully appreciate its connection to the larger society. We teach our students to see that sociology is all around us. It's in our families, our careers, our media, our jobs, our classrooms, our goals, our interests, our desires, and even our minds. Sociology can be found at the neighborhood pub, in conversation with the clerk at 7-Eleven, on a date, and in the highest offices of government. It's with us when we're alone and when we're in a group of people. Sociology focuses on questions of global significance as well as private concerns. For instance, sociologists study how some countries create and maintain dominance over others and also why we find some people more attractive than others. Sociology is an invitation to understand yourself within the context of your historical and cultural circumstances.

We have compiled this collection of short articles, chapters, and excerpts with the intent of providing comprehensive examples of the power of sociology for helping us to make sense of our lives and our times. The readings are organized in a format that demonstrates

- the uniqueness of the sociological perspective
- tools of sociological analysis
- the significance of different cultures in a global world
- social factors that influence identity development and self-management
- social rules about family, relationships, and belonging
- the influence of social institutions and organizations on everyday life
- the significance of socioeconomic class, gender, and racial/ethnic backgrounds in everyday life
- the significance of social demographics, such as aging populations and migration
- the power of social groups and social change

In general, our intent is to demonstrate the significance of sociology in everyday life and to show that what seems "obvious" is often not-so-obvious when subjected to rigorous sociological analysis. The metaphor of "architecture" used in the title for this reader illustrates the sociological idea that as social beings, we are constantly building and rebuilding our own social environment. The sociological promise is that if we understand these processes and how they affect us, we will be able to make more informed choices about how to live our lives and engage in our communities.

As in the first eight editions of the reader, the selections in this edition are intended to be vivid, provocative, and eye-opening examples of the practice of sociology. The readings represent a variety of styles. Some use common or everyday experiences and phenomena (such as drug use, employment, athletic performance, religious devotion, eating fast food, and the balance of work and family) to illustrate the relationship between the individual and society. Others focus on important social issues or problems (medical social control, race relations, poverty, educational inequalities, sexuality, immigration, global economics, environmental degradation, or political extremism) or on specific historical events (massacres during war, drug scares, and 9/11). Some were written quite recently; others are sociological classics. In addition to accurately representing the sociological perspective and providing rigorous coverage of the discipline, we hope the selections are thought-provoking, generate lots of discussion, and are enjoyable to read.

There are 41 selections in this reader, and 12 of them are new. These new readings focus on current, important social issues such as the pace of life in urban societies, gift exchange, media manipulation of statistics, status performance and race, parenting among same-sex couples, marriage promotion, cyberbrides, sexual regulation and consumer culture in high schools, gender technology in historical context, racism among well-intended white people, and working-class social movement tactics.

Most of the new readings are based on research studies that were written in the past 5 years. In recent editions of this reader, we have increased the number of selections drawn from contemporary social research. In doing so, we hope to provide you with illustrations of the ways in which social researchers combine theories and empirical studies to gain a better understanding of social patterns and processes. Although the professional language of some of these selections may seem challenging for introductory readers, we are confident that you will find them highly relevant and come away with a sense of being immersed in the most significant details of contemporary sociology.

To help you get the most out of these selections, we've written brief introductions that provide the sociological context for each chapter. We also included reflection points that can be used for comparing and contrasting the readings in each section and across sections. For those of you who are also reading the accompanying textbook, these introductions will furnish a quick intellectual link between the readings and information in the textbook. We have also included in these introductions brief instructions on what to look for when you read the selections in a given chapter. After each reading, you will find a set of discussion questions to ponder. Many of these questions ask you to apply a specific author's conclusions to some contemporary issue in society or to your own life experiences. It is our hope that these questions will generate a lot of classroom debate and help you see the sociological merit of the readings.

A website established for this ninth edition includes do-it-yourself reviews and tests for students, web-based activities designed to enhance learning, and a chat room where students and teachers can post messages and debate matters of sociological significance. The site can be accessed via the Pine Forge website at www.pineforge.com.

Books like these are enormous projects. We would like to thank David Repetto, Laureen Gleason, Erin Livingston, and the rest of the staff at SAGE for their useful advice and assistance in putting this reader together. It's always a pleasure to work with this very professional group. Thanks again to Jennifer Hamann for her assistance with reading selections and editing. Michelle Robertson joins us in this edition as a contributing editor, and we are especially grateful for her input.

Enjoy!

David M. Newman
Department of Sociology/Anthropology
DePauw University
Greencastle, IN 46135
E-mail: dnewman@depauw.edu

Jodi O'Brien
Department of Sociology
Seattle University
Seattle, WA 98122
E-mail: jobrien@seattleu.edu

About the Editors

David M. Newman (PhD, University of Washington) is Professor of Sociology at DePauw University. In addition to the introductory course, he teaches courses in research methods, family, social psychology, deviance, and mental illness. He has won teaching awards at both the University of Washington and DePauw University. His other written work includes *Identities and Inequalities: Exploring the Intersections of Race, Class, Gender, and Sexuality* (2012) and *Families: A Sociological Perspective* (2008).

Jodi O'Brien (PhD, University of Washington) is Professor of Sociology at Seattle University. She teaches courses in social psychology, sexuality, inequality, and classical and contemporary theory. She writes and lectures on the cultural politics of transgressive identities and communities. Her other books include *Everyday Inequalities* (Basil Blackwell), *Social Prisms: Reflections on Everyday Myths and Paradoxes* (SAGE), and *The Production of Reality: Essays and Readings on Social Interaction* (5th edition, SAGE).

PART I

The Individual and Society

Taking a New Look at a Familiar World

The primary claim of sociology is that our everyday feelings, thoughts, and actions are the product of a complex interplay between massive social forces and personal characteristics. We can't understand the relationship between individuals and their societies without understanding the connection between both. As C. Wright Mills discusses in the introductory article, the "sociological imagination" is the ability to see the impact of social forces on our private lives. When we develop a sociological imagination, we gain an awareness that our lives unfold at the intersection of personal biography and social history. The sociological imagination encourages us to move beyond individualistic explanations of human experiences to an understanding of the mutual influence between individuals and society. So rather than study what goes on within people, sociologists study what goes on between and among people as individuals, groups, organizations, or entire societies. Sociology teaches us to look beyond individual personalities and focus instead on the influence of social phenomena in shaping our ideas of who we are and what we think we can do.

Peter Berger, another well-known sociologist, invites us to consider the uniqueness of the sociological enterprise. According to Berger, the sociologist is driven by an insatiable curiosity to understand the social conditions that shape human behavior. The sociologist is also prepared to be surprised, disturbed, and sometimes even bored by what he or she discovers. In this regard, the sociologist is driven to make sense of the seemingly obvious with the understanding that once explored, it may not be so obvious after all. One example of the nonobvious is the influence that social institutions have on our behavior. It's not always easy to see this influence. We have a tendency to see people's behavior in individualistic, sometimes even biological, terms. This tendency toward individualistic explanations is particularly pronounced in U.S. society.

The influence of social institutions on our personal lives is often felt most forcefully when we are compelled to obey the commands of someone who is in a position of institutional authority. The social institution with the most explicit hierarchy of authority is the military. In "The My Lai Massacre: A Military Crime of Obedience," Herbert Kelman and V. Lee Hamilton describe a specific example of a crime in which the individuals involved attempted to deny responsibility for their actions by claiming that they were following the orders of a military officer who had the legitimate right to command them. This incident occurred in the midst of the Vietnam War. Arguably, people do things under such trying conditions that they wouldn't ordinarily do, even—as in this case—kill defenseless people. Kelman and Hamilton make a key sociological point by showing that these soldiers were not necessarily psychological misfits who were especially mean or violent. Instead, the researchers argue, they were ordinary people caught up in tense circumstances that made obeying the brutal commands of an authority seem like the normal and morally acceptable thing to do.

Something to Consider as You Read

As you read these selections, consider the effects of social context and situation on behavior. Even though it might appear extreme, how might the behavior of these soldiers be similar to other examples of social influence? Consider occasions in which you have done something publicly that you didn't feel right about personally. How do you explain your behavior? How might a sociologist explain your behavior?

The Sociological Imagination

C. Wright Mills

(1959)

"The individual can . . . know his own chances in life only by becoming aware of those of all individuals in his circumstances."

Nowadays men often feel that their private lives are a series of traps. They sense that within their everyday worlds, they cannot overcome their troubles, and in this feeling, they are often quite correct: What ordinary men are directly aware of and what they try to do are bounded by the private orbits in which they live; their visions and their powers are limited to the close-up scenes of job, family, neighborhood; in other milieux, they move vicariously and remain spectators. And the more aware they become, however vaguely, of ambitions and of threats which transcend their immediate locales, the more trapped they seem to feel.

Underlying this sense of being trapped are seemingly impersonal changes in the very structure of continent-wide societies. The facts of contemporary history are also facts about the success and the failure of individual men and women. When a society is industrialized, a peasant becomes a worker; a feudal lord is liquidated or becomes a businessman. When classes rise or fall, a man is employed or unemployed; when the rate of investment goes up or down, a man takes new heart or goes broke. When wars happen, an insurance salesman becomes a rocket launcher; a store clerk, a radar man; a wife lives alone; a child grows up without a father. Neither the life of an individual nor the history of a society can be understood without understanding both.

Yet men do not usually define the troubles they endure in terms of historical change and institutional contradiction. The well-being they enjoy, they do not usually impute to the big ups and downs of the societies in which they live. Seldom aware of the intricate connection between the patterns of their own lives and the course of world history, ordinary men do not usually know what this connection means for the kinds of men they are becoming and for the kinds of history-making in which they might take part. They do not possess the quality of mind essential to grasp the interplay of man and society, of biography and history, of self and world. They cannot cope with their personal troubles in such ways as to control the structural transformations that usually lie behind them.

Surely it is no wonder. In what period have so many men been so totally exposed at so fast a pace to such earthquakes of change? That Americans have not known such catastrophic changes as have the men and women of other societies is due to historical facts that are now quickly becoming "merely history." The history that now affects every man is world history. Within this scene and this period, in the course of a single generation, one-sixth of mankind is transformed from all that is feudal and backward into all that is modern, advanced, and fearful. Political colonies are freed, new and less visible forms of imperialism installed. Revolutions occur; men feel the intimate grip of new kinds of authority. Totalitarian societies rise, and are smashed to bits—or succeed fabulously. After two centuries of ascendancy, capitalism is shown up as only one way to make society into an industrial apparatus. After two centuries of hope, even formal democracy is restricted to a quite small portion of mankind. Everywhere in the underdeveloped world, ancient ways of life are broken up and vague expectations become urgent demands. Everywhere in the overdeveloped world, the means of authority and of violence become total in scope and bureaucratic in form. Humanity itself now lies before us, the supernation at either pole concentrating its most coordinated and massive efforts upon the preparation of World War Three.

The very shaping of history now outpaces the ability of men to orient themselves in accordance with cherished values. And which values? Even when they do not panic, men often sense that older ways of feeling and thinking have collapsed and that newer beginnings are ambiguous to the point of moral stasis. Is it any wonder that ordinary men feel they cannot cope with the larger worlds with which they are so suddenly confronted? That they cannot understand the meaning of their epoch for their own lives? That—in defense of selfhood—they become morally insensible, trying to remain altogether private men? Is it any wonder that they come to be possessed by a sense of the trap?

It is not only information that they need—in this Age of Fact, information often dominates their attention and overwhelms their capacities to assimilate it. It is not only the skills of reason that they need—although their struggles to acquire these often exhaust their limited moral energy.

What they need, and what they feel they need, is a quality of mind that will help them to use information and to develop reason in order to achieve lucid summations of what is going on in the world and of what may be happening within themselves. It is this quality, I am going to contend, that journalists and scholars, artists and publics, scientists and editors are coming to expect of what may be called the sociological imagination.

The sociological imagination enables its possessor to understand the larger historical scene in terms of its meaning for the inner life and the external career of a variety of individuals. It enables him to take into account how individuals, in the welter of their daily experience, often become falsely conscious of their social positions. Within that welter, the framework of modern society is sought, and within that framework the psychologies of a variety of men and women are formulated. By such means the personal uneasiness of individuals is focused upon explicit troubles and the indifference of publics is transformed into involvement with public issues.

The first fruit of this imagination—and the first lesson of the social science that embodies it—is the idea that the individual can understand his own experience and gauge his own fate only by locating himself within his period, that he can know his own chances in life only by becoming aware of those of all individuals in his circumstances. In many ways it is a terrible lesson; in many ways a magnificent one. We do not know the limits of man's capacities for supreme effort or willing degradation, for agony or glee, for pleasurable brutality or the sweetness of reason. But in our time we have come to know that the limits of "human nature" are frighteningly broad. We have come to know that every individual lives, from one generation to the next, in some society; that he lives out a biography, and that he lives it out within some historical sequence. By the fact of his living he contributes, however minutely, to the shaping of this society and to the course of its history, even as he is made by society and by its historical push and shove.

The sociological imagination enables us to grasp history and biography and the relations between the two within society. That is its task and its promise. To recognize this task and this promise is the mark of the classic social analyst. It is characteristic of Herbert Spencer—turgid, polysyllabic, comprehensive; of E. A. Ross—graceful, muckraking, upright; of Auguste Comte and Emile Durkheim; of the intricate and subtle Karl Mannheim. It is the quality of all that is intellectually excellent in Karl Marx; it is the clue to Thorstein Veblen's brilliant and ironic insight, to Joseph Schumpeter's many-sided constructions of reality; it is the basis of the psychological sweep of W. E. H. Lecky no less than of the profundity and clarity of Max Weber. And it is the signal of what is best in contemporary studies of man and society.

No social study that does not come back to the problems of biography, of history, and of their intersections within a society has completed its intellectual journey. Whatever the specific problems of the classic social analysts, however limited or however broad the features of social reality they have examined, those who have been imaginatively aware of the promise of their work have consistently asked three sorts of questions:

1. What is the structure of this particular society as a whole? What are its essential components, and how are they related to one another? How does it differ from other varieties of social order? Within it, what is the meaning of any particular feature for its continuance and for its change?

2. Where does this society stand in human history? What are the mechanics by which it is changing? What is its place within and its meaning for the development of humanity as a whole? How does any particular feature we are examining affect, and how is it affected by, the historical period in which it moves? And this period—what are its essential features? How does it differ from other periods? What are its characteristic ways of history making?

3. What varieties of men and women now prevail in this society and in this period? And what varieties are coming to prevail? In what ways are they selected and formed, liberated and repressed, made sensitive and blunted? What kinds of "human nature" are revealed in the conduct and character we observe in this society in this period? And what is the meaning for "human nature" of each and every feature of the society we are examining?

Whether the point of interest is a great power state or a minor literary mood, a family, a prison, a creed—these are the kinds of questions the best social analysts have asked. They are the intellectual pivots of classic studies of man in society—and they are the questions inevitably raised by any mind possessing the sociological imagination. For that imagination is the capacity to shift from one perspective to another—from the political to the psychological; from examination of a single family to comparative assessment of the national budgets of the world; from the theological school to the military establishment; from considerations of an oil industry to studies of contemporary poetry. It is the capacity to range from the most impersonal and remote transformations to the most intimate features of the human self—and to see the relations between the two. Back of its use there is always the urge to know the social and historical meaning of the individual in the society and in the period in which he has his quality and his being.

That, in brief, is why it is by means of the sociological imagination that men now hope to grasp what is going on in the world, and to understand what is happening in themselves as minute points of the intersections of biography and history within society. In large part, contemporary man's self-conscious view of himself as at least an outsider, if not a permanent stranger, rests upon an absorbed realization of social relativity and of the transformative power of history. The sociological imagination is the most fruitful form of this self-consciousness. By its use men whose mentalities have swept only a series of limited orbits often come to feel as if suddenly awakened in a house with which they had only supposed themselves to be familiar. Correctly or incorrectly, they often come to feel that they can now provide themselves with adequate summations, cohesive assessments, comprehensive orientations. Older decisions that once appeared sound now seem to them products of a mind unaccountably dense. Their capacity for astonishment is made lively again. They acquire a new way of thinking, they experience a transvaluation of values: in a word, by their reflection and by their sensibility, they realize the cultural meaning of the social sciences.

Perhaps the most fruitful distinction with which the sociological imagination works is between "the personal troubles of milieu" and "the public issues of social structure." This distinction is an essential tool of the sociological imagination and a feature of all classic work in social science.

Troubles occur within the character of the individual and within the range of his immediate relations with others; they have to do with his self and with those limited areas of social life of which he is directly and personally aware. Accordingly, the statement and the resolution of troubles properly lie within the individual as a biographical entity and within the scope of his immediate milieu—the social setting that is directly open to his personal experience and to some extent his willful activity. A trouble is a private matter: values cherished by an individual are felt by him to be threatened.

Issues have to do with matters that transcend these local environments of the individual and the range of his inner life. They have to do with the organization of many such milieux into the institutions of an historical society as a whole, with the ways in which various milieux overlap and interpenetrate to form the larger structure of social and historical life. An issue is a public matter: some value cherished by publics is felt to be threatened. Often there is a debate about what that value really is and about what it is that really threatens it. This debate is often without focus if only because it is the very nature of an issue, unlike even widespread trouble, that it cannot very well be defined in terms of the immediate and everyday environments of ordinary men. An issue, in fact, often involves a crisis in institutional arrangements, and often too it involves what Marxists call "contradictions" or "antagonisms."

In these terms, consider unemployment. When, in a city of 100,000, only one man is unemployed, that is his personal trouble, and for its relief we properly look to the character of the man, his skills, and his immediate opportunities. But when in a nation of 50 million employees, 15 million men are unemployed, that is an issue, and we may not hope to find its solution within the range of opportunities open to any one individual. The very structure of opportunities has collapsed. Both the correct statement of the problem and the range of possible solutions require us to consider the economic and political institutions of the society, and not merely the personal situation and character of a scatter of individuals.

Consider war. The personal problem of war, when it occurs, may be how to survive it or how to die in it with honor; how to make money out of it; how to climb into the higher safety of the military apparatus; or how to contribute to the war's termination. In short, according to one's values, to find a set of milieux and within it to survive the war or make one's death in it meaningful. But the structural issues of war have to do with its causes; with what types of men it throws up into command; with its effects upon economic and political, family, and religious institutions, with the unorganized irresponsibility of a world of nation-states.

Consider marriage. Inside a marriage a man and a woman may experience personal troubles, but when the divorce rate during the first four years of marriage is 250 out of every 1,000 attempts, this is an indication of a structural issue having to do with the institutions of marriage and the family and other institutions that bear upon them.

Or consider the metropolis—the horrible, beautiful, ugly, magnificent sprawl of the great city. For many upper-class people, the personal solution to "the problem of the city" is to have an apartment with private garage under it in the heart of the city, and forty miles out, a house by Henry Hill, garden by Garrett Eckbo, on a hundred acres of private land. In these two controlled environments—with a small staff at each end and a private helicopter connection—most people could solve many of the problems of personal milieux caused by the facts of the city. But all this, however splendid, does not solve the public issues that the structural fact of the city poses. What should be done with this wonderful monstrosity? Break it all up into scattered units, combining residence and work? Refurbish it as it stands? Or, after evacuation, dynamite it and build new cities according to new plans in new places? What should those plans be? And who is to decide and to accomplish whatever choice is made? These are structural issues; to confront them and to solve them requires us to consider political and economic issues that affect innumerable milieux.

Insofar as an economy is so arranged that slumps occur, the problem of unemployment becomes incapable of personal solution. Insofar as war is inherent in the nation-state system and in the uneven industrialization of the world, the ordinary individual in his restricted milieu will be powerless—with or without psychiatric aid—to solve the troubles this system or lack of system imposes upon him. Insofar as the family as an institution turns women into darling little slaves and men into their chief providers and unweaned dependents, the problem of a satisfactory marriage remains incapable of purely private solution. Insofar as the overdeveloped megalopolis and the overdeveloped automobile are built-in features of the overdeveloped society, the issues of urban living will not be solved by personal ingenuity and private wealth.

What we experience in various and specific milieux, I have noted, is often caused by structural changes. Accordingly, to understand the changes of many personal milieux we are required to look beyond them. And the number and variety of such structural changes increase as the institutions within which we live become more embracing and more intricately connected with one another. To be aware of the idea of social structure and to use it with sensibility is to be capable of tracing such linkages among a great variety of milieux. To be able to do that is to possess the sociological imagination. . . .

THINKING ABOUT THE READING

Consider the political, economic, familial, and cultural circumstances into which you were born. Make a list of some of these circumstances and also some of the major historical events that have occurred in your lifetime. How do you think these historical and social circumstances may have affected your personal biography? Can you think of ways in which your actions have influenced the course of other people's lives? Identify some famous people and consider how the intersection of history and biography led them to their particular position. How might the outcome have differed if some of the circumstances in their lives were different?

Invitation to Sociology

Peter Berger

(1963)

We would say then that the sociologist (that is, the one we would really like to invite to our game) is a person intensively, endlessly, shamelessly interested in the doings of men. His natural habitat is all the human gathering places of the world, wherever men* come together. The sociologist may be interested in many other things. But his consuming interest remains in the world of men, their institutions, their history, their passions. He will naturally be interested in the events that engage men's ultimate beliefs, their moments of tragedy and grandeur and ecstasy. But he will also be fascinated by the commonplace, the everyday. He will know reverence, but this reverence will not prevent him from wanting to see and to understand. He may sometimes feel revulsion or contempt. But this also will not deter him from wanting to have his questions answered. The sociologist, in his quest for understanding, moves through the world of men without respect for the usual lines of demarcation. Nobility and degradation, power and obscurity, intelligence and folly—these are equally *interesting* to him, however unequal they may be in his personal values or tastes. Thus his questions may lead him to all possible levels of society, the best and the least known places, the most respected and the most despised. And, if he is a good sociologist, he will find himself in all these places because his own questions have so taken possession of him that he has little choice but to seek for answers.

We could say that the sociologist, but for the grace of his academic title, is the man who must listen to gossip despite himself, who is tempted to look through keyholes, to read other people's mail, to open closed cabinets. What interests us is the curiosity that grips any sociologist in front of a closed door behind which there are human voices. If he is a good sociologist, he will want to open that door, to understand these voices. Behind each closed door he will anticipate some new facet of human life not yet perceived and understood.

The sociologist will occupy himself with matters that others regard as too sacred or as too distasteful for dispassionate investigation. He will find rewarding the company of priests or of prostitutes, depending not on his personal preferences but on the questions he happens to be asking at the moment. He will also concern himself with matters that others may find much too boring. He will be interested in the human interaction that goes with warfare or with great intellectual discoveries, but also in the relations between people employed in a restaurant or between a group of little girls playing with their dolls. His main focus of attention is not the ultimate significance of what men do, but the action in itself, as another example of the infinite richness of human conduct.

In these journeys through the world of men the sociologist will inevitably encounter other professional Peeping Toms. Sometimes these will resent his presence, feeling that he is poaching on their preserves. In some places the sociologist will meet up with the economist, in others with the political scientist, in yet others with the psychologist or the ethnologist. Yet chances are that the questions

*To be understood as people or persons.

that have brought him to these same places are different from the ones that propelled his fellow trespassers. The sociologist's questions always remain essentially the same: "What are people doing with each other here?" "What are their relationships to each other?" "How are these relationships organized in institutions?" "What are the collective ideas that move men and institutions?" In trying to answer these questions in specific instances, the sociologist will, of course, have to deal with economic or political matters, but he will do so in a way rather different from that of the economist or the political scientist. The scene that he contemplates is the same human scene that these other scientists concern themselves with. But the sociologist's angle of vision is different.

Much of the time the sociologist moves in sectors of experience that are familiar to him and to most people in his society. He investigates communities, institutions and activities that one can read about every day in the newspapers. Yet there is another excitement of discovery beckoning in his investigations. It is not the excitement of coming upon the totally unfamiliar, but rather the excitement of finding the familiar becoming transformed in its meaning. The fascination of sociology lies in the fact that its perspective makes us see in a new light the very world in which we have lived all our lives. This also constitutes a transformation of consciousness. Moreover, this transformation is more relevant existentially than that of many other intellectual disciplines, because it is more difficult to segregate in some special compartment of the mind. The astronomer does not live in the remote galaxies, and the nuclear physicist can, outside his laboratory, eat and laugh and marry and vote without thinking about the insides of the atom. The geologist looks at rocks only at appropriate times, and the linguist speaks English with his wife. The sociologist lives in society, on the job and off it. His own life, inevitably, is part of his subject matter. Men being what they are, sociologists too manage to segregate their professional insights from their everyday affairs. But it is a rather difficult feat to perform in good faith.

The sociologist moves in the common world of men, close to what most of them would call real. The categories he employs in his analyses are only refinements of the categories by which other men live—power, class, status, race, ethnicity. As a result, there is a deceptive simplicity and obviousness about some sociological investigations. One reads them, nods at the familiar scene, remarks that one has heard all this before and don't people have better things to do than to waste their time on truisms—until one is suddenly brought up against an insight that radically questions everything one had previously assumed about this familiar scene. This is the point at which one begins to sense the excitement of sociology.

Let us take a specific example. Imagine a sociology class in a Southern college where almost all the students are white Southerners. Imagine a lecture on the subject of the racial system of the South. The lecturer is talking here of matters that have been familiar to his students from the time of their infancy. Indeed, it may be that they are much more familiar with the minutiae of this system than he is. They are quite bored as a result. It seems to them that he is only using more pretentious words to describe what they already know. Thus he may use the term "caste," one commonly used now by American sociologists to describe the Southern racial system. But in explaining the term he shifts to traditional Hindu society, to make it clearer. He then goes on to analyze the magical beliefs inherent in caste tabus, the social dynamics of *commensalism* and *connubium,* the economic interests concealed within the system, the way in which religious beliefs relate to the tabus, the effects of the caste system upon the industrial development of the society and vice versa—all in India. But suddenly India is not very far away at all. The lecture then goes back to its Southern theme. The familiar now seems not quite so familiar anymore. Questions are raised that are new, perhaps raised angrily, but raised all the same. And at least some of the students have begun to understand that there are functions involved in this business of race that they have not read about in the newspapers (at least not those in their hometowns) and that their parents have not told them—partly, at least, because neither the newspapers nor the parents knew about them.

It can be said that the first wisdom of sociology is this—things are not what they seem. This too is a deceptively simple statement. It ceases to be simple after a while. Social reality turns out to have many layers of meaning. The discovery of each new layer changes the perception of the whole.

Anthropologists use the term "culture shock" to describe the impact of a totally new culture upon a newcomer. In an extreme instance such shock will be experienced by the Western explorer who is told, halfway through dinner, that he is eating the nice old lady he had been chatting with the previous day—a shock with predictable physiological if not moral consequences. Most explorers no longer encounter cannibalism in their travels today. However, the first encounters with polygamy or with puberty rites or even with the way some nations drive their automobiles can be quite a shock to an American visitor. With the shock may go not only disapproval or disgust but a sense of excitement that things can *really* be that different from what they are at home. To some extent, at least, this is the excitement of any first travel abroad. The experience of sociological discovery could be described as "culture shock" minus geographical displacement. In other words, the sociologist travels at home—with shocking results. He is unlikely to find that he is eating a nice old lady for dinner. But the discovery, for instance, that his own church has considerable money invested in the missile industry or that a few blocks from his home there are people who engage in cultic orgies may not be drastically different in emotional impact. Yet we would not want to imply that sociological discoveries are always or even usually outrageous to moral sentiment. Not at all. What they have in common with exploration in distant lands, however, is the sudden illumination of new and unsuspected facets of human existence in society. This is the excitement and . . . the humanistic justification of sociology.

People who like to avoid shocking discoveries, who prefer to believe that society is just what they were taught in Sunday School, who like the safety of the rules and the maxims of what Alfred Schuetz has called the "world-taken-for-granted," should stay away from sociology. People who feel no temptation before closed doors, who have no curiosity about human beings, who are content to admire scenery without wondering about the people who live in those houses on the other side of that river, should probably also stay away from sociology. They will find it unpleasant or, at any rate, unrewarding. People who are interested in human beings only if they can change, convert or reform them should also be warned, for they will find sociology much less useful than they hoped. And people whose interest is mainly in their own conceptual constructions will do just as well to turn to the study of little white mice. Sociology will be satisfying, in the long run, only to those who can think of nothing more entrancing than to watch men and to understand things human.

It may now be clear that we have, albeit deliberately, understated the case in the title of this chapter. To be sure, sociology is an individual pastime in the sense that it interests some men and bores others. Some like to observe human beings, others to experiment with mice. The world is big enough to hold all kinds and there is no logical priority for one interest as against another. But the word "pastime" is weak in describing what we mean. Sociology is more like a passion. The sociological perspective is more like a demon that possesses one, that drives one compellingly, again and again, to the questions that are its own. An introduction to sociology is, therefore, an invitation to a very special kind of passion.

THINKING ABOUT THE READING

Peter Berger claims that sociologists are tempted to listen to gossip, peek through keyholes, and look at other people's mail. This can be interpreted to mean that the sociologist has an insatiable curiosity about other people. What are some other behaviors and situations that might capture the attention of the sociologist? How does the sociologist differ from the psychologist or the economist or the historian? Are these fields of study likely to be in competition with sociology or to complement it?

The My Lai Massacre

A Military Crime of Obedience

Herbert Kelman and V. Lee Hamilton

(1989)

March 16, 1968, was a busy day in U.S. history. Stateside, Robert F. Kennedy announced his presidential candidacy, challenging a sitting president from his own party—in part out of opposition to an undeclared and disastrous war. In Vietnam, the war continued. In many ways, March 16 may have been a typical day in that war. We will probably never know. But we do know that on that day a typical company went on a mission—which may or may not have been typical—to a village called Son (or Song) My. Most of what is remembered from that mission occurred in the subhamlet known to Americans as My Lai 4.

The My Lai massacre was investigated and charges were brought in 1969 and 1970. Trials and disciplinary actions lasted into 1971. Entire books have been written about the army's year-long cover-up of the massacre (for example, Hersh, 1972), and the cover-up was a major focus of the army's own investigation of the incident. Our central concern here is the massacre itself—a crime of obedience—and public reactions to such crimes, rather than the lengths to which many went to deny the event. Therefore this account concentrates on one day: March 16, 1968.

Many verbal testimonials to the horrors that occurred at My Lai were available. More unusual was the fact that an army photographer, Ronald Haeberle, was assigned the task of documenting the anticipated military engagement at My Lai—and documented a massacre instead. Later, as the story of the massacre emerged, his photographs were widely distributed and seared the public conscience. What might have been dismissed as unreal or exaggerated was depicted in photographs of demonstrable authenticity. The dominant image appeared on the cover of *Life*: piles of bodies jumbled together in a ditch along a trail—the dead all apparently unarmed. All were Oriental, and all appeared to be children, women, or old men. Clearly there had been a mass execution, one whose image would not quickly fade.

So many bodies (over twenty in the cover photo alone) are hard to imagine as the handiwork of one killer. These were not. They were the product of what we call a crime of obedience. Crimes of obedience begin with orders. But orders are often vague and rarely survive with any clarity the transition from one authority down a chain of subordinates to the ultimate actors. The operation at Son My was no exception.

"Charlie" Company, Company C, under Lt. Col. Frank Barker's command, arrived in Vietnam in December 1967. As the army's investigative unit, directed by Lt. Gen. William R. Peers, characterized the personnel, they "contained no significant deviation from the average" for the time. Seymour S. Hersh (1970) described the "average" more explicitly: "Most of the men in Charlie Company had volunteered for the draft; only a few had gone to college for even one year. Nearly half were black, with a few Mexican-Americans. Most were eighteen to twenty-two years old. The favorite reading matter of Charlie Company, like that of other line infantry units in Vietnam, was comic books" (p. 18). The action at My Lai, like that throughout Vietnam, was fought by a cross-section of those Americans who either believed in the war or lacked the social resources to avoid participating in it. Charlie Company was indeed average for that time, that place, and that war.

Two key figures in Charlie Company were more unusual. The company's commander, Capt. Ernest Medina, was an upwardly mobile Mexican-American who wanted to make the army his career, although he feared that he might never advance beyond captain because of his lack of formal education. His eagerness had earned him a nickname among his men: "Mad Dog Medina." One of his admirers was the platoon leader Second Lt. William L. Calley, Jr., an undistinguished, five-foot-three-inch junior-college dropout who had failed four of the seven courses in which he had enrolled his first year. Many viewed him as one of those "instant officers" made possible only by the army's then-desperate need for manpower. Whatever the cause, he was an insecure leader whose frequent claim was "I'm the boss." His nickname among some of the troops was "Surfside 5½," a reference to the swashbuckling heroes of a popular television show, "Surfside 6."

The Son My operation was planned by Lieutenant Colonel Barker and his staff as a search-and-destroy mission with the objective of rooting out the Forty-eighth Viet Cong Battalion from their base area of Son My village. Apparently no written orders were ever issued. Barker's superior, Col. Oran Henderson, arrived at the staging point the day before. Among the issues he reviewed with the assembled officers were some of the weaknesses of prior operations by their units, including their failure to be appropriately aggressive in pursuit of the enemy. Later briefings by Lieutenant Colonel Barker and his staff asserted that no one except Viet Cong was expected to be in the village after 7 A.M. on the following day. The "innocent" would all be at the market. Those present at the briefings gave conflicting accounts of Barker's exact orders, but he conveyed at least a strong suggestion that the Son My area was to be obliterated. As the army's inquiry reported: "While there is some conflict in the testimony as to whether LTC Barker ordered the destruction of houses, dwellings, livestock, and other foodstuffs in the Song My area, the preponderance of the evidence indicates that such destruction was implied, if not specifically directed, by his orders of 15 March" (Peers Report, in Goldstein et al., 1976, p. 94).

Evidence that Barker ordered the killing of civilians is even more murky. What does seem clear, however, is that—having asserted that civilians would be away at the market—he did not specify what was to be done with any who might nevertheless be found on the scene. The Peers Report therefore considered it "reasonable to conclude that LTC Barker's minimal or nonexistent instructions concerning the handling of noncombatants created the potential for grave misunderstandings as to his intentions and for interpretation of his orders as authority to fire, without restriction, on all persons found in target area" (Goldstein et al., 1976, p. 95). Since Barker was killed in action in June 1968, his own formal version of the truth was never available.

Charlie Company's Captain Medina was briefed for the operation by Barker and his staff. He then transmitted the already vague orders to his own men. Charlie Company was spoiling for a fight, having been totally frustrated during its months in Vietnam—first by waiting for battles that never came, then by incompetent forays led by inexperienced commanders, and finally by mines and booby traps. In fact, the emotion-laden funeral of a sergeant killed by a booby trap was held on March 15, the day before My Lai. Captain Medina gave the orders for the next day's action at the close of that funeral. Many were in a mood for revenge.

It is again unclear what was ordered. Although all participants were alive by the time of the trials for the massacre, they were either on trial or probably felt under threat of trial. Memories are often flawed and self-serving at such times. It is apparent that Medina relayed to the men at least some of Barker's general message—to expect Viet Cong resistance, to burn, and to kill livestock. It is not clear that he ordered the slaughter of the inhabitants, but some of the men who heard him thought he had. One of those who claimed to have heard such orders was Lt. William Calley.

As March 16 dawned, much was expected of the operation by those who had set it into motion. Therefore a full complement of "brass" was present in helicopters overhead, including Barker, Colonel Henderson, and their superior, Major General Koster (who went on to become commandant of West Point before the story of My Lai broke). On the ground, the troops were to carry with them one reporter and one photographer to immortalize the anticipated battle.

The action for Company C began at 7:30 as their first wave of helicopters touched down near the subhamlet of My Lai 4. By 7:47 all of Company C was present and set to fight. But instead of the Viet Cong Forty-eighth Battalion, My Lai was filled with the old men, women, and children who were supposed to have gone to market. By this time, in their version of the war, and with whatever orders they thought they had heard, the men from Company C were nevertheless ready to find Viet Cong everywhere. By nightfall, the official tally was 128 VC killed and three weapons captured, although later, unofficial body counts ran as high as 500. The operation at Son My was over. And by nightfall, as Hersh reported: "the Viet Cong were back in My Lai 4, helping the survivors bury the dead. It took five days. Most of the funeral speeches were made by the Communist guerrillas. Nguyen Bat was not a Communist at the time of the massacre, but the incident changed his mind. 'After the shooting,' he said, 'all the villagers became Communists'" (1970, p. 74). To this day, the memory of the massacre is kept alive by markers and plaques designating the spots where groups of villagers were killed, by a large statue, and by the My Lai Museum, established in 1975 (Williams, 1985).

But what could have happened to leave American troops reporting a victory over Viet Cong when in fact they had killed hundreds of noncombatants? It is not hard to explain the report of victory; that is the essence of a cover-up. It is harder to understand how the killings came to be committed in the first place, making a cover-up necessary.

Mass Executions and the Defense of Superior Orders

Some of the atrocities on March 16, 1968, were evidently unofficial, spontaneous acts: rapes, tortures, killings. For example, Hersh (1970) describes Charlie Company's Second Platoon as entering "My Lai 4 with guns blazing" (p. 50); more graphically, Lieutenant "Brooks and his men in the second platoon to the north had begun to systematically ransack the hamlet and slaughter the people, kill the livestock, and destroy the crops. Men poured rifle and machine-gun fire into huts without knowing—or seemingly caring—who was inside" (pp. 49–50).

Some atrocities toward the end of the action were part of an almost casual "mopping-up," much of which was the responsibility of Lieutenant LaCross's Third Platoon of Charlie Company. The Peers Report states: "The entire 3rd Platoon then began moving into the western edge of My Lai (4), for the mop-up operation. . . . The squad . . . began to burn the houses in the southwestern portion of the hamlet" (Goldstein et al., 1976, p. 133). They became mingled with other platoons during a series of rapes and killings of survivors for which it was impossible to fix responsibility. Certainly to a Vietnamese all GIs would by this point look alike: "Nineteen-year-old Nguyen Thi Ngoc Tuyet watched a baby trying to open her slain mother's blouse to nurse. A soldier shot the infant while it was struggling with the blouse, and then slashed it with his bayonet." Tuyet also said she saw another baby hacked to death by GIs wielding their bayonets. "Le Tong, a twenty-eight-year-old rice farmer, reported seeing one woman raped after GIs killed her children. Nguyen Khoa, a thirty-seven-year-old peasant, told of a thirteen-year-old girl who was raped before being killed. GIs then attacked Khoa's wife, tearing off her clothes. Before they could rape her, however, Khoa said, their six-year-old son, riddled with bullets, fell and saturated her with blood. The GIs left her alone" (Hersh, 1970, p. 72). All of Company C was implicated in a pattern of death and destruction throughout the hamlet, much of which seemingly lacked rhyme or reason.

But a substantial amount of the killing was organized and traceable to one authority: the First Platoon's Lt. William Calley. Calley was originally charged with 109 killings, almost all of them mass executions at the trail and other locations. He stood trial for 102 of these killings, was convicted of 22 in 1971, and at first received a life sentence. Though others—both superior and subordinate to Calley—were brought to trial, he was the only one convicted for the My Lai crimes. Thus, the only actions of My Lai for which anyone was ever convicted were mass executions, ordered and committed. We suspect that there are commonsense reasons why this one type of killing was singled out. In the midst of rapidly moving events with people running about, an execution of stationary targets is literally a still life that stands out and whose participants are clearly visible. It can be proven that specific people committed specific deeds. An execution, in contrast to the shooting of someone on the run, is also more likely to meet the legal definition of an act resulting from intent—with malice aforethought. Moreover, American military law specifically forbids the killing of unarmed civilians or military prisoners, as does the Geneva Convention between nations. Thus common sense, legal standards, and explicit doctrine all made such actions the likeliest target for prosecution.

When Lieutenant Calley was charged under military law it was for violation of the Uniform Code of Military Justice (UCMJ) Article 118 (murder). This article is similar to civilian codes in that it provides for conviction if an accused:

> without justification or excuse, unlawfully kills a human being, when he—
>
> 1. has a premeditated design to kill;
> 2. intends to kill or inflict great bodily harm;
> 3. is engaged in an act which is inherently dangerous to others and evinces a wanton disregard of human life; or
> 4. is engaged in the perpetration or attempted perpetration of burglary, sodomy, rape, robbery, or aggravated arson. (Goldstein et al., 1976, p. 507)

For a soldier, one legal justification for killing is warfare; but warfare is subject to many legal limits and restrictions, including, of course, the inadmissibility of killing unarmed noncombatants or prisoners whom one has disarmed. The pictures of the trail victims at My Lai certainly portrayed one or the other of these. Such an action would be illegal under military law; ordering another to commit such an action would be illegal; and following such an order would be illegal.

But following an order may provide a second and pivotal justification for an act that would be murder when committed by a civilian. American military law assumes that the subordinate is inclined to follow orders, as that is the normal obligation of the role. Hence, legally, obedient subordinates are protected from unreasonable expectations regarding their capacity to evaluate those orders:

> An order requiring the performance of a military duty may be inferred to be legal. An act performed manifestly beyond the scope of authority, or pursuant to an order that a man of ordinary sense and understanding would know to be illegal, or in a wanton manner in the discharge of a lawful duty, is not excusable. (Par. 216, Subpar. *d*, Manual for Courts Martial, United States, 1969 Rev.)

Thus what *may* be excusable is the good-faith carrying out of an order, as long as that order appears to the ordinary soldier to be a legal one. In military law, invoking superior orders moves the question from one of the action's consequences—the body count—to one of evaluating the actor's motives and good sense.

In sum, if anyone is to be brought to justice for a massacre, common sense and legal codes decree that the most appropriate targets are those who make themselves executioners. This is the kind of target the government selected in prosecuting Lieutenant Calley with the greatest fervor. And in a military context, the most promising way in which one can redefine one's undeniable deeds into acceptability is to invoke superior orders. This is what Calley did in attempting to avoid conviction. Since the core legal issues involved points of mass execution—the ditches and trail where America's image of My Lai was formed—we review these events in greater detail.

The day's quiet beginning has already been noted. Troops landed and swept unopposed into the village. The three weapons eventually reported as the haul from the operation were picked up from three apparent Viet Cong who fled the village when the troops arrived and were pursued and killed by helicopter gunships. Obviously the Viet Cong did frequent the area. But it appears that by about 8:00 A.M. no one who met the troops was aggressive, and no one was armed. By the laws of war Charlie Company had no argument with such people.

As they moved into the village, the soldiers began to gather its inhabitants together. Shortly after 8:00 A.M. Lieutenant Calley told Pfc. Paul Meadlo that "you know what to do with" a group of villagers Meadlo was guarding. Estimates of the numbers in the group ranged as high as eighty women, children, and old men, and Meadlo's own estimate under oath was thirty to fifty people. As Meadlo later testified, Calley returned after ten or fifteen minutes: "He [Calley] said, 'How come they're not dead?' I said, 'I didn't know we were supposed to kill them.' He said, 'I want them dead.' He backed off twenty or thirty feet and started shooting into the people—the Viet Cong—shooting automatic. He was beside me. He burned four or five magazines. I burned off a few, about three. I helped shoot 'em" (Hammer, 1971, p. 155). Meadlo himself and others testified that Meadlo cried as he fired; others reported him later to be sobbing and "all broke up." It would appear that to Lieutenant Calley's subordinates something was unusual, and stressful, in these orders.

At the trial, the first specification in the murder charge against Calley was for this incident; he was accused of premeditated murder of "an unknown number, not less than 30, Oriental human beings, males and females of various ages, whose names are unknown, occupants of the village of My Lai 4, by means of shooting them with a rifle" (Goldstein et al., 1976, p. 497).

Among the helicopters flying reconnaissance above Son My was that of CWO Hugh Thompson. By 9:00 or soon after, Thompson had noticed some horrifying events from his perch. As he spotted wounded civilians, he sent down smoke markers so that soldiers on the ground could treat them. They killed them instead. He reported to headquarters, trying to persuade someone to stop what was going on. Barker, hearing the message, called down to Captain Medina. Medina, in turn, later claimed to have told Calley that it was "enough for today." But it was not yet enough.

At Calley's orders, his men began gathering the remaining villagers—roughly seventy-five individuals, mostly women and children—and herding them toward a drainage ditch. Accompanied by three or four enlisted men, Lieutenant Calley executed several batches of civilians who had been gathered into ditches. Some of the details of the process were entered into testimony in such accounts as Pfc. Dennis Conti's: "A lot of them, the people, were trying to get up and mostly they was just screaming and pretty bad shot up. . . . I seen a woman tried to get up. I seen Lieutenant Calley fire. He hit the side of her head and blew it off" (Hammer, 1971, p. 125).

Testimony by other soldiers presented the shooting's aftermath. Specialist Four Charles Hall, asked by Prosecutor Aubrey Daniel how he knew the people in the ditch were dead, said: "There was blood coming from them. They were just scattered all over the ground in the ditch, some in piles and some scattered out 20, 25 meters perhaps up the ditch. . . . They were very old people, very young children, and mothers. . . . There was blood all over them" (Goldstein et al., 1976, pp. 501–502). And Pfc.

Gregory Olsen corroborated the general picture of the victims: "They were—the majority were women and children, some babies. I distinctly remember one middle-aged Vietnamese male dressed in white right at my feet as I crossed. None of the bodies were mangled in any way. There was blood. Some appeared to be dead, others followed me with their eyes as I walked across the ditch" (Goldstein et al., 1976, p. 502).

The second specification in the murder charge stated that Calley did "with premeditation, murder an unknown number of Oriental human beings, not less than seventy, males and females of various ages, whose names are unknown, occupants of the village of My Lai 4, by means of shooting them with a rifle" (Goldstein et al., 1976, p. 497). Calley was also charged with and tried for shootings of individuals (an old man and a child); these charges were clearly supplemental to the main issue at trial—the mass killings and how they came about.

It is noteworthy that during these executions more than one enlisted man avoided carrying out Calley's orders, and more than one, by sworn oath, directly refused to obey them. For example, Pfc. James Joseph Dursi testified, when asked if he fired when Lieutenant Calley ordered him to: "No I just stood there. Meadlo turned to me after a couple of minutes and said 'Shoot! Why don't you shoot! Why don't you fire!' He was crying and yelling. I said, 'I can't! I won't!' And the people were screaming and crying and yelling. They kept firing for a couple of minutes, mostly automatic and semi-automatic" (Hammer, 1971, p. 143). . . .

Disobedience of Lieutenant Calley's own orders to kill represented a serious legal and moral threat to a defense *based* on superior orders, such as Calley was attempting. This defense had to assert that the orders seemed reasonable enough to carry out; that they appeared to be legal orders. Even if the orders in question were not legal, the defense had to assert that an ordinary individual could not and should not be expected to see the distinction. In short, if what happened was "business as usual," even though it might be bad business, then the defendant stood a chance of acquittal. But under direct command from "Surfside 5½," some ordinary enlisted men managed to refuse, to avoid, or at least to stop doing what they were ordered to do. As "reasonable men" of "ordinary sense and understanding," they had apparently found something awry that morning; and it would have been hard for an officer to plead successfully that he was more ordinary than his men in his capacity to evaluate the reasonableness of orders.

Even those who obeyed Calley's orders showed great stress. For example, Meadlo eventually began to argue and cry directly in front of Calley. Pfc. Herbert Carter shot himself in the foot, possibly because he could no longer take what he was doing. We were not destined to hear a sworn version of the incident, since neither side at the Calley trial called him to testify.

The most unusual instance of resistance to authority came from the skies. CWO Hugh Thompson, who had protested the apparent carnage of civilians, was Calley's inferior in rank but was not in his line of command. He was also watching the ditch from his helicopter and noticed some people moving after the first round of slaughter—chiefly children who had been shielded by their mothers' bodies. Landing to rescue the wounded, he also found some villagers hiding in a nearby bunker. Protecting the Vietnamese with his own body, Thompson ordered his men to train their guns on the Americans and to open fire if the Americans fired on the Vietnamese. He then radioed for additional rescue helicopters and stood between the Vietnamese and the Americans under Calley's command until the Vietnamese could be evacuated. He later returned to the ditch to unearth a child buried, unharmed, beneath layers of bodies. In October 1969, Thompson was awarded the Distinguished Flying Cross for heroism at My Lai, specifically (albeit inaccurately) for the rescue of children hiding in a bunker "between Viet Cong forces and advancing friendly forces" and for the rescue of a wounded child "caught in the intense crossfire" (Hersh, 1970, p. 119). Four months earlier, at the Pentagon, Thompson had identified Calley as having been at the ditch.

By about 10:00 A.M., the massacre was winding down. The remaining actions consisted largely of isolated rapes and killings, "clean-up" shootings of the wounded, and the destruction of the village by fire. We have already seen some examples of these more indiscriminate and possibly less premeditated acts. By the 11:00 A.M. lunch break, when the exhausted men of Company C were relaxing, two young girls wandered back from a hiding place only to be invited to share lunch. This surrealist touch illustrates the extent to which the soldiers' action had become dissociated from its meaning. An hour earlier, some of these men were making sure that not even a child would escape the executioner's bullet. But now the job was done and it was time for lunch—and in this new context it seemed only natural to ask the children who had managed to escape execution to join them. The massacre had ended. It remained only for the Viet Cong to reap the political rewards among the survivors in hiding.

The army command in the area knew that something had gone wrong. Direct commanders, including Lieutenant Colonel Barker, had firsthand reports, such as Thompson's complaints. Others had such odd bits of evidence as the claim of 128 Viet Cong dead with a booty of only three weapons. But the cover-up of My Lai began at once. The operation was reported as a victory over a stronghold of the Viet Cong Forty-eighth. . . .

William Calley was not the only man tried for the event at My Lai. The actions of over thirty soldiers and civilians were scrutinized by investigators; over half of these had to face charges or disciplinary action of some sort. Targets of investigation included Captain Medina, who was tried, and various higher-ups, including General Koster. But Lieutenant Calley was the only person convicted, the only person to serve time.

The core of Lieutenant Calley's defense was superior orders. What this meant to him—in contrast to what it meant to the judge and jury—can be gleaned from his responses to a series of questions from his defense attorney, George Latimer, in which Calley sketched out his understanding of the laws of war and the actions that constitute doing one's duty within those laws:

Latimer: Did you receive any training which had to do with the obedience to orders?

Calley: Yes, sir.

Latimer: . . . what were you informed [were] the principles involved in that field?

Calley: That all orders were to be assumed legal, that the soldier's job was to carry out any order given him to the best of his ability.

Latimer: . . . what might occur if you disobeyed an order by a senior officer?

Calley: You could be court-martialed for refusing an order and refusing an order in the face of the enemy, you could be sent to death, sir.

Latimer: [I am asking] whether you were required in any way, shape or form to make a determination of the legality or illegality of an order?

Calley: No, sir. I was never told that I had the choice, sir.

Latimer: If you had a doubt about the order, what were you supposed to do?

Calley: . . . I was supposed to carry the order out and then come back and make my complaint. (Hammer, 1971, pp. 240–241)

Lieutenant Calley steadfastly maintained that his actions within My Lai had constituted, in his mind, carrying out orders from Captain Medina. Both his own actions and the orders he gave to others (such as the instruction to Meadlo to "waste 'em") were entirely in response to superior orders. He denied any intent to kill individuals and any but the most passing awareness of distinctions among the individuals: "I was ordered to go in there and destroy the enemy. That was my job on that day. That was the mission I was given. I did not sit down and think in terms of men, women, and children. They were all classified the same, and that was the classification that we dealt with, just as enemy soldiers." When Latimer asked if in his own opinion Calley had acted "rightly and according to your understanding of your directions and orders," Calley replied, "I felt then and I still do that I acted as I was directed, and I carried out the orders that I was given, and I do not feel wrong in doing so, sir" (Hammer, 1971, p. 257).

His court-martial did not accept Calley's defense of superior orders and clearly did not share his interpretation of his duty. The jury evidently reasoned that, even if there had been orders to destroy everything in sight and to "waste the Vietnamese," any reasonable person would have realized that such orders were illegal and should have refused to carry them out. The defense of superior orders under such conditions is inadmissible under international and military law. The U.S. Army's *Law of Land Warfare* (Dept. of the Army, 1956), for example, states that "the fact that the law of war has been violated pursuant to an order of a superior authority, whether military or civil, does not deprive the act in question of its character of a war crime, nor does it constitute a defense in the trial of an accused individual, unless he did not know and could not reasonably have been expected to know that the act was unlawful" and that "members of the armed forces are bound to obey only lawful orders" (in Falk et al., 1971, pp. 71–72).

The disagreement between Calley and the court-martial seems to have revolved around the definition of the responsibilities of a subordinate to obey, on the one hand, and to evaluate, on the other. This tension . . . can best be captured via the charge to the jury in the Calley court-martial, made by the trial judge, Col. Reid Kennedy. The forty-one pages of the charge include the following:

> Both combatants captured by and noncombatants detained by the opposing force . . . have the right to be treated as prisoners. . . . Summary execution of detainees or prisoners is forbidden by law. . . . I therefore instruct you . . . that if unresisting human beings were killed at My Lai (4) while within the effective custody and control of our military forces, their deaths cannot be considered justified. . . . Thus if you find that Lieutenant Calley received an order directing him to kill unresisting Vietnamese within his control or within the control of his troops, *that order would be an illegal order.*

A determination that an order is illegal does not, of itself, assign criminal responsibility to the person following the order for acts done in compliance with it. Soldiers are taught to follow orders, and special attention is given to obedience of orders on the battlefield. Military effectiveness depends on obedience to orders. On the other hand, the obedience of a soldier is not the obedience of an automaton. A soldier is a reasoning agent, obliged to respond, not as a machine, but as a person. The law takes these factors into account in assessing criminal responsibility for acts done in compliance with illegal orders.

> The acts of a subordinate done in compliance with an unlawful order given him by his superior are excused and impose no criminal liability upon him unless the superior's order is one which a man of *ordinary sense and understanding* would, under the circumstances, know to be unlawful, or if the order in question is actually known to the accused to be unlawful. (Goldstein et al., 1976, pp. 525–526; emphasis added)

By this definition, subordinates take part in a balancing act, one tipped toward obedience but tempered by "ordinary sense and understanding."

A jury of combat veterans proceeded to convict William Calley of the premeditated murder of no less than twenty-two human beings. (The army, realizing some unfortunate connotations in referring to the victims as "Oriental human beings," eventually referred to them as "human beings.") Regarding the first specification in the murder charge, the bodies on the trail, [Calley] was convicted of premeditated murder of not less than one person. (Medical testimony had been able to pinpoint only one person whose wounds as revealed in Haeberle's photos were sure to be immediately fatal.) Regarding the second specification, the bodies in the ditch, Calley was convicted of the premeditated murder of not less than twenty human beings. Regarding additional specifications that he had killed an old man and a child, Calley was convicted of premeditated murder in the first case and of assault with intent to commit murder in the second.

Lieutenant Calley was initially sentenced to life imprisonment. That sentence was reduced: first to twenty years, eventually to ten (the latter by Secretary of Defense Callaway in 1974). Calley served three years before being released on bond. The time was spent under house arrest in his apartment, where he was able to receive visits from his girlfriend. He was granted parole on September 10, 1975.

Sanctioned Massacres

The slaughter at My Lai is an instance of a class of violent acts that can be described as sanctioned massacres (Kelman, 1973): acts of indiscriminate, ruthless, and often systematic mass violence, carried out by military or paramilitary personnel while engaged in officially sanctioned campaigns, the victims of which are defenseless and unresisting civilians, including old men, women, and children. Sanctioned massacres have occurred throughout history. Within American history, My Lai had its precursors in the Philippine war around the turn of the century (Schirmer, 1971) and in the massacres of American Indians. Elsewhere in the world, one recalls the Nazis' "final solution" for European Jews, the massacres and deportations of Armenians by Turks, the liquidation of the kulaks and the great purges in the Soviet Union, and more recently the massacres in Indonesia and Bangladesh, in Biafra and Burundi, in South Africa and Mozambique, in Cambodia and Afghanistan, in Syria and Lebanon. . . .

The occurrence of sanctioned massacres cannot be adequately explained by the existence of psychological forces—whether these be characterological dispositions to engage in murderous violence or profound hostility against the target—so powerful that they must find expression in violent acts unhampered by moral restraints. Instead, the major instigators for this class of violence derive from the policy process. The question that really calls for psychological analysis is why so many people are willing to formulate, participate in, and condone policies that call for the mass killings of defenseless civilians. Thus it is more instructive to look not at the motives for violence but at the conditions under which the usual moral inhibitions against violence become weakened. Three social processes that tend to create such conditions can be identified: authorization, routinization, and dehumanization. Through authorization, the situation becomes so defined that the individual is absolved of the responsibility to make personal moral choices. Through routinization, the action becomes so organized that there is no opportunity for raising moral questions. Through dehumanization, the actors' attitudes toward the target and toward themselves become so structured that it is neither necessary nor possible for them to view the relationship in moral terms.

Authorization

Sanctioned massacres by definition occur in the context of an authority situation, a situation in which, at least for many of the participants, the moral principles that generally govern human relationships do not apply. Thus, when acts of violence are explicitly ordered, implicitly encouraged, tacitly approved, or at least permitted by legitimate authorities, people's readiness to commit or condone them is enhanced. That such acts are authorized seems to carry automatic justification for them. Behaviorally, authorization obviates the necessity of making judgments or choices. Not only do normal moral principles become inoperative, but—particularly when the actions are explicitly ordered—a different kind of morality, linked to the duty to obey superior orders, tends to take over.

In an authority situation, individuals characteristically feel obligated to obey the orders of the authorities, whether or not these correspond with their personal preferences. They see themselves as having no choice as long as they accept the legitimacy of the orders and of the authorities who give them. Individuals differ considerably in the degree to which—and the conditions under which—they are prepared to challenge the legitimacy of an order on the grounds that the order itself is illegal, or that those giving it have overstepped their authority, or that it stems from a policy that violates fundamental societal values. Regardless of such individual differences, however, the basic structure of a situation of legitimate authority requires subordinates to respond in terms of their role obligations rather than their personal preferences; they can openly disobey only by challenging the legitimacy of the authority. Often people obey without question even though the behavior they engage in may entail great personal sacrifice or great harm to others.

An important corollary of the basic structure of the authority situation is that actors often do not see themselves as personally responsible for the consequences of their actions. Again, there are individual differences, depending on actors' capacity and readiness to evaluate the legitimacy of orders received. Insofar as they see themselves as having had no choice in their actions, however, they do not feel personally responsible for them. They were not personal agents, but merely extensions of the authority. Thus, when their actions cause harm to others, they can feel relatively free of guilt. A similar mechanism operates when a person engages in antisocial behavior that was not ordered by the authorities but was tacitly encouraged and approved by them—even if only by making it clear that such behavior will not be punished. In this situation, behavior that was formerly illegitimate is legitimized by the authorities' acquiescence.

In the My Lai massacre, it is likely that the structure of the authority situation contributed to the massive violence in both ways—that is, by conveying the message that acts of violence against Vietnamese villagers were *required*, as well as the message that such acts, even if not ordered, were *permitted* by the authorities in charge. The actions at My Lai represented, at least in some respects, responses to explicit or implicit orders. Lieutenant Calley indicated, by orders and by example, that he wanted large numbers of villagers killed. Whether Calley himself had been ordered by his superiors to "waste" the whole area, as he claimed, remains a matter of controversy. Even if we assume, however, that he was not explicitly ordered to wipe out the village, he had reason to believe that such actions were expected by his superior officers. Indeed, the very nature of the war conveyed this expectation. The principal measure of military success was the "body count"—the number of enemy soldiers killed—and any Vietnamese killed by the U.S. military was commonly defined as a "Viet Cong." Thus, it was not totally bizarre for Calley to believe that what he was doing at My Lai was to increase his body count, as any good officer was expected to do.

Even to the extent that the actions at My Lai occurred spontaneously, without reference to superior orders, those committing them had reason to assume that such actions might be tacitly approved of by the military authorities. Not only had they failed to punish such acts in most cases, but the very strategies and tactics that the authorities consistently devised were based on the proposition that the civilian population of South Vietnam—whether "hostile" or "friendly"—was expendable. Such policies as search-and-destroy missions, the establishment of free-shooting zones, the use of antipersonnel weapons, the bombing of entire villages if they were suspected of harboring guerrillas, the forced migration of masses of the rural population, and the defoliation of vast forest areas helped legitimize acts of massive violence of the kind occurring at My Lai.

Some of the actions at My Lai suggest an orientation to authority based on unquestioning obedience to superior orders, no matter how destructive the actions these orders call for. Such obedience is specifically fostered in the course of military training and reinforced by the structure of the military authority situation. It also reflects, however, an ideological orientation that may be more widespread in the general population. . . .

Routinization

Authorization processes create a situation in which people become involved in an action without considering its implications and without really making a decision. Once they have taken the initial step, they are in a new psychological and social situation in which the pressures to continue are powerful. As Lewin (1947) has pointed out, many forces that might originally have kept people out of a situation reverse direction once they have made a commitment (once they have gone through the "gate region") and now serve to keep them in the situation. For example, concern about the criminal nature of an action, which might originally have inhibited a person from becoming involved, may now lead to deeper involvement in efforts to justify the action and to avoid negative consequences.

Despite these forces, however, given the nature of the actions involved in sanctioned massacres, one might still expect moral scruples to intervene; but the likelihood of moral resistance is greatly reduced by transforming the action into routine, mechanical, highly programmed operations. Routinization fulfills two functions. First, it reduces the necessity of making decisions, thus minimizing the occasions in which moral questions may arise. Second, it makes it easier to avoid the implications of the action, since the actor focuses on the details of the job rather than on its meaning. The latter effect is more readily achieved among those who participate in sanctioned massacres from a distance—from their desks or even from the cockpits of their bombers.

Routinization operates both at the level of the individual actor and at the organizational level. Individual job performance is broken down into a series of discrete steps, most of them carried out in automatic, regularized fashion. It becomes easy to forget the nature of the product that emerges from this process. When Lieutenant Calley said of My Lai that it was "no great deal," he probably implied that it was all in a day's work. Organizationally, the task is divided among different offices, each of which has responsibility for a small portion of it. This arrangement diffuses responsibility and limits the amount and scope of decision making that is necessary. There is no expectation that the moral implications will be considered at any of these points, nor is there any opportunity to do so. The organizational processes also help further legitimize the actions of each participant. By proceeding in routine fashion—processing papers, exchanging memos, diligently carrying out their assigned tasks—the different units mutually reinforce each other in the view that what is going on must be perfectly normal, correct, and legitimate. The shared illusion that they are engaged in a legitimate enterprise helps the participants assimilate their activities to other purposes, such as the efficiency of their performance, the productivity of their unit, or the cohesiveness of their group (see Janis, 1972).

Normalization of atrocities is more difficult to the extent that there are constant reminders of the true meaning of the enterprise. Bureaucratic inventiveness in the use of language helps to cover up such meaning. For example, the SS had a set of *Sprachregelungen*, or "language rules," to govern descriptions of their extermination program. As Arendt (1964) points out, the term *language rule* in itself was "a code name; it meant what in ordinary language would be called a lie" (p. 85). The code names for killing and liquidation were "final solution," "evacuation," and "special treatment." The war in Indochina produced its own set of euphemisms, such as "protective reaction," "pacification," and "forced-draft urbanization and modernization." The use of euphemisms allows participants in sanctioned massacres to differentiate their actions from ordinary killing and destruction and thus to avoid confronting their true meaning.

Dehumanization

Authorization processes override standard moral considerations; routinization processes reduce the likelihood that such considerations will arise. Still, the inhibitions against murdering one's fellow human beings are generally so strong that the victims must also be stripped of their human status if they are to be subjected to systematic killing. Insofar as they are dehumanized, the usual principles of morality no longer apply to them.

Sanctioned massacres become possible to the extent that the victims are deprived in the perpetrators' eyes of the two qualities essential to being perceived as fully human and included in the moral compact that governs human relationships: *identity*—standing as independent, distinctive individuals, capable of making choices and entitled to live their own lives—and *community*—fellow membership in an interconnected network of individuals who care for each other and respect each other's individuality and rights (Kelman, 1973; see also Bakan, 1966, for a related distinction between "agency" and "communion"). Thus, when a group of people is defined entirely in terms of a category to which they belong, and when this category is excluded from the human family, moral restraints against killing them are more readily overcome.

Dehumanization of the enemy is a common phenomenon in any war situation. Sanctioned massacres, however, presuppose a more extreme degree of dehumanization, insofar as the killing is not in direct response to the target's threats or provocations. It is not what they have done that marks such victims for death but who they are—the category to which they happen to belong. They are the victims of policies that regard their systematic destruction as a desirable end or an acceptable means. Such extreme dehumanization becomes possible when the target group can readily be identified as a separate category of people who have historically been stigmatized and excluded by the victimizers; often the victims belong to a distinct racial, religious, ethnic, or political group regarded as inferior or sinister. The traditions, the habits, the images, and the vocabularies for dehumanizing such groups are already well established and can be drawn upon when the groups are selected for massacre. Labels help deprive the victims of identity and community, as in the epithet "gooks" that was commonly used to refer to Vietnamese and other Indochinese peoples.

The dynamics of the massacre process itself further increase the participants' tendency to dehumanize their victims. Those who participate as part of the bureaucratic apparatus increasingly come to see their victims as bodies to be counted and entered into their reports, as faceless figures that will determine their productivity rates and promotions. Those who participate in the massacre directly—in the field, as it were—are reinforced in their perception of the victims as less than human by observing their very victimization. The only way they can justify what is being done to these people—both by others and by themselves—and the only way they can extract some degree of meaning out of the absurd events in which they find themselves participating (see Lifton, 1971, 1973) is by coming to believe that the victims are subhuman and deserve to be rooted out. And thus the process of dehumanization feeds on itself.

REFERENCES

Arendt, H. (1964). *Eichmann in Jerusalem: A report on the banality of evil.* New York: Viking Press.

Bakan, D. (1966). *The duality of human existence.* Chicago: Rand McNally.

Department of the Army. (1956). *The law of land warfare* (Field Manual, No. 27–10). Washington, DC: U.S. Government Printing Office.

Falk, R. A., Kolko, G., & Lifton, R. J. (Eds.). (1971). *Crimes of war.* New York: Vintage Books.

French, P. (Ed.). (1972). *Individual and collective responsibility: The massacre at My Lai.* Cambridge, MA: Schenkman.

Goldstein, J., Marshall, B., & Schwartz, J. (Eds.). (1976). *The My Lai massacre and its cover-up: Beyond the reach of law?* (The Peers Report with a supplement and introductory essay on the limits of law). New York: Free Press.

Hammer, R. (1971). *The court-martial of Lt. Calley.* New York: Coward, McCann, & Geoghegan.

Hersh, S. (1970). *My Lai 4: A report on the massacre and its aftermath.* New York: Vintage Books.

______. (1972). *Cover-up.* New York: Random House.

Janis, I. L. (1972). *Victims of groupthink: A psychological study of foreign-policy decisions and fiascoes.* Boston: Houghton Mifflin.

Kelman, H. C. (1973). Violence without moral restraint: Reflections on the dehumanization of victims and victimizers. *Journal of Social Issues, 29*(4), 25–61.

Lewin, K. (1947). Group decision and social change. In T. M. Newcomb & E. L. Hartley (Eds.), *Readings in social psychology.* New York: Holt.

Lifton, R. J. (1971). Existential evil. In N. Sanford, C. Comstock, & Associates, *Sanctions for evil: Sources of social destructiveness.* San Francisco: Jossey-Bass.

______. (1973). *Home from the war—Vietnam veterans: Neither victims nor executioners.* NewYork: Simon & Schuster.

Manual for courts martial, United States (Rev. ed.). (1969). Washington, DC: U.S. Government Printing Office.

Schirmer, D. B. (1971, April 24). My Lai was not the first time. *New Republic,* pp. 18–21.

Williams, B. (1985, April 14–15). "I will never forgive," says My Lai survivor. *Jordan Times* (Amman), p. 4.

THINKING ABOUT THE READING

According to Kelman and Hamilton, social processes can create conditions under which usual restraints against violence are weakened. What social processes were in evidence during the My Lai massacre? The incident they describe provides us with an uncomfortable picture of human nature. Do you think most people would have reacted the way the soldiers at My Lai did? Are we all potential massacrers? Does the phenomenon of obedience to authority go beyond the tightly structured environment of the military? Can you think of incidents in your own life when you've done something—perhaps harmed or humiliated another person—because of the powerful influence of others? How might Kelman and Hamilton explain the actions of the individuals who carried out the hijackings and attacks of September 11, 2001, or of the American soldiers who abused Iraqi prisoners in their custody?

Seeing and Thinking Sociologically

2

Where is society located? This is an intriguing question. Society shapes our behavior and beliefs through social institutions such as religion, law, education, economics, and family. At the same time, we shape society through our interactions with one another and our participation in social institutions. In this way, we can say that society exists as an objective entity that transcends us. But it is also a construction that is created, reaffirmed, and altered through everyday interactions and behavior. Humans are social beings. We constantly look to others to help define and interpret the situations in which we find ourselves. Other people can influence what we see, feel, think, and do. But it's not just other people who influence us. We also live in a *society,* which consists of socially recognizable combinations of individuals—relationships, groups, and organizations—as well as the products of human action—statuses, roles, culture, and institutions. When we behave, we do so in a social context that consists of a combination of institutional arrangements, cultural influences, and interpersonal expectations. Thus, our behavior in any given situation is our own, but the reasons we do what we do are rooted in these more complex social factors.

The social context in which we reside—urban, rural, suburban, exurban, institutional—greatly influences and shapes our social interaction and individual experiences. In "The Metropolis and Mental Life," Georg Simmel uses the urban environment to investigate the social forces that affect a person's individuality and relationships. He shows how the urban environment challenges our sense of independence and individuality and pushes us toward a very calculative and rational approach to life. The path we take as individuals in constructing the architecture of our social environment is one shaped by these influential social forces of time and place.

As social beings, the mutual exchange between society and the individual is especially evident in our social interactions with others. In "Gift and Exchange," Zygmut Bauman explores the "gift-exchange" choice that individuals face on a daily basis in their social interaction. While he observes that self-interest drives our exchanges in certain moments, the needs and rights of other individuals control our exchanges at other times. In short, depending on the context, our social needs may pull us toward belonging or individuality, but in the end, we find that one without the other leaves us feeling unsettled in our social environment.

This social structure provides us with a sense of order in our daily lives. But sometimes that order breaks down. In "Culture of Fear," Barry Glassner shows us how the news media functions to *create* a culture that the public takes for granted. He focuses, in particular, on the emotion of fear in U.S. society. We constantly hear horror stories about such urgent social problems as deadly diseases, violent strangers, and out-of-control teens. But Glassner points out that the terrified public concern over certain issues is often inflated by the media and largely unwarranted. Ironically, when we live in a culture of fear, our most serious problems often go ignored.

Something to Consider as You Read

When reading the selections in this section, consider how "society" is defined and characterized. Is it something "out there," some invisible force that makes us behave in certain ways? What does it look like? How do you know it when you see it? Sociologists view society as the beliefs, practices, rules, and institutions that shape our lives. For example, consider certain practices of "gift exchange" that you use in your own life. How did you learn these practices? Why do you do them? What meaning do they have for you? Consider further the "rules of the road" that you follow when driving in traffic. Where did those come from? Why do people follow them? Sociologists explore these kinds of questions to understand the underlying patterns of structures that make up what we call "society."

The Metropolis and Mental Life

Georg Simmel

(1903)

The deepest problems of modern life flow from the attempt of the individual to maintain the independence and individuality of his existence against the sovereign powers of society, against the weight of the historical heritage and the external culture and technique of life.

This intellectualistic quality which is thus recognized as a protection of the inner life against the domination of the metropolis, becomes ramified into numerous specific phenomena. The metropolis has always been the seat of money economy because the many-sidedness and concentration of commercial activity have given the medium of exchange an importance which it could not have acquired in the commercial aspects of rural life. But money economy and the domination of the intellect stand in the closest relationship to one another. They have in common a purely matter-of-fact attitude in the treatment of persons and things in which a formal justice is often combined with an unrelenting hardness. The purely intellectualistic person is indifferent to all things personal because, out of them, relationships and reactions develop which are not to be completely understood by purely rational methods—just as the unique element in events never enters into the principle of money. Money is concerned only with what is common to all, i.e., with the exchange value which reduces all quality and individuality to a purely quantitative level. All emotional relationships between persons rest on their individuality, whereas intellectual relationships deal with persons as with numbers, that is, as with elements which, in themselves, are indifferent, but which are of interest only insofar as they offer something objectively perceivable. It is in this very manner that the inhabitant of the metropolis reckons with his merchant, his customer, and with his servant, and frequently with the persons with whom he is thrown into obligatory association. These relationships stand in distinct contrast with the nature of the smaller circle in which the inevitable knowledge of individual characteristics produces, with an equal inevitability, an emotional tone in conduct, a sphere which is beyond the mere objective weighting of tasks performed and payments made. What is essential here as regards the economic-psychological aspect of the problem is that in less advanced cultures production was for the customer who ordered the product so that the producer and the purchaser knew one another. The modern city, however, is supplied almost exclusively by production for the market, that is, for entirely unknown purchasers who never appear in the actual field of vision of the producers themselves. Thereby, the interests of each party acquire a relentless matter-of-factness, and its rationally calculated economic egoism need not fear any divergence from its set path because of the imponderability of personal relationships. This is all the more the case in the money economy which dominates the metropolis in which the last remnants of domestic production and direct barter of goods have been eradicated and in which the amount of production on direct personal order is reduced daily. Furthermore, this psychological intellectualistic attitude and the money economy are in such close integration that no one is able to say whether it was the former that effected the latter or vice versa. What is certain is only that the form of life in the metropolis is the soil which nourishes this interaction most fruitfully, a point which I shall attempt to demonstrate only with the statement of the most outstanding English constitutional historian to the effect that through the entire course of English history London has never acted as the heart of England but often as its intellect and always as its money bag.

In certain apparently insignificant characters or traits of the most external aspects of life are to be found a number of characteristic mental tendencies. The modern mind has become more and more a calculating one. The calculating exactness of practical life which has resulted from a money economy corresponds to the ideal of natural science, namely that of transforming the world into an arithmetical problem and of fixing every one of its parts in a mathematical formula. It has been money economy which has thus filled the daily life of so many people with weighing, calculating, enumerating and the reduction of qualitative values to quantitative terms. Because of the character of calculability which money has there has come into the relationships of the elements of life a precision and a degree of certainty in the definition of the equalities and inequalities and an unambiguousness in agreements and arrangements, just as externally this precision has been brought about through the general diffusion of pocket watches. It is, however, the conditions of the metropolis which are cause as well as effect for this essential characteristic. The relationships and concerns of the typical metropolitan resident are so manifold and complex that, especially as a result of the agglomeration of so many persons with such differentiated interests, their relationships and activities intertwine with one another into a many-membered organism. In view of this fact, the lack of the most exact punctuality in promises and performances would cause the whole to break down into an inextricable chaos. If all the watches in Berlin suddenly went wrong in different ways even only as much as an hour, its entire economic and commercial life would be derailed for some time. Even though this may seem more superficial in its significance, it transpires that the magnitude of distances results in making all waiting and the breaking of appointments an ill-afforded waste of time. For this reason the technique of metropolitan life in general is not conceivable without all of its activities and reciprocal relationships being organized and coordinated in the most punctual way into a firmly fixed framework of time which transcends all subjective elements. But here too there emerge those conclusions which are in general the whole task of this discussion, namely, that every event, however restricted to this superficial level it may appear, comes immediately into contact with the depths of the soul, and that the most banal externalities are, in the last analysis, bound up with the final decisions concerning the meaning and the style of life. Punctuality, calculability, and exactness, which are required by the complications and extensiveness of metropolitan life are not only most intimately connected with its capitalistic and intellectualistic character but also color the content of life and are conductive to the exclusion of those irrational, instinctive, sovereign human traits and impulses which originally seek to determine the form of life from within instead of receiving it from the outside in a general, schematically precise form. Even though those lives which are autonomous and characterized by these vital impulses are not entirely impossible in the city, they are, none the less, opposed to it *in abstracto*.

The same factors which, in the exactness and the minute precision of the form of life, have coalesced into a structure of the highest impersonality, have, on the other hand, an influence in a highly personal direction. There is perhaps no psychic phenomenon which is so unconditionally reserved to the city as the blasé outlook. It is at first the consequence of those rapidly shifting stimulations of the nerves which are thrown together in all their contrasts and from which it seems to us the intensification of metropolitan intellectuality seems to be derived. On that account it is not likely that stupid persons who have been hitherto intellectually dead will be blasé. Just as an immoderately sensuous life makes one blasé because it stimulates the nerves to their utmost reactivity until they finally can no longer produce any reaction at all, so, less harmful stimuli, through the rapidity and the contradictoriness of their shifts, force the nerves to make such violent responses, tear them about so brutally that they exhaust their last reserves of strength and, remaining in the same milieu, do not have time for new reserves to form. This incapacity to react to new stimulations with the required amount of energy constitutes in fact that blasé attitude which every child of a large city evinces when compared with the products of the more peaceful and more stable milieu.

Combined with this physiological source of the blasé metropolitan attitude there is another which derives from a money economy. The essence of the blasé attitude is an indifference toward the distinctions between things. Not in the sense that they are not perceived, as is the case of mental dullness, but rather that the meaning and the value of the distinctions between things, and therewith of the things themselves, are experienced as meaningless. They appear to the blasé person in a homogeneous, flat and gray color with no one of them worthy of being preferred to another. This psychic mood is the correct subjective reflection of a complete money economy to the extent that money takes the place of all the manifoldness of things and expresses all qualitative distinctions between them in the distinction of "how much." To the extent that money, with its colorlessness and its indifferent quality, can become a common denominator of all values it becomes the frightful leveler—it hollows out the core of things, their peculiarities, their specific values and their uniqueness and incomparability in a way which is beyond repair. They all float with the same specific gravity in the constantly moving stream of money. They all rest on the same level and are distinguished only by their amounts. In individual cases this coloring, or rather this decoloring of things, through their equation with money, may be imperceptibly small. In the relationship, however, which the wealthy person has to objects which can be bought for money, perhaps indeed in the total character which, for this reason, public opinion now recognizes in these objects, it takes on very considerable proportions. This is why the metropolis is the seat of commerce and it is in it that the purchasability of things appears in quite a different aspect than in simpler economies. It is also the peculiar seat of the blasé attitude. In it is brought to a peak, in a certain way, that achievement in the concentration of purchasable things which stimulates the individual to the highest degree of nervous energy. Through the mere quantitative intensification of the same conditions this achievement is transformed into its opposite, into this peculiar adaptive phenomenon—the blasé attitude—in which the nerves reveal their final possibility of adjusting themselves to the content and the form of metropolitan life by renouncing the response to them. We see that the self-preservation of certain types of personalities is obtained at the cost of devaluing the entire objective world, ending inevitably in dragging the personality downward into a feeling of its own valuelessness.

Cities are above all the seat of the most advanced economic division of labor. They produce such extreme phenomena as the lucrative vocation of the *quatorzieme* in Paris. These are persons who may be recognized by shields on their houses and who hold themselves ready at the dinner hour in appropriate costumes so they can be called upon on short notice in case thirteen persons find themselves at the table. Exactly in the measure of its extension the city offers to an increasing degree the determining conditions for the division of labor. It is a unit which, because of its large size, is receptive to a highly diversified plurality of achievements while at the same time the agglomeration of individuals and their struggle for the customer forces the individual to a type of specialized accomplishment in which he cannot be so easily exterminated by the other. The decisive fact here is that in the life of a city, struggle with nature for the means of life is transformed into a conflict with human beings and the gain which is fought for is granted, not by nature, but by man. For here we find not only the previously mentioned source of specialization but rather the deeper one in which the seller must seek to produce in the person to whom he wishes to sell ever new and unique needs. The necessity to specialize one's product in order to find a source of income which is not yet exhausted and also to specialize a function which cannot be easily supplanted is conducive to differentiation, refinement and enrichment of the needs of the public which obviously must lead to increasing personal variation within this public.

All this leads to the narrower type of intellectual individuation of mental qualities to which the city gives rise in proportion to its size. There is a whole series of causes for this. First of all there is the difficulty of giving one's own personality a certain status within the framework of metropolitan life. Where quantitative increase of value and energy has reached its limits, one seizes on qualitative distinctions, so that, through taking advantage of the existing sensitivity to differences, the attention of the social world can, in some way, be won for oneself. This leads ultimately to the strangest eccentricities, to specifically metropolitan extravagances of self-distanciation, of caprice, of fastidiousness, the meaning of which is no longer to be found in the content of such activity itself but rather in its being a form of "being different"—of making oneself noticeable. For many types of persons these are still the only means of saving for oneself, through the attention gained from others, some sort of self-esteem and the sense of filling a position. In the same sense there operates an apparently insignificant factor which in its effects however is perceptibly cumulative, namely, the brevity and rarity of meetings which are allotted to each individual as compared with social intercourse in a small city. For here we find the attempt to appear to-the-point, clear-cut and individual with extraordinarily greater frequency than where frequent and long association assures to each person an unambiguous conception of the other's personality.

This appears to me to be the most profound cause of the fact that the metropolis places emphasis on striving for the most individual forms of personal existence—regardless of whether it is always correct or always successful. The development of modern culture is characterized by the predominance of what one can call the objective spirit over the subjective; that is, in language as well as in law, in the technique of production as well as in art, in science as well as in the objects of domestic environment, there is embodied a sort of spirit (*Geist*), the daily growth of which is followed only imperfectly and with an even greater lag by the intellectual development of the individual. If we survey for instance the vast culture which during the last century has been embodied in things and in knowledge, in institutions and comforts, and if we compare them with the cultural progress of the individual during the same period—at least in the upper classes—we would see a frightful difference in rate of growth between the two which represents, in many points, rather a regression of the culture of the individual with reference to spirituality, delicacy and idealism. This discrepancy is in essence the result of the success of the growing division of labor. For it is this which requires from the individual an ever more one-sided type of achievement which, at its highest point, often permits his personality as a whole to fall into neglect. In any case this overgrowth of objective culture has been less and less satisfactory for the individual. Perhaps less conscious than in practical activity and in the obscure complex of feelings which flow from him, he is reduced to a negligible quantity. He becomes a single cog as over against the vast overwhelming organization of things and forces which gradually take out of his hands everything connected with progress, spirituality and value. The operation of these forces results in the transformation of the latter from a subjective form into one of purely objective existence. It need only be pointed out that the metropolis is the proper arena for this type of culture which has outgrown every personal element. Here in buildings and in educational institutions, in the wonders and comforts of space-conquering technique, in the formations of social life and in the concrete institutions of the State is to be found such a tremendous richness of crystallizing, depersonalized cultural accomplishments that the personality can, so to speak, scarcely maintain itself in the face of it. From one angle life is made infinitely more easy in the sense that stimulations, interests, and the taking up of time and attention, present themselves from all sides and carry it in a stream which scarcely requires any individual efforts for its ongoing. But from another angle, life is composed more and more of these impersonal cultural elements and existing goods and values which seek to suppress peculiar personal interests and incomparabilities. As a result, in order that this most personal element be saved, extremities and peculiarities and individualizations must be produced and they must be over-exaggerated merely to be brought into the awareness even of the individual himself. The atrophy of individual culture through the hypertrophy of objective culture lies at the root of the bitter hatred which the preachers of the most extreme individualism, in the footsteps of Nietzsche, directed against the metropolis. But it is also the explanation of why indeed they are so passionately loved in the metropolis and indeed appear to its residents as the saviors of their unsatisfied yearnings.

When both of these forms of individualism which are nourished by the quantitative relationships of the metropolis, i.e., individual independence and the elaboration of personal peculiarities, are examined with reference to their historical position, the metropolis attains an entirely new value and meaning in the world history of the spirit. The eighteenth century found the individual in the grip of powerful bonds which had become meaningless—bonds of a political, agrarian, guild and religious nature—delimitations which imposed upon the human being at the same time an unnatural form and for a long time an unjust inequality. In this situation arose the cry for freedom and equality—the belief in the full freedom of movement of the individual in all his social and intellectual relationships which would then permit the same noble essence to emerge equally from all individuals as Nature had placed it in them and as it had been distorted by social life and historical development. Alongside of this liberalistic ideal there grew up in the nineteenth century from Goethe and the Romantics, on the one hand, and from the economic division of labor on the other, the further tendency, namely, that individuals who had been liberated from their historical bonds sought now to distinguish themselves from one another. No longer was it the "general human quality" in every individual but rather his qualitative uniqueness and irreplaceability that now became the criteria of his value. In the conflict and shifting interpretations of these two ways of defining the position of the individual within the totality is to be found the external as well as the internal history of our time. It is the function of the metropolis to make a place for the conflict and for the attempts at unification of both of these in the sense that its own peculiar conditions have been revealed to us as the occasion and the stimulus for the development of both. Thereby they attain a quite unique place, fruitful with an inexhaustible richness of meaning in the development of the mental life. They reveal themselves as one of those great historical structures in which conflicting life-embracing currents find themselves with equal legitimacy. Because of this, however, regardless of whether we are sympathetic or antipathetic with their individual expressions, they transcend the sphere in which a judge-like attitude on our part is appropriate. To the extent that such forces have been integrated, with the fleeting existence of a single cell, into the root as well as the crown of the totality of historical life to which we belong—it is our task not to complain or to condone but only to understand.

THINKING ABOUT THE READING

According to Simmel, the individual is both liberated and repressed by the social forces of modern urban life. How does the objectivism of the urban environment lead individuals to feel alienation from or superficial relationships with those around them? How does it make the urban individual more calculative and rational in contrast with the sentimentality of the rural individual? On the other hand, how does the metropolis allow the urban individual more autonomy and flexibility to define themselves without the cultural pressures found in smaller communities?

Where have you lived in your lifetime—a small town or a large city? Perhaps both? Consider how the social forces of the space you live in affect your inner meaning and development of self. Have you developed a "blasé outlook" in the midst of the disorienting nature of large urban environments? Do you find more sentimentality and familiarity in your interactions within small town communities "where everyone knows your name"?

Conversely, can you give any examples of spaces within urban life that allow for the sentimentality and familiarity that Simmel associates with rural areas? What social forces might impede the social intercourse and personal relationships that Simmel suggests are indicative of rural life?

Gift and Exchange

Zygmunt Bauman

(1990)

Reminders from my creditors pile up on my desk. Some bills are most urgent; still, there are things I must buy—my shoes are in tatters, I cannot work late without a desk lamp, and one needs to eat every day . . . What can I do?

I can go to my brother and ask him for a loan. I'll explain my situation to him. Most probably, he'll grumble a little, and preach to me about the virtue of foresight, prudence and planning, of not living above one's means; but he'll reach into his pocket in the end and count his money. If he finds some, he'll give me what I need. Or at least a part he can afford.

Alternatively, I may go to my bank manager. But it would make no sense to explain to him how dreadfully I suffer. What does he care? The only question he'll ask me is what guarantee I can offer that the loan will be repaid. He'd like to know whether I have a regular income large enough to afford the repayment of the loan together with the interest. So I'll have to show him my salary slips; if I had a property, I'd have to offer it as collateral—a second mortgage, perhaps. If the manager is satisfied that I am not too excessive a risk and that the loan is likely to be duly repaid (with handsome interest, of course), he will lend me the money.

Depending on where I turn to solve my problem, I can expect two very different kinds of treatment. Two different sets of questions, referring obviously to two different conceptions of my right (my entitlement) to receive assistance. My brother is unlikely to inquire about my solvency; for him, the loan is not a matter of choice between good and bad business. What counts for him is that I am his brother; being his brother and being in need, I have sufficient claim to his help. My need is his obligation. The bank manager, on the other hand, could not care less who I am and whether I need the money I ask for. The one thing he would wish to know is whether the loan is likely to be a sensible, profitable business transaction for him or the bank he represents. In no way is he obliged, morally or otherwise, to lend me money. Were my brother to refuse my request, *he* would have to prove to *me* that he was unable to offer me the loan. With the bank manager, it would be the other way around: if I wanted him to offer me a loan, it is *I* who would have to prove to *him* that I am capable of prompt repayment of my debts.

Human interaction succumbs to the pressure of two principles which all too often contradict each other: the principle of *equivalent exchange*, and the principle of the *gift*. In the case of equivalent exchange, self-interest rules supreme. The other partner of the interaction may be recognized as an autonomous person, a legitimate subject of needs and rights—and yet those needs and rights are viewed first and foremost as constraints and obstacles to the full satisfaction of one's own interests. One is guided above all by concern with "just" payment in exchange for the services one renders to the needs of the other. "How much will I be paid?" "What is in it for me?" "Could I be better off doing something else?" "Have I not been cheated?" These and similar questions are what one asks about the prospective action in order to evaluate its desirability and to establish the order of preference between alternative choices. One bargains about the meaning of *equivalence*. One deploys all the resources one can lay one's hand on in order to obtain the best deal possible and tilt the transaction in one's own favor. Not so in the case of a gift. Here, the needs and the rights of others are the main—perhaps the only—motive for action. Rewards, even if they come in the end, are not a factor in the calculation of desirability of action. The concept of equivalence is altogether inapplicable. The goods are given away, the services are extended merely because the other person needs them and, being the person it is, has the right for the needs to be respected.

The two kinds of treatment which we discussed at the beginning of this article offer an example of the daily manifestations of the gift-exchange choice. As a first approximation, we may call my relationship with my brother (in which the gift motive prevailed) a *personal* one, and the relationship between myself and my bank manager (in which the exchange attitude came to the fore) *impersonal.* What happens in the framework of a personal relationship depends almost entirely on who we, the partners, are—and very little on what either of us has done, is doing or will do: on our *quality,* not *performance.* We are brothers, and hence we are obliged to assist each other in need. It does not matter (or, at least, it should not matter) whether the need in question arises from bad luck, miscalculation or improvidence. It matters even less whether the sum offered to bail me out is "secure"—that means, whether my performance is such that it warrants the hope of repayment. In an impersonal relationship, the opposite is true. It is only performance which counts, not the quality. It does not matter who I am, only what I am likely to do. My partner will be interested in my past record, as a basis on which to judge the likelihood of my future behavior.

Talcott Parsons, one of the most influential post-war sociologists, considered the opposition between quality and performance as one of the four major oppositions among which each conceivable pattern of human relationship must choose. He gave these oppositions the name of *pattern variables.* Another pair of opposite options to choose between is, according to Parsons, that of *universalism* and *particularism.* Pondering over my request, my brother might think of many things, yet universal principles, like legal regulations, codes of conduct or current interest rates were not, in all probability, among them. For him, I was not a "specimen of a category," a case to which some universal rule may apply. I was a particular, unique case—his brother. Whatever he was going to do he would do because I was such a unique person, unlike any other, and hence the question "What would he do in other similar circumstances?" simply did not arise. It was again very different with the bank manager. For him, I was just one member of a large category of past, current and prospective borrowers. Having dealt with so many others "like me" before, the bank manager surely worked out universal rules to be applied for all similar cases in the future. The outcome of my particular application would depend therefore on whatever the universal rules say about the credibility of my case.

The next pattern variable also sets the two cases under consideration in opposition to each other. The relationship between me and my brother is *diffuse;* the relationship between me and the bank manager is *specific.* The generosity of my brother was not just a one-off whim; it was not an attitude improvised specifically for the distress I reported in this one conversation. His brotherly predisposition towards me spills over into everything which concerns me—everything is also a matter of his concern. There is nothing in the life of either of us which does not matter for the other. If my brother was inclined to be helpful in this particular case, it is because he is generally well disposed towards me and interested in everything I do and might have done. His understanding and care would not be confined to financial matters. Not so with the bank manager; his conduct was specifically geared to this application there and then; his reaction to my request and his final decision were based entirely on the facts of the case and bore no relation whatsoever to other aspects of my life or personality. Most things important to me he rightly saw as irrelevant as far as the application for the loan was concerned, and hence left them out of consideration.

The fourth opposition crowns, so to speak, the other three (though one could argue as well that it underlies them and, indeed, makes them possible). This is the opposition, in Parsons' words, between *affectivity* and *affective neutrality.* This means that some interactions are infused with emotions—compassion, sympathy or love. Some others are "cool," detached, unemotional. Impersonal relations do not arouse in the actors any feelings other than a passionate urge to achieve a successful transaction. The actors themselves are not objects of emotions; they are neither liked nor disliked. If they strike a hard bargain, try to cheat, prevaricate or avoid commitments, some of the impatience with the unduly slow progress of the transaction may rub off on the attitude towards them; again, some affection may be born if they cooperate in the transaction with zeal and goodwill—when they are the sort of person "it is a pleasure to do business with." By and large, however, emotions are not an indispensable part of impersonal interactions, while they are the very factors which make personal interaction plausible.

My brother and I both feel deeply about our relationship. In all probability, we like each other. It is more than probable, however, that we empathize with each other and have a fellow-feeling for each other: we tend to put ourselves in each other's position, to understand each other's predicament, to imagine the joy or the agony of the other partner, feel good about the joy, and suffer because of the agony. This is hardly the case in my relationship with the bank manager. We meet too seldom and know each other too little to "read out" the partner's feeling. Were we able to do so, we still wouldn't do it—unless the feelings we wished to discover or anticipate in the partner bear a direct relationship to the success of the transaction at hand (I'd try to avoid making the bank manager angry; instead, I'd wish to elicit his or her good humor through jokes or flattery, hoping to relax the defenses—counting on human weaknesses). Otherwise, feelings are neither here nor there. Moreover, they might be positively harmful if allowed to interfere with judgment; if, for instance, my bank manager decides to offer me the loan because of feeling pity and compassion for my misery, and for this reason disregards my financial recklessness, which may easily cause a loss to the bank he or she represents.

While emotion is an indispensable accompaniment to personal relations, it would be out of place in impersonal ones. In the latter, dispassion and cool calculation are the rules one may disregard only at one's own peril. The unemotional stance taken by my partners in an impersonal transaction may well hurt my feelings, particularly when the situation which prompted me to turn to them in the first place has caused me pain and anguish. Unreasonably, I'll then be inclined to blame the unemotional attitude, so jarringly at odds with my excitement, on the "heartlessness and insensitivity of the bureaucrats." This is not an image which will assist the success of impersonal transactions. Hence we hear time and again of "listening banks" and "banks who like to say yes"; banks consider it profitable to conceal the impersonality of their attitude towards clients (that is, their interest in clients' money, not in their private problems and feelings) and therefore promise what they cannot and do not intend to deliver: to conduct impersonal transactions in a mood geared to personal ones.

Perhaps the most crucial distinction between personal and impersonal contexts of interaction lies in the factors on which the actors rely for the success of their action. We all depend on the actions of so many people of whom we know, if anything, very little; much too little to base our plans and our hopes on their personal characteristics, like reliability, trustworthiness, honesty, industry and so on. With so little knowledge at our disposal, a transaction would be downright impossible were it not for the opportunity to settle the issue in an impersonal manner: the chance to appeal not to the personal traits or aptitudes of the partners (which we do not know anyhow), but to the universal rules which apply to all cases of the same category, whoever happens to be our partner at the moment. Under conditions of limited personal knowledge, appealing to rules is the only way to make communication possible. Imagine what an incredibly large, unwieldy volume of knowledge you would need to amass if all your transactions with others were based solely on your properly researched estimate of their personal qualities. The much more realistic alternative is to get hold of the few general rules that guide the interchange, and trust that the partner will do the same and observe the same rules.

Most things in life are indeed organized in such a way as to enable partners to interact without any or with little personal information about each other. It would be quite impossible for me, ignorant about medical science as I am, to assess the healing ability and dedication of the consultants to whom I turn for help; fortunately, however, their competence to deal with my ailment has been confirmed, testified and certified by the British Medical Association, which accepted them as members, and by the hospital management, which employed them. I can, therefore, limit myself to reporting the fact of my case and assume—trust—that in exchange I'll receive the service which the case warrants and requires. When trying to make sure that the train I have boarded is scheduled to travel to the town I wish to reach, I may safely ask people dressed in the British Railroad uniform, without worrying about putting their love of truth to the test. I let into my house someone showing me a gas board inspector's card without going through the checks and inquiries I'd normally apply to a complete stranger. In all these and similar cases some people, personally unknown to me (like those sitting on the board of the Medical Association or the gas board), took upon themselves the task of vouching for the competent, rule-conforming conduct of people whose credentials they endorse. By so doing, they have made it possible for me to accept the services of such people on *trust*.

And yet it is precisely because so many of our transactions are performed in an impersonal context that the need of personal relationships becomes so poignant and acute.

Our craving for "deep and wholesome" personal relationships grows in intensity the wider and less penetrable is the network of impersonal dependencies in which we are entangled. I am an employee in the company where I earn my salary, a customer in the many shops where I buy things I need or believe I need, a passenger on the bus or on the train which takes me to and from my place of work, a viewer in the theatre, a voter in the party which I support, a patient in the doctor's surgery, and so many other things in so many other places. Everywhere I feel that only a small section of myself is present. I must constantly watch myself not to allow the rest of myself to interfere, as its other aspects are irrelevant and unwelcome in this particular context. And thus nowhere do I feel truly myself; nowhere am I fully at home. All in all, I begin to feel like a collection of the many different roles I play, each one among different people and in a different place. Is there something to connect them? Who am I in the end—the true, the real "I"?

As Georg Simmel observed a long time ago, in the densely populated, variegated world we inhabit the individuals tend to fall back upon themselves in the never-ending search for sense and unity. Once focused on ourselves rather than on the world outside, this overwhelming thirst for unity and coherence is articulated as the search for *self-identity*.

The German sociologist Niklas Luhmann presented the search for self-identity as the primary and the most powerful cause of our overwhelming need of love—of loving and being loved. Being loved means being treated by the other person as unique, as unlike any other; it means that the loving persons accept that the loved ones need not invoke universal rules in order to justify the images they hold of themselves or their demands; it means that the loving person accepts and confirms the sovereignty of myself, my right to decide for myself and to choose myself on my own authority; it means that he or she agrees with my emphatic and stubborn statement "Here is what I am, what I do, and where I stand."

Being loved means, in other words, being *understood*—or at least "understood" in the sense in which we use it whenever we say, "I want you to understand me!" or ask with anguish, "Do you understand me? Do you really *understand* me?" This craving for being understood is a desperate call to someone to put himself or herself in my shoes, to see things from my point of view, to accept without further proof that I indeed have such a point of view, which ought to be respected for the simple reason that it is mine. What I am after when craving to be understood is a confirmation that my own, private experience—my inner motives, my image of ideal life, my image of myself, my misery or joy—are *real*. I want a *validation* of my self-portrayal. I find such a validation in my acceptance by another person; in the other person's approval of what I would otherwise suspect of being just a figment of my imagination, my idiosyncrasy, the product of my fantasy running wild. I hope to achieve such a validation through my partner's willingness to listen seriously and with sympathy when I am talking about myself; my partner, in Luhmann's words, should "lower the threshold of relevance": my partner ought to accept everything I say as relevant and worth listening to and thinking about.

As a matter of fact, there is a paradox in my wishes. On one hand, I want myself to be a unique whole, and not just a collection of the roles I put on when "out there," only to take them off the moment I move from one place (or one company) to another. Thus I want to be unlike anybody else, similar to no one but myself—not to be just one of the many cogs in someone else's wheel. On the other hand, I know that nothing exists just because I have imagined it. I know the difference between fantasy and reality, and I know that whatever truly exists must surely exist for others as much as it does for me (remember the knowledge of everyday life each of us has and without which life in society is inconceivable; one of the crucial items of this knowledge is the belief that experiences are *shared*, that the world looks to others the same way as it looks to ourselves). And so the more I succeed in developing a truly unique self, in making my experience unique, the more I need a social confirmation of my experience. It seems, at least at first sight, that such a confirmation can only be had through love. The outcome of the paradox is that, in our complex society in which most human needs are attended to in an impersonal way, the need of a loving relationship is deeper than at any other time. This means also that the burden love must carry is formidable—and so are the pressures, tensions and obstacles the lovers must fight and conquer.

What makes a love relationship particularly vulnerable and fragile is the need for *reciprocity*. If I want to be loved, the partner I select will in all probability ask me to reciprocate—to respond with love. And this means, as we have already noticed, that I should return the services of my lover: act in such a way as to confirm the reality of my partner's experience; to understand at the same time as I am seeking to be understood myself. Ideally, each partner will strive to find meaning in the other partner's world. But the two realities (mine and my partner's) are surely not identical; worse still, they have but a few, if any, common points. When two people meet for the first time, both have behind them a long life of their own which was not shared with the other. Two distinct biographies would in all probability have produced two fairly distinct sets of experiences and expectations. Now they must be renegotiated. At least in some respects the two sets are likely to be found mutually contradictory. It is improbable that I and my partner will be ready to admit right away that both sets, in their entirety, are equally real and acceptable and do not need corrections and compromises. One set, or even both, will have to give way, be trimmed or even surrendered for the sake of a lasting relationship. And yet such a surrender defies the very purpose of love and the very need love is expected to satisfy. If renegotiation indeed takes place, if both partners see it through, the rewards are great. But the road to the happy end is thorny, and much patience and looking forward is needed to travel it unscathed.

The American sociologist Richard Sennett coined the term, *"destructive Gemeinschaft"* for a relationship in which both partners obsessively pursue the right to *intimacy*, to open oneself up to the partner, to share with the partner the whole, most private truth about one's inner life, to be absolutely sincere—that is, to hide nothing, however upsetting the information may be for the partner. In Sennett's view, stripping one's soul bare in front of the partner thrusts an enormous burden on the latter's shoulders, as the partner is asked to give agreement to things which do not necessarily arouse enthusiasm, and to be equally sincere and honest in reply. Sennett does not believe that a lasting relationship, and particularly a lasting *loving* relationship, can be erected on the wobbly ground of mutual intimacy. The odds are overwhelming that the partners will make demands of each other which they cannot meet (or, rather, do not wish to meet, considering the price); they will suffer and feel tormented and frustrated—and more often than not they will decide to call it a day, stop trying and withdraw. One or other of the partners will choose to opt out, and to seek satisfaction of his or her need of self-confirmation elsewhere.

Once again we find out, therefore, that the fragility of the love relationship—the destructiveness of the communion sought by the partners in love—is caused first and foremost by the requirement of reciprocity. Paradoxically, my love would be sustainable and safe only if I did not expect it to be reciprocated. Strange as it may seem, the least vulnerable is love as a gift: I am prepared to accept my beloved's world, to put myself in that world and try to comprehend it from inside—without expecting a similar service in exchange. . . . I need no negotiation, agreement or contract. Once aimed in both directions, however, intimacy makes negotiation and compromise inevitable. And it is precisely the negotiation and compromise which one or both partners may be too impatient, or too self-concerned, to bear lightly. With love being such a difficult and costly achievement, it is no wonder one finds demand for a substitute for love: for someone who would perform the function of love (that is, supply confirmation of inner experience, having first patiently absorbed a full, intimate confession) without demanding reciprocity in exchange. Herein lies the secret of the astounding success and popularity of psychoanalytic sessions, psychological counseling, marriage guidance, etc. For the right to open oneself up, make one's innermost feelings known to another person, and in the end receive the longed-for approval of one's identity, one need only pay money. Monetary payment transforms the analyst's or the therapist's relation to their patients or clients into an impersonal one. And so one can be loved without loving. One can be concerned with oneself, and have the concerns shared, without giving a single thought to the people whose services have been bought, and who have therefore taken upon themselves the obligation of sharing as a part of a business transaction.

Another, perhaps less vulnerable substitute for love (more precisely, for the function of identity-approval) is offered by the consumer market. The market puts on display a wide range of "identities" from which one can select one's own. Commercial advertisements take pains to show the commodities they try to sell in their social context, that means as a part of a particular *lifestyle*, so that the prospective customer can consciously purchase symbols of such self-identity as he or she would wish to possess. The market also offers identity-making tools, which can be used differentially, to produce results which differ somewhat from each other and are in this way personalized. Through the market, one can put together various elements of the complete identikit of a DIY, customized self. One can learn how to express oneself as a modern, liberated, carefree woman; or as a thoughtful, reasonable, caring housewife; or as an aspiring, self-confident tycoon; or as an easy-going, likeable fellow; or as an outdoor, physically fit, macho man; or as a romantic, dreamy and love-hungry creature; or as any mixture of all these. The advantage of market-promoted identities is that they come complete with their social approval, and so the agony of seeking confirmation is spared. Identikits and lifestyle symbols are introduced by people with authority, and supported by the information that very many people approve of them by using them or by "switching to them." Social approval therefore does not need to be negotiated—it has been, so to speak, built into the marketed product from the start.

With such alternatives widely available and growing in popularity, the effort required by the drive to solve the self-identity problem through reciprocal love has an ever smaller chance of success. As we have seen before, negotiating approval is a tormenting experience for the partners in love. Success is not possible without long and dedicated effort. It needs self-sacrifice on both sides. The effort and the sacrifice would perhaps be made more frequently and with greater zeal were it not for the availability of "easy" substitutes. With the substitutes being easy to obtain (the only sacrifice needed is to *part with* a quantity of money) and aggressively peddled by the sellers, there is, arguably, less motivation for a laborious, time-consuming and frequently frustrating effort. Resilience withers when confronted with alluringly "foolproof" and less demanding marketed alternatives. Often the first hurdle, the first setback in the developing and vulnerable love partnership would be enough for one or both partners to wish to slow down, or to leave the track altogether. Often the substitutes are first sought with the intention to "complement," and hence to strengthen or resuscitate, the failing love relationship; sooner or later, however, the substitutes unload that relationship of its original function and drain off the energy which prompted the partners to seek its resurrection in the first place.

A love relationship is thus exposed to a twofold danger. It may collapse under the pressure of inner tensions. Or it may retreat before, or turn itself into, another type of relationship—one which bears many or all the marks of an impersonal relationship: that of *exchange.*

We have observed a typical form of exchange relationship when considering bank customers' transactions with the bank manager. We have noted that the only thing which counted there was the passing of a particular object, or a service, from one side of the transaction to the other; an object was changing hands. The living persons involved in the transactions did not do much more than play the role of carriers or mediators; they prompted and facilitated the circulation of goods. Only apparently was their gaze fixed on their respective partners. In fact, they assigned relevance solely to the object of exchange, while granting the other persons a secondary, derivative importance—as holders or gatekeepers of the goods they wanted. They saw "through" their partners, straight into the goods themselves. The last thing the partners would consider would be the tender feelings or spiritual cravings of their counterparts (that is, unless the mood of the partner should influence the successful completion of the exchange). To put it bluntly, both partners acted *selfishly,* the supreme motive of their action was to give away as little as possible and to get as much as possible; both pursued their own self-interest, concentrating their thought solely on the task at hand. Their aims were, therefore, at cross purposes. We may say that in transactions of impersonal exchange, the interests of the actors are in *conflict.*

Nothing in an exchange transaction is done simply for the sake of the other; nothing about the partner is important unless it may be used to secure a better bargain in the transaction. The actors are therefore naturally suspicious of each other's motives. They fear being cheated. They feel they need to remain wide awake, wary and vigilant. They cannot afford to look the other way, lose their attention for a single moment. They want protection against the selfishness of the other side; they would not, of course, expect the other side to act selflessly, but they insist on a fair deal—that is, on whatever they consider to be an equivalent exchange. Hence the exchange relationships call for a *binding rule,* a law, and an *authority* entrusted with the task of adjudicating the fairness of the transaction and capable of imposing its decision by force in the case of transgression. Various consumer associations, consumer watchdogs, ombudsmen, etc. are established out of this urge for protection. They take upon themselves the difficult task of monitoring the fairness of exchange. They also press the authorities for laws which would restrain the freedom of the stronger side to exploit the ignorance or naivety of the weaker one.

Seldom are the two counterparts of a transaction in a truly equal position: those who produce or sell the goods know much more of the quality of their product than the buyers and users are ever likely to learn. They may well push the product to gullible customers under false pretences, unless constrained by a law like the Trade Descriptions Act. The more complex and technically sophisticated the goods, the less their buyers are able to judge their true quality and value. To avoid being deceived, the prospective buyers have to resort to the help of independent, that is, disinterested, authorities; they would press for a law which clearly states their rights and allows them to make up for their relatively inferior position by taking their case to court.

It is, however, precisely because the partners enter exchange relationships only as functions of exchange, as conveyers of the goods, and consequently remain "invisible" to each other, that they feel much less overwhelmed and tied down than in the case of love relationships. They are much less involved. They do not take upon themselves cumbersome duties, or obligations other than the promise to abide by the terms of the transaction. Aspects of their selves which are not relevant to the transaction at hand are unaffected and retain their autonomy. All in all, they feel that their freedom has not been compromised, and their future choices will not be constrained by the bond they enter. Exchange is relatively "inconsequential" as it is confined to a transaction entered into and finished here and now and restricted in time and space. Neither does it involve the whole of one's personality.

Love and exchange are two extremes of a continuous line along which all human relations may be plotted. In the form we have described them here they seldom appear in your or my experience. We have discussed them in a pure form, as models. Most relationships are "impure," and mix the two models in varying proportions. Most love relationships contain elements of business-like bargaining for the fair rate of exchange in the "I'll do this if you do that" style. Except for a chance encounter or one-off transaction, the actors in exchange relationships seldom remain indifferent to each other for long, and sooner or later more is involved than just money and goods. Each extreme model, however, retains its relative identity even if submerged in a mixed relationship. Each carries its own set of expectations, its own image of the perfect state of affairs—and hence orients the conduct of the actors in its own specific direction. Much of the ambiguity of the relationships we enter with other people can be accounted for by reference to the tensions and contradictions between the two extreme, complementary yet incompatible, sets of expectations. The model-like, pure relationships seldom appear in life, where the ambivalence of human relationships is the rule.

Our dreams and cravings seem to be torn between two needs it well might be impossible to gratify at the same time, yet equally difficult to satisfy when pursued separately. These are the needs of *belonging* and of *individuality.* The first need prompts us to seek strong and secure ties with others. We express this need whenever we speak or think of togetherness or of community. The second need sways us towards privacy, a state in which we are immune to pressures and free from demands, do whatever we think is worth doing, "are ourselves." Both needs are pressing and powerful; the pressure of each grows the less the given need is satisfied. On the other hand, the nearer one need comes to its satisfaction, the more painfully we feel the neglect of the other. We find out that community without privacy feels more like oppression than belonging. And that privacy without community feels more like loneliness than "being oneself."

THINKING ABOUT THE READING

According to Bauman, what is the gift-exchange choice that underlies human relationships? How does he apply these principles to love relationships? Do you see this gift-exchange choice manifested in your social interactions? How do some of Talcott Parsons's pattern variables factor into the principles of equivalent exchange and the gift?

Culture of Fear

Barry Glassner

(1999)

Why are so many fears in the air, and so many of them unfounded? Why, as crime rates plunged throughout the 1990s, did two-thirds of Americans believe they were soaring? How did it come about that by mid-decade 62 percent of us described ourselves as "truly desperate" about crime—almost twice as many as in the late 1980s, when crime rates were higher? Why, on a survey in 1997, when the crime rate had already fallen for a half dozen consecutive years, did more than half of us disagree with the statement "This country is finally beginning to make some progress in solving the crime problem"?[1]

In the late 1990s the number of drug users had decreased by half compared to a decade earlier; almost two-thirds of high school seniors had never used any illegal drugs, even marijuana. So why did a majority of adults rank drug abuse as the greatest danger to America's youth? Why did nine out of ten believe the drug problem is out of control, and only one in six believe the country was making progress?[2]

Give us a happy ending and we write a new disaster story. In the late 1990s the unemployment rate was below 5 percent for the first time in a quarter century. People who had been pounding the pavement for years could finally get work. Yet pundits warned of imminent economic disaster. They predicted inflation would take off, just as they had a few years earlier—also erroneously—when the unemployment rate dipped below 6 percent.[3]

We compound our worries beyond all reason. Life expectancy in the United States has doubled during the twentieth century. We are better able to cure and control diseases than any other civilization in history. Yet we hear that phenomenal numbers of us are dreadfully ill. In 1996 Bob Garfield, a magazine writer, reviewed articles about serious diseases published over the course of a year in the *Washington Post,* the *New York Times,* and *USA Today.* He learned that, in addition to 59 million Americans with heart disease, 53 million with migraines, 25 million with osteoporosis, 16 million with obesity, and 3 million with cancer, many Americans suffer from more obscure ailments such as temporomandibular joint disorders (10 million) and brain injuries (2 million). Adding up the estimates, Garfield determined that 543 million Americans are seriously sick—a shocking number in a nation of 266 million inhabitants. "Either as a society we are doomed, or someone is seriously double-dipping," he suggested.[4]

Garfield appears to have underestimated one category of patients: for psychiatric ailments his figure was 53 million. Yet when Jim Windolf, an editor of the *New York Observer,* collated estimates for maladies ranging from borderline personality disorder (10 million) and sex addiction (11 million) to less well-known conditions such as restless leg syndrome (12 million) he came up with a figure of 152 million. "But give the experts a little time," he advised. "With another new quantifiable disorder or two, everybody in the country will be officially nuts."[5]

Indeed, Windolf omitted from his estimates new-fashioned afflictions that have yet to make it into the *Diagnostic and Statistical Manual of Mental Disorders* of the American Psychiatric Association: ailments such as road rage, which afflicts more than half of Americans, according to a psychologist's testimony before a congressional hearing in 1997.[6]

The scope of our health fears seems limitless. Besides worrying disproportionately about legitimate ailments and prematurely about would-be diseases, we continue to fret over already refuted dangers. Some still worry, for instance, about "flesh-eating bacteria," a bug first rammed into our consciousness in 1994 when the U.S. news media picked up on a screamer headline in a British tabloid, "Killer Bug Ate My Face." The bacteria, depicted as more brutal than anything seen in modern times, was said to be spreading faster than the pack of photographers outside the home of its latest victim. In point of fact, however, we were not "terribly vulnerable" to these "superbugs," nor were they "medicine's worst nightmares," as voices in the media warned.

Group A strep, a cyclical strain that has been around for ages, had been dormant for half a century or more before making a comeback. The British pseudoepidemic had resulted in a total of about a dozen deaths in the previous year. Medical experts roundly rebutted the scares by noting that of 20 to 30 million strep infections each year in the United States fewer than 1 in 1,000 involve serious strep A complications, and only 500 to 1,500 people suffer the flesh-eating syndrome, whose proper name is necrotizing fasciitis. Still the fear persisted. Years after the initial scare, horrifying news stories continued to appear, complete with grotesque pictures of victims. A United Press International story in 1998 typical of the genre told of a child in Texas who died of the "deadly strain" of bacteria that the reporter warned "can spread at a rate of up to one inch per hour."[7]

Killer Kids

When we are not worrying about deadly diseases we worry about homicidal strangers. Every few months for the past several years it seems we discover a new category of people to fear: government thugs in Waco, sadistic cops on Los Angeles freeways and in Brooklyn police stations, mass-murdering youths in small towns all over the country. A single anomalous event can provide us with multiple groups of people to fear. After the 1995 explosion at the federal building in Oklahoma City, first we panicked about Arabs. "Knowing that the car bomb indicates Middle Eastern terrorists at work, it's safe to assume that their goal is to promote free-floating fear and a measure of anarchy, thereby disrupting American life," a *New York Post* editorial asserted. "Whatever we are doing to destroy Mideast terrorism, the chief terrorist threat against Americans, has not been working," wrote A. M. Rosenthal in the *New York Times*.[8]

When it turned out that the bombers were young white guys from middle America, two more groups instantly became spooky: right-wing radio talk show hosts who criticize the government—depicted by President Bill Clinton as "purveyors of hatred and division"—and members of militias. No group of disgruntled men was too ragtag not to warrant big, prophetic news stories.[9]

We have managed to convince ourselves that just about every young American male is a potential mass murderer—a remarkable achievement, considering the steep downward trend in youth crime throughout the 1990s. Faced year after year with comforting statistics, we either ignore them—adult Americans estimate that people under eighteen commit about half of all violent crimes when the actual number is 13 percent—or recast them as "The Lull Before the Storm" (*Newsweek* headline). "We know we've got about six years to turn this juvenile crime thing around or our country is going to be living with chaos," Bill Clinton asserted in 1997, even while acknowledging that the youth violent crime rate had fallen 9.2 percent the previous year.[10]

The more things improve the more pessimistic we become. Violence-related deaths at the nation's schools dropped to a record low during the 1996–97 academic year (19 deaths out of 54 million children), and only one in ten public schools reported *any* serious crime. Yet *Time* and *U.S. News & World Report* both ran headlines in 1996 referring to "Teenage Time Bombs." In a nation of "Children Without Souls" (another *Time* headline that year), "America's beleaguered cities are about to be victimized by a paradigm shattering wave of ultraviolent, morally vacuous young people some call 'the superpredators,'" William Bennett, the former Secretary of Education, and John

DiIulio, a criminologist, forecast in a book published in 1996.[11]

Instead of the arrival of superpredators, violence by urban youths continued to decline. So we went looking elsewhere for proof that heinous behavior by young people was "becoming increasingly more commonplace in America" (CNN). After a sixteen-year-old in Pearl, Mississippi, and a fourteen-year-old in West Paducah, Kentucky, went on shooting sprees in late 1997, killing five of their classmates and wounding twelve others, these isolated incidents were taken as evidence of "an epidemic of seemingly depraved adolescent murderers" (Geraldo Rivera). Three months later in March 1998 all sense of proportion vanished after two boys ages eleven and thirteen killed four students and a teacher in Jonesboro, Arkansas. No longer, we learned in *Time*, was it "unusual for kids to get back at the world with live ammunition." When a child psychologist on NBC's "Today" show advised parents to reassure their children that shootings at schools are rare, reporter Ann Curry corrected him. "But this is the fourth case since October," she said.[12]

Over the next couple of months young people failed to accommodate the trend hawkers. None committed mass murder. Fear of killer kids remained very much in the air nonetheless. In stories on topics such as school safety and childhood trauma, reporters recapitulated the gory details of the killings. And the news media made a point of reporting every incident in which a child was caught at school with a gun or making a death threat. In May, when a fifteen-year-old in Springfield, Oregon, did open fire in a cafeteria filled with students, killing two and wounding twenty-three others, the event felt like a continuation of a "disturbing trend" (*New York Times*). The day after the shooting, on National Public Radio's "All Things Considered," the criminologist Vincent Schiraldi tried to explain that the recent string of incidents did not constitute a trend, that youth homicide rates had declined by 30 percent in recent years, and more than three times as many people were killed by lightning than by violence at schools. But the show's host, Robert Siegel, interrupted him. "You're saying these are just anomalous events?" he asked, audibly peeved. The criminologist reiterated that *anomalous* is precisely the right word to describe the events, and he called it "a grave mistake" to imagine otherwise.

Yet given what had happened in Mississippi, Kentucky, Arkansas, and Oregon, could anyone doubt that today's youths are "more likely to pull a gun than make a fist," as Katie Couric declared on the "Today" show?[13]

Roosevelt Was Wrong

We had better learn to doubt our inflated fears before they destroy us. Valid fears have their place; they cue us to danger. False and overdrawn fears only cause hardship.

Even concerns about real dangers, when blown out of proportion, do demonstrable harm. Take the fear of cancer. Many Americans overestimate the prevalence of the disease, underestimate the odds of surviving it, and put themselves at greater risk as a result. Women in their forties believe they have a 1 in 10 chance of dying from breast cancer, a Dartmouth study found. Their real lifetime odds are more like 1 in 250. Women's heightened perception of risk, rather than motivating them to get checkups or seek treatment, can have the opposite effect. A study of daughters of women with breast cancer found an inverse correlation between fear and prevention: the greater a daughter's fear of the disease the less frequent her breast self-examination. Studies of the general population—both men and women—find that large numbers of people who believe they have symptoms of cancer delay going to a doctor, often for several months. When asked why, they report they are terrified about the pain and financial ruin cancer can cause as well as poor prospects for a cure. The irony of course is that early treatment can prevent precisely those horrors they most fear.[14]

Still more ironic, if harder to measure, are the adverse consequences of public panics. Exaggerated perceptions of the risks of cancer at least produce beneficial by-products, such as bountiful funding for research and treatment of this leading cause of death. When it comes to large-scale panics, however, it is difficult to see how potential victims benefit from the frenzy. Did panics a few years ago over sexual assaults on children by preschool teachers and priests leave children better off? Or did they prompt teachers and clergy to maintain excessive distance from children in their care, as social scientists and journalists who have studied the panics suggest? How well can care givers do their jobs when regulatory agencies, teachers' unions, and archdioceses explicitly prohibit them from any physical contact with children, even kindhearted hugs?[15]

Was it a good thing for children and parents that male day care providers left the profession for fear of being falsely accused of sex crimes? In an article in the *Journal of American Culture,* sociologist Mary DeYoung has argued that day care was "refeminized" as a result of the panics. "Once again, and in the time-honored and very familiar tradition of the family, the primary responsibility for the care and socialization of young children was placed on the shoulders of low-paid women," she contends.[16]

We all pay one of the costs of panics: huge sums of money go to waste. Hysteria over the ritual abuse of children cost billions of dollars in police investigations, trials, and imprisonments. Men and women went to jail for years "on the basis of some of the most fantastic claims ever presented to an American jury," as Dorothy Rabinowitz of the *Wall Street Journal* demonstrated in a series of investigative articles for which she became a Pulitzer Prize finalist in 1996. Across the nation expensive surveillance programs were implemented to protect children from fiends who reside primarily in the imaginations of adults.[17]

The price tag for our panic about overall crime has grown so monumental that even law-and-order zealots find it hard to defend. The criminal justice system costs Americans close to $100 billion a year, most of which goes to police and prisons. In California we spend more on jails than on higher education. Yet increases in the number of police and prison cells do not correlate consistently with reductions in the number of serious crimes committed. Criminologists who study reductions in homicide rates, for instance, find little difference between cities that substantially expand their police forces and prison capacity and others that do not.[18]

The turnabout in domestic public spending over the past quarter century, from child welfare and antipoverty programs to incarceration, did not even produce reductions in *fear* of crime. Increasing the number of cops and jails arguably has the opposite effect: it suggests that the crime problem is all the more out of control.[19]

Panic-driven public spending generates over the long term a pathology akin to one found in drug addicts. The more money and attention we fritter away on our compulsions, the less we have available for our real needs, which consequently grow larger. While fortunes are being spent to protect children from dangers that few ever encounter, approximately 11 million children lack health insurance, 12 million are malnourished, and rates of illiteracy are increasing.[20]

I do not contend, as did President Roosevelt in 1933, that "the only thing we have to fear is fear itself." My point is that we often fear the wrong things. In the 1990s middle-income and poorer Americans should have worried about unemployment insurance, which covered a smaller share of workers than twenty years earlier. Many of us have had friends or family out of work during economic downturns or as a result of corporate restructuring. Living in a nation with one of the largest income gaps of any industrialized country, where the bottom 40 percent of the population is worse off financially than their counterparts two decades earlier, we might also have worried about income inequality. Or poverty. During the mid- and late 1990s 5 million elderly Americans had no food in their homes, more than 20 million people used emergency food programs each year, and one in five children lived in poverty—more than a quarter million of them homeless. All told, a larger proportion of Americans were poor than three decades earlier.[21]

One of the paradoxes of a culture of fear is that serious problems remain widely ignored even though they give rise to precisely the dangers that the populace most abhors. Poverty, for example, correlates strongly with child abuse, crime, and drug abuse. Income inequality is also associated with adverse outcomes for society as a whole. The larger the gap between rich and poor in a society, the higher its overall death rates from heart disease, cancer, and murder. Some social scientists argue that extreme inequality also threatens political stability in a nation such as the United States, where we think of ourselves not as "haves and have nots" but as "haves and will haves." "Unlike the citizens of most other nations, Americans have always been united less by a shared past than by the shared dreams of a better future. If we lose that common future," the Brandeis University economist Robert Reich has suggested, "we lose the glue that holds our nation together."[22] . . .

Two Easy Explanations

In the following discussion I will try to answer two questions: Why are Americans so fearful lately, and why are our fears so often misplaced? To both questions the same two-word answer is commonly given . . . [One] popular explanation blames the news media. We have so many fears, many of them off-base, the argument goes, because the media bombard us with sensationalistic stories designed to increase ratings. This explanation, sometimes called the media-effects theory . . . contains sizable kernels of truth. When researchers from Emory University computed the levels of coverage of various health dangers in popular magazines and newspapers they discovered an inverse relationship: much less space was devoted to several of the major causes of death than to some uncommon causes. The leading cause of death, heart disease, received approximately the same amount of coverage as the eleventh-ranked cause of death, homicide. They found a similar inverse relationship in coverage of risk factors associated with serious illness and death. The lowest-ranking risk factor, drug use, received nearly as much attention as the second-ranked risk factor, diet and exercise.[23]

Disproportionate coverage in the news media plainly has effects on readers and viewers. When Esther Madriz, a professor at Hunter College, interviewed women in New York City about their fears of crime, they frequently responded with the phrase "I saw it in the news." The interviewees identified the news media as both the source of their fears and the reason they believed those fears were valid. Asked in a national poll why they believe the country has a serious crime problem, 76 percent of people cited stories they had seen in the media. Only 22 percent cited personal experience.[24]

When professors Robert Blendon and John Young of Harvard analyzed forty-seven surveys about drug abuse conducted between 1978 and 1997, they too discovered that the news media, rather than personal experience, provide Americans with their predominant fears. Eight out of ten adults say that drug abuse has never caused problems in their family, and the vast majority report relatively little direct experience with problems related to drug abuse. Widespread concern about drug problems emanates, Blendon and Young determined, from scares in the news media, television in particular.[25]

Television news programs survive on scares. On local newscasts, where producers live by the dictum "if it bleeds, it leads," drug, crime, and disaster stories make up most of the news portion of the broadcasts. Evening newscasts on the major networks are somewhat less bloody, but between 1990 and 1998, when the nation's murder rate declined by 20 percent, the number of murder stories on network newscasts increased 600 percent (*not* counting stories about O.J. Simpson).[26]

After the dinnertime newscasts the networks broadcast newsmagazines, whose guiding principle seems to be that no danger is too small to magnify into a national nightmare. Some of the risks reported by such programs would be merely laughable were they not hyped with so much fanfare: "Don't miss *Dateline* tonight or YOU could be the next victim!" Competing for ratings with drama programs and movies during prime-time evening hours, newsmagazines feature story lines that would make a writer for "Homicide" or "ER" wince.[27]

"It can happen in a flash. Fire breaks out on the operating table. The patient is surrounded by flames," Barbara Walters exclaimed on ABC's "20/20" in 1998. The problem—oxygen from a face mask ignited by a surgical instrument—occurs "more often than you might think," she cautioned in her introduction, even though reporter Arnold Diaz would note later, during the actual report, that out of 27 million surgeries each year the situation arises only about a hundred times. No matter, Diaz effectively nullified the reassuring numbers as soon as they left his mouth. To those who "may say it's too small a risk to worry about" he presented distraught victims: a woman with permanent scars on her face and a man whose son had died.[28]

The gambit is common. Producers of TV newsmagazines routinely let emotional accounts trump objective information. In 1994 medical authorities attempted to cut short the brouhaha over flesh-eating bacteria by publicizing the fact that an American is fifty-five times more likely to be struck by lightning than die of the suddenly celebrated microbe. Yet TV journalists brushed this fact aside with remarks like, "whatever the statistics, it's devastating to the victims" (Catherine Crier on "20/20"), accompanied by stomach-turning videos of disfigured patients.[29]

Sheryl Stolberg, then a medical writer for the *Los Angeles Times*, put her finger on what makes the TV newsmagazines so cavalier: "Killer germs are perfect for prime time," she wrote. "They are invisible, uncontrollable, and, in the case of Group A strep, can invade the body in an unnervingly simple manner, through a cut or scrape." Whereas print journalists only described in words the actions of "billions of bacteria" spreading "like underground fires" throughout a person's body, TV newsmagazines made use of special effects to depict graphically how these "merciless killers" do their damage.[30]

In Praise of Journalists

Any analysis of the culture of fear that ignored the news media would be patently incomplete, and of the several institutions most culpable for creating and sustaining scares the news media are arguably first among equals. They are also the most promising candidates for positive change. Yet by the same token critiques such as Stolberg's presage a crucial shortcoming in arguments that blame the media. Reporters not only spread fears, they also debunk them and criticize one another for spooking the public. A wide array of groups, including businesses, advocacy organizations, religious sects, and political parties, promote and profit from scares. News organizations are distinguished from other fear-mongering groups because they sometimes bite the scare that feeds them.

A group that raises money for research into a particular disease is not likely to negate concerns about that disease. A company that sells alarm systems is not about to call attention to the fact that crime is down. News organizations, on the other hand, periodically allay the very fears they arouse to lure audiences. Some newspapers that ran stories about child murderers, rather than treat every incident as evidence of a shocking trend, affirmed the opposite. After the schoolyard shooting in Kentucky the *New York Times* ran a sidebar alongside its feature story with the headline "Despite Recent Carnage, School Violence Is Not on Rise." Following the Jonesboro killings they ran a similar piece, this time on a recently released study showing the rarity of violent crimes in schools.[31]

Several major newspapers parted from the pack in other ways. *USA Today* and the *Washington Post*, for instance, made sure their readers knew that what should worry them is the availability of guns. *USA Today* ran news stories explaining that easy access to guns in homes accounted for increases in the number of juvenile arrests for homicide in rural areas during the 1990s. While other news outlets were respectfully quoting the mother of the thirteen-year-old Jonesboro shooter, who said she did not regret having encouraged her son to learn to fire a gun ("it's like anything else, there's some people that can drink a beer and not become an alcoholic"), *USA Today* ran an op-ed piece proposing legal parameters for gun ownership akin to those for the use of alcohol and motor vehicles. And the paper published its own editorial in support of laws that require gun owners to lock their guns or keep them in locked containers. Adopted at that time by only fifteen states, the laws had reduced the number of deaths among children in those states by 23 percent.[32]

The *Washington Post,* meanwhile, published an excellent investigative piece by reporter Sharon Walsh showing that guns increasingly were being marketed to teenagers and children. Quoting advertisements and statistics from gun manufacturers and the National Rifle Association, Walsh revealed that by 1998 the primary market for guns—white males—had been saturated and an effort to market to women had failed. Having come to see children as its future, the gun industry has taken to running ads like the one Walsh found in a Smith & Wesson catalog: "Seems like only yesterday that your father brought you here for the first time," reads the copy beside a photo of a child aiming a handgun, his father by his side. "Those sure were the good times—just you, dad and his Smith & Wesson."[33]

As a social scientist I am impressed and somewhat embarrassed to find that journalists, more often than media scholars, identify the jugglery involved in making small hazards appear huge and huge hazards disappear from sight. Take, for example, the scare several years ago over the Ebola virus. Another *Washington Post* reporter, John Schwartz, identified a key bit of hocus-pocus used to sell that scare. Schwartz called it "the Cuisinart Effect," because it involves the mashing together of images and story lines from fiction and reality. A report by *Dateline NBC* on death in Zaire, for instance, interspersed clips from *Outbreak,* a movie whose plot involves a lethal virus that threatens to kill the entire U.S. population. Alternating between Dustin Hoffman's character exclaiming, "We can't stop it!" and real-life science writer Laurie Garrett, author of *The Coming Plague,* proclaiming that "HIV is not an aberration . . . it's part of a trend," *Dateline*'s report gave the impression that swarms of epidemics were on their way.[34]

Another great journalist-debunker, Malcolm Gladwell, noted that the book that had inspired *Outbreak,* Richard Preston's *The Hot Zone,* itself was written "in self-conscious imitation of a sci-fi thriller." In the real-world incident that occasioned *The Hot Zone,* monkeys infected in Zaire with a strain of Ebola virus were quarantined at a government facility in Reston, Virginia. The strain turned out not to be lethal in humans, but neither Preston in his book nor the screenwriters for *Outbreak* nor TV producers who sampled from the movie let that anticlimax interfere with the scare value of their stories. Preston speculates about an airborne strain of Ebola being carried by travelers from African airports to European, Asian, and American cities. In *Outbreak* hundreds of people die from such an airborne strain before a cure is miraculously discovered in the nick of time to save humanity. In truth, Gladwell points out in a piece in *The New Republic,* an Ebola strain that is both virulent to humans and airborne is unlikely to emerge and would mutate rapidly if it did, becoming far less potent before it had a chance to infect large numbers of people on a single continent, much less throughout the globe. "It is one of the ironies of the analysis of alarmists such as Preston that they are all too willing to point out the limitations of human beings, but they neglect to point out the limitations of microscopic life forms," Gladwell notes.[35]

Such disproofs of disease scares appear rather frequently in general-interest magazines and newspapers, including in publications where one might not expect to find them. The *Wall Street Journal,* for instance, while primarily a business publication and itself a retailer of fears about governmental regulators, labor unions, and other corporate-preferred hobgoblins, has done much to demolish medical myths. Among my personal favorites is an article published in 1996 titled "Fright by the Numbers," in which reporter Cynthia Crossen rebuts a cover story in *Time* magazine on prostate cancer. One in five men will get the disease, *Time* thundered. "That's scary. But it's also a lifetime risk—the accumulated risk over some 80 years of life," Crossen responds. A forty-year-old's chance of coming down with (not dying of) prostate cancer in the next ten years is 1 in 1,000, she goes on to report. His odds rise to 1 in 100 over twenty years. Even by the time he's seventy, he has only a 1 in 20 chance of *any* kind of cancer, including prostate.[36]

In the same article Crossen counters other alarmist claims as well, such as the much-repeated pronouncement that one in three Americans is obese. The number actually refers to how many are overweight, a less serious condition. Fewer are *obese* (a term that is less than objective itself), variously defined as 20 to 40 percent above ideal body weight as determined by current standards.[37]

Morality and Marketing

To blame the media is to oversimplify the complex role that journalists play as both proponents and doubters of popular fears. . . . Why do news organizations and their audiences find themselves drawn to one hazard rather than another?

Mary Douglas, the eminent anthropologist who devoted much of her career to studying how people interpret risk, pointed out that every society has an almost infinite quantity of potential dangers from which to choose. Societies differ both in the types of dangers they select and the number. Dangers get selected for special emphasis, Douglas showed, either because they offend the basic moral principles of the society or because they enable criticism of disliked groups and institutions. In *Risk and Culture,* a book she wrote with Aaron Wildavsky, the authors give an example from fourteenth-century Europe. Impure water had been a health danger long before that time, but only after it became convenient to accuse Jews of poisoning the wells did people become preoccupied with it.

Or take a more recent institutional example. In the first half of the 1990s U.S. cities spent at least $10 billion to purge asbestos from public schools, even though removing asbestos from buildings posed a greater health hazard than leaving it in place. At a time when about one-third of the nation's schools were in need of extensive repairs the money might have been spent to renovate dilapidated buildings. But hazards posed by seeping asbestos are morally repugnant. A product that was supposed to protect children from fires might be giving them cancer. By directing our worries and dollars at asbestos we express outrage at technology and industry run afoul.[38]

From a psychological point of view extreme fear and outrage are often projections. Consider, for example, the panic over violence against children. By failing to provide adequate education, nutrition, housing, parenting, medical services, and child care over the past couple of decades we have done the nation's children immense harm. Yet we project our guilt onto a cavalcade of bogeypeople—pedophile preschool teachers, preteen mass murderers, and homicidal au pairs, to name only a few.[39]

When Debbie Nathan, a journalist, and Michael Snedeker, an attorney, researched the evidence behind publicized reports in the 1980s and early 1990s of children being ritually raped and tortured they learned that although seven out of ten Americans believed that satanic cults were committing these atrocities, few of the incidents had actually occurred. At the outset of each ritual-abuse case the children involved claimed they had not been molested. They later changed their tunes at the urging of parents and law enforcement authorities. The ghastly tales of abuse, it turns out, typically came from the parents themselves, usually the mothers, who had convinced themselves they were true. Nathan and Snedeker suggest that some of the mothers had been abused themselves and projected those horrors, which they had trouble facing directly, onto their children. Other mothers, who had not been victimized in those ways, used the figure of ritually abused children as a medium of protest against male dominance more generally. Allegations of children being raped allowed conventional wives and mothers to speak out against men and masculinity without having to fear they would seem unfeminine. "The larger culture," Nathan and Snedeker note, "still required that women's complaints about inequality and sexual violence be communicated through the innocent, mortified voice of the child."

Diverse groups used the ritual-abuse scares to diverse ends. Well-known feminists such as Gloria Steinem and Catharine MacKinnon took up the cause, depicting ritually abused children as living proof of the ravages of patriarchy and the need for fundamental social reform.[40]

This was far from the only time feminist spokeswomen have mongered fears about sinister breeds of men who exist in nowhere near the high numbers they allege. Another example occurred a few years ago when teen pregnancy was much in the news. Feminists helped popularize the frightful but erroneous statistic that two out of three teen mothers had been seduced and abandoned by adult men. The true figure is more like one in ten, but some feminists continued to cultivate the scare well after the bogus stat had been definitively debunked.[41] . . .

Final Thoughts

The short answer to why Americans harbor so many misbegotten fears is that immense power and money await those who tap into our moral insecurities and supply us with symbolic substitutes. . . .

(1) Statements of alarm by newscasters; [and] (2) glorification of wannabe experts are two telltale tricks of the fear mongers' trade. [Other tactics include] (3) the use of poignant anecdotes in place of scientific evidence; (4) the christening of isolated incidents as trends; and (5) depletions of entire categories of people as innately dangerous.

If journalists would curtail such practices, there would be fewer anxious and misinformed Americans. Ultimately, though, neither the ploys that narrators use nor what Cantril termed "the sheer dramatic excellence" of their presentations fully accounts for why people in 1938 swallowed a tall tale about martians taking over New Jersey or why people today buy into tales about perverts taking over cyberspace, unionizing employees taking over workplaces, heroin dealers taking over middle-class suburbs, and so forth.[42] . . .

Fear mongers have knocked the optimism out of us by stuffing us full of negative presumptions about our fellow citizens and social institutions. But the United States is a wealthy nation. We have the resources to feed, house, educate, insure, and disarm our communities if we resolve to do so.

There should be no mystery about where much of the money and labor can be found—in the culture of fear itself. We waste tens of billions of dollars and person-hours every year on largely mythical hazards like road rage, on prison cells occupied by people who pose little or no danger to others, on programs designed to protect young people from dangers that few of them ever face, on compensation for victims of metaphorical illnesses, and on technology to make airline travel—which is already safer than other means of transportation—safer still.

We can choose to redirect some of those funds to combat serious dangers that threaten large numbers of people. At election time we can choose candidates that proffer programs rather than scares.[43]

Or we can go on believing in martian invaders.

Notes

1. Crime data here and throughout are from reports of the Bureau of Justice Statistics unless otherwise noted. Fear of crime: Esther Madriz, *Nothing*

Bad Happens to Good Girls (Berkeley: University of California Press, 1997), ch. 1; Richard Morin, "As Crime Rate Falls, Fears Persist," *Washington Post* National Edition, 16 June 1997, p. 35; David Whitman, "Believing the Good News," *U.S. News & World Report*, 5 January 1998, pp. 45–46.

2. Eva Bertram, Morris Blachman et al., *Drug War Politics* (Berkeley: University of California Press, 1996), p. 10; Mike Males, *Scapegoat Generation* (Monroe, ME: Common Courage Press, 1996), ch. 6; Karen Peterson, "Survey: Teen Drug Use Declines," *USA Today*, 19 June 1998, p. A6; Robert Blendon and John Young, "The Public and the War on Illicit Drugs," *Journal of the American Medical Association* 279 (18 March 1998): 827–32. In presenting these statistics and others I am aware of a seeming paradox: I criticize the abuse of statistics by fearmongering politicians, journalists, and others but hand down precise-sounding numbers myself. Yet to eschew all estimates because some are used inappropriately or do not withstand scrutiny would be as foolhardy as ignoring all medical advice because some doctors are quacks. Readers can be assured I have interrogated the statistics presented here as factual. As notes throughout the book make clear, I have tried to rely on research that appears in peer-reviewed scholarly journals. Where this was not possible or sufficient, I traced numbers back to their sources, investigated the research methodology utilized to produce them, or conducted searches of the popular and scientific literature for critical commentaries and conflicting findings.

3. Bob Herbert, "Bogeyman Economics," *New York Times*, 4 April 1997, p. A15; Doug Henwood, "Alarming Drop in Unemployment," *Extra*, September 1994, pp. 16–17; Christopher Shea, "Low Inflation and Low Unemployment Spur Economists to Debate 'Natural Rate' Theory," *Chronicle of Higher Education*, 24 October 1997, p. A13.

4. Bob Garfield, "Maladies by the Millions," *USA Today*, 16 December 1996, p. A15.

5. Jim Windolf, "A Nation of Nuts," *Wall Street Journal*, 22 October 1997, p. A22.

6. Andrew Ferguson, "Road Rage," *Time*, 12 January 1998, pp. 64–68; Joe Sharkey, "You're Not Bad, You're Sick. It's in the Book," *New York Times*, 28 September 1997, pp. Nl, 5.

7. Malcolm Dean, "Flesh-eating Bugs Scare," *Lancet* 343 (4 June 1994): 1418; "Flesh-eating Bacteria," *Science* 264 (17 June 1994): 1665; David Brown, "The Flesh-eating Bug," *Washington Post* National Edition, 19 December 1994, p. 34; Sarah Richardson, "Tabloid Strep," *Discover* (January 1995): 71; Liz Hunt, "What's Bugging Us," *The Independent*, 28 May 1994, p. 25; Lisa Seachrist, "The Once and Future Scourge," *Science News* 148 (7 October 1995): 234–35. Quotes are from Bernard Dixon, "A Rampant Non-epidemic," *British Medical Journal* 308 (11 June 1994): 1576–77; and Michael Lemonick and Leon Jaroff, "The Killers All Around," *Time*, 12 September 1994, pp. 62–69. More recent coverage: "Strep A Involved in Baby's Death," UPI, 27 February 1998; see also, e.g., Steve Carney, "Miracle Mom," *Los Angeles Times*, 4 March 1998, p. A6; KTLA, "News at Ten," 28 March 1998.

8. Jim Naureckas, "The Jihad That Wasn't," *Extra*, July 1995, pp. 6–10, 20 (contains quotes). See also Edward Said, "A Devil Theory of Islam," *Nation*, 12 August 1996, pp. 28–32.

9. Lewis Lapham, "Seen but Not Heard," *Harper's*, July 1995, pp. 29–36 (contains Clinton quote). See also Robin Wright and Ronald Ostrow, "Illusion of Immunity Is Shattered," *Los Angeles Times*, 20 April 1995, pp. Al, 18; Jack Germond and Jules Witcover, "Making the Angry White Males Angrier," column syndicated by Tribune Media Services, May 1995; and articles by James Bennet and Michael Janofsky in the *New York Times*, May 1995.

10. Tom Morganthau, "The Lull Before the Storm?" *Newsweek*, 4 December 1995, pp. 40–42; Mike Males, "Wild in Deceit," *Extra*, March 1996, pp. 7–9; *Progressive*, July 1997, p. 9 (contains Clinton quote); Robin Templeton, "First, We Kill All the 11-Year-Olds," *Salon*, 27 May 1998.

11. Statistics from "Violence and Discipline Problems in U.S. Public Schools: 1996–97," National Center on Education Statistics, U.S. Department of Education, Washington, DC, March 1998; CNN, "Early Prime," 2 December 1997; and Tamar Lewin, "Despite Recent Carnage, School Violence Is Not on Rise," *New York Times*, 3 December 1997, p. A14. Headlines: *Time*, 15 January 1996; *U.S. News & World Report*, 25 March 1996; Margaret Carlson, "Children Without Souls," *Time*, 2 December 1996, p. 70. William J. Bennett, John J. Dilulio, and John Walters, *Body Count* (New York: Simon & Schuster, 1996).

12. CNN, "Talkback Live," 2 December 1997; CNN, "The Geraldo Rivera Show," 11 December 1997; Richard Lacayo, "Toward the Root of Evil," *Time*, 6 April 1998, pp. 38–39; NBC, "Today," 25 March 1998. See also Rick Bragg, "Forgiveness, After 3 Die in Shootings in Kentucky," *New York Times*, 3 December 1997, p. A14; Maureen Downey, "Kids and Violence," 28 March 1998, *Atlanta Journal and Constitution*, p. A12.

13. Jocelyn Stewart, "Schools Learn to Take Threats More Seriously," *Los Angeles Times,* 2 May 1998, pp. Al, 17; "Kindergarten Student Faces Gun Charges," *New York Times,* 11 May 1998, p. A11; Rick Bragg, "Jonesboro Dazed by Its Darkest Day" and "Past Victims Relive Pain as Tragedy Is Repeated," *New York Times,* 18 April 1998, p. A7, and idem, 25 May 1998, p. A8. Remaining quotes are from Tamar Lewin, "More Victims and Less Sense in Shootings," *New York Times,* 22 May 1998, p. A20; NPR, "All Things Considered," 22 May 1998; NBC, "Today," 25 March 1998. See also Mike Males, "Who's Really Killing Our Schoolkids," *Los Angeles Times,* 31 May 1998, pp. M1, 3; Michael Sniffen, "Youth Crime Fell in 1997, Reno Says," Associated Press, 20 November 1998.

14. Overestimation of breast cancer: William C. Black et al., "Perceptions of Breast Cancer Risk and Screening Effectiveness in Women Younger Than 50," *Journal of the National Cancer Institute* 87 (1995): 720–31; B. Smith et al., "Perception of Breast Cancer Risk Among Women in Breast and Family History of Breast Cancer," *Surgery* 120 (1996): 297–303. Fear and avoidance: Steven Berman and Abraham Wandersman, "Fear of Cancer and Knowledge of Cancer," *Social Science and Medicine* 31 (1990): 81–90; S. Benedict et al., "Breast Cancer Detection by Daughters of Women with Breast Cancer," *Cancer Practice* 5 (1997): 213–19; M. Muir et al., "Health Promotion and Early Detection of Cancer in Older Adults," *Cancer Oncology Nursing Journal 7* (1997): 82–89. For a conflicting finding see Kevin McCaul et al., "Breast Cancer Worry and Screening," *Health Psychology* 15 (1996): 430–33.

15. Philip Jenkins, *Pedophiles and Priests* (New York: Oxford University Press, 1996), see esp. ch. 10; Debbie Nathan and Michael Snedeker, *Satan's Silence* (New York: Basic Books, 1995), see esp. ch. 6; Jeffrey Victor, "The Danger of Moral Panics," *Skeptic* 3 (1995): 44–51. See also Noelle Oxenhandler, "The Eros of Parenthood," *Family Therapy Networker* (May 1996): 17–19.

16. Mary DeYoung, "The Devil Goes to Day Care," *Journal of American Culture* 20 (1997): 19–25.

17. Dorothy Rabinowitz, "A Darkness in Massachusetts," *Wall Street Journal,* 30 January 1995, p. A20 (contains quote); "Back in Wenatchee" (unsigned editorial), *Wall Street Journal,* 20 June 1996, p. A18; Dorothy Rabinowitz, "Justice in Massachusetts," *Wall Street Journal,* 13 May 1997, p. A19. See also Nathan and Snedeker, *Satan's Silence;* James Beaver, "The Myth of Repressed Memory," *Journal of Criminal Law and Criminology* 86 (1996): 596–607; Kathryn Lyon, *Witch Hunt* (New York: Avon, 1998); Pam Belluck, "'Memory' Therapy Leads to a Lawsuit and Big Settlement," *New York Times,* 6 November 1997, pp. A1, 10.

18. Elliott Currie, *Crime and Punishment in America* (New York: Metropolitan, 1998); Tony Pate et al., *Reducing Fear of Crime in Houston and Newark* (Washington, DC: Police Foundation, 1986); Steven Donziger, *The Real War on Crime* (New York: HarperCollins, 1996); Christina Johns, *Power, Ideology and the War on Drugs* (New York: Praeger, 1992); John Irwin et al., "Fanning the Flames of Fear," *Crime and Delinquency* 44 (1998): 32–48.

19. Steven Donziger, "Fear, Crime and Punishment in the U.S.," *Tikkun* 12 (1996): 24–27, 77.

20. Peter Budetti, "Health Insurance for Children," *New England Journal of Medicine* 338 (1998): 541–42; Eileen Smith, "Drugs Top Adult Fears for Kids' Well-being," *USA Today,* 9 December 1997, p. D1. Literacy statistic: Adult Literacy Service.

21. "The State of America's Children," report by the Children's Defense Fund, Washington, DC, March 1998; "Blocks to Their Future," report by the National Law Center on Homelessness and Poverty, Washington, DC, September 1997; reports released in 1998 from the National Center for Children in Poverty, Columbia University, New York; Douglas Massey, "The Age of Extremes," *Demography* 33 (1996): 395–412; Notes Trudy Lieberman, "Hunger in America," *Nation,* 30 March 1998, pp. 11–16; David Lynch, "Rich Poor World," *USA Today,* 20 September 1996, p. B1; Richard Wolf, "Good Economy Hasn't Helped the Poor," *USA Today,* 10 March 1998, p. A3; Robert Reich, "Broken Faith," *Nation,* 16 February 1998, pp. 11–17.

22. Inequality and mortality studies: Bruce Kennedy et al., "Income Distribution and Mortality," *British Medical Journal* 312 (1996): 1004–7; Ichiro Kawachi and Bruce Kennedy, "The Relationship of Income Inequality to Mortality," *Social Science and Medicine* 45 (1997): 1121–27. See also Barbara Chasin, *Inequality and Violence in the United States* (Atlantic Highlands, NJ: Humanities Press, 1997). Political stability: John Sloan, "The Reagan Presidency, Growing Inequality, and the American Dream," *Policy Studies Journal* 25 (1997): 371–86 (contains Reich quotes and "will haves" phrase). On both topics see also Philippe Bourgois, *In Search of Respect: Selling Crack in El Barrio* (Cambridge: Cambridge University Press, 1996); William J. Wilson, *When Work Disappears* (New York, Knopf,

1996); Richard Gelles, "Family Violence," *Annual Review of Sociology* 11 (1985): 347–67; Sheldon Danziger and Peter Gottschalk, *America Unequal* (Cambridge, MA: Harvard University Press, 1995); Claude Fischer et al., *Inequality by Design* (Princeton, NJ: Princeton University Press, 1996).

23. Karen Frost, Erica Frank et al., "Relative Risk in the News Media," *American Journal of Public Health* 87 (1997): 842–45. Media-effects theory: Nancy Signorielli and Michael Morgan, eds., *Cultivation Analysis* (Newbury Park, CA: Sage, 1990); Jennings Bryant and Dolf Zillman, eds., *Media Effects* (Hillsdale, NJ: Erlbaum, 1994); Ronald Jacobs, "Producing the News, Producing the Crisis," *Media, Culture and Society* 18 (1996): 373–97.

24. Madriz, *Nothing Bad Happens to Good Girls,* see esp. pp. 111–14; David Whitman and Margaret Loftus, "Things Are Getting Better? Who Knew," *U.S. News & World Report,* 16 December 1996, pp. 30–32.

25. Blendon and Young, "War on Illicit Drugs." See also Ted Chiricos et al., "Crime, News and Fear of Crime," *Social Problems* 44 (1997): 342–57.

26. Steven Stark, "Local News: The Biggest Scandal on TV," *Washington Monthly* (June 1997): 38–41; Barbara Bliss Osborn, "If It Bleeds, It Leads," *Extra,* September–October 1994, p. 15; Jenkins, *Pedophiles and Priests,* pp. 68–71; "It's Murder," *USA Today,* 20 April 1998, p. D2; Lawrence Grossman, "Does Local TV News Need a National Nanny?" *Columbia Journalism Review* (May 1998): 33.

27. Regarding fearmongering by newsmagazines, see also Elizabeth Jensen et al., "Consumer Alert," *Brill's Content* (October 1998): 130–47.

28. ABC, "20/20," 16 March 1998.

29. Thomas Maugh, "Killer Bacteria a Rarity," *Los Angeles Times,* 3 December 1994, p. A29; Ed Siegel, "Roll Over, Ed Murrow," *Boston Globe,* 21 August 1994, p. 14. Crier quote from ABC's "20/20," 24 June 1994.

30. Sheryl Stolberg, "'Killer Bug' Perfect for Prime Time," *Los Angeles Times,* 15 June 1994, pp. A1, 30–31. Quotes from Brown, "Flesh-eating Bug"; and Michael Lemonick and Leon Jaroff, "The Killers All Around," *Time,* 12 September 1994, pp. 62–69.

31. Lewin, "More Victims and Less Sense"; Tamar Lewin, "Study Finds No Big Rise in Public-School Crimes," *New York Times,* 25 March 1998, p. A18.

32. "Licensing Can Protect," *USA Today,* 7 April 1998, p. A11; Jonathan Kellerman, "Few Surprises When It Comes to Violence," *USA Today,* 27 March 1998, p. A13; Gary Fields, "Juvenile Homicide Arrest Rate on Rise in Rural USA," *USA Today,* 26 March 1998, p. A11; Karen Peterson and Glenn O'Neal, "Society More Violent, So Are Its Children," *USA Today,* 25 March 1998, p. A3; Scott Bowles, "Armed, Alienated and Adolescent," *USA Today,* 26 March 1998, p. A9. Similar suggestions about guns appear in Jonathan Alter, "Harnessing the Hysteria," *Newsweek,* 6 April 1998, p. 27.

33. Sharon Walsh, "Gun Sellers Look to Future—Children," *Washington Post,* 28 March 1998, pp. A1, 2.

34. John Schwartz, "An Outbreak of Medical Myths," *Washington Post* National Edition, 22 May 1995, p. 38.

35. Richard Preston, *The Hot Zone* (New York: Random House, 1994); Malcolm Gladwell, "The Plague Year," *New Republic,* 17 July 1995, p. 40.

36. Erik Larson, "A False Crisis: How Workplace Violence Became a Hot Issue," *Wall Street Journal,* 13 October 1994, pp. A1, 8; Cynthia Crossen, "Fright By the Numbers," *Wall Street Journal,* 11 April 1996, pp. B1, 8. See also G. Pascal Zachary, "Junk History," *Wall Street Journal,* 19 September 1997, pp. A1, 6.

37. On variable definitions of obesity see also Werner Cahnman, "The Stigma of Obesity," *Sociological Quarterly* 9 (1968): 283–99; Susan Bordo, *Unbearable Weight* (Berkeley: University of California Press, 1993); Joan Chrisler, "Politics and Women's Weight," *Feminism and Psychology* 6 (1996): 181–84.

38. Mary Douglas and Aaron Wildavsky, *Risk and Culture* (Berkeley: University of California Press, 1982), see esp. pp. 6–9; Mary Douglas, *Risk and Blame* (London: Routledge, 1992). See also Mary Douglas, *Purity and Danger* (New York: Praeger, 1966). Asbestos and schools: Peter Cary, "The Asbestos Panic Attack," *U.S. News & World Report,* 20 February 1995, pp. 61–64; Children's Defense Fund, "State of America's Children."

39. See Marina Warner, "Peroxide Mug-shot," *London Review of Books,* 1 January 1998, pp. 10–11.

40. Nathan and Snedeker, *Satan's Silence* (quote from p. 240). See also David Bromley, "Satanism: The New Cult Scare," in James Richardson et al., eds., *The Satanism Scare* (Hawthorne, NY: Aldine de Gruyter, 1991), pp. 49–71.

41. Of girls ages fifteen to seventeen who gave birth, fewer than one in ten were unmarried and had been made pregnant by men at least five years older. See Steven Holmes, "It's Awful, It's Terrible, It's . . . Never Mind," *New York Times,* 6 July 1997, p. E3.

42. CNN, "Crossfire," 27 August 1995 (contains Huffington quote); Ruth Conniff, "Warning:

Feminism Is Hazardous to Your Health," *Progressive,* April 1997, pp. 33–36 (contains Sommers quote). See also Susan Faludi, *Backlash* (New York: Crown, 1991); Deborah Rhode, "Media Images, Feminist Issues," *Signs* 20 (1995): 685–710; Paula Span, "Did Feminists Forget the Most Crucial Issues?" *Los Angeles Times,* 28 November 1996, p. E8.

43. See Katha Pollitt, "Subject to Debate," *Nation,* 26 December 1994, p. 788, and idem, 20 November 1995, p. 600.

THINKING ABOUT THE READING

Glassner originally wrote this piece over a decade ago. What are some contemporary examples of cultural fears that he might include if he were writing this today? How do you determine if the fear is a cultural myth or something that should be taken seriously as a social problem? According to Glassner, how are these cultural myths created, and why are we so inclined to believe in them? Do you think a culture less organized by the medium of television would be more or less likely to support such myths?

PART II

The Construction of Self and Society

Building Reality

The Social Construction of Knowledge

Sociologists often talk about reality as a social construction. What they mean is that truth and knowledge are discovered, communicated, reinforced, and changed by members of society. Truth doesn't just fall from the sky and hit us on the head. What is considered truth or knowledge is specific to a given culture. All cultures have specific rules for determining what counts as good and right and true. As social beings, we respond to our interpretations and definitions of situations, not to the situations themselves. We learn from our cultural environment what sorts of ideas and interpretations are reasonable and expected. Thus, we make sense of situations and events in our lives by applying culturally shared definitions and interpretations. In this way, we distinguish fact from fantasy, truth from fiction, myth from reality. This process of interpretation or "meaning making" is tied to interpersonal interaction, group membership, culture, history, power, economics, and politics.

Discovering patterns and determining useful knowledge are the goals of any academic discipline. The purpose of an academic field such as sociology is to provide the public with useful and relevant information about how society works. This task is typically accomplished through systematic social research—experiments, field research, unobtrusive observation, and surveys. But gathering trustworthy data can be difficult. People sometimes lie or have difficulty recalling past events in their lives. Sometimes the simple fact of observing people's behavior changes that behavior. Sometimes the information needed to answer questions about important, controversial issues is hard to obtain without raising ethical issues.

Moreover, sometimes the characteristics and phenomena we're interested in understanding are difficult to observe and measure. Unlike other disciplines (say, the natural sciences), sociologists deal with concepts that can't be seen or touched. In "Concepts, Indicators, and Reality," Earl Babbie gives us a brief introduction to some of the problems researchers face when they try to transform important but abstract concepts into *indicators* (things that researchers can systematically quantify so they can generate statistical information). In so doing, he shows us that although sociologists provide us with useful empirical findings about the world in which we live, an understanding of the measurement difficulties they face will provide us with the critical eye of an informed consumer as we go about digesting research information.

In a similar vein, in "Missing Numbers," Joel Best shows us how, in discussions of social issues such as school shootings, relevant and obtainable statistics that can help us better understand the issue are often ignored by the media. He details several reasons for these "missing numbers" and warns us about the risk of misunderstanding certain events and societal patterns without the use of accurate statistics. As Best acknowledges, the process of gathering data and generating statistics is never a perfect one, but our potential to be competent and informed citizens of our socially constructed reality is affected by our ability to recognize these missing numbers.

Something to Consider as You Read

Babbie's and Best's comments remind us that even scientists must make decisions about how to interpret information. Thus, scientists, working within academic communities, define truth and knowledge. This knowledge is often significant and useful, but we need to remember that it is the construction of a group of people following particular rules, not something that is just "out there." As you read these selections, think about the kind of information you would need or would want that might convince you to question some truth that you have always taken for granted.

Concepts, Indicators, and Reality

Earl Babbie

(1986)

Measurement is one of the fundamental aspects of social research. When we describe science as logical/empirical, we mean that scientific conclusions should (1) make sense and (2) correspond to what we can observe. It is the second of these characteristics I want to explore in this essay.

Suppose we are interested in learning whether education really reduces prejudice. To do that, we must be able to measure both prejudice and education. Once we've distinguished prejudiced people from unprejudiced people and educated people from uneducated people, we'll be in a position to find out whether the two variables are related.

Social scientific measurement operates in accordance with the following implicit model:

- Prejudice exists as a *variable*: some people are more prejudiced than others.
- There are numerous *indicators* of prejudice.
- None of the indicators provides a perfect reflection of prejudice as it "really" is, but they can point to it at least approximately.
- We should try to find better and better indicators of prejudice—indicators that come ever closer to the "real thing."

This model applies to all of the variables social scientists study. Take a minute to look through the following list of variables commonly examined in social research.

Arms race	Tolerance
Religiosity	Fascism
Urbanism	Parochialism
TV watching	Maturity
Susceptibility	Solidarity
Stereotyping	Instability
Anti-Semitism	Education
Voting	Liberalism
Dissonance	Authoritarianism
Pessimism	Race
Anxiety	Happiness
Revolution	Powerlessness
Alienation	Mobility
Social class	Consistency
Age	Delinquency
Self-esteem	Compassion
Idealism	Democracy
Prestige	Influence

Even if you've never taken a course in social science, many of these terms are at least somewhat familiar to you. Social scientists study things that are of general interest to everyone. The nuclear arms race affects us all, for example, and it is a special concern for many of us. Differences in *religiosity* (some of us are more religious than others) are also of special interest to some people. As our country has evolved from small towns to large cities, we've all thought and talked more about *urbanism*—the good and bad associated with city life. Similar interests can be identified for all of the other terms.

My point is that you've probably thought about many of the variables mentioned in the list. Those you are familiar with undoubtedly have the quality of reality for you: that is, you know they exist. Religiosity, for example, is real. Regardless of whether you're in favor of it, opposed to it, or don't care much one way or the other, you at least know that religiosity exists. Or does it?

This is a particularly interesting question for me, since my first book, *To Comfort and to Challenge* (with Charles Glock and Benjamin Ringer), was about this subject. In particular, we wanted to know why some people were more religious than others (the sources of religiosity) and what impact differences in religiosity had on other aspects of life (the consequences of religiosity). Looking for the sources and consequences of a particular variable is a conventional social scientific undertaking; the first step is to develop a measure of that variable. We had to develop methods for distinguishing religious people, nonreligious people, and those somewhere in between.

The question we faced was, if religiosity is real, how do we know that? How do we distinguish religious people from nonreligious people? For most contemporary Americans, a number of answers come readily to mind. Religious people go to church, for example. They believe in the tenets of their faith. They pray. They read religious materials, such as the Bible, and they participate in religious organizations.

Not all religious people do all of these things, of course, and a great deal depends on their particular religious affiliation, if any. Christians believe in the divinity of Jesus; Jews do not. Moslems believe Mohammed's teachings are sacred; Jews and Christians do not. Some signs of religiosity are to be found in seemingly secular realms. Orthodox Jews, for example, refrain from eating pork; Seventh-Day Adventists don't drink alcohol.

In our study, we were interested in religiosity among a very specific group: Episcopal church members in America. To simplify our present discussion, let's look at that much narrower question: How can you distinguish religious from nonreligious Episcopalians in America?

As I've indicated above, we are likely to say that religious people attend church, whereas nonreligious people do not. Thus, if we know someone who attends church every week, we're likely to think of that person as religious; indeed, religious people joke about church members who only attend services on Easter and at Christmas. The latter are presumed to be less religious.

Of course, we are speaking rather casually here, so let's see whether church attendance would be an adequate measure of religiosity for Episcopalians and other mainstream American Christians. Would you be willing to equate religiosity with church attendance? That is, would you be willing to call religious everyone who attended church every week, let's say, and call nonreligious everyone who did not?

I suspect that you would not consider equating church attendance with religiosity a wise policy. For example, consider a political figure who attends church every Sunday, sits in the front pew, puts a large contribution in the collection plate with a flourish, and by all other evidence seems only interested in being known as a religious person for the political advantage that may entail. Let's add that the politician in question regularly lies and cheats, exhibits no Christian compassion toward others, and ridicules religion in private. You'd probably consider it inappropriate to classify that person as religious.

Now imagine someone confined to a hospital bed, who spends every waking minute reading in the Bible, leading other patients in prayer, raising money for missionary work abroad—but never going to church. Probably this would fit your image of a religious person.

These deviant cases illustrate that, while church attendance is somehow related to religiosity, it is not a sufficient indicator in and of itself. So how can we distinguish religious from nonreligious people?

Prayer is a possibility. Presumably, people who pray a lot are more religious than those who don't. But wouldn't it matter what they prayed for? Suppose they were only praying for money. How about the Moslem extremist praying daily for the extermination of the Jews? How about the athlete praying for an opponent to be hit by a truck? Like church attendance, prayer seems to have something to do with religiosity, but we can't simply equate the two.

We might consider religious beliefs. Among Christians, for example, it would seem to make sense that a person who believes in God is more religious than one who does not. However, this would require that we consider the person who says, "I'll believe anything they say just as long as I don't rot in Hell" more religious than, say, a concerned theologian who completes a lifetime of concentrated and devoted study of humbly concluding that who or what God is cannot be known with certainty. We'd probably decide that this was a misclassification.

Without attempting to exhaust all the possible indicators of religiosity, I hope it's clear that we would never find a single measure that will satisfy us as tapping the real essence of religiosity. In recognition of this, social researchers use a combination of indicators to create a *composite measure*—an index or a scale—of variables such as religiosity. Such a measure might include all of the indicators discussed so far: church attendance, prayer, and beliefs.

While composite measures are usually a good idea, they do not really solve the dilemma I've laid out. With a little thought, we could certainly imagine circumstances in which a "truly" religious person nonetheless didn't attend church, pray, or believe, and we could likewise imagine a nonreligious person who did all of those things. In either event, we would have demonstrated the imperfection of the composite measure.

Recognition of this often leads people to conclude that variables like religiosity are simply beyond empirical measurement. This conclusion is true and false and even worse.

The conclusion is false in that we can make any measurement we want. For example, we can ask people if they attend church regularly and call that a measure of religiosity just as easily as Yankee Doodle called the feather in his hat macaroni. In our case, moreover, most people would say that what we've measured is by no means irrelevant to religiosity.

The conclusion is true in that no empirical measurement—single or composite—will satisfy all of us as having captured the essence of religiousness. Since that can never happen, we can never satisfactorily measure religiosity.

The situation is worse than either of these comments suggests in that the reason we can't measure religiosity is that it doesn't exist! Religiosity isn't real. Neither is prejudice, love, alienation, or any of those other variables. Let's see why.

There's a very old puzzle I'm sure you're familiar with: when a tree falls in the forest, does it make a sound if no one is there to hear it? High school and college students have struggled with that one for centuries. There's no doubt that the unobserved falling tree will still crash through the branches of its neighbors, snap its own limbs into pieces, and slam against the ground. But would it make a sound?

If you've given this any thought before, you've probably come to the conclusion that the puzzle rests on the ambiguity of the word *sound.* Where does sound occur? In this example, does it occur in the falling tree, in the air, or in the ear of the beholder? We can be reasonably certain that the falling tree generates turbulent waves in the air; if those waves in the air strike your ear, you will experience something we call *hearing.* We say you've heard a sound. But do the waves in the air per se qualify as sound?

The answer to this central question is necessarily arbitrary. We can have it be whichever way we want. The truth is that (1) a tree fell; (2) it created waves in the air; and (3) if the waves reached someone's ear, they would cause an experience for that person. Humans created the idea of *sound* in the context of that whole process. Whenever waves in the air cause an experience by way of our ears, we use the term *sound* to identify that experience. We're usually not too precise about where the sound happens: in the tree, in the air, or in our ears.

Our imprecise use of the term *sound* produces the apparent dilemma. So what's the truth? What's really the case? Does it make a sound or not? The truth is that (1) a tree fell; (2) it created waves in the air; and

(3) if the waves reached someone's ear, they would cause an experience for that person. That's it. That's the final and ultimate truth of the matter.

I've belabored this point, because it sets the stage for understanding a critical issue in social research—one that often confuses students. To move in the direction of that issue, let's shift from sound to sight for a moment. Here's a new puzzle for you: are the tree's leaves green if no one is there to see them? Take a minute to think about that, and then continue reading.

Here's how I'd answer the question. The tree's leaves have a certain physical and chemical composition that affects the reflection of light rays off of them; specifically, they only reflect the green portion of the light spectrum. When rays from that portion of the light spectrum hit our eyes, they create an experience we call the color green.

"But are the leaves green if no one sees them?" you may ask. The answer to that is whatever we want it to be, since we haven't specified where the color green exists: in the physical/chemical composition of the leaf, in the light rays reflected from the leaf, or in our eyes.

While we are free to specify what we mean by the color green in this sense, nothing we do can change the ultimate truth, the ultimate reality of the matter. The truth is that (1) the leaves have a certain physical and chemical composition; (2) they reflect only a portion of the light spectrum; and (3) that portion of the light spectrum causes an experience if it hits our eyes. That's the ultimate truth of the universe in this matter.

By the same token, the truth about religiosity is that (1) some people to go church more than others; (2) some pray more than others; (3) some believe more than others; and so forth. This is observably the case.

At some point, our ancestors noticed that the things we're discussing were not completely independent of one another. People who went to church seemed to pray more, on the whole, than people who didn't go to church. Moreover, those who went to church and prayed seemed to believe more of the church's teachings than did those who neither went to church nor prayed. The observation of relationships such as these led them to conclude literally that "there is more here than meets the eye." The term *religiosity* was created to represent the *concept* that all the concrete observables seemed to have in common. People gradually came to believe that the concepts were real and the "indicators" only pale reflections.

We can never find a "true" measure of religiosity, prejudice, alienation, love, compassion, or any other such concepts, since none of them exists except in our minds. Concepts are "figments of our imaginations." I do not mean to suggest that concepts are useless or should be dispensed with. Life as we know it depends on the creation and use of concepts, and science would be impossible without them. Still, we should recognize that they are fictitious, then we can trade them in for more useful ones whenever appropriate.

THINKING ABOUT THE READING

Define the following terms: *poverty*, *happiness*, *academic effort*, and *love*. Now consider what indicators you would use to determine people's levels of each of these concepts. The indicator must be something that will allow you to clearly determine whether or not someone is in a particular state (such as poor or not poor; happy or not happy; in love or not in love). For example, you might decide that blushing in the presence of someone is one indicator of being in love or that the number of hours a person spends studying for a test is an indicator of academic effort. What's wrong with simply asking people if they're poor, if they're in love, if they're happy, or if they work hard? Consider the connection between how a concept is defined and how it can be measured. Is it possible that sociology sometimes uses concepts that seem meaningless because they are easier to "see" and measure?

Missing Numbers

Joel Best

(2004)

CBS News anchor Dan Rather began his evening newscast on March 5, 2001, by declaring: "School shootings in this country have become an epidemic." That day, a student in Santee, California, had killed two other students and wounded thirteen more, and media coverage linked this episode to a disturbing trend. Between December 1997 and May 1998, there had been three heavily publicized school shooting incidents: in West Paducah, Kentucky (three dead, five wounded); Jonesboro, Arkansas (five dead, ten wounded); and Springfield, Oregon (two dead and twenty-one wounded at the school, after the shooter had killed his parents at home). The following spring brought the rampage at Columbine High School in Littleton, Colorado, in which two students killed twelve fellow students and a teacher, before shooting themselves. Who could doubt Rather's claim about an epidemic?

And yet the word *epidemic* suggests a widespread, growing phenomenon. Were school shootings indeed on the rise? Surprisingly, a great deal of evidence indicated that they were not:

Since school shootings are violent crimes, we might begin by examining trends in criminality documented by the Federal Bureau of Investigation. The *Uniform Crime Reports,* the FBI's tally of crimes reported to the police, showed that the overall crime rate, as well as the rates for such major violent crimes as homicide, robbery, and aggravated assault, fell during the 1990s.

Similarly, the National Crime Victimization Survey (which asks respondents whether anyone in their household has been a crime victim) revealed that victimization rates fell during the 1990s; in particular, reports of teenagers being victimized by violent crimes at school dropped.

Other indicators of school violence also showed decreases. The Youth Risk Behavior Survey conducted by the U.S. Centers for Disease Control and Prevention found steadily declining percentages of high school students who reported fighting or carrying weapons on school property during the 1990s.

Finally, when researchers at the National School Safety Center combed media reports from the school years 1992–1993 through 2000–2001, they identified 321 violent deaths that had occurred at schools. Not all of these incidents involved student-on-student violence; they included, for example, 16 accidental deaths and 56 suicides, as well as incidents involving nonstudents, such as a teacher killed by her estranged husband (who then shot himself) and a nonstudent killed on a school playground during a weekend. Even if we include all 321 of these deaths, however, the average fell from 48 violent deaths per year during the school years 1992–1993 through 1996–1997 to 32 per year from 1997–1998 through 2000–2001. If we eliminate accidental deaths and suicides, the decline remains, with the average falling from 31 deaths per year in the earlier period to 24 per year in the later period (which included all of the heavily publicized incidents mentioned earlier). While violent deaths are tragedies, they are also rare. Tens of millions of children attend school; for every million students, fewer than one violent death per year occurs in school.

In other words, a great deal of statistical evidence was available to challenge claims that the country was experiencing a sudden epidemic of school shootings. The FBI's *Uniform Crime Reports* and the National Crime Victimization Survey in particular are standard sources for reporters who examine crime trends; the media's failure to incorporate findings from these sources in their coverage of school shootings is striking.

Although it might seem that statistics appear in every discussion of every social issue, in some cases—such as the media's coverage of school shootings—relevant, readily available statistics are ignored. We might think of these as *missing numbers*. This [reading] examines several reasons for missing numbers, including overwhelming examples, incalculable concepts, uncounted phenomena, forgotten figures, and legendary numbers. It asks why potentially relevant statistics don't figure in certain public debates and tries to assess the consequences of their absence.

The Power of Examples

Why are numbers missing from some debates over social problems and social policies? One answer is that a powerful example can overwhelm discussion of an issue. The 1999 shootings at Columbine High School are a case in point. The high death toll ensured that Columbine would be a major news story. Moreover, the school's location in a suburb of a major city made it easy for reporters to reach the scene. As it took some hours to evacuate the students and secure the building, the press had time to arrive and capture dramatic video footage that could be replayed to illustrate related stories in the weeks that followed. The juxtaposition of a terrible crime in a prosperous suburban community made the story especially frightening—if this school shooting could happen at Columbine, surely such crimes could happen anywhere. In addition, the Columbine tragedy occurred in the era of competing twenty-four-hour cable news channels; their decisions to run live coverage of several funeral and memorial services and to devote broadcast time to extended discussions of the event and its implications helped to keep the story alive for weeks.

For today's media, a dramatic event can become more than simply a news story in its own right; reporters have become attuned to searching for the larger significance of an event so that they can portray newsworthy incidents as instances of a widespread pattern or problem. Thus, Columbine, when coupled with the earlier, heavily publicized school shooting stories of 1997–1998, came to exemplify the problem of school violence. And, commentators reasoned, if a larger problem existed, it must reflect underlying societal conditions; that is, school shootings needed to be understood as a trend, wave, or epidemic with identifiable causes. Journalists have been identifying such crime waves since at least the nineteenth century—and, for nearly as long, criminologists have understood that crime waves are not so much patterns in criminal behavior as they are patterns in media coverage. All of the available statistical evidence suggested that school violence had declined from the early 1990s to the late 1990s; there was no actual wave of school shootings. But the powerful images from Columbine made that evidence irrelevant. One terrible example was "proof that school shootings were epidemic."

The power of examples is widely recognized. A reporter preparing a story about any broad social condition—say, homelessness—is likely to begin by illustrating the problem with an example, perhaps a particular homeless person. Journalists (and their editors) prefer interesting, compelling examples that will intrigue their audience. And advocates who are trying to promote particular social policies learn to help journalists by guiding them to examples that can be used to make specific points. Thus, activists calling for increased services for the homeless might showcase a homeless family, perhaps a mother of young children whose husband has been laid off by a factory closing and who cannot find affordable housing. In contrast, politicians seeking new powers to institutionalize the homeless mentally ill might point to a deranged, violent individual who seems to endanger passersby. The choice of examples conveys a sense of a social problem's nature.

The problem with examples—whether they derive from dramatic events, contemporary legends, or the strategic choices of journalists or advocates—is that they probably aren't especially typical. Examples compel when they have emotional power, when they frighten or disturb us. But atypical examples usually distort our understanding of a social problem; when we concentrate on the dramatic exception, we tend to overlook the more common, more typical—but more mundane—cases. Thus, Democrats used to complain about Republican President Ronald Reagan's fondness for repeating the story of a "welfare queen" who had supposedly collected dozens of welfare checks using false identities.[1] Using such colorful examples to typify welfare fraud implies that welfare recipients are undeserving or don't really need public assistance. Defenders of welfare often countered Reagan's anecdotes with statistics showing that recipients were deserving (as evidenced by the small number of able-bodied adults without dependent children who received benefits) or that criminal convictions for fraud were relatively few.[2] The danger is that the powerful but atypical example—the homeless intact family, the welfare queen—will warp our vision of a social problem, thereby reducing a complicated social condition to a simple, melodramatic fable.

Statistics, then, offer a way of checking our examples. If studies of the homeless find few intact families (or individuals who pose threats of violence), or if studies of welfare recipients find that fraud involving multiple false identities is rare, then we should recognize the distorting effects of atypical examples and realize that the absence of numbers can damage our ability to grasp the actual dimensions of our problems.

The Incalculable

Sometimes numbers are missing because phenomena are very hard to count. Consider another crime wave. During the summer of 2002, public concern turned to kidnapped children. Attention first focused on the case of an adolescent girl abducted from her bedroom one night—a classic melodramatic example of a terrible crime that seemingly could happen to anyone. As weeks passed without a sign of the girl, both the search and the accompanying news coverage continued. Reports of other cases of kidnapped or murdered children began linking these presumably unrelated crimes to the earlier kidnapping, leading the media to begin talking about an epidemic of abductions.

This issue had a history, however. Twenty years earlier, activists had aroused national concern about the problem of missing children by coupling frightening examples to large statistical estimates. One widespread claim alleged that nearly two million children went missing each year, including fifty thousand kidnapped by strangers. Later, journalists and social scientists exposed these early estimates as being unreasonably high. As a result, in 2002, some reporters questioned the claims of a new abduction epidemic; in fact, they argued, the FBI had investigated more kidnappings the previous year, which suggested that these crimes were actually becoming less common.

Both sets of claims—that kidnappings were epidemic and that they were declining—were based on weak evidence. Missing-children statistics can never be precise because missing children are so difficult to count. We encounter problems of definition:

What is a child—that is, what is the upper age limit for being counted?

What do we mean by missing? How long must a child be missing to be counted—a few minutes, one day, seventy-two hours?

What sorts of absences should be counted? Wandering off and getting lost? Running away? Being taken by a relative during a family dispute? Is a child who is with a noncustodial parent at a known location considered missing?

People need to agree about what to count before they can start counting, but not everyone agrees about the answers to these questions.

Obviously, the answers chosen will affect the numbers counted; using a broad definition means that more missing children will be counted.

A second set of problems concerns reporting. Parents of missing children presumably call their local law enforcement agency—usually a police or sheriff's department. But those authorities may respond in different ways. Some states require them to forward all missing-children reports to a statewide clearinghouse, which is supposed to contact all law enforcement agencies in the state in order to facilitate the search. The clearinghouses—and some departments—may notify the National Crime Information Center, a branch of the FBI that compiles missing-persons reports. Some reports also reach the National Center for Missing and Exploited Children (the federally funded group best known for circulating pictures of missing children) or FBI investigators (who claim jurisdiction over a few, but by no means most, kidnappings). Authorities in the same jurisdiction do not necessarily handle all missing-children reports the same way; the case of a six-year-old seen being dragged into a strange car is likely to be treated differently than a report of a sixteen-year-old who has run away. We can suspect that the policies of different agencies will vary significantly. The point is that the jurisdiction from which a child disappears and the particulars of the case probably affect whether a particular missing-child report finds its way into various agencies' records.

It is thus very difficult to make convincing comparisons of the numbers of missing children from either time to time or place to place. Reporters who noted that fewer child-kidnapping reports were filed with the FBI in 2002 than in 2001, and who therefore concluded that the problem was declining, mistakenly assumed that the FBI's records were more complete and authoritative than they actually were. Some things—like missing children—are very difficult to count, which should make us skeptical about the accuracy of statistics that claim to describe the situation.

Our culture has a particularly difficult time assigning values to certain types of factors. Periodically, for example, the press expresses shock that a cost-benefit analysis has assigned some specific value to individual lives. Such revelations produce predictably outraged challenges: how can anyone place a dollar value on a human life—aren't people's lives priceless? The answer to that question depends on when and where it is asked. Americans' notion that human life is priceless has a surprisingly short history. Only a century ago, the parents of a child killed by a streetcar could sue the streetcar company for damages equal to the child's economic value to the family (basically, the child's expected earnings until adulthood); today, of course, the parents would sue for the (vastly greater) value of their pain and suffering. Even the dollar value of a child's life varies across time and space.

The Uncounted

A third category of missing numbers involves what is deliberately uncounted, records that go unkept. Consider the U.S. Bureau of the Census's tabulations of religious affiliation: there are none. In fact, the census asks no questions about religion. Arguments about the constitutionally mandated separation of church and state, as well as a general sense that religion is a touchy subject, have led the Census Bureau to omit any questions about religion when it surveys the citizenry (in contrast to most European countries, where such questions are asked).

Thus, anyone trying to estimate the level of religious activity in the United States must rely on less accurate numbers, such as church membership rolls or individuals' reports of their attendance at worship services. The membership rolls of different denominations vary in what they count: Are infants counted once baptized, or does one become an enrolled member only in childhood or even adulthood? Are individuals culled from the rolls if they stop attending or actively participating in religious activities? Such variation makes it difficult to compare the sizes of different faiths. Surveys other than the census sometimes ask people how often they attend religious services, but we have good reason to suspect that respondents overreport attendance (possibly to make a good impression on the interviewers). The result is that, for the United States, at least, it is difficult to accurately measure the population's religious preferences or level of involvement. The policy of not asking questions about religion through the census means that such information simply does not exist.

The way choices are phrased also creates uncounted categories. Since 1790, each census has asked about race or ethnicity, but the wording of the questions—and the array of possible answers—has changed. The 2000 census, for example, was the first to offer respondents the chance to identify themselves as multiracial. Proponents of this change had argued that many Americans have family trees that include ancestors of different races and that it was unreasonable to force people to place themselves within a single racial category.

But some advocates had another reason for promoting this change. When forced to choose only one category, people who knew that their family backgrounds included people of different ethnicities had to oversimplify; most probably picked the option that fit the largest share of their ancestors. For example, an individual whose grandparents included three whites and one Native American was likely to choose "white." In a society in which a group's political influence depends partly on its size, such choices could depress the numbers of people of American Indian ancestry (or any other relatively small, heavily intermarried group) identified by the census. Native American activists favored letting people list themselves as being of more than one race because they believed that this would help identify a larger Native American population and presumably increase that group's political clout. In contrast, African American activists tended to be less enthusiastic about allowing people to identify themselves as multiracial. Based in part on the legacy of segregation, which sometimes held that having a single black ancestor was sufficient to warrant being considered nonwhite, people with mixed black and white ancestry (who account for a majority of those usually classified as African Americans) had tended to list themselves as "black." If large numbers of these individuals began listing more than one racial group, black people might risk losing political influence.

As is so often the case, attitudes toward altering the census categories depended on whether one expected to win or lose by the change. The reclassification had the expected effect, even though only 2.4 percent of respondents to the 2000 census opted to describe themselves as multiracial. The new classification boosted the numbers of people classified as Native Americans: although only 2.5 million respondents listed themselves under the traditional one-ethnicity category, adding those who identified themselves as part-Indian raised the total to 4.1 million—a 110 percent increase since 1990. However, relatively small numbers of people (fewer than eight hundred thousand) listed their race as both white and black, compared to almost 34 million identified as black.[3]

Sometimes only certain cases go uncounted. Critics argue that the official unemployment rate, which counts only those without full-time work who have actively looked for a job during the previous four weeks, is too low. They insist that a more accurate count would include those who want to work but have given up looking as well as those who want full-time work but have had to settle for part-time jobs—two groups that, taken together, actually outnumber the officially unemployed. Of course, every definition draws such distinctions between what does—and doesn't—count.

The lesson is simple. Statistics depend on collecting information. If questions go unasked, or if they are asked in ways that limit responses, or if measures count some cases but exclude others, information goes ungathered, and missing numbers result. Nevertheless, choices regarding which data to collect and how to go about collecting the information are inevitable. If we want to describe America's racial composition in a way that can be understood, we need to distill incredible diversity into a few categories. The cost of classifying anything into a particular set of categories is that some information is inevitably lost: distinctions seem sharper; what may have been arbitrary cut-offs are treated as meaningful; and, in particular, we tend to lose sight of the choices and uncertainties that went into creating our categories.

There are a variety of ways to ensure that things remain uncounted. The simplest is to not collect the information (for instance, don't ask census respondents any questions about religion). But, even when the data exist, it is possible to avoid compiling information (by simply not doing the calculations necessary to produce certain statistics), to refuse to publish the information, or even to block access to it. More subtly, both data collection and analysis can be time-consuming and expensive; in a society where researchers depend on others for funding, decisions not to fund certain research can have the effect of relegating those topics to the ranks of the uncounted.

The Forgotten

Another form of missing numbers is easy to overlook—these are figures, once public and even familiar, that we no longer remember or don't bother to consider. Consider the number of deaths from measles. In 1900, the death rate from measles was 13.3 per 100,000 in the population; measles ranked among the top ten diseases causing death in the United States. Over the course of a century, however, measles lost its power to kill; first more effective treatments and then vaccination eliminated measles as a major medical threat. Nor was this an exceptional case. At the beginning of the twentieth century, many of the leading causes of death were infectious diseases; influenza/pneumonia, tuberculosis, diphtheria, and typhoid/typhoid fever also ranked in the top ten. Most of those formerly devastating diseases have been brought under something approaching complete control in the United States through the advent of vaccinations and antibiotics. The array of medical threats has changed.

Forgotten numbers have the potential to help us put things in perspective, if only we can bring ourselves to remember them. When we lose sight of the past, we have more trouble assessing our current situation. However, people who are trying to draw attention to social problems are often reluctant to make comparisons with the past. After all, such comparisons may reveal considerable progress. During the twentieth century, for example, Americans' life expectancies increased dramatically. In 1900, a newborn male could expect to live forty-six years; a century later, male life expectancy had risen to seventy-three. The increase for females was even greater—from age forty-eight to eighty. During the same period, the proportion of Americans completing high school rose from about 6 percent to about 85 percent. Many advocates seem to fear that talking about long-term progress invites complacency about contemporary society, and they prefer to focus on short-run trends—especially if the numbers seem more compelling because they show things getting worse.

Similarly, comparing our society to others can help us get a better sense of the size and shape of our problems. Again, in discussions of social issues, such comparisons tend to be made selectively, in ways that emphasize the magnitude of our contemporary problems. Where data suggest that the United States lags behind other nations, comparative statistics are commonplace, but we might suspect that those trying to promote social action will be less likely to present evidence showing America to advantage. (Of course, those resisting change may favor just such numbers.) Comparisons across time and space are recalled when they help advocates make their points, but otherwise they tend to be ignored, if not forgotten.

Legendary Numbers

One final category deserves mention. It does not involve potentially relevant numbers that are missing, but rather includes irrelevant or erroneous figures that somehow find their way into discussions of social issues. Recently, for example, it became fairly common for journalists to compare various risks against a peculiar standard: the number of people killed worldwide each year by falling coconuts (the annual coconut-death figure usually cited was 150). Do 150 people actually die in this way? It might seem possible—coconuts are hard and heavy, and they fall a great distance, so being bonked on the head presumably might be fatal. But who keeps track of coconut fatalities? The answer: no one. Although it turns out that the medical literature includes a few reports of injuries—not deaths—inflicted by falling coconuts, the figure of 150 deaths is the journalistic equivalent of a contemporary legend. It gets passed along as a "true fact," repeated as something that "everybody knows."

Other legendary statistics are attributed to presumably authoritative sources. A claim that a World Health Organization study had determined that blondness was caused by a recessive gene and that blonds would be extinct within two hundred years was carried by a number of prominent news outlets, which presumably ran the story on the basis of one another's coverage, without bothering to check with the World Health Organization (which denied the story).

Legendary numbers can become surprisingly well established. Take the claim that fifty-six is the average age at which a woman becomes widowed. In spite of its obvious improbability (after all, the average male lives into his seventies, married men live longer than those who are unmarried, and husbands are only a few years older on average than their wives), this statistic has circulated for more than twenty years. It appeared in a television commercial for financial services, in materials distributed to women's studies students, and in countless newspaper and magazine articles; its origins are long lost. Perhaps it has endured because no official agency collects data on age at widowhood, making it difficult to challenge such a frequently repeated figure. Nevertheless, demographers—using complicated equations that incorporate age-specific death rates, the percentage of married people in various age cohorts, and age differences between husbands and wives—have concluded that the average age at which women become widows has, to no one's surprise, been rising steadily, from sixty-five in 1970 to about sixty-nine in 1988.[4]

Even figures that actually originate in scientists' statements can take on legendary qualities. In part, this reflects the difficulties of translating complex scientific ideas into what are intended to be easy-to-understand statements. For example, the widely repeated claim that individuals need to drink eight glasses of water each day had its origin in an analysis that did in fact recommend that level of water intake. But the analysis also noted that most of this water would ordinarily come from food (bread, for example, is 35 percent water, and meats and vegetables contain even higher proportions of water). However, the notion that food contained most of the water needed for good health was soon forgotten, in favor of urging people to consume the entire amount through drinking. Similarly, the oft-repeated statements that humans and chimpanzees have DNA that is 98 percent similar—or, variously, 98.4, 99, or 99.44 percent similar—may seem precise, but they ignore the complex assumptions involved in making such calculations and imply that this measure is more meaningful than it actually is.

Widely circulated numbers are not necessarily valid or even meaningful. In the modern world, with ready access to the Internet and all manner of electronic databases, even figures that have been thoroughly debunked can remain in circulation; they are easy to retrieve and disseminate but almost impossible to eradicate. The problem is not one of missing numbers—in such cases, the numbers are all too present. What is absent is the sort of evidence needed to give the statistics any credibility.

The attraction of legendary numbers is that they seem to give weight or authority to a claim. It is far less convincing to argue, "That's not such an important cause of death! Why, I'll bet more people are killed each year by falling coconuts!" than to flatly compare 150 coconut deaths to whatever is at issue. Numbers are presumed to be factual; numbers imply that someone has actually counted something. Of course, if that is true, it should be possible to document the claim—which cannot be done for legendary numbers.

A related phenomenon is that some numbers, if not themselves fanciful, come to be considered more meaningful than they are. We see this particularly in the efforts of bureaucrats to measure the unmeasurable. A school district, for example, might want to reward good teaching. But what makes a good teacher? Most of us can look back on our teachers and identify some as better than others. But what made them better? Maybe they helped us when we were having trouble, encouraged us, or set high standards. My reasons for singling out some of my teachers as especially good might be very different from the reasons you would cite. Teachers can be excellent in many ways, and there's probably no reliable method of translating degree of excellence into a number. How can we measure good teaching or artistic genius? Even baseball fans—those compulsive record-keepers and lovers of statistics—can argue about the relative merits of different athletes, and baseball has remarkably complete records of players' performances.

But that sort of soft appeal to the immeasurability of performance is unlikely to appease politicians or an angry public demanding better schools. So educational bureaucrats—school districts and state education departments—insist on measuring "performance." In recent years, the favored measure has been students' scores on standardized tests. This is not completely unreasonable—one could argue that, overall, better teaching should lead to students learning more and, in turn, to higher test scores. But test scores are affected by many things besides teachers' performance, including students' home lives. And our own memories of our "best teachers" probably don't depend on how they shaped our performances on standardized tests.

However imperfect test scores might be as an indicator of the quality of teaching, they do offer a nice quantitative measure—this student got so many right, the students in this class scored this well, and so on. No wonder bureaucrats gravitate toward such measures—they are precise (and it is relatively inexpensive to get the information), even if it isn't clear just what they mean. The same thing happens in many settings. Universities want their professors to do high-quality research and be good teachers, but everyone recognizes that these qualities are hard to measure. Thus, there is a tremendous temptation to focus on things that are easy to count: How many books or articles has a faculty member published? (Some departments even selectively weigh articles in different journals, depending on some measure of each journal's influence.) Are a professor's teaching evaluation scores better than average?

The problem with such bureaucratic measures is that we lose sight of their limitations. We begin by telling ourselves that we need some way of measuring teaching quality and that this method—whatever its flaws—is better than nothing. Even if some resist adopting the measure at first, over time inertia sets in, and people come to accept its use. Before long, the measure is taken for granted, and its flaws tend to be forgotten. The criticism of being an imperfect measure can be leveled at many of the numbers discussed. If pressed, a statistic's defenders will often acknowledge that the criticism is valid, that the measure is flawed. But, they ask, what choice do we have? How else can we measure—quickly, cheaply, and more or less objectively—good teaching (or whatever else concerns us)? Isn't an imperfect statistic better than none at all? They have a point. But we should never blind ourselves to a statistic's shortcomings; once we forget a number's limitations, we give it far more power and influence than it deserves. We need to remember that a clear and direct measure would be preferable and that our imperfect measure is—once again—a type of missing number.

What's Missing?

When people use statistics, they assume—or, at least, they want their listeners to assume—that the numbers are meaningful. This means, at a minimum, that someone has actually counted something and that they have done the counting in a way that makes sense. Statistical information is one of the best ways we have of making sense of the world's complexities, of identifying patterns amid the confusion. But bad statistics give us bad information.

This [reading] argues that some statistics are bad not so much because the information they contain is bad but because of what is missing—what has not been counted. Numbers can be missing in several senses: a powerful example can make us forget to look for statistics; things can go uncounted because they are considered difficult or impossible to count or because we decide not to count them. In other cases, we count, but something gets lost in the process: things once counted are forgotten, or we brandish numbers that lack substance.

In all of these cases, something is missing. Understanding that helps us recognize what counts as a good statistic. Good statistics are not only products of people counting; the quality of statistics also depends on people's willingness and ability to count thoughtfully and on their decisions about what, exactly, ought to be counted so that the resulting numbers will be both accurate and meaningful.

This process is never perfect. Every number has its limitations; every number is a product of choices that inevitably involve compromise. Statistics are intended to help us summarize, to get an overview of part of the world's complexity. But some information is always sacrificed in the process of choosing what will be counted and how. Something is, in short, always missing. In evaluating statistics, we should not forget what has been lost, if only because this helps us understand what we still have.

In addition, a recent three-hundred-page book designed to help statistics instructors reach their students includes a seventeen-page chapter entitled "Lying with Statistics," which addresses issues of bias; see Andrew Gelman and Deborah Nolan, *Teaching Statistics: A Bag of Tricks* (New York: Oxford University Press, 2002), pp. 147–163.

Notes

1. Compare Kiron K. Skinner, Annelise Anderson, and Martin Anderson, *Reagan, in His Own Hand* (New York: Free Press, 2001), pp. 241, 459; and Tip O'Neill, *Man of the House* (New York: Random House, 1987), pp. 347–348.

2. In response, welfare critics offered their own numbers suggesting that abuses were widespread. One analysis suggested: "A major source of the variations among estimates [for welfare fraud] . . . has been the estimators' use of very different definitions of improprieties" (John A. Gardiner and Theodore R. Lyman, *The Fraud Control Game* [Bloomington: Indiana University Press, 1984], p. 2).

3. On the 2000 census, see Margo Anderson and Stephen E. Fienberg, "Census 2000 and the Politics of Census Taking," *Society* 39 (November 2001): 17–25; and Eric Schmitt, "For 7 Million People in Census, One Race Category Isn't Enough," *New York Times*, March 13, 2001, p. A1. For historical background, see Petersen, *Ethnicity Counts.*

4. Robert Schoen and Robert Weinick "The Slowing Metabolism of Marriage," *Demography* 30 (1993): 737–746.

THINKING ABOUT THE READING

Statistics can provide empirical evidence that tells us what is happening around crucial social issues and the noteworthy trends in these issues over time. For example, statistics on demographic characteristics—race/ethnicity, gender, social class, age, religion, and sexuality—can help us understand the experiences of different groups in our society. Indeed, fully understanding these statistics allows us to become more competent and informed citizens. As Best noted, we want accuracy in our numbers, because bad statistics give us bad information. Take one of the several different reasons he lists for these missing numbers and critically discuss who may benefit and who may lose in these cases. Furthermore, Best suggests that statistics offer us a way to check our assumptions. Consider an assumption you have made about individuals, groups, or institutions and how you could verify this assumption with statistical evidence. Do a search for statistics that provide you with evidence that confirms or perhaps challenges this assumption. Review this statistical evidence and ask critical questions about what, if anything, is missing from these numbers that may affect your understanding of the topic.

Building Order

Culture and History

Culture provides members of a society with a common bond and a set of shared rules and beliefs for making sense of the world in similar ways. Shared cultural knowledge makes it possible for people to live together in a society. Sociologists refer to shared cultural expectations as *social norms.* Norms are the rules and standards that govern all social encounters and the mechanisms that provide order in our day-to-day lives. Shared norms make it possible to know what to expect from others and what others can expect from us. When norms are violated, we are reminded of the boundaries of social behavior. These violations lead us to notice otherwise taken-for-granted rules about what is considered right and wrong.

When we examine the social influences on our behavior, things that were once familiar and taken for granted suddenly become unfamiliar and curious. During the course of our lives, we are rarely forced to examine *why* we do the common things we do; we just do them. But if we take a step back and examine our common customs and behaviors, they begin to look as strange as the "mystical" rituals of some far-off, exotic land. It is for this reason that Horace Miner's article, "Body Ritual among the Nacirema," has become a classic in sociology and anthropology. As you read this selection, consider the process of using the sociological imagination to understand your own life and the lives of others. When you think about other cultures, how can you be sure that your perceptions, as an outsider, are not as bizarre as Miner's perspective on the Nacirema? When done well, sociological research helps us to understand different points of view and different cultural contexts from the perspective of insiders.

Cultural clashes can be quite confusing and painful for newly arrived immigrants from countries with vastly different cultural traditions. In the article "The Melting Pot," Anne Fadiman examines the experiences of Hmong refugees in the United States. Hundreds of thousands of Hmong people have fled Laos since that country fell to communist forces in 1975. Most have settled in the United States. Virtually every element of Hmong culture and tradition stands in stark contrast to the highly modernized culture of U.S. society. The Hmong have been described in the U.S. media as simplistic, primitive, and throwbacks to the Stone Age. This article vividly portrays the everyday conflicts immigrants face as they straddle two vastly different cultures.

What happens when an industry is transported from one culture to another, especially if the cultures are very different? In the third reading for the section, James L. Watson addresses this question in the context of the transportation of the fast food industry, specifically McDonald's, to Hong Kong. McDonald's epitomizes Western cultural patterns of fast food consumption and other norms of consumer capitalism. What happens when this industry arrives in Asia? Watson observes that the result is cultural changes that reflect both globalization processes and also local resistance to those processes. Accordingly, this article demonstrates the importance of looking closely at local cultures in studying the effects of globalization.

Something to Consider as You Read

How do cultural practices provide social order? Where is this order located? In our minds? In our interactions with others? Think about what happens to your own sense of order when you become immersed in a different culture. What are some of the challenges you might face in trying to maintain your own cultural beliefs and practices while living in a completely different culture? Are some cultural practices easier to export than others? Why? As you read and compare these selections, think about why some cultures consider their ways to be better and more "real" than others. Do you think this ethnocentrism is a hallmark of all cultures or just some? As processes of globalization increase, what are the consequences for local cultures?

Body Ritual among the Nacirema

Horace Miner

(1956)

The anthropologist has become so familiar with the diversity of ways in which different peoples behave in similar situations that he is not apt to be surprised by even the most exotic customs. In fact, if all of the logically possible combinations of behavior have not been found somewhere in the world, he is apt to suspect that they must be present in some yet undescribed tribe. This point has, in fact, been expressed with respect to clan organization by Murdock (1949, p. 71). In this light, the magical beliefs and practices of the Nacirema present such unusual aspects that it seems desirable to describe them as an example of the extremes to which human behavior can go.

Professor Linton first brought the ritual of the Nacirema to the attention of anthropologists twenty years ago (1936, p. 326), but the culture of this people is still very poorly understood. They are a North American group living in the territory between the Canadian Cree, the Yaqui and Tarahumara of Mexico, and the Carib and Arawak of the Antilles. Little is known of their origin, although tradition states that they came from the east. According to Nacirema mythology, their nation was originated by a culture hero, Notgnihsaw, who is otherwise known for two great feats of strength—the throwing of a piece of wampum across the river Pa-To-Mac and the chopping down of a cherry tree in which the Spirit of Truth resided.

Nacirema culture is characterized by a highly developed market economy which has evolved in a rich natural habitat. While much of the people's time is devoted to economic pursuits, a large part of the fruits of these labors and a considerable portion of the day are spent in ritual activity. The focus of this activity is the human body, the appearance and health of which loom as a dominant concern in the ethos of the people. While such a concern is certainly not unusual, its ceremonial aspects and associated philosophy are unique.

The fundamental belief underlying the whole system appears to be that the human body is ugly and that its natural tendency is to debility and disease. Incarcerated in such a body, man's only hope is to avert these characteristics through the use of the powerful influences of ritual and ceremony. Every household has one or more shrines devoted to this purpose. The more powerful individuals in this society have several shrines in their houses and, in fact, the opulence of a house is often referred to in terms of the number of such ritual centers it possesses. Most houses are of wattle and daub construction, but the shrine rooms of the more wealthy are walled with stone. Poorer families imitate the rich by applying pottery plaques to their shrine walls.

While each family has at least one such shrine, the rituals associated with it are not family ceremonies but are private and secret. The rites are normally only discussed with children, and then only during the period when they are being initiated into these mysteries. I was able, however, to establish sufficient rapport with the natives to examine these shrines and to have the rituals described to me.

The focal point of the shrine is a box or chest which is built into the wall. In this chest are kept the many charms and magical potions without which no native believes he could live. These preparations are secured from a variety of specialized practitioners. The most powerful of these are the medicine men, whose assistance must be rewarded with substantial gifts.

However, the medicine men do not provide the curative potions for their clients, but decide what the ingredients should be and then write them down in an ancient and secret language. This writing is understood only by the medicine men and by the herbalists who, for another gift, provide the required charm.

The charm is not disposed of after it has served its purpose, but is placed in the charm-box of the household shrine. As these magical materials are specific for certain ills, and the real or imagined maladies of the people are many, the charm-box is usually full to overflowing. The magical packets are so numerous that people forget what their purposes were and fear to use them again. While the natives are very vague on this point, we can only assume that the idea in retaining all the old magical materials is that their presence in the charm-box, before which the body rituals are conducted, will in some way protect the worshipper.

Beneath the charm-box is a small font [fountain]. Each day every member of the family, in succession, enters the shrine room, bows his head before the charm-box, mingles different sorts of holy water in the font, and proceeds with a brief rite of ablution. The holy waters are secured from the Water Temple of the community, where the priests conduct elaborate ceremonies to make the liquid ritually pure.

In the hierarchy of magical practitioners, and below the medicine men in prestige, are specialists whose designation is best translated "holy-mouth-men." The Nacirema have an almost pathological horror of and fascination with the mouth, the condition of which is believed to have a supernatural influence on all social relationships. Were it not for the rituals of the mouth, they believe that their teeth would fall out, their gums bleed, their jaws shrink, their friends desert them, and their lovers reject them. They also believe that a strong relationship exists between oral and moral characteristics. For example, there is a ritual ablution of the mouth for children which is supposed to improve their moral fiber.

The daily body ritual performed by everyone includes a mouth-rite. Despite the fact that these people are so punctilious about care of the mouth, this rite involves a practice which strikes the uninitiated stranger as revolting. It was reported to me that the ritual consists of inserting a small bundle of hog hairs into the mouth, along with certain magical powders, and then moving the bundle in a highly formalized series of gestures.

In addition to the private mouth-rite, the people seek out a holy-mouth-man once or twice a year. These practitioners have an impressive set of paraphernalia, consisting of a variety of augers, awls, probes, and prods. The use of these objects in the exorcism of the evils of the mouth involves almost unbelievable ritual torture of the client. The holy-mouth-man opens the client's mouth and, using the above-mentioned tools, enlarges any holes which decay may have created in the teeth. Magical materials are put into these holes. If there are no naturally occurring holes in the teeth, large sections of one or more teeth are gouged out so that the supernatural substance can be applied. In the client's view, the purpose of these ministrations is to arrest decay and to draw friends. The extremely sacred and traditional character of the rite is evident in the fact that the natives return to the holy-mouth-man year after year, despite the fact that their teeth continue to decay.

It is to be hoped that, when a thorough study of the Nacirema is made, there will be careful inquiry into the personality structure of these people. One has but to watch the gleam in the eye of a holy-mouth-man, as he jabs an awl into an exposed nerve, to suspect that a certain amount of sadism is involved. If this can be established, a very interesting pattern emerges, for most of the population shows definite masochistic tendencies. It was to these that Professor Linton referred in discussing a distinctive part of the daily body ritual which is performed only by men. This part of the rite involves scraping and lacerating the surface of the face with a sharp instrument. Special women's rites are performed only four times during each lunar month, but what they lack in frequency is made up in barbarity. As part of this ceremony, women bake their heads in small ovens for about an hour. The theoretically interesting point is that what seems to be a preponderantly masochistic people have developed sadistic specialists.

The medicine men have an imposing temple, or *latipso,* in every community of any size. The more elaborate ceremonies required to treat very sick patients can only be performed at this temple. These ceremonies involve not only the thaumaturge but a permanent group of vestal maidens who move sedately about the temple chambers in distinctive costume and headdress.

The *latipso* ceremonies are so harsh that it is phenomenal that a fair proportion of the really sick natives who enter the temple ever recover. Small children whose indoctrination is still incomplete have been known to resist attempts to take them to the temple because "that is where you go to die." Despite this fact, sick adults are not only willing but eager to undergo the protracted ritual purification, if they can afford to do so. No matter how ill the supplicant or how grave the emergency, the guardians of many temples will not admit a client if he cannot give a rich gift to the custodian. Even after one has gained admission and survived the ceremonies, the guardians will not permit the neophyte to leave until he makes still another gift.

The supplicant entering the temple is first stripped of all his or her clothes. In everyday life the Nacirema avoids exposure of his body and its natural functions. Bathing and excretory acts are performed only in the secrecy of the household shrine, where they are ritualized as part of the body-rites. Psychological shock results from the fact that body secrecy is suddenly lost upon entry into the *latipso.* A man, whose own wife has never seen him in an excretory act, suddenly finds himself naked and assisted by a vestal maiden while he performs his natural functions into a sacred vessel. This sort of ceremonial treatment is necessitated by the fact that the excreta are used by a diviner to ascertain the course and nature of the client's sickness. Female clients, on the other hand, find their naked bodies are subjected to the scrutiny, manipulation, and prodding of the medicine men.

Few supplicants in the temple are well enough to do anything but lie on their hard beds. The daily ceremonies, like the rites of the holy-mouth-men, involve discomfort and torture. With ritual precision, the vestals awaken their miserable charges each dawn and roll them about on their beds of pain while performing ablutions, in the formal movements of which the maidens are highly trained. At other times they insert magic wands in the supplicant's mouth or force him to eat substances which are supposed to be healing. From time to time the medicine men come to their clients and jab magically treated needles into their flesh. The fact that these temple ceremonies may not cure, and may even kill the neophyte, in no way decreases the people's faith in the medicine men.

There remains one other kind of practitioner, known as a "listener." This witch-doctor has the power to exorcise the devils that lodge in the heads of people who have been bewitched. The Nacirema believe that parents bewitch their own children. Mothers are particularly suspected of putting a curse on children while teaching them the secret body rituals. The counter-magic of the witch-doctor is unusual in its lack of ritual. The patient simply tells the "listener" all his troubles and fears, beginning with the earliest difficulties he can remember. The memory displayed by the Nacirema in these exorcism sessions is truly remarkable. It is not uncommon for the patient to bemoan the rejection he felt upon being weaned as a babe, and a few individuals even see their troubles going back to the traumatic effects of their own birth.

In conclusion, mention must be made of certain practices which have their base in native esthetics but which depend upon the pervasive aversion to the natural body and its functions. There are ritual fasts to make fat people thin and ceremonial feasts to make thin people fat. Still other rites are used to make women's breasts larger if they are small, and smaller if they are large. General dissatisfaction with breast shape is symbolized in the fact that the ideal form is virtually outside the range of human variation. A few women afflicted with almost inhuman hypermammary development are so idolized that they make a handsome living by simply going from village to village and permitting the natives to stare at them for a fee.

Reference has already been made to the fact that excretory functions are ritualized, routinized, and relegated to secrecy. Natural reproductive functions are similarly distorted. Intercourse is taboo as a topic and scheduled as an act. Efforts are made to avoid pregnancy by the use of magical materials or by limiting intercourse to certain phases of the moon. Conception is actually very infrequent. When pregnant, women dress so as to hide their condition. Parturition takes place in secret, without friends or relatives to assist, and the majority of women do not nurse their infants.

Our review of the ritual life of the Nacirema has certainly shown them to be a magic-ridden people. It is hard to understand how they have managed to exist so long under the burdens which they have imposed upon themselves. But even such exotic customs as these take on real meaning when they are viewed with the insight provided by Malinowski when he wrote (1948, p. 70):

> Looking from far and above, from our high places of safety in the developed civilization, it is easy to see all the crudity and irrelevance of magic. But without its power and guidance early man could not have mastered his practical difficulties as he has done, nor could man have advanced to the higher stages of civilization.

REFERENCES

Linton, R. (1936). *The study of man.* New York: Appleton-Century.

Malinowski, B. (1948). *Magic, science, and religion.* Glencoe, IL: Free Press.

Murdock, G. P. (1949). *Social structure.* New York: Macmillan.

THINKING ABOUT THE READING

What do you think of this culture? Do their ways seem very foreign or are there some things that seem familiar? This article was written more than 50 years ago and, of course, much has changed since then. How might you update this description of the Nacirema to account for current values and rituals? Imagine you are an anthropologist from a culture completely unfamiliar with Western traditions. Using your own life as a starting point, think of common patterns of work, leisure, learning, intimacy, eating, sleeping, and so forth. Are there some customs that distinguish your group (religious, racial, ethnic, friendship, etc.) from others? See if you can find the reasons why these customs exist, which customs serve an obvious purpose (e.g., health), and which might seem arbitrary and silly to an outside observer.

The Melting Pot

Anne Fadiman

(1997)

The Lee family—Nao Kao, Foua, Chong, Zoua, Cheng, May, Yer, and True—arrived in the United States on December 18, 1980. Their luggage consisted of a few clothes, a blue blanket, and a wooden mortar and pestle that Foua had chiseled from a block of wood in Houaysouy. They flew from Bangkok to Honolulu, and then to Portland, Oregon, where they were to spend two years before moving to Merced. Other refugees told me that their airplane flights—a mode of travel that strained the limits of the familiar Hmong concept of migration—had been fraught with anxiety and shame: they got airsick, they didn't know how to use the bathroom but were afraid to soil themselves, they thought they had to pay for their food but had no money, they tried to eat the Wash'n Dris. The Lees, though perplexed, took the novelties of the trip in stride. Nao Kao remembers the airplane as being "just like a big house."

Their first week in Portland, however, was miserably disorienting. Before being placed by a local refugee agency in a small rented house, they spent a week with relatives, sleeping on the floor. "We didn't know anything so our relatives had to show us everything," Foua said. "They knew because they had lived in America for three or four months already. Our relatives told us about electricity and said the children shouldn't touch those plugs in the wall because they could get hurt. They told us that the refrigerator is a cold box where you put meat. They showed us how to open the TV so we could see it. We had never seen a toilet before and we thought maybe the water in it was to drink or cook with. Then our relatives told us what it was, but we didn't know whether we should sit or whether we should stand on it. Our relatives took us to the store but we didn't know that the cans and packages had food in them. We could tell what the meat was, but the chickens and cows and pigs were all cut up in little pieces and had plastic on them. Our relatives told us the stove is for cooking the food, but I was afraid to use it because it might explode. Our relatives said in America the food you don't eat you just throw away. In Laos we always fed it to the animals and it was strange to waste it like that. In this country there were a lot of strange things and even now I don't know a lot of things and my children have to help me, and it still seems like a strange country."

Seventeen years later, Foua and Nao Kao use American appliances, but they still speak only Hmong, celebrate only Hmong holidays, practice only the Hmong religion, cook only Hmong dishes, sing only Hmong songs, play only Hmong musical instruments, tell only Hmong stories, and know far more about current political events in Laos and Thailand than about those in the United States. When I first met them, during their eighth year in this country, only one American adult, Jeanine Hilt, had ever been invited to their home as a guest. It would be hard to imagine anything further from the vaunted American ideal of assimilation, in which immigrants are expected to submerge their cultural differences in order to embrace a shared national identity. *E pluribus unum:* from many, one.

During the late 1910s and early 1920s, immigrant workers at the Ford automotive plant in Dearborn, Michigan, were given free, compulsory "Americanization" classes. In addition to English lessons, there were lectures on work habits, personal hygiene, and table manners. The first sentence they memorized was "I am a good American." During their graduation ceremony they gathered next to a gigantic wooden pot, which their teachers stirred with ten-foot ladles. The students walked through a door into the pot, wearing traditional costumes from their countries of origin and singing songs in their native languages. A few minutes later, the door in the pot opened, and the students walked out again, wearing suits and ties, waving American flags, and singing "The Star-Spangled Banner."

The European immigrants who emerged from the Ford Motor Company melting pot came to the United States because they hoped to assimilate into mainstream American society. The Hmong came to the United States for the same reason they had left China in the nineteenth century: because they were trying to *resist* assimilation. As the anthropologist Jacques Lemoine has observed, "they did not come to our countries only to save their lives, they rather came to save their selves, that is, their Hmong ethnicity." If their Hmong ethnicity had been safe in Laos, they would have preferred to remain there, just as their ancestors—for whom migration had always been a problem-solving strategy, not a footloose impulse—would have preferred to remain in China. Unlike the Ford workers who enthusiastically, or at least uncomplainingly, belted out the "The Star-Spangled Banner" (of which Foua and Nao Kao know not a single word), the Hmong are what sociologists call "involuntary migrants." It is well known that involuntary migrants, no matter what pot they are thrown into, tend not to melt.

What the Hmong wanted here was to be left alone to be Hmong: clustered in all-Hmong enclaves, protected from government interference, self-sufficient, and agrarian. Some brought hoes in their luggage. General Vang Pao has said, "For many years, right from the start, I tell the American government that we need a little bit of land where we can grow vegetables and build homes like in Laos . . . I tell them it does not have to be the best land, just a little land where we can live." This proposal was never seriously considered. "It was just out of the question," said a spokesman for the State Department's refugee program. "It would cost too much, it would be impractical, but most of all it would set off wild protests from [other Americans] and from other refugees who weren't getting land for themselves." . . .

Just as newly arrived immigrants in earlier eras had been called "FOBs"—Fresh Off the Boat—some social workers nicknamed the incoming Hmong, along with the other Southeast Asian refugees who entered the United States after the Vietnamese War, "JOJs": Just Off the Jet. Unlike the first waves of Vietnamese and Cambodian refugees, most of whom received several months of vocational and language training at regional "reception centers," the Hmong JOJs, who arrived after the centers had closed, were all sent directly to their new homes. (Later on, some were given "cultural orientation" training in Thailand before flying to the United States. Their classes covered such topics as how to distinguish a one-dollar bill from a ten-dollar bill and how to use a peephole.) The logistical details of their resettlement were contracted by the federal government to private nonprofit groups known as VOLAGs, or national voluntary resettlement agencies, which found local sponsors. Within their first few weeks in this country, newly arrived families were likely to deal with VOLAG officials, immigration officials, public health officials, social service officials, employment officials, and public assistance officials. The Hmong are not known for holding bureaucrats in high esteem. As one proverb puts it, "To see a tiger is to die; to see an official is to become destitute." In a study of adaptation problems among Indochinese refugees, Hmong respondents rated "Difficulty with American Agencies" as a more serious problem than either "War Memories" or "Separation from Family." Because many of the VOLAGs had religious affiliations, the JOJs also often found themselves dealing with Christian ministers, who, not surprisingly, took a dim view of shamanistic animism. A sponsoring pastor in Minnesota told a local newspaper, "It would be wicked to just bring them over and feed and clothe them and let them go to hell. The God who made us wants them to be converted. If anyone thinks that a gospel-preaching church would bring them over and not tell them about the Lord, they're out of their mind." The proselytizing backfired. According to a study of Hmong mental health problems, refugees sponsored by this pastor's religious organization were significantly more likely, when compared to other refugees, to require psychiatric treatment.

The Hmong were accustomed to living in the mountains, and most of them had never seen snow. Almost all their resettlement sites had flat topography and freezing winters. The majority were sent to cities, including Minneapolis, Chicago, Milwaukee, Detroit, Hartford, and Providence, because that was where refugee services—health care, language classes, job training, public housing—were concentrated. To encourage assimilation, and to avoid burdening any one community with more than its "fair share" of refugees, the Immigration and Naturalization Service adopted a policy of dispersal rather than clustering. Newly arrived Hmong were assigned to fifty-three cities in twenty-five different states: stirred into the melting pot in tiny, manageable portions, or, as John Finck, who worked with Hmong at the Rhode Island Office of Refugee Resettlement, put it, "spread like a thin layer of butter throughout the country so they'd disappear." In some places, clans were broken up. In others, members of only one clan were resettled, making it impossible for young people, who were forbidden by cultural taboo from marrying within their own clan, to find local marriage partners. Group solidarity, the cornerstone of Hmong social organization for more than two thousand years, was completely ignored.

Although most Hmong were resettled in cities, some nuclear families, unaccompanied by any of their extended relations, were placed in isolated rural areas. Disconnected from traditional supports, these families exhibited unusually high levels of anxiety, depression, and paranoia. In one such case, the distraught and delusional father of the Yang family—the only Hmong family sponsored by the First Baptist Church of Fairfield, Iowa—attempted to hang himself in the basement of his wooden bungalow along with his wife and four children. His wife changed her mind at the last minute and cut the family down, but she acted too late to save their only son. An Iowa grand jury declined to indict either parent, on the grounds that the father was suffering from Post-Traumatic Stress Disorder, and the mother, cut off from all sources of information except her husband, had no way to develop an independent version of reality.

Reviewing the initial resettlement of the Hmong with a decade's hindsight, Lionel Rosenblatt, the former United States Refugee Coordinator in Thailand, conceded that it had been catastrophically mishandled. "We knew at the start their situation was different, but we just couldn't make any special provisions for them," he said. "I still feel it was no mistake to bring the Hmong here, but you look back now and say, 'How could we have done it so shoddily?'" Eugene Douglas, President Reagan's ambassador-at-large for refugee affairs, stated flatly, "It was a kind of hell they landed into. Really, it couldn't have been done much worse."

The Hmong who sought asylum in the United States were, of course, not a homogeneous lump. A small percentage, mostly the high-ranking military officers who were admitted first, were multilingual and cosmopolitan, and a larger percentage had been exposed in a desultory fashion to some aspects of American culture and technology during the war or while living in Thai refugee camps. But the experience of tens of thousands of Hmong was much like the Lees'. It is possible to get some idea of how monumental the task of adjustment was likely to be by glancing at some of the pamphlets, audiotapes, and videos that refugee agencies produced for Southeast Asian JOJs. For example, "Your New Life in the United States," a handbook published by the Language and Orientation Resource Center in Washington, D.C., included the following tips:

Learn the meaning of "WALK"–"DON'T WALK" signs when crossing the street.

To send mail, you must use stamps.

To use the phone:

1) Pick up the receiver
2) Listen for dial tone
3) Dial each number separately
4) Wait for person to answer after it rings
5) Speak.

The door of the refrigerator must be shut.

Never put your hand in the garbage disposal.

Do not stand or squat on the toilet since it may break.

Never put rocks or other hard objects in the tub or sink since this will damage them.

Always ask before picking your neighbor's flowers, fruit, or vegetables.

In colder areas you must wear shoes, socks, and appropriate outerwear. Otherwise, you may become ill.

Always use a handkerchief or a kleenex to blow your nose in public places or inside a public building.

Never urinate in the street. This creates a smell that is offensive to Americans. They also believe that it causes disease.

Spitting in public is considered impolite and unhealthy. Use a kleenex or handkerchief.

Picking your nose or your ears in public is frowned upon in the United States.

The customs they were expected to follow seemed so peculiar, the rules and regulations so numerous, the language so hard to learn, and the emphasis on literacy and the decoding of other unfamiliar symbols so strong, that many Hmong were overwhelmed. Jonas Vangay told me, "In America, we are blind because even though we have eyes, we cannot see. We are deaf because even though we have ears, we cannot hear." Some newcomers wore pajamas as street clothes; poured water on electric stoves to extinguish them; lit charcoal fires in their living rooms; stored blankets in their refrigerators; washed rice in their toilets; washed their clothes in swimming pools; washed their hair with Lestoil; cooked with motor oil and furniture polish; drank Clorox; ate cat food; planted crops in public parks; shot and ate skunks, porcupines, woodpeckers, robins, egrets, sparrows, and a bald eagle; and hunted pigeons with crossbows in the streets of Philadelphia.

If the United States seemed incomprehensible to the Hmong, the Hmong seemed equally incomprehensible to the United States. Journalists seized excitedly on a label that is still trotted out at regular intervals: "the most primitive refugee group in America." (In an angry letter to the *New York Times,* in which that phrase had appeared in a 1990 news article, a Hmong computer specialist observed, "Evidently, we were not too primitive to fight as proxies for United States troops in the war in Laos.") Typical phrases from newspaper and magazine stories in the late seventies and eighties included "low-caste hill tribe," "Stone Age," "emerging from the mists of time," "like Alice falling down a rabbit hole." Inaccuracies were in no short supply. A 1981 article in the *Christian Science Monitor* called the Hmong language "extremely simplistic"; declared that the Hmong, who have been sewing *paj ntaub* [embroidered cloth] with organic motifs for centuries, make "no connection between a picture of a tree and a real tree"; and noted that "the Hmong have no oral tradition of literature. . . . Apparently no folk tales exist." Some journalists seemed to shed all inhibition, and much of their good sense as well, when they were loosed on the Hmong. . . .

Timothy Dunnigan, a linguistic anthropologist who has taught a seminar at the University of Minnesota on the media presentation of Hmong and Native Americans, once remarked to me, "The kinds of metaphorical language that we use to describe the Hmong say far more about us, and our attachment to our own frame of reference, than they do about the Hmong." . . .

It could not be denied that the Hmong were genuinely mysterious—far more so, for instance, than the Vietnamese and Cambodians who were streaming into the United States at the same time. Hardly anyone knew how to pronounce the word "Hmong." Hardly anyone—except the anthropology graduate students who suddenly realized they could write dissertations on patrilineal exogamous clan structures without leaving their hometowns—knew what role the Hmong had played during the war, or even what war it had been, since our government had succeeded all too well in keeping the Quiet War quiet. Hardly anyone knew they had a rich history, a complex culture, an efficient social system, and enviable family values. They were therefore an ideal blank surface on which to project xenophobic fantasies. . . .

Not everyone who wanted to make the Hmong feel unwelcome stopped at slander. In the words of the president of a youth center in Minneapolis, his Hmong neighbors in the mid-eighties were "prime meat for predators." In Laos, Hmong houses had no locks. Sometimes they had no doors. Cultural taboos against theft and intra-community violence were poor preparation for life in the high-crime, inner-city neighborhoods in which most Hmong were placed. Some of the violence directed against them had nothing to do with their ethnicity; they were simply easy marks. But a good deal of it was motivated by resentment, particularly in urban areas, for what was perceived as preferential welfare treatment.

In Minneapolis, tires were slashed and windows smashed. A high school student getting off a bus was hit in the face and told to "go back to China." A woman was kicked in the thighs, face, and kidneys, and her purse, which contained the family's entire savings of $400, was stolen; afterwards, she forbade her children to play outdoors, and her husband, who had once commanded a fifty-man unit in the Armée Clandestine, stayed home to guard the family's belongings. In Providence, children were beaten walking home from school. In Missoula, teenagers were stoned. In Milwaukee, garden plots were vandalized and a car was set on fire. In Eureka, California, two burning crosses were placed on a family's front lawn. In a random act of violence near Springfield, Illinois, a twelve-year-old boy was shot and killed by three men who forced his family's car off Interstate 55 and demanded money. His father told a reporter, "In a war, you know who your enemies are. Here, you don't know if the person walking up to you will hurt you."

In Philadelphia, anti-Hmong muggings, robberies, beatings, stonings, and vandalism were so commonplace during the early eighties that the city's Commission on Human Relations held public hearings to investigate the violence. One source of discord seemed to be a $100,000 federal grant for Hmong employment assistance that had incensed local residents, who were mostly unemployed themselves and believed the money should have been allocated to American citizens, not resident aliens. . . .

One thing stands out in all these accounts: the Hmong didn't fight back. . . .

Although on the battlefield the Hmong were known more for their fierceness than for their long livers, in the United States many were too proud to lower themselves to the level of the petty criminals they encountered, or even to admit they had been victims. An anthropologist named George M. Scott, Jr., once asked a group of Hmong in San Diego, all victims of property damage or assault, why they had not defended themselves or taken revenge. Scott wrote, "several Hmong victims of such abuse, both young and old, answered that to have done so, besides inviting further, retaliatory abuse, would have made them feel 'embarrassed' or ashamed. . . . In addition, the current president of Lao Family [a Hmong mutual assistance organization], when asked why his people did not 'fight back' when attacked here as they did in Laos, replied simply, 'because nothing here is worth defending to us.'"

In any case, Hmong who were persecuted by their neighbors could exercise a time-honored alternative to violence: flight. . . . Between 1982 and 1984, three quarters of the Hmong population of Philadelphia simply left town and joined relatives in other cities. During approximately the same period, one third of all the Hmong in the United States moved from one city to another. When they decided to relocate, Hmong families often lit off without notifying their sponsors, who were invariably offended. If they couldn't fit one of their possessions, such as a television set, in a car or bus or U-Haul, they left it behind, seemingly without so much as a backward glance. Some families traveled alone, but more often they moved in groups. When there was an exodus from Portland, Oregon, a long caravan of overloaded cars motored together down Interstate 5, bound for the Central Valley of California. With this "secondary migration," as sociologists termed it, the government's attempt to stir the Hmong evenly into the melting pot was definitively sabotaged.

Although local violence was often the triggering factor, there were also other reasons for migrating. In 1982, when all refugees who had lived in the United States for more than eighteen months stopped receiving Refugee Cash Assistance—the period of eligibility had previously been three years—many Hmong who had no jobs and no prospects moved to states that provided welfare benefits to two-parent families. Their original host states were often glad to get rid of them. For a time, the Oregon Human Resources Department, strapped by a tight state budget, sent refugees letters that pointedly detailed the levels of welfare benefits available in several other states. California's were among the highest. Thousands of Hmong also moved to California because they had heard it was an agricultural state where they might be able to farm. But by far the most important reason for relocating was reunification with other members of one's clan. Hmong clans are sometimes at odds with each other, but within a clan, whose thousands of members are regarded as siblings, one can always count on support and sympathy. A Hmong who tries to gain acceptance to a kin group other than his own is called a *puav,* or bat. He is rejected by the birds because he has fur and by the mice because he has wings. Only when a Hmong lives among his own subspecies can he stop flitting restlessly from group to group, haunted by the shame of not belonging.

The Hmong may have been following their venerable proverb, "There's always another mountain," but in the past, each new mountain had yielded a living. Unfortunately, the most popular areas of secondary resettlement all had high unemployment rates, and they got higher. . . .

By 1985, at least eighty percent of the Hmong in Merced, Fresno, and San Joaquin counties were on welfare.

That didn't halt the migration. Family reunification tends to have a snowball effect. The more Thaos or Xiongs there were in one place, the more mutual assistance they could provide, the more cultural traditions they could practice together, and the more stable their community would be. Americans, however, tended to view secondary migration as an indication of instability and dependence. . . .

Seeing that the Hmong were redistributing themselves as they saw fit, and that they were becoming an economic burden on the places to which they chose to move, the federal Office of Refugee Resettlement tried to slow the migratory tide. The 1983 Highland Lao Initiative, a three-million-dollar "emergency effort" to bolster employment and community stability in Hmong communities outside California, offered vocational training, English classes, and other enticements for the Hmong to stay put. Though the initiative claimed a handful of modest local successes, the California migration was essentially unstoppable. By this time, most Hmong JOJs were being sponsored by relatives in America rather than by voluntary organizations, so the government no longer had geographic control over their placements. The influx therefore came—and, in smaller increments, is still coming—from Thailand as well as from other parts of America. Therefore, in addition to trying to prevent the Hmong from moving to high-welfare states, the Office of Refugee Resettlement started trying to encourage the ones who were already there to leave. Spending an average of $7,000 per family on moving expenses, job placement, and a month or two of rent and food subsidies, the Planned

Secondary Resettlement Program, which was phased out in 1994, relocated about 800 unemployed Hmong families from what it called "congested areas" to communities with "favorable employment opportunities"—i.e., unskilled jobs with wages too low to attract a full complement of local American workers.

Within the economic limitations of blue-collar labor, those 800 families have fared well. Ninety-five percent have become self-sufficient. They work in manufacturing plants in Dallas, on electronics assembly lines in Atlanta, in furniture and textile factories in Morganton, North Carolina. More than a quarter of them have saved enough money to buy their own houses, as have three quarters of the Hmong families who live in Lancaster County, Pennsylvania, where the men farm or work in food-processing plants, and the women work for the Amish, sewing quilts that are truthfully advertised as "locally made." Elsewhere, Hmong are employed as grocers, carpenters, poultry processors, machinists, welders, auto mechanics, tool and die makers, teachers, nurses, interpreters, and community liaisons. In a survey of Minnesota employers, the respondents were asked "What do you think of the Hmong as workers?" Eighty-six percent rated them "very good." . . .

Some younger Hmong have become lawyers, doctors, dentists, engineers, computer programmers, accountants, and public administrators. Hmong National Development, an association that promotes Hmong self-sufficiency, encourages this small corps of professionals to serve as mentors and sponsors for other Hmong who might thereby be induced to follow suit. The cultural legacy of mutual assistance has been remarkably adaptive. Hundreds of Hmong students converse electronically, trading gossip and information—opinions on the relevance of traditional customs, advice on college admissions, personal ads—via the Hmong Channel on the Internet Relay Chat system. . . . There is also a Hmong Homepage on the World Wide Web (http://www.stolaf.edu/people/cdr/hmong/) and several burgeoning Hmong electronic mailing lists, including Hmongnet, Hmongforum, and Hmong Language Users Group.

The M.D.s and J.D.s and digital sophisticates constitute a small, though growing, minority. Although younger, English-speaking Hmong who have been educated in the United States have better employment records than their elders, they still lag behind most other Asian-Americans. . . .

For the many Hmong who live in high-unemployment areas, questions of advancement are often moot. They have no jobs at all. This is the reason the Hmong are routinely called this country's "least successful refugees." It is worth noting that the standard American tests of success that they have flunked are almost exclusively economic. If one applied social indices instead—such as rates of crime, child abuse, illegitimacy, and divorce—the Hmong would probably score better than most refugee groups (and also better than most Americans), but those are not the forms of success to which our culture assigns its highest priority. Instead, we have trained the spotlight on our best-loved index of failure, the welfare rolls. In California, Minnesota, and Wisconsin, where, not coincidentally, benefits tend to be relatively generous and eligibility requirements relatively loose, the percentages of Hmong on welfare are approximately forty-five, forty, and thirty-five (an improvement over five years ago, when they were approximately sixty-five, seventy, and sixty). The cycle of dependence that began with rice drops in Laos and reinforced with daily handouts at Thai refugee camps has been completed here in the United States. The conflicting structures of the Hmong culture and the American welfare system make it almost impossible for the average family to become independent. . . .

Few things gall the Hmong more than to be criticized for accepting public assistance. For one thing, they feel they deserve the money. Every Hmong has a different version of what is commonly called "The Promise": a written or verbal contract, made by CIA personnel in Laos, that if they fought for the Americans, the Americans would aid them if the Pathet Lao won the war. After risking their lives to rescue downed American pilots, seeing their villages flattened by incidental American bombs, and being forced to flee their country because they had supported the "American War," the Hmong expected a hero's welcome here. According to many of them, the first betrayal came when the American airlifts rescued only the officers from Long Tieng, leaving nearly everyone else behind. The second betrayal came in the Thai camps, when the Hmong who wanted to come to the United States were not all automatically admitted. The third betrayal came when they arrived here and found they were ineligible for veterans' benefits. The fourth betrayal came when Americans condemned them for what the Hmong call "eating welfare." The fifth betrayal came when the Americans announced that the welfare would stop.

Aside from some older people who consider welfare a retirement benefit, most Hmong would prefer almost any other option—if other options existed. What right-thinking Hmong would choose to be yoked to one of the most bureaucratic institutions in America? . . .

In a study of Indochinese refugees in Illinois, the Hmong exhibited the highest degree of "alienation from their environment." According to a Minnesota study, Hmong refugees who had lived in the United States for a year and a half had "very high levels of depression, anxiety, hostility, phobia, paranoid ideation, obsessive compulsiveness and feelings of inadequacy." (Over the next decade, some of these symptoms moderated, but the refugees' levels of anxiety, hostility, and paranoia showed little or no improvement.) The study that I found most disheartening was the 1987 California Southeast Asian Mental Health Needs Assessment, a statewide epidemiological survey funded by the Office of Refugee Resettlement and the National Institute of Mental Health. It was shocking to look at the bar graphs comparing the Hmong with the Vietnamese, the Chinese-Vietnamese, the Cambodians, and the Lao—all of whom, particularly the Cambodians, fared poorly compared to the general population—and see how the Hmong stacked up: Most depressed. Most psychosocially dysfunctional. Most likely to be severely in need of mental health treatment. Least educated. Least literate. Smallest percentage in labor force. Most likely to cite "fear" as a reason for immigration and least likely to cite "a better life." . . .

"Full" of both past trauma and past longing, the Hmong have found it especially hard to deal with present threats to their old identities. I once went to a conference on Southeast Asian mental health at which a psychologist named Evelyn Lee, who was born in Macao, invited six members of the audience to come to the front of the auditorium for a role-playing exercise. She cast them as a grandfather, a father, a mother, an eighteen-year-old son, a sixteen-year-old daughter, and a twelve-year-old daughter. "Okay," she told them, "line up according to your status in your old country." Ranking themselves by traditional notions of age and gender, they queued up in the order I've just mentioned, with the grandfather standing proudly at the head of the line. "Now they come to America," said Dr. Lee. "Grandfather has no job. Father can only chop vegetables. Mother didn't work in the old country, but here she gets a job in a garment factory. Oldest daughter works there too. Son drops out of high school because he can't learn English. Youngest daughter learns the best English in the family and ends up at U.C. Berkeley. Now you line up again." As the family reshuffled, I realized that its power structure had turned completely upside down, with the twelve-year-old girl now occupying the head of the line and the grandfather standing forlornly at the tail.

Dr. Lee's exercise was an eloquent demonstration of what sociologists call "role loss." Of all the stresses in the Hmong community, role loss . . . may be the most corrosive to the ego. . . .

And in this country the real children have assumed some of the power that used to belong to their elders. The status conferred by speaking English and understanding American conventions is a phenomenon familiar to most immigrant groups, but the Hmong, whose identity has always hinged on tradition, have taken it particularly hard. . . .

Although Americanization may bring certain benefits—more job opportunities, more money, less cultural dislocation—Hmong parents are likely to view any earmarks of assimilation as an insult and a threat. "In our families, the kids eat hamburger and bread," said Dang Moua sadly, "whereas the parents prefer hot soup with vegetables, rice, and meat like tripes or liver or kidney that the young ones don't want." . . .

Sukey Waller, Merced's maverick psychologist, once recalled a Hmong community meeting she had attended. "An old man of seventy or eighty stood up in the front row," she said, "and he asked one of the most poignant questions I have ever heard: 'Why, when what we did worked so well for two hundred years, is everything breaking down?'" When Sukey told me this, I understood why the man had asked the question, but I thought he was wrong. Much has broken down, but not everything. Jacques Lemoine's analysis of the postwar hegira—that the Hmong came to the West to save not only their lives but their ethnicity—has been at least partially confirmed in the United States. I can think of no other group of immigrants whose culture, in its most essential aspects, has been so little eroded by assimilation. Virtually all Hmong still marry other Hmong, marry young, obey the taboo against marrying within their own clans, pay brideprices, and have large families. Clan and lineage structures are intact, as is the ethic of group solidarity and mutual assistance. On most weekends in Merced, it is possible to hear a death drum beating at a Hmong funeral or a *txiv neeb's* gong and rattle sounding at a healing ceremony. Babies wear strings on their wrists to protect their souls from abduction by *dabs*. People divine their fortunes by interpreting their dreams. (If you dream of opium, you will have bad luck; if you dream you are covered with excrement, you will have good luck; if you dream you have a snake on your lap, you will become pregnant.) Animal sacrifices are common, even among Christian converts, a fact I first learned when May Ying Xiong told me that she would be unavailable to interpret one weekend because her family was sacrificing a cow to safeguard her niece during an upcoming open-heart operation. When I said, "I didn't know your family was so religious," she replied, "Oh yes, we're Mormon." . . .

I was able to see the whole cycle of adjustment to American life start all over again during one of my visits to Merced. When I arrived at the Lees' apartment, I was surprised to find it crammed with people I'd never met before. These turned out to be a cousin of Nao Kao's named Joua Chai Lee, his wife, Yeng Lor, and their nine children, who ranged in age from eight months to twenty-five years. They had arrived from Thailand two weeks earlier, carrying one piece of luggage for all eleven of them. In it were packed some clothes, a bag of rice, and, because Joua is a *txiv neeb's* assistant, a set of rattles, a drum, and a pair of divinatory water-buffalo horns. The cousins were staying with Foua and Nao Kao until they found a place of their own. The two families had not seen each other in more than a decade, and there was a festive atmosphere in the little apartment, with small children dashing around in their new American sneakers and the four barefooted adults frequently throwing back their heads and laughing. Joua said to me, via May Ying's translation, "Even though there are a lot of us, you can spend the night here too." May Ying explained to me later that Joua didn't really expect me to lie down on the floor with twenty of his relatives. It was simply his way, even though he was in a strange country where he owned almost nothing, of extending a face-saving bit of Hmong hospitality.

I asked Joua what he thought of America. "It is really nice but it is different," he said, "It is very flat. You cannot tell one place from another. There are many things I have not seen before, like that"—a light switch—"and that"—a telephone—"and that"—an air conditioner. "Yesterday our relatives took us somewhere in a car and I saw a lady and I thought she was real but she was fake." This turned out to have been a mannequin at the Merced Mall.

"I couldn't stop laughing all the way home," he said. And remembering how funny his mistake had been, he started to laugh again.

Then I asked Joua what he hoped for his family's future here. "I will work if I can," he said, "but I think I probably cannot. As old as I am, I think I will not be able to learn one word of English. If my children put a heart to it, they will be able to learn English and get really smart. But as for myself, I have no hope."

THINKING ABOUT THE READING

Why has it been so difficult for Hmong refugees to adjust to life in the United States? How do the experiences of younger Hmong compare to those of their elders? Why are the Hmong such a popular target of anti-immigrant violence and persecution? Why is the U.S. government so unwilling to grant the Hmong their wish to be "left alone"? In other words, why is there such a strong desire to assimilate them into American culture? On a more general level, why is there such distaste in this society when certain ethnic groups desire to retain their traditional way of life? Consider the differences that might emerge between different generations within immigrant families. What aspects of culture are the most difficult to maintain through the generations?

McDonald's in Hong Kong

Consumerism, Dietary Change, and the Rise of a Children's Culture

James L. Watson

(1997)

Transnationalism and the Fast Food Industry*

Does the roaring success of McDonald's and its rivals in the fast food industry mean that Hong Kong's local culture is under siege? Are food chains helping to create a homogenous, "global" culture better suited to the demands of a capitalist world order? Hong Kong would seem to be an excellent place to test the globalization hypothesis, given the central role that cuisine plays in the production and maintenance of a distinctive local identity. Man Tso-chuen's great-grandchildren are today avid consumers of Big Macs, pizza, and Coca-Cola; does this somehow make them less "Chinese" than their grandfather?

The people of Hong Kong have embraced American-style fast foods, and by so doing they might appear to be in the vanguard of a worldwide culinary revolution. But they have not been stripped of their cultural traditions, nor have they become "Americanized" in any but the most superficial of ways. Hong Kong in the late 1990s constitutes one of the world's most heterogeneous cultural environments. Younger people, in particular, are fully conversant in transnational idioms, which include language, music, sports, clothing, satellite television, cyber-communications, global travel, and—of course—cuisine. It is no longer possible to distinguish what is local and what is not. In Hong Kong, the transnational *is* the local.

Eating Out: A Social History of Consumption

By the time McDonald's opened its first Hong Kong restaurant in 1975, the idea of fast food was already well established among local consumers. Office workers, shop assistants, teachers, and transport workers had enjoyed various forms of take-out cuisine for well over a century; an entire industry had emerged to deliver mid-day meals direct to workplaces. In the 1960s and 1970s thousands of street vendors produced snacks and simple meals on demand, day or night. Time has always been money in Hong Kong; hence, the dual keys to success in the catering trade were speed and convenience. Another essential characteristic was that the food, based primarily on rice or noodles, had to be hot. Even the most cosmopolitan of local consumers did not (and many still do not) consider cold foods, such as sandwiches and salads, to be acceptable meals. Older people in South China associate cold food with offerings to the dead and are understandably hesitant to eat it.

The fast food industry in Hong Kong had to deliver hot items that could compete with traditional purveyors of convenience foods (noodle shops, dumpling stalls, soup carts, portable grills).

*Seven of the world's ten busiest McDonald's restaurants are located in Hong Kong. When McDonald's first opened in 1975, few thought it would survive more than a few months. By January 1, 1997, Hong Kong had 125 outlets, which means that there was one McDonald's for every 51,200 residents, compared to one for every 30,000 people in the United States.

McDonald's mid-1970s entry corresponded to an economic boom associated with Hong Kong's conversion from a low-wage, light-industrial outpost to a regional center for financial services and high-technology industries. McDonald's takeoff thus paralleled the rise of a new class of highly educated, affluent consumers who thrive in Hong Kong's ever-changing urban environment—one of the most stressful in the world. These new consumers eat out more often than their parents and have created a huge demand for fast, convenient foods of all types. In order to compete in this market, McDonald's had to offer something different. That critical difference, at least during the company's first decade of operation, was American culture packaged as all-American, middle-class food.

* * *

Mental Categories: Snack Versus Meal

As in other parts of East Asia, McDonald's faced a serious problem when it began operation in Hong Kong: Hamburgers, fries, and sandwiches were perceived as snacks (Cantonese *siu sihk*, literally "small eats"); in the local view these items did not constitute the elements of a proper meal. This perception is still prevalent among older, more conservative consumers who believe that hamburgers, hot dogs, and pizza can never be "filling." Many students stop at fast food outlets on their way home from school; they may share hamburgers and fries with their classmates and then eat a full meal with their families at home. This is not considered a problem by parents, who themselves are likely to have stopped for tea and snacks after work. Snacking with friends and colleagues provides a major opportunity for socializing (and transacting business) among southern Chinese. Teahouses, coffee shops, bakeries, and ice cream parlors are popular precisely because they provide a structured yet informal setting for social encounters. Furthermore, unlike Chinese restaurants and banquet halls, snack centers do not command a great deal of time or money from customers.

Contrary to corporate goals, therefore, McDonald's entered the Hong Kong market as a purveyor of snacks. Only since the late 1980s has its fare been treated as the foundation of "meals" by a generation of younger consumers who regularly eat non-Chinese food. Thanks largely to McDonald's, hamburgers and fries are now a recognized feature of Hong Kong's lunch scene. The evening hours remain, however, the weak link in McDonald's marketing plan; the real surprise was breakfast, which became a peak traffic period.

The mental universe of Hong Kong consumers is partially revealed in the everyday use of language. Hamburgers are referred to, in colloquial Cantonese, as *han bou bao*—*han* being a homophone for "ham" and *bao* the common term for stuffed buns or bread rolls. *Bao* are quintessential snacks, and however excellent or nutritious they might be, they do not constitute the basis of a satisfying (i.e., filling) meal. In South China that honor is reserved for culinary arrangements that rest, literally, on a bed of rice (*fan*). Foods that accompany rice are referred to as *sung*, probably best translated as "toppings" (including meat, fish, and vegetables). It is significant that hamburgers are rarely categorized as meat (*yuk*); Hong Kong consumers tend to perceive anything that is served between slices of bread (Big Macs, fish sandwiches, hot dogs) as *bao*. In American culture the hamburger is categorized first and foremost as a meat item (with all the attendant worries about fat and cholesterol content), whereas in Hong Kong the same item is thought of primarily as bread.

From Exotic to Ordinary: McDonald's Becomes Local

Following precedents in other international markets, the Hong Kong franchise promoted McDonald's basic menu and did not introduce items that would be more recognizable to Chinese consumers (such as rice dishes, tropical fruit, soup noodles). Until recently the food has been indistinguishable from that served in Mobile, Alabama, or Moline, Illinois. There are, however, local preferences: the best-selling items in many outlets are fish sandwiches and plain hamburgers; Big Macs tend to be the favorites of children and teenagers. Hot tea and hot chocolate outsell coffee, but Coca-Cola remains the most popular drink.

McDonald's conservative approach also applied to the breakfast menu. When morning service was introduced in the 1980s, American-style items such as eggs, muffins, pancakes, and hash brown potatoes were not featured. Instead, the local outlets served the standard fare of hamburgers and fries for breakfast. McDonald's initial venture into the early morning food market was so successful that Mr. Ng [managing director of McDonald's Hong Kong], hesitated to introduce American-style breakfast items, fearing that an abrupt shift in menu might alienate consumers who were beginning to accept hamburgers and fries as a regular feature of their diet. The transition to eggs, muffins, and hash browns was a gradual one, and today most Hong Kong customers order breakfasts that are similar to those offered in American outlets. But once established, dietary preferences change slowly: McDonald's continues to feature plain hamburgers (but not the Big Mac) on its breakfast menu in most Hong Kong outlets.

Management decisions of the type outlined above helped establish McDonald's as an icon of popular culture in Hong Kong. From 1975 to approximately 1985, McDonald's became the "in" place for young people wishing to associate themselves with the laid-back, nonhierarchical dynamism they perceived American society to embody. The first generation of consumers patronized McDonald's precisely because it was *not* Chinese and was *not* associated with Hong Kong's past as a backward-looking colonial outpost where (in their view) nothing of consequence ever happened. Hong Kong was changing and, as noted earlier, a new consumer culture was beginning to take shape. McDonald's caught the wave of this cultural movement and has been riding it ever since.

Today, McDonald's restaurants in Hong Kong are packed—wall-to-wall—with people of all ages, few of whom are seeking an American cultural experience. Twenty years after Mr. Ng opened his first restaurant, eating at McDonald's has become an ordinary, everyday experience for hundreds of thousands of Hong Kong residents. The chain has become a local institution in the sense that it has blended into the urban landscape; McDonald's outlets now serve as rendezvous points for young and old alike.

What's in a Smile? Friendliness and Public Service

American consumers expect to be served "with a smile" when they order fast food, but this is not true in all societies. In Hong Kong people are suspicious of anyone who displays what is perceived to be an excess of congeniality, solicitude, or familiarity. The human smile is not, therefore, a universal symbol of openness and honesty. "If you buy an apple from a hawker and he smiles at you," my Cantonese tutor once told me, "you know you're being cheated."

Given these cultural expectations, it was difficult for Hong Kong management to import a key element of the McDonald's formula—service with a smile—and make it work. Crew members were trained to treat customers in a manner that approximates the American notion of "friendliness." Prior to the 1970s, there was not even an indigenous Cantonese term to describe this form of behavior. The traditional notion of friendship is based on loyalty to close associates, which by definition cannot be extended to strangers. Today the concept of *public* friendliness is recognized—and verbalized—by younger people in Hong Kong, but the term many of them use to express this quality is "friendly," borrowed directly from English. McDonald's, through its television advertising, may be partly responsible for this innovation, but to date it has had little effect on workers in the catering industry.

During my interviews it became clear that the majority of Hong Kong consumers were uninterested in public displays of congeniality from service personnel. When shopping for fast food most people cited convenience, cleanliness, and table space as primary considerations; few even mentioned service except to note that the food should be delivered promptly. Counter staff in Hong Kong's fast food outlets (including McDonald's) rarely make great efforts to smile or to behave in a manner Americans would interpret as friendly. Instead, they project qualities that are admired in the local culture: competence, directness, and unflappability. In a North American setting the facial expression that Hong Kong employees use to convey these qualities would likely be interpreted as a deliberate attempt to be rude or indifferent. Workers who smile on the job are assumed to be enjoying themselves at the consumer's (and management's) expense: In the words of one diner I overheard while standing in a queue, "They must be playing around back there. What are they laughing about?"

Consumer Discipline?

[A] hallmark of the American fast food business is the displacement of labor costs from the corporation to the consumers. For the system to work, consumers must be educated—or "disciplined"—so that they voluntarily fulfill their side of an implicit bargain: We (the corporation) will provide cheap, fast service, if you (the customer) "earn" your own tray, seat yourself, and help clean up afterward. Time and space are also critical factors in the equation: Fast service is offered in exchange for speedy consumption and a prompt departure, thereby making room for others. This system has revolutionized the American food industry and has helped to shape consumer expectations in other sectors of the economy. How has it fared in Hong Kong? Are Chinese customers conforming to disciplinary models devised in Oak Brook, Illinois?

The answer is both yes and no. In general Hong Kong consumers have accepted the basic elements of the fast food formula, but with "localizing" adaptations. For instance, customers generally do not bus their own trays, nor do they depart immediately upon finishing. Clearing one's own table has never been an accepted part of local culinary culture, owing in part to the low esteem attaching to this type of labor. During McDonald's first decade in Hong Kong, the cost of hiring extra cleaners was offset by low wages. A pattern was thus established, and customers grew accustomed to leaving without attending to their own rubbish. Later, as wages escalated in the late 1980s and early 1990s. McDonald's tried to introduce self-busing by posting announcements in restaurants and featuring the practice in its television advertisements. As of February 1997, however, little had changed. Hong Kong consumers have ignored this aspect of consumer discipline.

What about the critical issues of time and space? Local managers with whom I spoke estimated that the average eating time for most Hong Kong customers was between 20 and 25 minutes, compared to 11 minutes in the United States fast food industry. This estimate confirms my own observations of McDonald's consumers in Hong Kong's central business districts (Victoria and Tsimshatsui). A survey conducted in the New Territories city of Yuen Long—an old market town that has grown into a modern urban center—revealed that local McDonald's consumers took just under 26 minutes to eat.

Perhaps the most striking feature of the American-inspired model of consumer discipline is the queue. Researchers in many parts of the world have reported that customers refuse, despite "education" campaigns by the chains involved, to form neat lines in front of cashiers. Instead, customers pack themselves into disorderly scrums and jostle for a chance to place their orders. Scrums of this nature were common in Hong Kong when McDonald's opened in 1975. Local managers discouraged this practice by stationing queue monitors near the registers during busy hours and, by the 1980s, orderly lines were the norm at McDonald's. The disappearance of the scrum corresponds to a general change in Hong Kong's public culture as a new generation of residents, the children of refugees, began to treat the territory as their home. Courtesy toward strangers was largely unknown in the 1960s: Boarding a bus during rush hour could be a nightmare and transacting business at a bank teller's window required brute strength. Many people credit McDonald's with being the first public institution in Hong Kong to enforce queuing, and thereby helping to create a more "civilized" social order. McDonald's did not, in fact, introduce the queue to Hong Kong, but this belief is firmly lodged in the public imagination.

Hovering and the Napkin Wars

Purchasing one's food is no longer a physical challenge in Hong Kong's McDonald's but finding a place to sit is quite another matter. The traditional practice of "hovering" is one solution: Choose a group of diners who appear to be on the verge of leaving and stake a claim to their table by hovering nearby, sometimes only inches away. Seated customers routinely ignore the intrusion; it would, in fact, entail a loss of face to notice. Hovering was the norm in Hong Kong's lower- to middle-range restaurants during the 1960s and 1970s, but the practice has disappeared in recent years. Restaurants now take names or hand out tickets at the entrance; warning signs, in Chinese and English, are posted: "Please wait to be seated." Customers are no longer allowed into the dining area until a table is ready.

Fast food outlets are the only dining establishments in Hong Kong where hovering is still tolerated, largely because it would be nearly impossible to regulate. Customer traffic in McDonald's is so heavy that the standard restaurant design has failed to reproduce American-style dining routines: Rather than ordering first and finding a place to sit afterward, Hong Kong consumers usually arrive in groups and delegate one or two people to claim a table while someone else joins the counter queues. Children make ideal hoverers and learn to scoot through packed restaurants, zeroing in on diners who are about to finish. It is one of the wonders of comparative ethnography to witness the speed with which Hong Kong children perform this reconnaissance duty. Foreign visitors are sometimes unnerved by hovering, but residents accept it as part of everyday life in one of the world's most densely populated cities. It is not surprising, therefore, that Hong Kong's fast food chains have made few efforts to curtail the practice.

Management is less tolerant of behavior that affects profit margins. In the United States fast food companies save money by allowing (or requiring) customers to collect their own napkins, straws, plastic flatware, and condiments. Self-provisioning is an essential feature of consumer discipline, but it only works if the system is not abused. In Hong Kong napkins are dispensed, one at a time, by McDonald's crew members who work behind the counter; customers who do not ask for napkins do not receive any. This is a deviation from the corporation's standard operating procedure and adds a few seconds to each transaction, which in turn slows down the queues. Why alter a well-tested routine? The reason is simple: napkins placed in public dispensers disappear faster than they can be replaced.

* * *

Buffets, like fast food outlets, depend upon consumers to perform much of their own labor in return for reduced prices. Abuse of the system—wasting food or taking it home—is taken for granted and is factored into the price of buffet meals. Fast food chains, by contrast, operate at lower price thresholds where consumer abuse can seriously affect profits.

Many university students of my acquaintance reported that they had frequently observed older people pocketing wads of paper napkins, three to four inches thick, in restaurants that permit self-provisioning. Management efforts to stop this behavior are referred to, in the Cantonese-English slang of Hong Kong youth, as the "Napkin Wars." Younger people were appalled by what they saw as the waste of natural resources by a handful of customers. As they talked about the issue, however, it became obvious that the Napkin Wars represented more—in their eyes—than a campaign to conserve paper. The sight of diners abusing public facilities reminded these young people of the bad old days of their parents and grandparents, when Hong Kong's social life was dominated by refugees who had little stake in the local community. During the 1960s and 1970s, economic insecurities were heightened by the very real prospect that Red Guards might take over the colony at any moment. The game plan was simple during those decades: Make money as quickly as possible and move on. In the 1980s a new generation of local-born youth began treating Hong Kong as home and proceeded to build a public culture better suited to their vision of life in a cosmopolitan city. In this new Hong Kong, consumers are expected to be sophisticated and financially secure, which means that it would be beneath their dignity to abuse public facilities. Still, McDonald's retains control of its napkins.

* * *

Children as Consumers

During the summer of 1994, while attending a business lunch in one of Hong Kong's fanciest hotels, I watched a waiter lean down to consult with a customer at an adjoining table. The object of his attention was a six-year-old child who studied the menu with practiced skill. His parents beamed as their prodigy performed; meanwhile, sitting across the table, a pair of grandparents sat bolt upright, scowling in obvious disapproval. Twenty years ago the sight of a child commanding such attention would have shocked the entire restaurant into silence. No one, save the immediate party (and this observer), even noticed in 1994.

Hong Kong children rarely ate outside their home until the late 1970s, and when they did, they were expected to eat what was put in front of them. The idea that children might actually order their own food or speak to a waiter would have outraged most adults; only foreign youngsters (notably the offspring of British and American expatriates) were permitted to make their preferences known in public. Today, Hong Kong children as young as two or three participate in the local economy as full-fledged consumers, with their own tastes and brand loyalties. Children now have money in their pockets and they spend it on personal consumption, which usually means snacks. In response, new industries and a specialized service sector has emerged to "feed" these discerning consumers. McDonald's was one of the first corporations to recognize the potential of the children's market; in effect, the company started a revolution by making it possible for even the youngest consumers to *choose* their own food.

* * *

Many Hong Kong children of my acquaintance are so fond of McDonald's that they refuse to eat with their parents or grandparents in Chinese-style restaurants or *dim sam* teahouses. This has caused intergenerational distress in some of Hong Kong's more conservative communities. In 1994, a nine-year-old boy, the descendant of illustrious ancestors who settled in the New Territories eight centuries ago, talked about his concerns as we consumed Big Macs, fries, and shakes at McDonald's: "A-bak [uncle], I like it here better than any place in the world. I want to come here every day." His father takes him to McDonald's at least twice a week, but his grandfather, who accompanied them a few times in the late 1980s, will no longer do so. "I prefer to eat *dim sam*," the older man told me later. "That place [McDonald's] is for kids." Many grandparents have resigned themselves to the new consumer trends and take their preschool grandchildren to McDonald's for mid-morning snacks—precisely the time of day that local teahouses were once packed with retired people. Cantonese grandparents have always played a prominent role in child minding, but until recently the children had to accommodate to the proclivities of their elders. By the 1990s grandchildren were more assertive and the mid-morning *dim sam* snack was giving way to hamburgers and Cokes.

* * *

Ronald McDonald and the Invention of Birthday Parties

Until recently most people in Hong Kong did not even know, let alone celebrate, their birthdates in the Western calendrical sense; dates of birth according to the lunar calendar were recorded for divinatory purposes but were not noted in annual rites. By the late 1980s, however, birthday parties, complete with cakes and candles, were the rage in Hong Kong. Any child who was anyone had to have a party, and the most popular venue was a fast food restaurant, with McDonald's ranked above all competitors. The majority of Hong Kong people live in overcrowded flats, which means that parties are rarely held in private homes.

Except for the outlets in central business districts, McDonald's restaurants are packed every Saturday and Sunday with birthday parties, cycled through at the rate of one every hour. A party hostess, provided by the restaurant, leads the children in games while the parents sit on the sidelines, talking quietly among themselves. For a small fee celebrants receive printed invitation cards, photographs, a gift box containing toys and a discount coupon for future trips to McDonald's. Parties are held in a special enclosure, called the Ronald Room, which is equipped with low tables and tiny stools—suitable only for children. Television commercials portray Ronald McDonald leading birthday celebrants on exciting safaris and expeditions. The clown's Cantonese name, Mak Dong Lou Suk-Suk ("Uncle McDonald"), plays on the intimacy of kinship and has helped transform him into one of Hong Kong's most familiar cartoon figures.

* * *

McDonald's as a Youth Center

Weekends may be devoted to family dining and birthday parties for younger children, but on weekday afternoons, from 3:00 to 6:00 P.M., McDonald's restaurants are packed with teenagers stopping for a snack on their way home from school. In many outlets 80 percent of the late afternoon clientele appear in school uniforms, turning the restaurants into a sea of white frocks, light blue shirts, and dark trousers. The students, aged between 10 and 17, stake out tables and buy snacks that are shared in groups. The noise level at this time of day is deafening; students shout to friends and dart from table to table. Few adults, other than restaurant staff, are in evidence. It is obvious that McDonald's is treated as an informal youth center, a recreational extension of school where students can unwind after long hours of study.

* * *

In contrast to their counterparts in the United States, where fast food chains have devised ways to discourage lingering, McDonald's in Hong Kong does not set a limit on table time. When I asked the managers of several Hong Kong outlets how they coped with so many young people chatting at tables that might otherwise be occupied by paying customers, they all replied that the students were "welcome." The obvious strategy is to turn a potential liability into an asset: "Students create a good atmosphere which is good for our business," said one manager as he watched an army of teenagers—dressed in identical school uniforms—surge into his restaurant. Large numbers of students also use McDonald's as a place to do homework and prepare for exams, often in groups. Study space of any kind, public or private, is hard to find in overcrowded Hong Kong.

* * *

Conclusions: Whose Culture Is It?

In what sense, if any, is McDonald's involved in these cultural transformations (the creation of a child-centered consumer culture, for instance)? Has the company helped to create these trends, or merely followed the market? Is this an example of American-inspired, transnational culture crowding out indigenous cultures?

The deeper I dig into the lives of consumers themselves, in Hong Kong and elsewhere, the more complex the picture becomes. Having watched the processes of culture change unfold for nearly thirty years, it is apparent to me that the ordinary people of Hong Kong have most assuredly *not* been stripped of their cultural heritage, nor have they become the uncomprehending dupes of transnational corporations. Younger people—including many of the grandchildren of my former neighbors in the New Territories—are avid consumers of transnational culture in all of its most obvious manifestations: music, fashion, television, and cuisine. At the same time, however, Hong Kong has itself become a major center for the *production* of transnational culture, not just a sinkhole for its *consumption.* Witness, for example, the expansion of Hong Kong popular culture into China, Southeast Asia, and beyond: "Cantopop" music is heard on radio stations in North China, Vietnam, and Japan; the Hong Kong fashion industry influences clothing styles in Los Angeles, Bangkok, and Kuala Lumpur; and, perhaps most significant of all, Hong Kong is emerging as a center for the production and dissemination of satellite television programs throughout East, Southeast, and South Asia.

A lifestyle is emerging in Hong Kong that can best be described as postmodern, postnationalist, and flamboyantly transnational. The wholesale acceptance and appropriation of Big Macs, Ronald McDonald, and birthday parties are small, but significant aspects of this redefinition of Chinese cultural identity. In closing, therefore, it seems appropriate to pose an entirely new set of questions: Where does the transnational end and the local begin? Whose culture is it, anyway? In places like Hong Kong the postcolonial periphery is fast becoming the metropolitan center, where local people are consuming and simultaneously producing new cultural systems.

* * *

THINKING ABOUT THE READING

According to Watson's observations, the people in Hong Kong have adopted some of the characteristics of McDonald's "fast food" culture and resisted others. How has the presence of McDonald's changed Chinese culture? How have the Chinese changed this "fast food" culture to better fit their own? What does this reading suggest about the relationship between globalization and local cultures? Consider other examples of globalized industries and the impact on local cultures.

Building Identity

Socialization

5

Sociology teaches us that humans don't develop in a social vacuum. Other people, cultural practices, historical events, and social institutions shape what we do and say, what we value, and who we become. Our self-concept, identity, and sense of self-worth are derived from our interactions with other people. We are especially tuned in to the reactions, real or imagined, of others.

Socialization is the process by which individuals learn their culture and learn to live according to the norms of their society. Through socialization, we learn how to perceive our world, gain a sense of our own identity, and discover how to interact appropriately with others. This learning process occurs within the context of several social institutions—schools, religious institutions, the media, and the family—and it extends beyond childhood. Adults must be resocialized into a new galaxy of norms, values, and expectations each time they leave or abandon current positions and enter new ones.

The conditions into which we are born shape our initial socialization in profound ways. Circumstances such as race, ethnicity, and social class are particularly significant factors in socialization processes. In "Life as the Maid's Daughter," sociologist Mary Romero describes a research interview with a young Chicana regarding her recollections of growing up as the daughter of a live-in maid for a white, upper-class family living in Los Angeles. Romero describes the many ways in which this girl learns to move between different social settings, adapt to different expectations, and occupy different social positions. This girl must constantly negotiate the boundaries of inclusion and exclusion, as she struggles between the socializing influence of her own ethnic group and that of the white, upper-class employers she and her mother live with. Through this juggling, she illustrates the ways in which we manage the different, often contradictory, identities that we take on in different situations.

Although ethnicity is an important factor in shaping our identities, the ways in which this works are complex. Zhou and Lee suggest that Asian American youth have a distinct culture and sense of identity that reflects their unique position as "in between" in a society that focuses on "black/white" racial identities. Immigration status, racism, and globalization are also significant features in shaping this unique identity.

In a popular sociological article, "Code of the Streets," Elijah Anderson looks at the complexities of socialization among young African American men living in the inner cities. Anderson is particularly interested in the development of "manhood" and respect among these young men and the relationship between violence and survival. To survive, one must adopt the "code"—a complex system of norms, dress, rituals, and expected behavior. The third reading for this section is a contemporary study focusing on inner-city girls and their relationship to the "code." Nikki Jones suggests that these girls are no more immune to violence than boys. She provides narrative descriptions of some of the ways in which street life shapes the behavior and identity of the young women in her study.

Something to Consider as You Read

According to sociologists, we are shaped by our cultural environment and by the influences of significant people and groups in our lives. Consider some of the people or groups whose opinions matter to you. Can you imagine them as a kind of audience in your head, observing and reacting to your behavior? Think about the desire to feel included. To what extent has this desire shaped your participation in a group that has had an impact on your self-image? How important are "role models" in the socialization process? If someone is managing conflicting identities and has no role models or others in similar situations, how might this conflict affect her or his sense of self and relationships with others? What do these readings suggest about the importance of being the "right person in the right place" even if that's not all you feel yourself to be? How do power and authority affect people's sense of self and their right to be whomever they want to be in any situation? How do social conditions shape our choices and opportunities?

Life as the Maid's Daughter

An Exploration of the Everyday Boundaries of Race, Class, and Gender

Mary Romero

(1995)

Introduction

. . . My current research attempts to expand the sociological understanding of the dynamics of race, class, and gender in the everyday routines of family life and reproductive labor. . . . I am lured to the unique setting presented by domestic service . . . and I turn to the realities experienced by the children of private household workers. This focus is not entirely voluntary. While presenting my research on Chicana private household workers, I was approached repeatedly by Latina/os and African Americans who wanted to share their knowledge about domestic service—knowledge they obtained as the daughters and sons of household workers. Listening to their accounts about their mothers' employment presents another reality to understanding paid and unpaid reproductive labor and the way in which persons of color are socialized into a class-based, gendered, racist social structure. The following discussion explores issues of stratification in everyday life by analyzing the life story of a maid's daughter. This life story illustrates the potential of the standpoint of the maid's daughter for generating knowledge about race, class, and gender. . . .

Social Boundaries Presented in the Life Story

The first interview with Teresa,[1] the daughter of a live-in maid, eventually led to a life history project. I am intrigued by Teresa's experiences with her mother's white, upper-middle-class employers while maintaining close ties to her relatives in Juarez, Mexico, and Mexican friends in Los Angeles. While some may view Teresa's life as a freak accident, living a life of "rags to riches," and certainly not a common Chicana/o experience, her story represents a microcosm of power relationships in the larger society. Life as the maid's daughter in an upper-middle-class neighborhood exemplifies many aspects of the Chicano/Mexicano experience as "racial ethnics" in the United States, whereby the boundaries of inclusion and exclusion are constantly changing as we move from one social setting and one social role to another.

Teresa's narrative contains descriptive accounts of negotiating boundaries in the employers' homes and in their community. As the maid's daughter, the old adage "Just like one of the family" is a reality, and Teresa has to learn when she must act like the employer's child and when she must assume the appropriate behavior as the maid's daughter. She has to recognize all the social cues and interpret social settings correctly—when to expect the same rights and privileges as the employer's children and when to fulfill the expectations and obligations as the maid's daughter. Unlike the employers' families, Teresa and her mother rely on different ways of obtaining knowledge. The taken-for-granted reality of the employers' families do not contain conscious experiences of negotiating race and class status, particularly not in the intimate setting of the home. Teresa's status is constantly changing in response to the wide range of social settings she encounters—from employers' dinner parties with movie stars and corporate executives to Sunday dinners with Mexican garment workers in Los Angeles and factory workers in El Paso. Since Teresa remains bilingual and bicultural throughout her life, her story reflects the constant struggle and resistance to maintain her Mexican identity, claiming a reality that is neither rewarded nor acknowledged as valid.

Teresa's account of her life as the maid's daughter is symbolic of the way that racial ethnics participate in the United States; sometimes we are included and other times excluded or ignored. Teresa's story captures the reality of social stratification in the United States, that is, a racist, sexist, and class-structured society upheld by an ideology of equality. I will analyze the experiences of the maid's daughter in an upper-middle-class neighborhood in Los Angeles to investigate the ways that boundaries of race, class, and gender are maintained or diffused in everyday life. I have selected various excerpts from the transcripts that illustrate how knowledge about a class-based and gendered, racist social order is learned, the type of information that is conveyed, and how the boundaries between systems of domination impact everyday life. I begin with a brief history of Teresa and her mother, Carmen.

Learning Social Boundaries: Background

Teresa's mother was born in Piedras Negras, a small town in Aguas Calientes in Mexico. After her father was seriously injured in a railroad accident, the family moved to a small town outside Ciudad Juarez. . . . By the time she was fifteen she moved to Juarez and took a job as a domestic, making about eight dollars a week. She soon crossed the border and began working for Anglo families in the country club area in El Paso. Like other domestics in El Paso, Teresa's mother returned to Mexico on weekends and helped support her mother and sisters. In her late twenties she joined several of her friends in their search for better-paying jobs in Los Angeles. The women immediately found jobs in the garment industry. Yet, after six months in the sweatshops, Teresa's mother went to an agency in search of domestic work. She was placed in a very exclusive Los Angeles neighborhood. Several years later Teresa was born. Her friends took care of the baby while Carmen continued working; childcare became a burden, however, and she eventually returned to Mexico. At the age of thirty-six Teresa's mother returned to Mexico with her newborn baby. Leaving Teresa with her grandmother and aunts, her mother sought work in the country club area. Three years later Teresa and her mother returned to Los Angeles.

Over the next fifteen years Teresa lived with her mother in the employer's (Smith) home, usually the two sharing the maid's room located off the kitchen. From the age of three until Teresa started school, she accompanied her mother to work. She continued to live in the Smiths' home until she left for college. All of Teresa's live-in years were spent in one employer's household. The Smiths were unable to afford a full-time maid, however, so Teresa's mother began doing day work throughout the neighborhood. After school Teresa went to whatever house her mother was cleaning and waited until her mother finished working, around 4 or 6 P.M., and then returned to the Smiths' home with her mother. Many prominent families in the neighborhood knew Teresa as the maid's daughter and treated her accordingly. While Teresa wanted the relationship with the employers to cease when she went to college and left the neighborhood, her mother continued to work as a live-in maid with no residence other than the room in the employer's home; consequently, Teresa's social status as the maid's daughter continued. . . .

One of the Family

As Teresa got older, the boundaries between insider and outsider became more complicated, as employers referred to her and Carmen as "one of the family." Entering into an employer's world as the maid's daughter, Teresa was not only subjected to the rules of an outsider but also had to recognize when the rules changed, making her momentarily an insider. While the boundaries dictating Carmen's work became blurred between the obligations of an employee and that of a friend or family member, Teresa was forced into situations in which she was expected to be just like one of the employer's children, and yet she remained the maid's daughter....

Living under conditions established by the employers made Teresa and her mother's efforts to maintain a distinction between their family life and an employer's family very difficult. Analyzing incidents in which the boundaries between the worker's family and employer's family were blurred highlights the issues that complicate the mother-daughter relationship. Teresa's account of her mother's hospitalization was the first of numerous conflicts between the two that stemmed from the live-in situation and their relationships with the employer's family. The following excerpt demonstrates the difficulty in interacting as a family unit and the degree of influence and power employers exerted over their daily lives:

> When I was about ten my mother got real sick. That summer, instead of sleeping downstairs in my mother's room when my mother wasn't there, one of the kids was gone away to college, so it was just Rosalyn, David and myself that were home. The other two were gone, so I was gonna sleep upstairs in one of the rooms. I was around eight or nine, ten I guess. I lived in the back room. It was a really neat room because Rosalyn was allowed to paint it. She got her friend who was real good, painted a big tree and clouds and all this stuff on the walls. So I really loved it and I had my own room. I was with the Smiths all the time, as my parents, for about two months. My mother was in the hospital for about a month. Then when she came home, she really couldn't do anything. We would all have dinner, the Smiths were really, really supportive. I went to summer school and I took math and English and stuff like that. I was in this drama class and I did drama and I got to do the leading role. Everybody really liked me and Ms. Smith would come and see my play. So things started to change when I got a lot closer to them and I was with them alone. I would go see my mother every day, and my cousin was there. I think that my cousin kind of resented all the time that the Smiths spent with me. I think my mother was really afraid that now that she wasn't there that they were going to steal me from her. I went to see her, but I could only stay a couple of hours and it was really weird. I didn't like seeing my mother in pain and she was in a lot of pain. I remember before she came home the Smiths said that they thought it would be a really good idea if I stayed upstairs and I had my own room now that my mother was going to be sick and I couldn't sleep in the same bed 'cause I might hurt her. It was important for my mother to be alone. And how did I feel about that? I was really excited about that [having her own room]—you know. They said, "Your mom she is probably not going to like it and she might get upset about it, but I think that we can convince her that it is ok." When my mom came home, she understood that she couldn't be touched and that she had to be really careful, but she wanted it [having her own room] to be temporary. Then my mother was really upset. She got into it with them and said, "No, I don't want it that way." She would tell me, "No, I want you to be down here. ¿Qué crees que eres hija de ellos? You're gonna be with me all the time, you can't do that." So I would tell Ms. Smith. She would ask me when we would go to the market together, "How does your mom seem, what does she feel, what does she say?" She would get me to relay that. I would say, "I think my mom is really upset about me moving upstairs. She doesn't like it and she just says no." I wouldn't tell her everything. They would talk to her and finally they convinced her, but my mom really, really resented it and was really angry about it. She was just generally afraid. All these times that my mother wasn't there, things happened and they would take me places with them, go out to dinner with them and their friends. So that was a real big change, in that I slept upstairs and had different rules. Everything changed. I was more independent. I did my own homework; they would open the back door and yell that dinner was ready—you know. Things were just real different.

The account illustrates how assuming the role of insider was an illusion because neither the worker's daughter nor the worker ever became a member of the white, middle-class family. Teresa was only allowed to move out of the maid's quarter, where she shared a bed with her mother, when two of the employer's children were leaving home, vacating two bedrooms. . . .

Teresa and Carmen did not experience the boundaries of insider and outsider in the same way. Teresa was in a position to assume a more active family role when employers made certain requests. Unlike her mother, she was not an employee and was not expected to clean and serve the employer. Carmen's responsibility for the housework never ceased, however, regardless of the emotional ties existing between employee and employers. She and her employers understood that, whatever family activity she might be participating in, if the situation called for someone to clean, pick up, or serve, that was Carmen's job. When the Smiths requested Teresa to sit at the dinner table with the family, they placed Teresa in a different class position than her mother, who was now expected to serve her daughter alongside her employer. Moving Teresa upstairs in a bedroom alongside the employer and their children was bound to drive a wedge between Teresa and Carmen. There is a long history of spatial deference in domestic service, including separate entrances, staircases, and eating and sleeping arrangements. Carmen's room reflected her position in the household. As the maid's quarter, the room was separated from the rest of the bedrooms and was located near the maid's central work area, the kitchen. The room was obviously not large enough for two beds because Carmen and Teresa shared a bed. Once Teresa was moved upstairs, she no longer shared the same social space in the employer's home as her mother. Weakening the bonds between the maid and her daughter permitted the employers to broaden their range of relationships and interaction with Teresa.

Carmen's feelings of betrayal and loss underline how threatening the employers' actions were. She understood that the employers were in a position to buy her child's love. They had already attempted to socialize Teresa into Euro-American ideals by planning Teresa's education and deciding what courses she would take. Guided by the importance they place on European culture, the employers defined the Mexican Spanish spoken by Teresa and her mother as inadequate and classified Castillan Spanish as "proper" Spanish. As a Mexican immigrant woman working as a live-in maid, Carmen was able to experience certain middle-class privileges, but her only access to these privileges was through her relationship with employers. Therefore, without the employers' assistance, she did not have the necessary connections to enroll Teresa in private schools or provide her with upper-middle-class experiences to help her develop the skills needed to survive in elite schools. Carmen only gained these privileges for her daughter at a price; she relinquished many of her parental rights to her employers. To a large degree the Smiths determined Carmen's role as a parent, and the other employers restricted the time she had to attend school functions and the amount of energy left at the end of the day to mother her own child.

Carmen pointed to the myth of "being like one of the family" in her comment, "¿Qué crees que eres hija de ellos? You're gonna be with me all the time, you can't do that." The statement underlines the fact that the bond between mother and daughter is for life, whereas the pseudofamily relationship with employers is temporary and conditional. Carmen wanted her daughter to understand that taking on the role of being one of the employer's family did not relinquish her from the responsibility of fulfilling her "real" family obligations. The resentment Teresa felt from her cousin who was keeping vigil at his aunt's hospital bed indicated that she had not been a dutiful daughter. The outside pressure from an employer did not remove her own family obligations and responsibilities. Teresa's relatives expected a daughter to be at her mother's side providing any assistance possible as a caretaker, even if it was limited to companionship. The employer determined Teresa's activity, however, and shaped her behavior into that of a middle-class child; consequently, she was kept away from the hospital and protected from the realities of her mother's illness. Furthermore, she was submerged into the employer's world, dining at the country club and interacting with their friends.

Her mother's accusation that Teresa wanted to be the Smiths' daughter signifies the feelings of betrayal or loss and the degree to which Carmen was threatened by the employer's power and authority. Yet Teresa also felt betrayal and loss and viewed herself in competition with the employers for her mother's time, attention, and love. In this excerpt Teresa accuses her mother of wanting to be part of employers' families and community:

> I couldn't understand it—you know—until I was about eighteen and then I said, "It is your fault. If I treat the Smiths differently, it is your fault. You chose to have me live in this situation. It was your decision to let me have two parents, and for me to balance things off, so you can't tell me that I said this. You are the one who wanted this." When I was about eighteen we got into a huge fight on Christmas. I hated the holidays because I hated spending them with the Smiths. My mother always worked. She worked on every holiday. She loved to work on the holidays! She would look forward to working. My mother just worked all the time! I think that part of it was that she wanted to have power and control over this community, and she wanted the network, and she wanted to go to different people's houses.

As employers, Mr. and Mrs. Smith were able to exert an enormous amount of power over the relationship between Teresa and her mother. Carmen was employed in an occupation in which the way to improve working conditions, pay, and benefits was through the manipulation of personal relationships with employers. Carmen obviously tried to take advantage of her relationship with the Smiths in order to provide the best for her daughter. The more intimate and interpersonal the relationship, the more likely employers were to give gifts, do favors, and provide financial assistance. Although speaking in anger and filled with hurt, Teresa accused her mother of choosing to be with employers and their families rather than with her own daughter. Underneath Teresa's accusation was the understanding that the only influence and status her mother had as a domestic was gained through her personal relationships with employers. Although her mother had limited power in rejecting the Smiths' demands, Teresa held her responsible for giving them too much control. Teresa argued that the positive relationship with the Smiths was done out of obedience to her mother and denied any familial feelings toward the employers. The web between employee and employers' families affected both mother and daughter, who were unable to separate the boundaries of work and family.

Maintaining Cultural Identity

A major theme in Teresa's narrative was her struggle to retain her Mexican culture and her political commitment to social justice. Rather than internalizing meaning attached to Euro-American practices and redefining Mexican culture and bilingualism as negative social traits, Teresa learned to be a competent social actor in both white, upper-middle-class environments and in working- and middle-class Chicano and Mexicano environments. To survive as a stranger in so many social settings, Teresa developed an acute skill for assessing the rules governing a particular social setting and acting accordingly. Her ability to be competent in diverse social settings was only possible, however, because of her life with the employers' children. Teresa and her mother maintained another life—one that was guarded and protected against any employer intrusion. Their other life was Mexican, not white, was Spanish speaking, not English speaking, was female dominated rather than male dominated, and was poor and working-class, not upper-middle-class. During the week Teresa and her mother visited the other Mexican maids in the neighborhoods, on weekends they occasionally took a bus into the Mexican barrio in Los Angeles to have dinner with friends, and every summer they spent a month in Ciudad Juarez with their family. . . .

Teresa's description of evening activity with the Mexican maids in the neighborhood provides insight into her daily socialization and explains how she learned to live in the employer's home without internalizing all their negative attitudes toward Mexican and working-class culture. Within the white, upper-class neighborhood in which they worked, the Mexican maids got together on a regular basis and cooked Mexican food, listened to Mexican music, and gossiped in Spanish about their employers. Treated as invisible or as confidants, the maids were frequently exposed to the intimate details of their employers' marriages and family life. The Mexican maids voiced their disapproval of the lenient child-rearing practices and parental decisions, particularly surrounding drug usage and the importance of material possessions:

> Raquel was the only one [maid] in the neighborhood who had her own room and own TV set. So everybody would go over to Raquel's. . . . This was my mother's support system. After hours, they would go to different people's [maid's] rooms depending on what their rooms had. Some of them had kitchens and they would go and cook all together, or do things like play cards and talk all the time. I remember that in those situations they would sit, and my mother would talk about the Smiths, what they were like. When they were going to negotiate for raises, when they didn't like certain things, I would listen and hear all the different discussions about what was going on in different houses. And they would talk, also, about the family relationships. The way they interacted, the kids did this and that. At the time some of the kids were smoking pot and they would talk about who was smoking marijuana. How weird it was that the parents didn't care. They would talk about what they saw as being wrong. The marriage relationship, or how weird it was they would go off to the beauty shop and spend all this money, go shopping and do all these weird things and the effect that it had on the kids.

The interaction among the maids points to the existence of another culture operating invisibly within a Euro-American and male-dominated community. The workers' support system did not include employers and addressed their concerns as mothers, immigrants, workers, and women. They created a Mexican-dominated domain for themselves. Here they ate Mexican food, spoke Spanish, listened to the Spanish radio station, and watched novellas on TV. Here Teresa was not a cultural artifact but, instead, a member of the Mexican community.

In exchanging gossip and voicing their opinions about the employers' lifestyles, the maids rejected many of the employers' priorities in life. Sharing stories about the employers' families allowed the Mexican immigrant women to be critical of white, upper-middle-class families and to affirm and enhance their own cultural practices and beliefs. The regular evening sessions with other working-class Mexican immigrant women were essential in preserving Teresa and her mother's cultural values and were an important agency of socialization for Teresa. For instance, the maids had a much higher regard for their duties and responsibilities as mothers than as wives or lovers. In comparison to their mistresses, they were not financially dependent on men, nor did they engage in the expensive and time-consuming activity of being an ideal wife, such as dieting, exercising, and maintaining a certain standard of beauty in their dress, makeup, and hairdos. Unlike the employers' daughters, who attended cotillions and were socialized to acquire success through marriage, Teresa was constantly pushed to succeed academically in order to pursue a career. The gender identity cultivated among the maids did not include dependence on men or the learned helplessness that was enforced in the employers' homes but, rather, promoted self-sufficiency. However, both white women employers and Mexican women employees were expected to be nurturing and caring. These traits were further reinforced when employers asked Teresa to babysit for their children or to provide them with companionship during their husbands' absences.

So, while Teresa observed her mother adapting to the employers' standards in her interaction with their children, she learned that her mother did not approve of their lifestyle and understood that she had another set of expectations to adhere to. Teresa attended the same schools as employers' children, wore similar clothes, and conducted most of her social life within the same socioeconomic class, but she remained the maid's daughter—and learned the limitations of that position. Teresa watched her mother uphold higher standards for her and apply a different set of standards to the employers' children; most of the time, however, it appeared to Teresa as if they had no rules at all.

Sharing stories about the Smiths and other employers in a female, Mexican, and worker-dominated social setting provided Teresa with a clear image of the people she lived with as employers rather than as family members. Seeing the employers through the eyes of the employees forced Teresa to question their kindness and benevolence and to recognize their use of manipulation to obtain additional physical and emotional labor from the employees. She became aware of the workers' struggles and the long list of grievances, including no annual raises, no paid vacations, no social security or health benefits, little if any privacy, and sexual harassment. Teresa was also exposed to the price that working-class immigrant women employed as live-in maids paid in maintaining white, middle-class, patriarchal communities. Employers' careers and lifestyles, particularly the everyday rituals affirming male privilege, were made possible through the labor women provided for men's physical, social, and emotional needs. Female employers depended on the maid's labor to assist in the reproduction of their gendered class status. Household labor was expanded in order to accommodate the male members of the employers' families and to preserve their privilege. Additional work was created by rearranging meals around men's work and recreation schedules and by waiting on them and serving them. Teresa's mother was frequently called upon to provide emotional labor for the wife, husband, mother, and father within an employer's family, thus freeing members to work or increase their leisure time.

Discussion

Teresa's account offers insight into the ways racial ethnic women gain knowledge about the social order and use the knowledge to develop survival strategies. As the college-educated daughter of an immigrant Mexican woman employed as a live-in maid, Teresa's experiences in the employers' homes, neighborhood, and school and her experiences in the homes of working-class Mexicano families and barrios provided her with the skills to cross the class and cultural boundaries separating the two worlds. The process of negotiating social boundaries involved an evaluation of Euro-American culture and its belief system in light of an intimate knowledge of white, middle-class families. Being in the position to compare and contrast behavior within different communities, Teresa debunked notions of "American family values" and resisted efforts toward assimilation. Learning to function in the employers' world was accomplished without internalizing its belief system, which defined ethnic culture as inferior. Unlike the employers' families, Teresa was not able to assume the taken-for-granted reality of her mother's employers because her experiences provided a different kind of knowledge about the social order.

While the employers' children were surrounded by positive images of their race and class status, Teresa faced negative sanctions against her culture and powerless images of her race. Among employers' families she quickly learned that her "mother tongue" was not valued and that her culture was denied. All the Mexican adults in the neighborhood were in subordinate positions to the white adults and were responsible for caring for and nurturing white children. Most of the female employers were full-time homemakers who enjoyed the financial security provided by their husbands, whereas the Mexican immigrant women in the neighborhood all worked as maids and were financially independent; in many cases they were supporting children, husbands, and other family members. By directly observing her mother serve, pick up after, and nurture employers and their families, Teresa learned about white, middle-class privileges. Her experiences with other working-class Mexicans were dominated by women's responsibility for their children and extended families. Here the major responsibility of mothering was financial; caring and nurturing were secondary and were provided by the extended family or children did without. Confronted with a working mother who was too tired to spend time with her, Teresa learned about the racial, class, and gender parameters of parenthood, including its privileges, rights, responsibilities, and obligations. She also learned that the role of a daughter included helping her mother with everyday household tasks and, eventually, with the financial needs of the extended family. Unlike her uncles and male cousins, Teresa was not exempt from cooking and housework, regardless of her financial contributions. Within the extended family Teresa was subjected to standards of beauty strongly weighted by male definitions of women as modest beings, many times restricted in her dress and physical movements. Her social worlds became clearly marked by race, ethnic, class, and gender differences.

Successfully negotiating movement from a white, male, and middle-class setting to one dominated by working-class, immigrant, Mexican women involved a socialization process that provided Teresa with the skills to be bicultural. Since neither setting was bicultural, Teresa had to become that in order to be a competent social actor in each. Being bicultural included having the ability to assess the rules governing each setting and to understand her ethnic, class, and gender position. Her early socialization in the employers' households was not guided by principles of creativity, independence, and leadership but, rather, was based on conformity and accommodation. Teresa's experiences in two different cultural groups allowed her to separate each and to fulfill the employers' expectations without necessarily internalizing the meaning attached to the act. Therefore, she was able to learn English without internalizing the idea that English is superior to Spanish or that monolingualism is normal. The existence of a Mexican community within the employers' neighborhood provided Teresa with a collective experience of class-based racism, and the maids' support system affirmed and enhanced their own belief system and culture. As Philomena Essed (1991, 294) points out, "The problem is not only how knowledge of racism is acquired but also what kind of knowledge is being transmitted."

Teresa's life story lends itself to a complex set of analyses because the pressures to assimilate were challenged by the positive interactions she experienced within her ethnic community. Like other bilingual persons in the United States, Teresa's linguistic abilities were shaped by the linguistic practices of the social settings she had access to. Teresa learned the appropriate behavior for each social setting, each marked by different class and cultural dynamics and in which women's economic roles and relationships to men were distinct. An overview of Teresa's socialization illustrates the process of biculturalism—a process that included different sets of standards and rules governing her actions as a woman, as a Chicana, and as the maid's daughter. . . .

Notes

This essay was originally presented as a paper at the University of Michigan, "Feminist Scholarship: Thinking through the Disciplines," 30 January 1992. I want to thank Abigail J. Stewart and Donna Stanton for their insightful comments and suggestions.

1. The names are pseudonyms.

REFERENCE

Essed, Philomena. 1991. *Understanding Everyday Racism.* Newbury Park, Calif.: Sage Publications.

THINKING ABOUT THE READING

Teresa's childhood is unique in that she and her mother lived in the household of her mother's employer, requiring them to conform to the expectations of the employers even when her mother was "not at work." Her childhood was shaped by the need to read signals from others to determine her position in various social settings. What were some of the different influences in Teresa's early socialization? Did she accept people's attempts to mold her, or did she resist? How did she react to her mother's employers' referring to her as "one of the family"? Teresa came from a poor family, but she spent her childhood in affluent households. With respect to socialization, what advantages do you think these experiences provided her? What were the disadvantages? How do you think these experiences would have changed if she was a *son* of a live-in maid, rather than a daughter? If she was a poor *white* girl rather than Latina?

The Making of Culture, Identity, and Ethnicity Among Asian American Youth

Min Zhou and Jennifer Lee

(2004)

Youth and Culture

In preindustrial societies, one's life course was roughly marked by two discrete stages—childhood and adulthood. However, in postindustrial societies, the duration of childhood has been prolonged and also includes the distinct, yet overlapping stages of adolescence and youth. Today, youth generally refer to those between the ages of 16 and 24 (and sometimes even 30). These young people are at the stage in their life cycles where they strive to find their own spaces, make their own choices, and form their own identities, while at the same time deterred by certain norms, rules, regulations, and social forces from accepting the myriad responsibilities that accompany full adulthood.

The delayed entrance into adulthood stems from two sources: on the one hand, societal constraints prohibit youth from partaking in certain adult activities; and on the other, youth also prolong this stage in their lives. For instance, while American youth may legally enter the labor force, those under the age of 18 are considered minors and therefore banned from participating in electoral politics or purchasing a pack of cigarettes. In addition, it is a criminal offense for an adult to have sex with a minor, even when the sex is consensual. Furthermore, while accorded the full rights of citizenship at the age of 18, those under the age of 21 are prohibited from buying or consuming alcoholic beverages and entering nightclubs that serve alcohol. While societal constraints play an active part in delaying the entrance into adulthood, American youth themselves are increasingly active participants in prolonging this stage of their lives. For instance, it is becoming increasingly more common (and to some extent necessary in today's economy) for those under the age of 25 to continue with school full time after graduating from high school. Consequently, seeking higher education delays the transition to adulthood, which is typically characterized by a stable job, marriage, home life, and parenthood.

Today, to be young is to be hip, cool, fun loving, carefree, and able to follow one's heart's desires. As a significant social group, this age cohort is inherently ambiguous as it juxtaposes and strives to find balance between the dialectics of parental influence and individual freedom, dependence and independence, innocence and responsibility, and ultimately adolescence and adulthood. It is precisely the tension arising from this ambiguity that drives the public misrepresentation of youth as deviant, delinquent, deficient, and rebellious or resistant.

Culture, on the other hand, is defined as the ways, forms, and patterns of life in which socially identifiable groups interact with their environments and express their symbolic and material existences. Young people experience the conditions of their lives, define them, and respond to them, and in the process, they produce unique cultural forms and practices that become the expressions and products of their own experiences (Brake, 1985). Thus, youth culture is broadly referred to as a particular way of life, combined with particular patterns of beliefs, values, symbols, and activities that are shared, lived, or expressed by young people (Frith, 1984). As social scientists, our goal is not only to identify young people's shared activities but also to uncover the values that underlie their activities and behavior. As early as 1942, Talcott Parsons coined the phrase "youth culture" to describe a distinctive world of youth structured by age and sex roles with a value system in opposition to the adult world of productive work, responsibility, and routine (p. 606). While youth culture may oppose the adult values of conformity to adult culture and responsibilities, it also serves as an invaluable problem-solving resource—the development and use of day-to-day practices to help make sense of and cope with youth's shared problems (Frith, 1984).

Contemporary research on youth culture reveals a tendency to move youth out of class-based categories, and instead, emphasize the diversity of youth cultures as well as the multidimensional nature of resistance. The resurgent literature on youth culture, especially since the 1980s, turns our attention to the distinctive characteristics of youth, with a particular emphasis on the impacts of class, race/ethnicity, gender, and geography on their cultural expressions, signs, symbols, and activities.

We highlight several significant conceptual advancements in the recent literature on youth culture as they help provide some contextual background for our understanding of the Asian American youth culture. Instead of placing youth within a class framework or portraying them as delinquents, new research on youth in the United States focuses on the interactive processes between various macro and micro social forces in the formation and practice of culture and identity. Going beyond analyses of class, contemporary work recognizes the diversity of youth cultures, examining differences across a wide range of social categories, including class, race/ethnicity, gender, sexuality, and geography, and sometimes even making further distinctions within these categories. For example, the experiences of white and nonwhite youth, boys and girls, and heterosexuals and homosexuals are now presumed to have distinct characteristics that interact with structural forces such as hegemony, racism, sexism, and homophobia. Their experiences shape their socialization, which in turn affect the ways in which members of each respective group express, represent, and identify themselves.

Intergenerational differences among Asians complicate intragroup dynamics and family relations. Most notably, native-born children and grandchildren of Asian ancestry feel a sense of ambivalence toward newer arrivals. Because about two-thirds of the Asian American population is first generation, native-born Asians must now confront renewed images of Asians as "foreigners." Resembling the new immigrants in phenotype, but not necessarily in behavior, language, and culture, the more "assimilated" native born find that they must actively and constantly distinguish themselves from the newer arrivals. The "immigrant shadow" looms large for Asian American youth and can weigh heavily on the identity formation of native-born youth. However, native- and foreign-born youth react differently to the "immigrant shadow." For instance, comments about one's "good English" or inquiries about where one comes from are often taken as insensitive at best, and offensive or even insulting at worst, to native-born Asian Americans. By stark contrast, similar encounters tend to be interpreted or felt more positively among foreign-born Asians. Different lived experiences between the native and foreign-born are thus not only generational but also cultural.

While recognizing the vast diversity among Asian Americans, we argue that intragroup dynamics and their consequences render the Asian American experience unique, and the imposed pan-ethnic category meaningful for analysis. However, we do not lose sight of the fact that "Asian American" is an imposed identity and most often, not adopted by either the first or second generation who are much more likely to identify with their national origins than other Americans.

Immigration

Immigration is the most immediate process that shapes the cultural formation of Asian American youth. As noted earlier, Asian Americans are a predominantly immigrant group, which in turn has an enormous impact on the experiences of Asian American youth. As they grow up, they are intimately influenced and often intensely constrained by the immigrant family, the ethnic community, and their parents' ancestral homeland. Research has illustrated how the immigrant family and ethnic community have been the primary sources of support as well as the primary sites of conflict (Zhou and Bankston, 1998).

Asian American children, despite their diverse origins, share certain common family experiences—most prominently the unduly familial obligation to obey their elders and repay parental sacrifices, along with the extraordinarily high parental expectations for educational and occupational achievement. Many Asian immigrant parents (especially those who had already secured middle-class status in their home countries) migrated to provide better opportunities for their children. As new immigrants, the first generation often endure difficulties associated with migration such as lack of English-language proficiency, American cultural literacy, and familiarity with the host society. Moreover, those who gave up their middle-class occupations often endure downward occupational mobility, relative deprivation, and discrimination from the host society (Lee, 2000). In their directed quest to achieve socioeconomic mobility, the first generation appears to their children as little more than one-dimensional hard workers who focus too much on material achievement and too little on leisure.

Although their children may feel that their immigrant parents have a narrow vision of success, the first generation are all too aware of their own limits in ensuring socioeconomic mobility for their children, and hence, turn to education as the surest path to move ahead. Thus, not only do they place an enormous amount of pressure on their children to excel in school, but they also provide the material means to assure success. For example, they move to neighborhoods with strong public schools, send their children to private after-school programs (including language programs, academic tutoring, and enrichment institutions in the ethnic community), spend time to seek out detailed academic information, and make decisions about schools and majors for their children (Zhou and Li, 2003).

Although the parents feel that they are doing what is best for their children, the children—whose frame of reference is "American"—see things differently. From their point of view, their parents appear rigid and "abnormal," that is, unacculturated, old-fashioned, and traditional disciplinarians who are incapable of having fun with them and unwilling to show respect for their individuality. The children view the immigrant family and ethnic community as symbols of the old world—strictly authoritarian, severely demanding, and overwhelmingly stifling.

At the same time, however, the children witness at first hand their parents' daily struggles as new immigrants trying to make it in America, and consequently, develop a unique respect and sensitivity toward them. One of the most prominent ways that they demonstrate their respect and sensitivity is through a subtle blend of conformity to and rebellion against their parents. Asian American youth are less likely to talk back or blatantly defy their parents than other American youth. For example, in her study of Nisei daughters during the years of Asian exclusion, Valerie Matsumoto depicts the tension between native-born daughters and their immigrant parents in the way they defined womanhood. However, rather than challenging their parents head-on, running away from the family, or leaving the ethnic community, the native-born daughters judiciously negotiated the roles assigned to them by their parents and the tightly knit community by creating various cultural forms—dances, dating, and courtship romances—to assert a gender identity that was simultaneously feminine, "Americanized," and "Japanese." Although this delicate balancing act may take a heavy emotional toll on Asian American youth, it is precisely their ambivalence toward their immigrant families that makes the youth culturally sensitive, which, in turn, expands their repertoire for cultural expression.

Racial Exclusion

Along with the experiences associated with immigration, racialization is a second important process that shapes the cultural formation of Asian American youth. The youth confront the consequences of racialization in a number of ways, the first of which is their encounter with racial exclusion. During the period of Asian exclusion, Asian Americans—who were considered an "inferior race"—were confined to ethnic enclaves. The youth who grew up in this era had few social, educational, or occupational options beyond the walls of the ethnic enclave and were barred from full participation in American life. Consequently, some Asian American youth turned their attention overseas to their parents' ancestral homeland for opportunities that were denied to them in the United States. For example, frustrated by their limited mobility options in America, native-born Chinese youth in the 1930s promoted the "Go West to China" movement to seek better opportunities in their ancestral homeland.

Cultural forms and expressions such as ethnic [newspaper] presses, dances, and beauty pageant contests not only affirmed ethnic identities but also inadvertently reinforced Asian Americans as the foreign "Other." While the days of racial segregation in ethnic enclaves have long since disappeared, the effects of racialization still remain a part of growing up in the United States for Asian American youth. For instance, today's Asian American youth develop an awareness of their nonwhite racial identity that functions as a marker of exclusion in some facets of American society, as Nazli Kibria (2002) describes in her study of second-generation Chinese and Korean Americans.

Racial Stereotyping

Another way in which Asian American youth face the consequences of racialization is through racial stereotyping. Excluded from fair representation in mainstream American media, most portrayals of Asian Americans have been either insidious stereotypes of the "foreign" *other* or celebrated images of the "super" *other,* setting Asian Americans apart from other Americans. So pervasive are racial stereotypes of Asian Americans that Asian American youth culture is, in part, produced within the context of counteracting these narrowly circumscribed, one-dimensional images, most prominently that of the model minority.

The celebrated "model minority" image of Asian Americans was born in the mid-1960s, at the peak of the civil rights and ethnic consciousness movements, but *before* the rising waves of immigration and refugee influx from Asia. Two articles published in 1966—"Success Story, Japanese-American Style," by William Petersen in the *New York Times Magazine* and "Success of One Minority Group in U.S.," by the *U.S. News and World Report* staff—marked a significant departure from the portrayal of Asian Americans as aliens and foreigners, and changed the way that the media depicted Asian immigrants and their descendants. Both articles extolled Japanese and Chinese Americans for their persistence in overcoming extreme hardship and discrimination in order to achieve success (unmatched even by U.S.-born whites) with "their own almost totally unaided effort" and "no help from anyone else." The press attributed their winning wealth and ability to get ahead in American society to hard work, family solidarity, discipline, delayed gratification, nonconfrontation, and disdain for welfare.

Although the image of the model minority may seem laudatory, it has far-reaching consequences that extend beyond Asian Americans. First, the model minority stereotype serves to buttress the myth that the United States is a country devoid of racism, and one that accords equal opportunity for all who take the initiative to work hard to get ahead. The image functions to blame those who lag behind and are not making it for their failure to work hard, their inability to delay gratification, and their inferior culture. Not only does the image thwart other racial/ethnic minorities' demands for social justice, it also pits minority groups against each other.

Perhaps one of the most devastating consequences of the model minority stereotype is its effect on Asian American youth who feel frustrated and burdened because others judge them by standards *different* from those of other American youth. Their rebellion against the image manifests itself in their adoption of cultural forms and styles that are influenced by other racial/ethnic minority youth, especially African American youth. For example, many Asian American youth adorn hip-hop style clothing, listen to rap music, and frequent hip-hop clubs. However, it would be a mistake to assume that Asian American youth simply imitate other youth cultures. While they borrow elements from other minority youth cultures, they consciously create unique Asian-American-style cultural forms and practices that manifest themselves in import car racing, DJ-ing and emceeing, raving, and film and theater.

A case in point is the recently released all-Asian-cast Hollywood film *Better Luck Tomorrow,* directed by Justin Lin (2003). Defiantly turning the model minority image on its head, the film exposes the dark side of living up to the meek, studious, overachiever stereotype. Moreover, it depicts a series of shocking scenarios that illustrates that Asian American adolescents are just as confused and disturbed as other disaffected American youth who are bored with life. The multifaceted portrayal of Asian American youth proves that they can be good-looking, smart, funny, susceptible to drugs and alcohol, prone to violence and other vices, capable of self-destructive behavior, and completely lacking in morals all at once. While the model minority stereotype still endures in the media and popular culture, Asian American youth use cultural expression to show that they are far more multidimensional, complex, and, in fact, normal than the stereotype allows.

Invisibility

Yet another way in which Asian American youth confront the consequences of racialization is with invisibility, or the lack of public/media exposure. While all American youth are marginalized in society, they are nevertheless highly visible in mainstream American culture. Stereotypical or not, images of white and black youth permeate the media—from high culture on television, film, theater, music, dance, and fashion, to low culture on the street. However, images of Asian American youth are virtually absent, and perceptibly so. Hence, their expression through cultural forms and practices is an avenue through which Asian American youth make their presence on the American scene.

For example, Asian Americans have never had a significant place in the American recording industry as performers, producers, or consumers. By contrast, whites and blacks, and more recently Latinos, have been highly visible in and targeted by the recording industry. To counter the invisibility, many Asian American cultural workers believe that "just by being there" is the first step to being recognized, but beyond that, Asian Americans need to take active steps to realize their full political and artistic potential (Wong, 1997). So invisible are Asian Americans in the recording industry that many Asian American cultural workers consider making a mark on this scene as an urgent political and revolutionary project.

Whereas cultural workers agree that Asians are invisible in the arts, they do not agree on the means by which to assert their presence. For instance, Fred Ho (1999), an Asian American jazz artist, argues that "just being there" is not enough. Moreover, Ho contends that because Asian American cultural work is revolutionary in nature, Asian Americans should be in control of the means of cultural production by establishing their own production companies. Only by establishing their own means of cultural production can Asian Americans problematize the ideology of assimilation, reject Eurocentric and essentialist forms, and draw upon the rich Asian, immigrant, and working-class traditions of Asian Americans.

Globalization and Transnationalism

Globalization and transnationalism provide yet another means through which Asian

Americans confront the effects of racialization. Advancements in digital technologies and the Internet allow today's generation of Asian American youth to communicate and stay in close contact with their respective ancestral homelands through visual imagery, music, sounds, and words beyond the imagination of earlier generations. Globalization beyond national borders widens the cultural space in which Asian American youth are able to maneuver at relative ease to create new opportunities for cultural production and expression.

Cultural forms such as styles, music, dances, desires, and dreams among South Asian American youth, for example, are shaped not only by American influences but also through diasporic influences to create an empowering sense of identity (Maira, 2002). The opportunities presented by globalization for the identity formation of Asian American youth are not reserved only for the native born.

In sum, Asian American youths' adoption of cultural forms are the products of both opportunities and constraints presented to them from the larger processes of immigration and racialization. Because many Asian Americans are foreign born, the experiences of Asian American youth are inextricably linked to immigrant adaptation, the immigrant family, and the ethnic community, thereby making their experiences distinctive from those of other American youth. And while other racial/ethnic minority youth also experience the racialization of their identities, the effects of racialization and the way they manifest themselves in cultural expression are distinctive. Racial/ethnic minorities in the United States have suffered and continue to suffer from exclusion and stereotypes, but the stereotypes of each group are different and affect minority youth differently. The stereotype of the Asian model minority constrains and frustrates youth, pigeonholing them into a one-dimensional, superhuman mold. In other realms, Asian American youth experience the pangs of invisibility and attempt to make their mark through unique cultural practices such as import car racing, dance, theater, and clothing styles. Finally, today's advancements in technology offer Asian American youth influences beyond America, and also broaden the scope of their cultural terrain.

Asian American Identity and an "Emergent Culture of Hybridity"

A diverse lot, today's Asian American youth often adopt a number of different identities—ethnic, hyphenated-American, pan-ethnic, or multiracial—and these identities are not necessarily mutually exclusive. However, unlike white American youth whose choice of ethnic identities is symbolic, the identity choices among Asian American youth are far more limited and consequential. Previous research indicates that nativity, generational status, bilingualism, gender, neighborhood context, and perceptions of discrimination are important factors in determining the identity choices among today's Asian American youth (Lee and Bean, 2004; Portes and Rumbaut, 2001; Xie and Goyette, 1997; Zhou and Bankston, 1998). Perhaps most importantly, as nonwhite racial/ethnic minorities, Asian American youth are subject to outsiders' ascription, meaning that how others perceive them has a profound effect on the way they choose to identify themselves. Joane Nagel (1994) has long noted that the choice of identities is a dialectical process that involves both internal and external opinions and processes, that is, what *you* think your identity is versus what *they* think your identity is.

Yet another distinctive feature of Asian American identity is that Asians are neither black nor white, but occupy a position in between, at least at this moment in time. In a society that has long been divided by an impenetrable black-white color line, it is not at all clear that today's Asian Americans see themselves (or for that matter that others see them) as either black *or* white. While Asian Americans may be considered "people of color," the degree to which they view themselves and are viewed by others as closer to black or white is highly ambiguous (Lee and Bean, 2004). On the one hand, they are minorities and therefore subject to racial discrimination and prejudice. On the other, some Asian ethnic groups have achieved social status on a par with—and in some arenas, superseding—whites. Consequently, the "in-between" and "dual status" of Asians may provide greater flexibility in the identity choices for Asian American youth, especially for Asian multiracials.

Although the centrality of race, outsiders' ascription, and the black-white color line has a powerful effect on the identity choices of Asian American youth, also relevant is the fact that most Asian American youth are either immigrants or the children of immigrants. Hence, their identities are inextricably bound by the experiences of immigration, the immigrant family, and the ethnic community, as well as their interactions with mainstream institutions such as schools.

Asian American youth feel they are both a part of and yet apart from mainstream America. Asian American youth have successfully carved out a unique cultural space for themselves that is, in part, a consequence of their constant negotiation between the traditions of their immigrant families and the marginalization and exclusion they experience from the larger society. This does not mean that Asian American youth divorce themselves entirely from mainstream American culture. However, the unique dual status of Asian American youth combined with their immigrant backgrounds prompt them to actively craft a culture of their own that is distinctive both from their ethnic communities and from other American institutions and youth. In doing so, they negotiate between "American" and "Asian" traits, which often results in an "emergent culture of hybridity" that mixes elements of both worlds. By carving out a space and culture of their own through "grass roots cultural production" (Bielby, 2004), Asian American youth have been able to adopt an identity apart from the restraints imposed by the roles and expectations of the family, the ethnic community, work, and school. Consequently, this culture offers young Asian Americans a collective identity—a reference group from which they can develop an individual identity (Brake, 1985).

REFERENCES

Bielby, William T., "Rock in a Hard Place: Grass-Roots Cultural Production in the Post-Elvis Era," *American Sociological Review 69* (2004): 1–13.

Brake, Michael, *Comparative Youth Culture: The Sociology of Youth Cultures and Youth Subcultures in America, Britain, and Canada* (New York: Routledge, 1985).

Frith, Simon, *The Sociology of Youth* (Ormskirk, Lancashire: Causeway Press, 1984).

Ho, Fred, "Identity: Beyond Asian American Jazz: My Musical and Political Changes in the Asian American Movement," *Leonardo Musk Journal 9* (1999): 45–51.

Kibria, Nazli, *Becoming Asian American: Second Generation Chinese and Korean Identity* (Baltimore: Johns Hopkins University Press, 2002).

Lee, Jennifer, "Striving for the American Dream: Struggle, Success, and Intergroup Conflict among Korean Immigrant Entrepreneurs," in *Contemporary Asian America: A Multidisciplinary Reader*, ed. Min Zhou and James V. Gatewood (New York: New York University Press, 2000), 278–296.

Lee, Jennifer and Frank D. Bean, "America's Changing Color Lines: Immigration, Race/Ethnicity, and Multiracial Identification" *Annual Review of Sociology 30* (2004): 221–242.

Lin, Justin (Director), *Better Luck Tomorrow*, Hollywood, CA: Paramount Pictures, 2003.

Maira, Sunaina Marr, *Desis in the House: Indian American Youth Culture in New York City* (Philadelphia: Temple University Press, 2002).

Nagel, Joane, "Constructing Ethnicity: Creating and Recreating Ethnic Identity and Culture," *Social Problems 41* (1994): 152–171.

Parsons, Talcott, "Age and Sex in the Social Structure of the United States," *American Sociological Review 7* (1942): 604–616.

Petersen, William, "Success Story, Japanese-American Style," *New York Times Magazine*, January 9, 1966, pp. 20–21, 33, 36, 38, 40–41, 43.

Portes, Alejandro and Rubén G. Rumbaut, *Legacies: The Story of the Immigrant Second Generation* (Berkeley: University of California Press, 2001).

U.S. News & World Report, "Success Story of One Minority in the U.S.," December 26, 1966, pp. 73–78.

Wong, Deborah, "Just Being There: Making Asian American Space in the Recording Industry," in *Music of Multicultural America: A Study of Twelve Musical Communities*, ed. Kip Lornell and Anne K. Rasmussen (London: Schirmer Books, 1997), 287–316.

Xie, Yu and Kimberly Goyette, "The Racial Identification of Biracial Children with One Asian Parent: Evidence from the 1990 Census," *Social Forces 76* (1997): 547–570.

Zhou, Min and Carl L. Bankston, III., *Growing Up American: How Vietnamese Children Adapt to Life in the United States* (New York: Russell Sage Foundation, 1998).

Zhou, Min and Xiyuan Li, "Ethnic Language Schools and the Development of Supplementary Education in the Immigrant Chinese Community in the United States," *New Directions for Youth Development: Understanding the Social Worlds of Immigrant Youth* (Winter 2003): 57–73.

THINKING ABOUT THE READING

Zhou and Lee suggest that Asian American youth have created their own distinctive culture and identity. What are some examples of this culture? What are some of the factors that contribute to the construction of Asian American youth identities? What do the authors mean when they say that Asian American youth occupy a position between "black and white"? What are some of the implications of this position for their sense of identity and culture?

Working 'the Code'

On Girls, Gender, and Inner-City Violence

Nikki Jones

(2008)

In mainstream American society, it is commonly assumed that women and girls shy away from conflict, are not physically aggressive, and do not fight like boys and men.

In this article, I draw on field research among African-American girls in the United States to argue that the circumstances of inner-city life have encouraged the development of uniquely situated femininities that simultaneously encourage and limit inner-city girls' use of physical aggression and violence. First, I begin by arguing that, in the urban environments that I studied, gender—being a girl—does not protect inner-city girls from much of the violence experienced by inner-city boys. In fact, teenaged boys and girls are both preoccupied with 'survival' as an ongoing project. I use my analysis of interviews with young people involved in violent incidents to demonstrate similarities in how young people work 'the code of the street' across perceived gender lines. This in-depth examination of young people's use of physical aggression and violence reveals that while young men and young women fight, survival is still a gendered project.

Race, Gender, and Inner-City Violence

Inner-city life has changed dramatically over the last century and especially over the last 30 years.

In his ethnographic account of life in inner-city Philadelphia, Elijah Anderson writes that the code of the street is 'a set of prescriptions and proscriptions, or informal rules, of behavior organized around a desperate search for respect that governs public social relations, especially violence among so many residents, particularly young men and women' (Anderson, 1999, p. 10). Furthermore, the code is 'a system of accountability that promises "an eye for an eye," or a certain "payback" for transgressions' (Anderson, 1999, p. 10). Fundamental elements of the code include respect and 'a credible reputation for vengeance that works to deter aggression' (Anderson, 1999, p. 10). According to Anderson, it is this complex relationship between masculinity, respect and violence that, at times, encourages poor, urban young men to risk their lives in order to be recognized and respected by others *as a man.*

Black feminist scholar Patricia Hill Collins considers Anderson's discussion of masculinity and the 'code of the street' in her recent analysis of the relationship between hegemonic (and racialized) masculinities and femininities, violence and dominance (Collins, 2004, pp. 188–212). Collins argues that the hyper-criminalization of urban spaces is exacerbated by the culture of the code. As young men from distressed urban areas cycle in and out of correctional facilities at historically remarkable rates, she argues, urban public schools, street concerns and homes have become a '... nexus of street, prison and youth culture,' which exerts 'a tremendous amount of pressure on Black men, especially young, working class men, to avoid being classified as "weak"' (Collins, 2004, p. 211).

What About Girls?

Over the last few decades, feminist criminologists and gender and crime scholars have examined women's and girls' experiences with aggression and violence with increasing complexity. Emphasizing how particular material circumstances influence women's and girls' relationship to violence shifts the focus from the consideration of dichotomous gender differences to the empirical examination of gender similarities and differences in experiences with violence among young women and men who live in poor, urban areas (Simpson, 1991). The analysis presented here follows in this tradition by recognizing the influence of shared life circumstances on young people's use of violence.

The young people from Philadelphia's inner-city neighborhoods that I encountered generally share similar life circumstances, yet how they respond to these structural and cultural circumstances—that is, how they work the code of the street—is also gendered in ways that reflect differences among inner-city girls' and boys' understanding of what you 'got to' do to 'survive.'

Methods

Each of the respondents featured in this study was enrolled in a city hospital-based violence intervention project that targeted youth aged 12 to 24 who [were] presented in the emergency department as a result of an intentional violent incident and were considered to be at either moderate or high risk for involvement in future violent incidents. As a consequence of patterns of racial segregation within the city, almost the entire population of young women and men who voluntarily enrolled in the hospital's violence intervention project were African-American.

My fieldwork for this study took place in three phases over 3 years (2001–2003). During the first phase of the study, which lasted about a year and a half, I conducted 'ride alongs' with intervention counselors who met with young people in their homes shortly after their initial visit to the emergency room. I also conducted a series of interviews with members of the intervention counseling staff. Most of the staff grew up in Philadelphia and were personally familiar with many of the neighborhoods we visited. During this time and throughout the study, I also observed interactions in the spaces and places that were significant in the lives of the young people I met. These spaces included trolley cars and buses (transportation to and from school), a neighborhood high school nicknamed 'the Prison on the Hill,' the city's family and criminal court, and various correctional facilities in the area. I also intentionally engaged in extended conversations with grandmothers and mothers, sisters, brothers, cousins and friends of the young people I visited and interviewed. I recorded this information in my field notes and used it to complement, supplement, test and, at times, verify the information collected during interviews.

Shared Circumstances, Shared Code

While the problem of inner-city violence is believed to impact boys and men only, my interviews with teenaged inner-city girls revealed that young women are regularly exposed to many of the same forms of violence that men are exposed to in their everyday lives and are deeply influenced by its normative order. In the inner-city neighborhoods I visited, which were often quite isolated from the rest of the city, I encountered young men and young women who could quickly recall a friend, relative or 'associate' who had been shot, robbed or stabbed. In the public high school I visited, I watched adolescent girls and boys begin their school day with the same ritual: they dropped their bags on security belts, stepped through a metal detector, and raised their arms and spread their legs for a police-style 'pat down' before entering the building. Repeatedly, I encountered teenaged girls who, like the young men they share space with in the inner city, had stories to tell about getting 'rolled on,' or getting 'jumped,' or about the 'fair one' gone bad. It is these shared circumstances of life that engender a shared understanding about how to survive in a setting where your safety is never guaranteed. In the following sections, I provide portraits of four young people involved in violent incidents in order to illustrate what was revealed to me during the course of field research and interviews: an appreciation of 'the code of the street' that cut across gender lines. The first two respondents, Billy and DeLisha, tell stories of recouping from a very public loss in a street fight. The second set of respondents, Danielle and Robert, highlight how even those who are averse to fighting must sometimes put forth a 'tough front' to deter potentially aggressive challenges in the future.

Billy and DeLisha: 'I'm Not Looking Over My Shoulder'

Billy was 'jumped' by a group of young men while in 'their' neighborhood, which is within walking distance of his own. He tells me this story as we sit in the living room of his row home. Billy recently reached his 20s, although he looks older than his age. He is White but shares a class background that is similar to many of the young people I interviewed. His block, like most of the others I visited during this study, is a collection of row homes in various states of disrepair. Billy spends more time here than he would like. He is unemployed and when asked how best the intervention project he enrolled in could help him his request was simple: I need a job. As we talk, I think that Billy is polite—he offers me a drink (a beer, which I decline) before we begin our interview—and even quiet. He recalls two violent battles within the last year, both of which ended with him in the emergency room, without wavering too far from a measured, even tone. The first incident he recalls for me happened in South Philadelphia. He was walking down the block, when he came across a group of guys on the corner, guys who he had 'trouble' with in the past. As he stood talking to an acquaintance, Billy was approached from behind and punched in the back of the head. The force of the punch was multiplied exponentially by brass knuckles, 'splitting [his] head open.' Billy was knocked out instantly, fell face-first toward the ground and split his nose on a concrete step. The thin scar from this street-fight remains several months later.

In contrast to Billy's even tone, DeLisha is loud. She is thin with a medium-brown complexion. Her retelling of the story of her injury is more like a re-enactment as the adrenaline, anxiety and excitement of the day return. She comes across as fiercely independent, especially for a 17-year-old girl. DeLisha, a young mother with a 1-year-old daughter, has been unable to rely on her own drug-addicted mother for much of her life. After years of this independence, she is convinced that she does not need anyone's help to 'make it' in life. While she has been a 'fighter' for as long as she can remember, she was never hurt before. Not in school. Not in her neighborhood, which is one of the most notorious in the city. And not like this. She had agreed to a fight with another neighborhood girl. The younger girl, pressured by her family and peers to win the battle, shielded a box-cutter from DeLisha's sight until the very last minute. When it seemed that she would lose, the girl flashed the box-cutter and slashed DeLisha across the hand, tearing past skin and muscle into a tendon on her arm.

During my interviews with Billy and DeLisha, I asked each of them how these very public losses, which also resulted in serious physical injuries, would influence their mobility within the neighborhood. Would they avoid certain people and places? Would or could they shrug their loss off or would they seek vengeance for their lost battle? Billy's and DeLisha's responses were strikingly similar in tone, nearly identical at some points, and equally revealing of two of the most basic elements of the code of the street: the commitment to maintaining a 'tough front' and 'payback.'

> Billy: I mean, just like I say, I walk around this neighborhood. I'm not looking over my shoulder. . . . I'm not going to walk [and] look around my shoulder because I've got people looking for me. I mean you want me . . . you know where I live. They can call me at any time they want. That's how, that's how I think. . . . I'm not going to sit around my own neighborhood and just say: 'Aww, I got to watch my back.' You want me? You got me.

> DeLisha: I'm not a scared type . . . I walk on the streets anytime I want to. I do anything I want to, anytime I want to do it. It's never been a problem walking on the street 3:00 in the morning. If I want to go home 3.00 in the morning, I'm going to go home. I'm not looking over my shoulder. My grandma never raised me to look over my shoulder. I'm not going to stop because of some little incident [being cut in the hand with a box cutter].

Billy and DeLisha's strikingly similar responses reveal their commitment to a shared 'system of accountability,' the code of the street, which, as Anderson argues, governs much of social life, especially violence, in distressed urban areas (Anderson, 1999). Billy and DeLisha hold themselves accountable to this system ('I'm not going to . . . ') and are also aware that others will hold them accountable for their behaviors and actions. Billy and DeLisha are acutely aware that someone who 'looks over their shoulder' while walking down the street is perceived as weak, a moving target, and both are determined to reject such a fate. Instead, Billy and DeLisha remain committed to managing their 'presentation of self' (Goffman, 1959) in a way that masks any signs of vulnerability.

In addition to their commitment to 'not looking over their shoulder,' Billy and DeLisha are also sensitive to the fact that the fights they were in were not 'fair.' These street-level injustices inform Billy and DeLisha's expectations for retaliation. Consistent with the code, both Billy and DeLisha—equally armed with long fight histories—realize the importance of 'payback' and consider future battles with their challengers to be inevitable. When I asked DeLisha if she anticipated another fight with the young woman who cut her, she replied with a strong yes, 'because I'm taking it there with her.' Billy was also equally committed to retaliation, telling me: ' . . . one by one, I will get them.'

Danielle and Robert: 'Sometimes You Got to Fight'

In *Code of the Street* (1999), Anderson demonstrates how important it is for young people to prove publicly that they are not someone to be 'messed with.' One of the ways that young people prove this to others is by engaging in fights in public, when necessary. The following statements from Danielle and Robert, two young people who are adept at avoiding conflicts, illustrate teenaged girls' and boys' shared understanding of the importance of demonstrating that one is willing to fight as a way to deter ongoing challenges to one's well-being:

> Danielle: 'cause sometimes you got to fight, not fight, but get into that type of battle to let them know that I'm not scared of you and you can't keep harassing me thinking that it's okay.

> Robert: . . . you know, if someone keep picking on you like that, you gonna have to do something to prove a point to them: that you not going to be scared of them . . . So, sometimes you do got to, you do got to fight. Cause you just got to tell them that you not scared of them.

Like DeLisha and Billy, Danielle, a recent high-school graduate, and Robert, who is in the 11th grade, offer nearly identical explanations of the importance of physically protecting one's own boundaries by demonstrating to others that you will fight, if necessary. While neither Danielle nor Robert identify as 'fighters,' both are convinced that sometimes you 'got to fight.' Again, this shared language reveals an awareness and commitment to a shared system of accountability, 'the code of the street,' which encourages young people—teenaged girls and boys—to present a 'tough front' as a way to discourage ongoing challenges to one's personal security. For the young people in this study, the value placed on maintaining a tough front or 'proving a point' cut across perceived gender lines.

In addition to possibly deterring future challenges, Anderson argues that presenting and ultimately proving oneself as someone who is not to be 'messed with' helps to build a young person's confidence and self-esteem: 'particularly for young men and perhaps increasingly among females . . . their identity, their self-respect, and their honor are often intricately tied up with the way they perform on the streets during and after such [violent] encounters' (Anderson, 1999, p. 76). Those young people who are able to perform well during these public encounters acquire a sense of confidence that will facilitate their movement throughout the neighborhood. This boost to one's sense of self is not restricted to young men; young women who can fight and win may also demonstrate a strong sense of pride and confidence in their ability to 'handle' potentially aggressive or violent conflicts, as illustrated by the following interview with Nicole.

Nicole: 'I Feel Like I Can Defend Myself'

My conversation with Nicole typifies the confidence expressed by teenaged girls who can fight and win. Nicole is a smart, articulate young woman who attended some community college courses while still a senior in high school. She planned to attend a state university to study engineering after graduation. While in high school, she tells me, she felt confident in her ability to walk the hallways of her sometimes chaotic public school: 'I feel like I can defend myself.' Unlike some young women who walk the hallways constantly testing others, Nicole's was a quiet confidence: 'I don't, like, I mean, when I'm walking around school or something, I don't walk around talking about "yeah, I beat this girl up."' Nicole could, in fact, claim that she didn't beat up just one girl but several, at the same time. Nicole explained to me how her most recent fight began:

> We [she and another young woman] had got into two arguments in the hallway and then her friends were holding her back. So I just said, 'Forget it. I'm just going to my class.' So I'm in class, I'm inside the classroom and I hear Nina say, 'Is this that bitch's class?' I came to the door and was like, 'Yes, this is my class.' And she puts her hands up [in fighting position] and she swings . . . And me and her was fighting, and then I got her on the wall, and then I felt somebody pulling my hair, and it turns out to be Jessica. Right? And then we fighting, and then I see Tasha, and it's me and all these three people and then they broke it all up.

Nicole's only injury in the fight came from the elbow of the school police officer who eventually ended the battle. As Nicole recalls this fight, and her performance in particular, I notice that she is smiling. This smile, together with the tone in which she tells the story of her earlier battle, makes it clear that she is proud of her ability to meet the challenge presented to her by these young women. Impressed at her ability to fight off three teenaged girls at the same time, I ask Nicole: 'How did you manage not to get jumped?' She quickly corrects my definition of the situation: 'No. I managed to beat them up.' After retelling her fight story, Nicole shakes her head from side to side and says: 'I had to end up beating them up. So sad.' I notice her sure smile return. 'You don't really look like you feel bad about that,' I say. 'I don't,' she replies.

The level of self-confidence that Nicole displays in this brief exchange contrasts with the passivity and submissiveness that is commonly expected of women and girls, especially white, middle-class women and girls (Collins, 2004). It is young men, not teenaged girls, who are expected to exude such confidence as they construct a 'tough front' to deter would-be challengers (Anderson, 1999). Nicole's confidence is also more than an expressive performance. Nicole knows that she is physically able to fight and win, when necessary, because she has done so in the past. For teenaged girls like Nicole and Sharmaine, whom I discuss below, this confidence is essential to their evaluation of how best to handle potential interpersonal conflicts in their everyday lives.

Sharmaine: '. . . I Have One Hand Left'

Sharmaine, an 8th grader, displayed a level of self-confidence similar to Nicole's after a fight with a boy in her classroom. Moments before the fight, the boy approached Sharmaine while she was looking out her classroom window, and 'whispered something' in her ear. Sharmaine knew that this boy liked her, but she thought she had made it quite clear that she did not like him. Sharmaine quickly told him to back off and then looked to her teacher for reinforcement. Her teacher, Sharmaine recalls, just laughed at the boy's advances. After he whispered in Sharmaine's ear a second time, she turned around and punched him in the face. Sharmaine later ended up in the emergency room with a jammed finger from the punch. I asked Sharmaine if she was concerned about him getting back at her when she returned to school. She tells me that someone in the emergency room asked her the same question. 'What did you say?' I ask. 'I told them no . . . because I have one hand left.'

For young women like Nicole and Sharmaine, the proven ability to defend themselves translates into a level of self-confidence that is not typically expected in girls and young women. Those girls who are confident in their ability to 'take care of themselves' become more mobile as they come to believe, as DeLisha says, that they can 'do anything [they] want to, anytime [they] want to do it.' Girls who are able to gain and maintain this level of self-confidence are able to challenge the real and imagined gendered boundaries on space and place in the inner city.

'Boys Got to Go Get Guns'

The need to be 'distinguished as a man'—a benchmark of hegemonic masculinity—often fosters adolescent boys' preoccupation with distinguishing themselves *from* women (Anderson, 1999; Collins, 2004, p. 210; Connell & Messerschmidt, 2005). This is a gendered preoccupation that was not revealed in urban adolescent girls' accounts of physical aggression and violence. The following statement from Craig, a young man who has deliberately checked his readiness to fight after being shot in the hip, illustrates how the need to 'be a man' influences young men's consideration of violence:

> Yeah, I don't fight no more. I can't fight [because of injury]. So, I really stop and think about stuff because it isn't even worth it . . . unless, I mean, you really want it [a fight] to happen . . . I'm going to turn the other cheek. But, I'm not going to be, like, wearing a skirt. That's the way you got to look at it.

While Craig is prepared to exit his life as a 'fighter,' he predicts that his newfound commitment to avoid fights will not stand up to the pressure of proving his manhood to a challenger. Craig is well aware of how another young man can communicate that he 'really want [a fight] to happen.' Once a challenger publicly escalates a battle in this way, young men like Craig have few choices. At this moment, a young man will have to demonstrate to his challenger, and his audience, that he isn't 'wearing a skirt.' Not only must he fight, he must also fight *like a man.*

Craig's admission is revealing of how a young man's concern with not being 'like' a woman influences his consideration of the appropriate use of physical aggression. While a similar type of preoccupation with intergender distinctions was not typically revealed in young women's accounts, I found that teenaged girls were generally aware of at least one significant difference in how young women and men were expected to work the code of the street. As is revealed in my conversation with Shante, a teenaged girl who was hit in the head with a brick by a neighborhood girl, young men are generally expected to use more serious or lethal forms of violence than girls or women.

I asked Shante what people in her neighborhood thought about girls fighting.

> 'Today,' she asked, 'you mean like people on the street?'
>
> 'Yeah.'
>
> 'If [a girl] get beat up, you just get beat up. That's on you.'
>
> 'Do you think it's different for boys?' I asked.
>
> 'Umm, boys got to go get guns. They got to blow somebody's head off. They got to shoot. They don't fight these days. They use guns.'

Shante's perception of what boys 'got to' do is informed by years of observation and experience. Shante has grown up in a neighborhood marked by violence. Days before this interview, she saw a young man get shot in the head. She tells me he was dead by the time he hit the sidewalk. When I asked Shante whether or not girls used guns, she could recall just one young woman from the neighborhood—the same young woman who hit Shante over the head with a brick—who had 'pistol whipped' another teenaged girl. While she certainly used the gun as a weapon, she didn't shoot her. These two incidents are actually quite typical of reported gender differences in the use of weapons in violent acts: boys and men are much more likely than girls and women to use guns to shoot and kill. Women and girls, like many of the young women I spoke with during this study, are far more likely to rely on knives and box-cutters, if they use a weapon at all (see also Miller, 1998 & 2001; Pastor, et al., 1996, p. 28). Those young women who did use a weapon, such as a knife or box-cutter, explained that they did so for protection. For example, Shante told me that she carried a razor blade, 'because she doesn't trust people.'

Takeya: 'A Good Girl'

In contrast to the commitment to protecting one's manhood, which Craig alludes to and Elijah Anderson describes in great detail (Anderson, 1999), the young women I spoke to did not suggest that they fought because that's what *women* do. Furthermore, while young women deeply appreciated the utility of a 'tough front,' they were unlikely to use phrases like 'I don't want to be wearing a skirt.' In fact, while young men like Craig work to prove their manhood by distinguishing themselves from women, many of the young women I spoke with—including the 'toughest' among them—embraced popular notions of femininity, 'skirts' and all. For many of the girls I interviewed, an appreciation of some aspects of hegemonic femininity modulated their involvement in violent interactions.

My conversation with Takeya sheds light on how inner-city girls attempt to reconcile the contradictory concerns that emerge from intersecting survival and gender projects. When I asked Takeya, a slim 13-year-old girl with a light brown complexion, about her fighting history, she replied, 'I'm not in no fights. I'm a good girl.' 'You are a good girl?' I asked. 'Yeah, I'm a good girl and I'm-a be a pretty girl at 18.'

Takeya's concern with being a 'pretty girl' reflects an appreciation of aspects of hegemonic femininity that place great value on beauty. Her understanding of what it means to be beautiful is also influenced by the locally placed value on skin color, hair texture and body figure. While brown skin and textured hair may not fit hegemonic (White, middle-class) conceptions of beauty, in this setting, a light-brown skinned complexion, 'straight' or 'good' hair, and a slim figure help to make one 'pretty' and 'good' (Banks, 2000). Yet, Takeya also knows that one's ability to stay pretty—to be a pretty girl at age 18—is directly influenced by one's involvement in interpersonal aggression or violence.

In order to be considered a 'pretty girl' by her peers, Takeya knows that she must avoid those types of interpersonal conflicts that tend to result in cuts and scratches to young women's faces, especially the ones that others consider beautiful (in *Code of the Street* [1999] Anderson writes that such visible scars often result in heightened status for the young women who leave their mark on pretty girls). Yet, Takeya is also aware that the culture of the code requires her to become an able fighter and to maintain a reputation as such. After expressing her commitment to being a 'good' girl, Takeya is sure to inform me that not only does she know how to fight, others also recognize her as an able fighter: 'I don't want you to think I don't know how to fight. I mean everybody always come get me [for fights]. [I'm] the number one [person they come to get].'

Takeya's simultaneous embrace of the culture of code and some aspects of normative femininity, Craig's concern with distinguishing himself from women, and Shante's convincing disclosure regarding what boys 'got to' do highlight how masculinity and femininity projects overlap and intersect with the project of survival for young people in distressed inner-city neighborhoods. Both Craig and Takeya appreciate fundamental elements of 'the code,' especially the importance of being known as an able fighter. Yet, Craig's use of physical aggression is likely to be encouraged by his commitment to a distinctive aspect of hegemonic masculinity: being distinguished from a girl. Meanwhile, Takeya's use of physical aggression and violence is tempered—though not extinguished—by seemingly typical 'female' concerns: being a 'good' and 'pretty' girl. In contrast to the project of accomplishing masculinity, which overlaps and, at times, contradicts the project of survival for young men, the project of accomplishing femininity can, at times, facilitate young women's struggle to survive in this setting.

Gender, Survival, and 'the Code'

I have argued that gender does not protect young women from much of the violence young men experience in distressed inner-city neighborhoods, and that given these shared circumstances, it becomes equally important for women and men to work 'the code of the street.' Like many adolescent boys, young women also recognize that reputation, respect and retaliation—the '3 Rs' of the code of the street—organize their social world (Anderson, 1999). Yet, as true as it is that, at times, young men and women work the code of the street in similar ways, it is also true that differences exist. These differences are rooted in the relationships between masculinity, femininity and the use of violence or aggression in distressed urban areas and emerge from overlapping and intersecting survival and gender projects.

In order to 'survive' in today's inner city, young women like DeLisha, Danielle, Shante and Takeya are encouraged to embrace some aspects of the 'code of the street' that organises much of inner-city life (Anderson, 1999). In doing so, these girls also embrace and accomplish some aspects of hegemonic masculinity that are embedded in the code. My analysis of interviews with teenaged girls and boys injured in intentional violent incidents reveals an appreciation of the importance of maintaining a tough front and demonstrating nerve across perceived gender lines. It is this appreciation of the cultural elements of the code that leads teenaged girls like Danielle to believe strongly that 'sometimes you got to fight.'

REFERENCES

Anderson, E. (1999). *Code of the street: Decency, violence and the moral life of the inner city.* New York: W.W. Norton.

Banks, I. (2000). *Hair matters: Beauty, power, and Black women's consciousness.* New York: New York University Press.

Collins, P. Hill. (2004). *Black sexual politics: African Americans, gender, and the new racism.* New York: Routledge.

Connell, R. W., & Messerschmidt, J. W. (2005). Hegemonic masculinity: Rethinking the concept. *Gender & Society,* 19(6), 829–859.

Goffman, E. (1959). *The presentation of self in everyday life.* New York: Anchor Books.

Miller, J. (1998). Up it up: Gender and the accomplishment of street robbery. *Criminology,* 36(1), 37–66.

Miller, J. (2001). *One of the guys: Girls, gangs, and gender.* Oxford: Oxford University Press.

Pastor, J., McCormick, J., & Fine, M. (1996). Makin' homes: An urban girl thing. In B. J. Ross Leadbeater & N. Way (Eds.), *Urban girls: Resisting stereotypes, creating identities.* New York: New York University Press.

Simpson, S. S. (1991). Caste, class, and violent crime: Explaining difference in female offending. *Criminology,* 29(1), 115–135.

THINKING ABOUT THE READING

This reading demonstrates that inner-city girls are also not as isolated from violence as is commonly thought. What are some of the reasons for their involvement with violence? What is the "code of the street"? How are violence and the "code" related to the ways in which these girls see themselves? How are they related to their survival? Does the way in which girls use and understand violence differ from the ways in which boys see it?

Supporting Identity

The Presentation of Self

Social behavior is highly influenced by the images we form of others. We typically form impressions of people based on an initial assessment of their social group membership (ethnicity, age, gender, etc.), their personal attributes (e.g., physical attractiveness), and the verbal and nonverbal messages they provide. These assessments are usually accompanied by a set of expectations we've learned to associate with members of certain social groups or people with certain attributes. Such judgments allow us to place people in broad categories and provide a degree of predictability in interactions.

While we are forming impressions of others, we are fully aware that they are doing the same thing with us. Early in life, most of us learn that it is to our advantage to have people think highly of us. In "The Presentation of Self in Everyday Life," Erving Goffman describes a process called *impression management,* in which we attempt to control and manipulate information about ourselves to influence the impressions others form of us. Impression management provides the link between the way we perceive ourselves and the way we want others to perceive us. We've all been in situations—a first date, a job interview, meeting a girlfriend's or boyfriend's family for the first time—in which we've felt compelled to "make a good impression." What we often fail to realize, however, is that personal impression management may be influenced by larger organizational and institutional forces.

Impression management is used to control the assessment others make of us in relation to certain group memberships. For example, initial impressions are often formed around key facets of identity like race and social class. In "Public Identities: Managing Race in Public Spaces," Karyn Lacy presents research based on interviews with middle-class blacks in Washington DC. Despite what the research literature on racial stigma theory has shown, these individuals use their agency and cultural capital to produce "public identities" and control interactions with whites that lead to positive outcomes in public arenas like shopping malls, real estate (house hunting), and the workplace. The use of public identities demonstrates how some middle class blacks can define their situation and avoid racial discrimination through the manipulation of their public interactions with whites.

The third reading is likely to raise considerable discussion. Why do college-age men go on "girl hunts"? Sociologist David Grazian uses Goffman's framework to explain this behavior. He suggests that the urban nightlife girl hunt scene is actually a ritual of male bonding and masculine identity building. Getting a girl is not really the goal. Hanging out and bonding with other men and reinforcing patterns of masculinity is the point of these ritual-like practices. If this is the case, then what are the implications for the perpetuation of cultural practices that build masculinity by objectifying women?

Something to Consider as You Read

As you read these selections on the presentation of self and identity, consider where people get their ideas about whom and what they can be in various settings. Consider a setting in which everyone present may be trying to create a certain impression, because that's what they all think everyone else wants. What would have to happen in order for the "impression script" to change in this setting? In what ways do material resources and authority influence the impression we're able to make? Are there certain types of people who needn't be concerned about the impressions they give off? Compare and contrast the readings on public identities and the "girl hunt." Are there similar cultural scripts operating in both these scenarios?

The Presentation of Self in Everyday Life

Selections

Erving Goffman

(1959)

Introduction

When an individual enters the presence of others, they commonly seek to acquire information about him or to bring into play information about him already possessed. They will be interested in his general socio-economic status, his conception of self, his attitude toward them, his competence, his trustworthiness, etc. Although some of this information seems to be sought almost as an end in itself, there are usually quite practical reasons for acquiring it. Information about the individual helps to define the situation, enabling others to know in advance what he will expect of them and what they may expect of him. Informed in these ways, the others will know how best to act in order to call forth a desired response from him.

For those present, many sources of information become accessible and many carriers (or "sign-vehicles") become available for conveying this information. If unacquainted with the individual, observers can glean clues from his conduct and appearance which allow them to apply their previous experience with individuals roughly similar to the one before them or, more important, to apply untested stereotypes to him. They can also assume from past experience that only individuals of a particular kind are likely to be found in a given social setting. They can rely on what the individual says about himself or on documentary evidence he provides as to who and what he is. If they know, or know of, the individual by virtue of experience prior to the interaction, they can rely on assumptions as to the persistence and generality of psychological traits as a means of predicting his present and future behavior.

* * *

The expressiveness of the individual (and therefore his capacity to give impressions) appears to involve two radically different kinds of sign activity: the expression that he *gives*, and the expression that he *gives off*. The first involves verbal symbols or their substitutes which he uses admittedly and solely to convey the information that he and the others are known to attach to these symbols. This is communication in the traditional and narrow sense. The second involves a wide range of action that others can treat as symptomatic of the actor, the expectation being that the action was performed for reasons other than the information conveyed in this way. As we shall have to see, this distinction has an only initial validity. The individual does of course intentionally convey misinformation by means of both of these types of communication, the first involving deceit, the second feigning.

Taking communication in both its narrow and broad sense, one finds that when the individual is in the immediate presence of others, his activity will have a promissory character. The others are likely to find that they must accept the individual on faith, offering him a just return while he is present before them in exchange for something whose true value will not be established until after he has left their presence. (Of course, the others also live by inference in their dealings with the physical world, but it is only in the world of social interaction that the objects about which they make inferences will purposely facilitate and hinder this inferential process.) The security that they justifiably feel in making inferences about the individual will vary, of course, depending on such factors as the amount of information they already possess about him, but no amount of such past evidence can entirely obviate the necessity of acting on the basis of inferences. As William I. Thomas suggested:

> It is also highly important for us to realize that we do not as a matter of fact lead our lives, make our decisions, and reach our goals in everyday life either statistically or scientifically. We live by inference. I am, let us say, your guest. You do not know, you cannot determine scientifically, that I will not steal your money or your spoons. But inferentially I will not and inferentially you have me as a guest.[1]

Let us now turn from the others to the point of view of the individual who presents himself before them. He may wish them to think highly of him, or to think that he thinks highly of them, or to perceive how in fact he feels toward them, or to obtain no clear-cut impression; he may wish to ensure sufficient harmony so that the interaction can be sustained, or to defraud, get rid of, confuse, mislead, antagonize, or insult them. Regardless of the particular objective which the individual has in mind and of his motive for having this objective, it will be in his interests to control the conduct of the others, especially their responsive treatment of him.[2] This control is achieved largely by influencing the definition of the situation which the others come to formulate, and he can influence this definition by expressing himself in such a way as to give them the kind of impression that will lead them to act voluntarily in accordance with his own plan. Thus, when an individual appears in the presence of others, there will usually be some reason for him to mobilize his activity so that it will convey an impression to others which it is in his interests to convey. . . .

I have said that when an individual appears before others his actions will influence the definition of the situation which they come to have. Sometimes the individual will act in a thoroughly calculating manner, expressing himself in a given way solely in order to give the kind of impression to others that is likely to evoke from them a specific response he is concerned to obtain. Sometimes the individual will be calculating in his activity but be relatively unaware that this is the case. Sometimes he will intentionally and consciously express himself in a particular way, but chiefly because the tradition of his group or social status require this kind of expression and not because of any particular response (other than vague acceptance or approval) that is likely to be evoked from those impressed by the expression. Sometimes the traditions of an individual's role will lead him to give a well-designed impression of a particular kind and yet he may be neither consciously nor unconsciously disposed to create such an impression. The others, in their turn, may be suitably impressed by the individual's efforts to convey something, or may misunderstand the situation and come to conclusions that are warranted neither by the individual's intent nor by the facts. In any case, in so far as the others act *as if* the individual had conveyed a particular impression, we may take a functional or pragmatic view and say that the individual has "effectively" projected a given definition of the situation and "effectively" fostered the understanding that a given state of affairs obtains. . . .

When we allow that the individual projects a definition of the situation when he appears before others, we must also see that the others, however passive their role may seem to be, will themselves effectively project a definition of the situation by virtue of their response to the individual and by virtue of any lines of action they initiate to him. Ordinarily the definitions of the situation projected by the several different participants are sufficiently attuned to one another so that open contradiction will not occur. I do not mean that there will be the kind of consensus that arises when each individual present candidly expresses what he really feels and honestly agrees with the expressed feelings of the others present.

This kind of harmony is an optimistic ideal and in any case not necessary for the smooth working of society. Rather, each participant is expected to suppress his immediate heartfelt feelings, conveying a view of the situation which he feels the others will be able to find at least temporarily acceptable. The maintenance of this surface of agreement, this veneer of consensus, is facilitated by each participant concealing his own wants behind statements which assert values to which everyone present feels obliged to give lip service. Further, there is usually a kind of division of definitional labor. Each participant is allowed to establish the tentative official ruling regarding matters which are vital to him but not immediately important to others, e.g., the rationalizations and justifications by which he accounts for his past activity. In exchange for this courtesy he remains silent or noncommittal on matters important to others but not immediately important to him. We have then a kind of interactional *modus vivendi.* Together the participants contribute to a single overall definition of the situation which involves not so much a real agreement as to what exists but rather a real agreement as to whose claims concerning what issues will be temporarily honored. Real agreement will also exist concerning the desirability of avoiding an open conflict of definitions of the situation.[3] I will refer to this level of agreement as a "working consensus." It is to be understood that the working consensus established in one interaction setting will be quite different in content from the working consensus established in a different type of setting. Thus, between two friends at lunch, a reciprocal show of affection, respect, and concern for the other is maintained. In service occupations, on the other hand, the specialist often maintains an image of disinterested involvement in the problem of the client, while the client responds with a show of respect for the competence and integrity of the specialist. Regardless of such differences in content, however, the general form of these working arrangements is the same.

* * *

Given the fact that the individual effectively projects a definition of the situation when he enters the presence of others, we can assume that events may occur within the interaction which contradict, discredit, or otherwise throw doubt upon this projection. When these disruptive events occur, the interaction itself may come to a confused and embarrassed halt. Some of the assumptions upon which the responses of the participants had been predicated become untenable, and the participants find themselves lodged in an interaction for which the situation has been wrongly defined and is now no longer defined. At such moments the individual whose presentation has been discredited may feel ashamed while the others present may feel hostile, and all the participants may come to feel ill at ease, nonplussed, out of countenance, embarrassed, experiencing the kind of anomy that is generated when the minute social system of face-to-face interaction breaks down. . . .

We find that preventive practices are constantly employed to avoid these embarrassments and that corrective practices are constantly employed to compensate for discrediting occurrences that have not been successfully avoided. When the individual employs these strategies and tactics to protect his own projections, we may refer to them as "defensive practices"; when a participant employs them to save the definition of the situation projected by another, we speak of "protective practices" or "tact." Together, defensive and protective practices comprise the techniques employed to safe-guard the impression fostered by an individual during his presence before others. It should be added that while we may be ready to see that no fostered impression would survive if defensive practices were not employed, we are less ready perhaps to see that few impressions could survive if those who received the impression did not exert tact in their reception of it.

In addition to the fact that precautions are taken to prevent disruption of projected definitions, we may also note that an intense interest in these disruptions comes to play a significant role in the social life of the group. Practical jokes and social games are played in which embarrassments which are to be taken unseriously are purposely engineered.[4] Fantasies are created in which devastating exposures occur. Anecdotes from the past—real, embroidered, or fictitious—are told and retold, detailing disruptions which occurred, almost occurred, or occurred and were admirably resolved. There seems to be no grouping which does not have a ready supply of these games, reveries, and cautionary tales, to be used as a source of humor, a catharsis for anxieties, and a sanction for inducing individuals to be modest in their claims and reasonable in their projected expectations. The individual may tell himself through dreams of getting into impossible positions. Families tell of the time a guest got his dates mixed and arrived when neither the house nor anyone in it was ready for him. Journalists tell of times when an all-too-meaningful misprint occurred, and the paper's assumption of objectivity or decorum was humorously discredited. Public servants tell of times a client ridiculously misunderstood form instructions, giving answers which implied an unanticipated and bizarre definition of the situation.[5] Seamen, whose home away from home is rigorously he-man, tell stories of coming back home and inadvertently asking mother to "pass the fucking butter."[6] Diplomats tell of the time a near-sighted queen asked a republican ambassador about the health of his king.[7]

To summarize, then, I assume that when an individual appears before others he will have many motives for trying to control the impression they receive of the situation. This report is concerned with some of the common techniques that persons employ to sustain such impressions and with some of the common contingencies associated with the employment of these techniques. It will be convenient to end this introduction with some definitions. . . . For the purpose of this report, interaction (that is, face-to-face interaction) may be roughly defined as the reciprocal influence of individuals upon one another's actions when in one another's immediate physical presence. An interaction may be defined as all the interaction which occurs throughout any one occasion when a given set of individuals are in one another's continuous presence; the term "an encounter" would do as well. A "performance" may be defined as all the activity of a given participant on a given occasion which serves to influence in any way any of the other participants. Taking a particular participant and his performance as a basic point of reference, we may refer to those who contribute the other performances as the audience, observers, or co-participants. The pre-established pattern of action which is unfolded during a performance and which may be presented or played through on other occasions may be called a "part" or "routine."[8] These situational terms can easily be related to conventional structural ones. When an individual or performer plays the same part to the same audience on different occasions, a social relationship is likely to arise. Defining social role as the enactment of rights and duties attached to a given status, we can say that a social role will involve one or more parts and that each of these different parts may be presented by the performer on a series of occasions to the same kinds of audience or to an audience of the same persons.

* * *

Performances

Front

I [use] the term "performance" to refer to all the activity of an individual which occurs during a period marked by his continuous presence before a particular set of observers and which has some influence on the observers. It will be convenient to label as "front" that part of the individual's performance which regularly functions in a general and fixed fashion to define the situation for those who observe the performance. Front, then, is the expressive equipment of a standard kind intentionally or unwittingly employed by the individual during his performance. For preliminary purposes, it will be convenient to distinguish and label what seem to be the standard parts of front.

First, there is the "setting," involving furniture, décor, physical layout, and other background items which supply the scenery and stage props for the spate of human action played out before, within, or upon it. A setting tends to stay put, geographically speaking, so that those who would use a particular setting as part of their performance cannot begin their act until they have brought themselves to the appropriate place and must terminate their performance when they leave it. It is only in exceptional circumstances that the setting follows along with the performers; we see this in the funeral cortège, the civic parade, and the dreamlike processions that kings and queens are made of. In the main, these exceptions seem to offer some kind of extra protection for performers who are, or who have momentarily become, highly sacred. . . .

It is sometimes convenient to divide the stimuli which make up personal front into "appearance" and "manner," according to the function performed by the information that these stimuli convey. "Appearance" may be taken to refer to those stimuli which function at the time to tell us of the performer's social statuses. These stimuli also tell us of the individual's temporary ritual state, that is, whether he is engaging in formal social activity, work, or informal recreation, whether or not he is celebrating a new phase in the season cycle or in his life-cycle. "Manner" may be taken to refer to those stimuli which function at the time to warn us of the interaction role the performer will expect to play in the oncoming situation. Thus a haughty, aggressive manner may give the impression that the performer expects to be the one who will initiate the verbal interaction and direct its course. A meek, apologetic manner may give the impression that the performer expects to follow the lead of others, or at least that he can be led to do so. . . .

Dramatic Realization

While in the presence of others, the individual typically infuses his activity with signs which dramatically highlight and portray confirmatory facts that might otherwise remain unapparent or obscure. For if the individual's activity is to become significant to others, he must mobilize his activity so that it will express *during the interaction* what he wishes to convey. In fact, the performer may be required not only to express his claimed capacities during the interaction but also to do so during a split second in the interaction. Thus, if a baseball umpire is to give the impression that he is sure of his judgment, he must forgo the moment of thought which might make him sure of his judgment; he must give an instantaneous decision so that the audience will be sure that he is sure of his judgment.[9] . . .

Similarly, the proprietor of a service establishment may find it difficult to dramatize what is actually being done for clients because the clients cannot "see" the overhead costs of the service rendered them. Undertakers must therefore charge a great deal for their highly visible product—a coffin that has been transformed into a casket—because many of the other costs of conducting a funeral are ones that cannot be readily dramatized.[10] Merchants, too, find that they must charge high prices for things that look intrinsically inexpensive in order to compensate the establishment for expensive things like insurance, slack periods, etc., that never appear before the customers' eyes. . . .

Idealization

. . . I want to consider here another important aspect of this socialization process—the tendency for performers to offer their observers an impression that is idealized in several different ways.

The notion that a performance presents an idealized view of the situation is, of course, quite common. Cooley's view may be taken as an illustration:

> If we never tried to seem a little better than we are, how could we improve or "train ourselves from the outside inward"? And the same impulse to show the world a better or idealized aspect of ourselves finds an organized expression in the various professions and classes, each of which has to some extent a cant or pose, which its members assume unconsciously, for the most part, but which has the effect of a conspiracy to work upon the credulity of the rest of the world. There is a cant not only of theology and of philanthropy, but also of law, medicine, teaching, even of science—perhaps especially of science, just now, since the more a particular kind of merit is recognized and admired, the more it is likely to be assumed by the unworthy.[11]

Thus, when the individual presents himself before others, his performance will tend to incorporate and exemplify the officially accredited values of the society, more so, in fact, than does his behavior as a whole.

To the degree that a performance highlights the common official values of the society in which it occurs, we may look upon it, in the manner of Durkheim and Radcliffe-Brown, as a ceremony—as an expressive rejuvenation and reaffirmation of the moral values of the community. Furthermore, insofar as the expressive bias of performances comes to be accepted as reality, then that which is accepted at the moment as reality will have some of the characteristics of a celebration. To stay in one's room away from the place where the party is given, or away from where the practitioner attends his client, is to stay away from where reality is being performed. The world, in truth, is a wedding.

One of the richest sources of data on the presentation of idealized performances is the literature on social mobility. In most societies there seems to be a major or general system of stratification, and in most stratified societies there is an idealization of the higher strata and some aspiration on the part of those in low places to move to higher ones. (One must be careful to appreciate that this involves not merely a desire for a prestigeful place but also a desire for a place close to the sacred center of the common values of the society.) Commonly we find that upward mobility involves the presentation of proper performances and that efforts to move upward and efforts to keep from moving downward are expressed in terms of sacrifices made for the maintenance of front. Once the proper sign-equipment has been obtained and familiarity gained in the management of it, then this equipment can be used to embellish and illumine one's daily performances with a favorable social style.

Perhaps the most important piece of sign-equipment associated with social class consists of the status symbols through which material wealth is expressed. American society is similar to others in this regard but seems to have been singled out as an extreme example of wealth-oriented class structure—perhaps because in America the license to employ symbols of wealth and financial capacity to do so are so widely distributed. . . .

Reality and Contrivance

. . . Some performances are carried off successfully with complete dishonesty, others with complete honesty; but for performances in general neither of these extremes is essential and neither, perhaps, is dramaturgically advisable.

The implication here is that an honest, sincere, serious performance is less firmly connected with the solid world than one might first assume. And this implication will be strengthened if we look again at the distance usually placed between quite honest performances and quite contrived ones. In this connection take, for example, the remarkable phenomenon of stage acting. It does take deep skill, long training, and psychological capacity to become a good stage actor. But this fact should not blind us to another one: that almost anyone can quickly learn a script well enough to give a charitable audience some sense of realness in what is being contrived before them. And it seems this is so because ordinary social intercourse is itself put together as a scene is put together, by the exchange of dramatically inflated actions, counteractions, and terminating replies. Scripts even in the hands of unpracticed players can come to life because life itself is a dramatically enacted thing. All the world is not, of course, a stage, but the crucial ways in which it isn't are not easy to specify. . . .

When the individual does move into a new position in society and obtains a new part to perform, he is not likely to be told in full detail how to conduct himself, nor will the facts of his new situation press sufficiently on him from the start to determine his conduct without his further giving thought to it. Ordinarily he will be given only a few cues, hints, and stage directions, and it will be assumed that he already has in his repertoire a large number of bits and pieces of performances that will be required in the new setting. The individual will already have a fair idea of what modesty, deference, or righteous indignation looks like, and can make a pass at playing these bits when necessary. He may even be able to play out the part of a hypnotic subject[12] or commit a "compulsive" crime[13] on the basis of models for these activities that he is already familiar with.

A theatrical performance or a staged confidence game requires a thorough scripting of the spoken content of the routine; but the vast part involving "expression given off" is often determined by meager stage directions. It is expected that the performer of illusions will already know a good deal about how to manage his voice, his face, and his body, although he—as well as any person who directs him—may find it difficult indeed to provide a detailed verbal statement of this kind of knowledge. And in this, of course, we approach the situation of the straightforward man in the street. Socialization may not so much involve a learning of the many specific details of a single concrete part—often there could not be enough time or energy for this. What does seem to be required of the individual is that he learn enough pieces of expression to be able to "fill in" and manage, more or less, any part that he is likely to be given. The legitimate performances of everyday life are not "acted" or "put on" in the sense that the performer knows in advance just what he is going to do, and does this solely because of the effect it is likely to have. The expressions it is felt he is giving off will be especially "inaccessible" to him.[14] But as in the case of less legitimate performers, the incapacity of the ordinary individual to formulate in advance the movements of his eyes and body does not mean that he will not express himself through these devices in a way that is dramatized and performed in his repertoire of actions. In short, we all act better than we know how.

When we watch a television wrestler gouge, foul, and snarl at his opponent we are quite ready to see that, in spite of the dust, he is, and knows he is, merely playing at being the "heavy," and that in another match he may be given the other role, that of clean-cut wrestler, and perform this with equal verve and proficiency. We seem less ready to see, however, that while such details as the number and character of the falls may be fixed beforehand, the details of the expressions and movements used do not come from a script but from command of an idiom, a command that is exercised from moment to moment with little calculation or forethought. . . .

Personality-Interaction-Society

In recent years there have been elaborate attempts to bring into one framework the concepts and findings derived from three different areas of inquiry: the individual personality, social interaction, and society. I would like to suggest here a simple addition to these interdisciplinary attempts.

When an individual appears before others, he knowingly and unwittingly projects a definition of the situation, of which a conception of himself is an important part. When an event occurs which is expressively incompatible with this fostered impression, significant consequences are simultaneously felt in three levels of social reality, each of which involves a different point of reference and a different order of fact.

First, the social interaction, treated here as a dialogue between two teams, may come to an embarrassed and confused halt; the situation may cease to be defined. Previous positions may become no longer tenable, and participants may find themselves without a charted course of action. The participants typically sense a false note in the situation and come to feel awkward, flustered, and, literally, out of countenance. In other words, the minute social system created and sustained by orderly social interaction becomes disorganized. These are the consequences that the disruption has from the point of view of social interaction.

Secondly, in addition to these disorganizing consequences for action at the moment, performance disruptions may have consequences of a more far-reaching kind. Audiences tend to accept the self projected by the individual performer during any current performance as a responsible representative of his colleague-grouping, of his team, and of his social establishment. Audiences also accept the individual's particular performance as evidence of his capacity to perform the routine and even as evidence of his capacity to perform any routine. In a sense these larger social units—teams, establishments, etc.—become committed every time the individual performs his routine; with each performance the legitimacy of these units will tend to be tested anew and their permanent reputation put at stake. This kind of commitment is especially strong during some performances. Thus, when a surgeon and his nurse both turn from the operating table and the anesthetized patient accidentally rolls off the table to his death, not only is the operation disrupted in an embarrassing way, but the reputation of the doctor, as a doctor and as a man, and also the reputation of the hospital may be weakened. These are the consequences that disruptions may have from the point of view of social structure.

Finally, we often find that the individual may deeply involve his ego in his identification with a particular part, establishment, and group, and in his self-conception as someone who does not disrupt social interaction or let down the social units which depend upon that interaction. When a disruption occurs, then, we may find that the self-conceptions around which his personality has been built may become discredited. These are consequences that disruptions may have from the point of view of individual personality.

Performance disruptions, then, have consequences at three levels of abstraction: personality, interaction, and social structure. While the likelihood of disruption will vary widely from interaction to interaction, and while the social importance of likely disruptions will vary from interaction to interaction, still it seems that there is no interaction in which the participants do not take an appreciable chance of being slightly embarrassed or a slight chance of being deeply humiliated. Life may not be much of a gamble, but interaction is. Further, insofar as individuals make efforts to avoid disruptions or to correct for ones not avoided, these efforts, too, will have simultaneous consequences at the three levels. Here, then, we have one simple way of articulating three levels of abstraction and three perspectives from which social life has been studied.

Staging and the Self

The general notion that we make a presentation of ourselves to others is hardly novel; what ought to be stressed in conclusion is that the very structure of the self can be seen in terms of how we arrange for such performances in our Anglo-American society. . . .

The self, then, as a performed character, is not an organic thing that has a specific location, whose fundamental fate is to be born, to mature, and to die; it is a dramatic effect arising diffusely from a scene that is presented, and the characteristic issue, the crucial concern, is whether it will be credited or discredited.

In analyzing the self then we are drawn from its possessor, from the person who will profit or lose most by it, for he and his body merely provide the peg on which something of collaborative manufacture will be hung for a time. And the means for producing and maintaining selves do not reside inside the peg; in fact these means are often bolted down in social establishments. There will be a back region with its tools for shaping the body, and a front region with its fixed props. There will be a team of persons whose activity on stage in conjunction with available props will constitute the scene from which the performed character's self will emerge, and another team, the audience, whose interpretive activity will be necessary for this emergence. The self is a product of all of these arrangements, and in all of its parts bears the marks of this genesis.

The whole machinery of self-production is cumbersome, of course, and sometimes breaks down, exposing its separate components: back region control; team collusion; audience tact; and so forth. But, well oiled, impressions will flow from it fast enough to put us in the grips of one of our types of reality—the performance will come off and the firm self accorded each performed character will appear to emanate intrinsically from its performer.

Let us turn now from the individual as character performed to the individual as performer. He has a capacity to learn, this being exercised in the task of training for a part. He is given to having fantasies and dreams, some that pleasurably unfold a triumphant performance, others full of anxiety and dread that nervously deal with vital discreditings in a public front region. He often manifests a gregarious desire for teammates and audiences, a tactful considerateness for their concerns; and he has a capacity for deeply felt shame, leading him to minimize the chances he takes of exposure.

These attributes of the individual *qua* performer are not merely a depicted effect of particular performances; they are psychobiological in nature, and yet they seem to arise out of intimate interaction with the contingencies of staging performances.

And now a final comment. In developing the conceptual framework employed in this report, some language of the stage was used. I spoke of performers and audiences; of routines and parts; of performances coming off or falling flat; of cues, stage settings, and backstage; of dramaturgical needs, dramaturgical skills, and dramaturgical strategies. Now it should be admitted that this attempt to press a mere analogy so far was in part a rhetoric and a maneuver. . . .

And so here the language and mask of the stage will be dropped. Scaffolds, after all, are to build other things with, and should be erected with an eye to taking them down.

This report is not concerned with aspects of theater that creep into everyday life. It is concerned with the structure of social encounters—the structure of those entities in social life that come into being whenever persons enter one another's immediate physical presence. The key factor in this structure is the maintenance of a single definition of the situation, this definition having to be expressed, and this expression sustained in the face of a multitude of potential disruptions.

A character staged in a theater is not in some ways real, nor does it have the same kind of real consequences as does the thoroughly contrived character performed by a confidence man; but the *successful* staging of either of these types of false figures involves use of *real* techniques—the same techniques by which everyday persons sustain their real social situations. Those who conduct face to face interaction on a theater's stage must meet the key requirement of real situations; they must expressively sustain a definition of the situation: but this they do in circumstances that have facilitated their developing an apt terminology for the interactional tasks that all of us share.

Notes

1. Quoted in E. H. Volkart, editor, *Social Behavior and Personality,* Contributions of W. I. Thomas to Theory and Social Research (New York: Social Science Research Council, 1951), p. 9.

2. Here I owe much to an unpublished paper by Tom Burns of the University of Edinburgh. He presents the argument that in all interaction a basic

underlying theme is the desire of each participant to guide and control the responses made by the others present. A similar argument has been advanced by Jay Haley in a recent unpublished paper, but in regard to a special kind of control, that having to do with defining the nature of the relationship of those involved in the interaction.

3. An interaction can be purposely set up as a time and place for voicing differences in opinion. But in such cases participants *must* be careful to agree not to disagree on the proper tone of voice, vocabulary, and degree of seriousness in which all arguments are to be phrased, and upon the mutual respect which disagreeing participants must carefully continue to express toward one another. This debaters' or academic definition of the situation may also be invoked suddenly and judiciously as a way of translating a serious conflict of views into one that can be handled within a framework acceptable to all present.

4. Goffman, *op. cit.*, pp. 319–27.

5. Peter Blau, "Dynamics of Bureaucracy" (Ph.D. dissertation, Department of Sociology, Columbia University, forthcoming, University of Chicago Press), pp. 127–29.

6. Walter M. Beattie, Jr., "The Merchant Seaman" (unpublished M.A. report, Department of Sociology, University of Chicago, 1950), p. 35.

7. Sir Frederick Ponsonby, *Recollections of Three Reigns* (New York: Dutton, 1952), p. 46.

8. For comments on the importance of distinguishing between a routine of interaction and any particular instance when this routine is played through, see John van Neumann and Oskar Morgenstern, *The Theory of Games and Economic Behaviour* (2nd ed.) (Princeton: Princeton University Press, 1947), p. 49.

9. See Babe Pinelli, as told to Joe King, *Mr. Ump* (Philadelphia: Westminster Press, 1953), p. 75.

10. Material on the burial business used throughout this report is taken from Robert W. Habenstein, "The American Funeral Director" (unpublished Ph.D. dissertation, Department of Sociology, University of Chicago, 1954). I owe much to Mr. Habenstein's analysis of a funeral as a performance.

11. Charles H. Cooley, *Human Nature and the Social Order* (New York: Scribner's, 1922), pp. 352–53.

12. This view of hypnosis is neatly presented by T. R. Sarbin, "Contributions to Role-Taking Theory. I: Hypnotic Behavior," *Psychological Review*, 57, pp. 255–70.

13. See D. R. Cressey, "The Differential Association Theory and Compulsive Crimes," *Journal of Criminal Law, Criminology and Police Science*, 45, pp. 29–40.

14. This concept derives from T. R. Sarbin, "Role Theory," in Gardner Lindzey, *Handbook of Social Psychology* (Cambridge: Addison-Wesley, 1954), Vol. 1, pp. 235–36.

THINKING ABOUT THE READING

According to Goffman, why must everyone engage in impression management? What are some of the reasons we do this? What does he mean by the terms *definition of the situation*, *working consensus*, and *preventative strategies*? Consider a situation in which you were particularly aware of your own self-presentation. Do you think Goffman is interested primarily in the interactions between people or in their individual psychology? What is the source of the "scripts" that people use to determine what role they should play in a given situation or performance?

Public Identities: Managing Race in Public Spaces

Karyn Lacy

(2007)

"They're trying to be like the whites instead of being who they are," Andrea Creighton, a forty-three-year-old information analyst with the federal government, told me when I asked whether she believed blacks had made it in the United States or still had a long way to go. Andrea is black, and she perceives irrepressible distinctions between middle-class blacks and whites, even though many aspects of her life appear to reflect membership in the suburban middle-class mainstream. She and her husband, Greg, have two teenage children: a girl, age seventeen, and a boy, age fifteen. They have lived on a quiet street in Sherwood Park, an upper-middle-class suburb of Washington, D.C., for seven years. Their four-bedroom home is an imposing red-brick-front colonial with shiny black shutters, nestled on an acre of neatly manicured lawn. The children are active members of the local soccer team, and Greg is one of the team's coaches. Andrea and her husband each drive midsize cars and have provided their daughter, who is old enough to drive unaccompanied by an adult, with her own car. At first blush, they seem nearly identical to their white middle-class counterparts. But unlike the nearly all-white neighborhood that the average middle-class white family calls home, the Creightons' upscale subdivision is predominantly black. Andrea and Greg are pleased that their children are growing up in a community filled with black professionals. The Creightons' residence in Sherwood Park is one indication of the kind of social differentiation Andrea employs to define her identity as a member of the black middle class. Though she shares many life-style characteristics with mainstream whites, she feels that middle-class blacks are not mirror images of middle-class whites, nor should they aspire to be.

In terms of occupational status, educational attainment, income, and housing, the top segment of the black middle class is equal to the white middle class. The key distinction between the white and black middle classes is thus a matter of degree. Middle-class whites fit the public image of the middle class and may therefore take their middle-class status for granted, but blacks who have "made it" must work harder, more deliberately, and more consistently to make their middle-class status known to others.

Instances of discrimination against blacks in stores, in the workplace, and in other public spaces occur every day, unobserved by potential sympathizers and unreported by black victims. As sociologist Joe Feagin's gripping study of the black middle-class experience shows, middle-class status does not automatically shield blacks from discrimination by whites in public spaces (Feagin 1991). His interviewees' reports of being denied seating in restaurants, accosted while shopping, and harassed by police officers lead Feagin to conclude that a middle-class status does not protect blacks from the threat of racial discrimination. Feagin's study documents the formal and informal mechanisms that contribute to persistent discrimination toward blacks in the public sphere. His perspective, which has been invaluable in shedding light on the dynamics of racial stratification in the United States, suggests that contemporary patterns of discrimination often prevent accomplished blacks from enjoying the taken-for-granted privileges associated with a middle-class status, such as a leisurely dinner out or a carefree shopping experience.

This study demonstrates that despite the ever-present possibility of stigmatization, not all middle-class blacks feel as overwhelmed by and as ill-equipped to grapple with perceived discrimination as racial stigma theory implies. Some perceive themselves as active agents capable of orchestrating public interactions with whites to their advantage in a variety of public settings. Study participants from Lakeview, Riverton, and Sherwood Park describe how the strategic deployment of cultural capital, including language, mannerisms, clothing, and credentials, allows them to create what I call *public identities* that effectively lessen or short-circuit potential discriminatory treatment.

White Americans typically equate race with class and then reflexively consign all blacks to the lowest class levels. The experiences of middle-class blacks in my study suggest that those who actively correct the misapprehensions of white strangers reduce the likelihood of discriminatory treatment. This invocation of a public identity is a deliberate, conscious act—one that entails psychological costs as well as rewards. As Charlotte, an elementary school teacher and Lakeview resident, explains, black people "have two faces," and learn to distinguish self-presentation strategies suitable in the white world from self-presentation strategies useful in the black world.

Most middle-class whites, on the other hand, pay little overt attention to their own race or class. For them, most activities such as shopping, working as a manager, or buying a house are routinized, psychologically neutral, and relatively conflict free. Public challenges to their class status are rare. Middle-class blacks face a different reality. When they leave the familiarity of their upscale suburban communities, many of the accoutrements associated with their middle-class lifestyle fade from view. Skin color persists. On occasions when race trumps class, blacks' everyday interactions with white store clerks, real estate agents, and office subordinates can become exercises in frustration or humiliation or both. Asserting public identities makes it possible for blacks to tip the balance of a public interaction so that class trumps race. Blacks who successfully bring their middle-class status firmly into focus pressure white strangers and workplace subordinates to adjust their own behaviors in light of this information. Public identities, then, are not so much prepared responses that permit individuals to skillfully avoid or ignore strangers or social deviants when in public as they are strategies for sustaining problem-free interactions involving strangers. The use of public identities allows some middle-class blacks to complete their shopping without being accosted by store clerks or security guards, to supervise workplace subordinates effectively, and to disarm hostile real estate agents.

Constructing Public Identities: Boundary-Work in the Public Sphere

A key component of the public identities asserted by middle-class blacks is based on class and involves differentiating themselves from lower-class blacks through what I call *exclusionary* boundary-work. Washington-area middle-class blacks are firm in their belief that it is possible to minimize the probability of encountering racial discrimination if they can successfully convey their middle-class status to white strangers. To accomplish this feat, interviewees attempt to erect exclusionary boundaries against a bundle of stereotypes commonly associated with lower-class blacks. Exclusionary boundary-work is most readily apparent when middle-class blacks are shopping or managing employees in the workplace. Middle-class blacks also engage in *inclusionary* boundary-work in order to blur distinctions between themselves and white members of the middle class by emphasizing areas of consensus and shared experience. Efforts to highlight overlaps with the white middle class are common when middle-class blacks engage in house-hunting activities.

The construction and assertion of public identities varies according to social context and the basis of perceived discrimination. In the context of shopping, the middle-class blacks in this study perceive that race bias is operational, that is, that there is a failure by others to distinguish them from the black poor. Specifically, they know that whites wrongly assume that blacks are poor and that the poor are likely to be shoplifters. Consequently, when shopping, these middle-class blacks confront the stereotype of the street-savvy black shoplifter, which white store clerks often apply to blacks as a group. To disassociate themselves from this negative image and signal that they "belong" in the store (i.e., that they have money, can afford the merchandise, and have no need to steal), study participants report that they dress with care. "People make decisions about you based on how you're dressed and what you look like," Michelle says. "Because I know that," she elaborates, "I choose my dress depending on what the environment is." Interviewees contend that their decisions to eschew clothing associated with urban popular culture—for example, oversized gold earrings, baggy jeans, and designer tennis shoes—maximize their chances of enjoying a trouble-free shopping experience and signal their respectability to white strangers. This kind of exclusionary boundary-work helps middle-class blacks establish *social differentiation*—they make clear to store personnel that they are *not* like the poor.

Evidence of social differentiation emerges in the workplace as well. Just as the professions have used educational credentials to limit membership and to bring legitimacy to their discipline, professional blacks underscore their authority as managers by highlighting credentials such as job title and professional status. Holding positions of power, interviewees believe, makes them impervious to workplace discrimination.

In the context of house-hunting, middle-class blacks perceive that class [bias], rather than race bias, operates. In order to maximize their range of residential options, public identities are constructed to be linked in an inclusionary manner with their white counterparts. With the dominant cultural code in mind, middle-class blacks rely on mainstream language and mannerisms to carry out interactions with real estate agents. In cases in which these interactions break down, respondents use their own resources and social networks to find an acceptable home on their own. Put simply, middle-class blacks engage in inclusionary boundary-work to establish *social unity*—to show that middle-class blacks are much like the white middle-class. These identity construction processes are mutually reinforcing in that they each help to affirm respondents' position as legitimate members of the American middle class.

Cultural Capital and Cultural Literacy

Cultural capital, a key signifier of middle-class status, constitutes the means by which public identities are staked out. Cultural capital theorists argue that an important mechanism in the reproduction of inequality is a lack of exposure to dominant cultural codes, behaviors, and practices (Bourdieu and Passeron 1977). Middle-class blacks have obviously secured a privileged position in the occupational structure. But cultural capital differs from such economic capital in that cultural capital indicates a "proficiency in and familiarity with dominant cultural codes and practices—for example, linguistic styles, aesthetic preferences, styles of interaction" (Aschaffenburg and Maas 1997). These signifiers of middle-class status are institutionalized and taken for granted as normative, hence the underlying assumption that groups that cannot activate cultural capital fall victim to systematic inequality.

The majority of the blacks in this study are first-generation middle-class or grew up in working-class families; therefore, they could not acquire cultural capital through the process outlined by Bourdieu. They were not in a position to inherit from their parents the ability to signal their class position to whites via mainstream cultural resources because their parents either did not have access to middle-class cultural resources or they had views about black-white interaction that were informed by Jim Crow laws and other pillars of racial segregation. The few interviewees who did grow up in the middle class question how much their parents, who went about their everyday lives almost exclusively in black communities, could have effectively prepared them to negotiate routine interactions with whites as equals.

The blacks in my study were not endowed with the cultural capital useful in managing interactions with whites through their families of origin. As children, they were compelled to figure out these negotiations on their own through their immersion in white colleges, workplaces, and educational institutions, without involving their parents or other adults. They did so through two socialization processes that facilitate the construction of public identities: improvisation and script-switching.

Improvisations Socialization

During childhood, the blacks in this study were socialized into a set of informal strategies that allowed them to negotiate on their own the racial discrimination they faced at that time. In contrast to their parents' strategies of avoidance, deference, and unwillingness to confront authority, interviewees were more likely to fight back surreptitiously. For example, they often challenged indirectly the authority of white teachers and authority figures. When these middle-class blacks employed improvisational strategies, they left the impression that they were obeying the rules when they were, in fact, circumventing rules and established practices. This phenomenon is typified by Brad, who told me how a white guidance counselor had discouraged his applying to college: "My high school counselor told me that I should not go to Michigan because I probably wouldn't make it, and I should go to a trade school. [That way] I would have a job, [and] I could support my family." He pauses, visibly upset. Then, with sarcasm, he adds, "She was great."

I asked him, "This was a black woman telling you this?" He answered, "Uh-uh, she was a white woman. Miss Blupper. I remember her name." Miss Blupper's lack of confidence in Brad's intellectual ability made him even more determined to go to college. Brad acquired on his own a knowledge of college rankings and the admissions process that his guidance counselor was unwilling to provide. He ended up graduating from high school early to attend the University of Michigan and went on to become a judge.

In addition to being discouraged from pursuing a college track by high school teachers, middle-class blacks frequently faced white teachers who were heavily invested in symbolically maintaining the racial boundaries that had been dismantled by desegregation policy. Looking back on his tenure as class president during his junior year at the predominantly white high school, Greg remembers that the tradition dictating who should escort the homecoming queen was abandoned by his white teacher when she realized that he was slotted to escort a white girl:

> My junior year in high school there was always the tradition that the juniors put on the prom for the seniors. . . . I was the class president, and I was 'spose to escort the queen. . . . Well, they made an exception that year. [He laughs.] Basically they said, well, they'd let me and my date lead the parade, and the queen and everybody else [were to] follow behind. Well, you know how I am, I'm saying, "What's up with this?" I'm 'spose to walk the queen, but the queen's white. They didn't want me walking a white queen. I guess they didn't want this black guy walking in with this white queen. So it's really funny that the girl that I happened to be dating at that time, she had naturally red hair, and [she was] just as white as almost snow. But she was black! Yeah, she was a black girl! [He laughs.] . . . So, anyway, I said, "I'll take her to the prom. I'll fake 'em all out." So I'm leading the prom, me [and the girl], we're going to the prom together. So nobody knew anything; so me and [the girl] showed up, and my, my, my, you talking about fine [attractive]!

Greg decided to "pay back" the teacher not by using official channels and reporting her to higher authorities or by insisting that in fact he would escort the white queen as tradition dictated, but by devising a scheme on his own that would both expose the absurdity of the black-white boundary and preserve his dignity. Brad also circumvented official channels in his quest for a college education.

By the time he entered high school, Greg possessed an insider's knowledge of mainstream culture; he knew whites would be baffled by the apparent racial identity of his fair-skinned date. He acquired this familiarity with dominant codes and practices through exposure to white cultural norms in integrated settings, settings that required him to manage interactions with whites. Greg improvised strategies for managing these interactions as the specific conflict arose, yet these incidents prepared him, as I will demonstrate, for later experiences with racial discrimination.

Script-Switching

Script-switching processes refer to the strategies middle-class blacks employ to demonstrate that they are knowledgeable about middle-class lifestyles and to communicate their social position to others.

Scholars now recognize that blacks and whites tend to "behave" different kinds of scripts. For example, Thomas Kochman observed that blacks tend to communicate in an "emotionally intense, dynamic, and demonstrative" style, whereas whites tend to communicate in a "more modest and emotionally constrained" style (Kochman 1981:106). Of course, Kochman's schema is a generalization of these racial groups. There are whites and blacks who do not fit neatly into the categories he lays out. But these exceptions do not erase the powerful impact of these stereotypes on everyday interactions across the color line. Because public interactions are governed by mainstream scripts, middle-class blacks are compelled to switch from black scripts to white scripts in public spaces. Thus, public interactions require a different presentation of self than those asserted in majority-black spaces. In short, the middle-class blacks sometimes downplay their racial identities in public interactions with whites. Jasmine is short, with a bouncy haircut in the shape of a trendy bob. She seems taller than she actually is because she is extroverted and somewhat bossy, whereas her husband Richard is quiet and shy. Jasmine, now forty-five, describes how she felt compelled to script-switch as a teenager when her parents enrolled her in a predominantly white high school:

> I remember wanting to do "the white thing" when I was there. I had iodine and baby oil, trying to get a tan, and why wasn't [my] hair blowing in the wind? They [the white girls] would be shaving their legs and that type of thing, and most African American girls aren't that particular. I felt I needed to be a part of them, I needed to do their thing. . . . To this day, I think I made the blend [between two cultures] pretty decent because I have plenty of friends who just hate going back to our high school reunion. They just see no purpose [in going], but I enjoyed it because I participated in everything. . . . I was homecoming queen, I was in their beauty pageant when no other black person would dare to be in their pageant. I was like, "If you can do it, I can do it!"

Charlotte, speaking with admiration of a worker in a predominantly white school system with very few other black teachers, outlined how a black male art teacher who declined to script-switch was harassed by the white principal. "[The] white principal can't *stand* him, and I think it's because he's this big, black guy, and he's loud. You know, 'Hey, how ya doing!' Kind of like that. He's real down to earth, and I think they're kind of envious of him, because he's been in books, he's been in the [*Washington*] *Post,* he's been on TV, and they're trying to get their little doctorates. And they're always demeaning him. . . . They are just awful to him."

According to Charlotte, this teacher is subjected to a different set of evaluation criteria than the other teachers working at the school. But because Charlotte and the few other black teachers have not been mistreated in the ways that she observes the black male teacher has been, Charlotte feels that the white principal is reacting not so much to the art teacher's race as to his refusal to display the appropriate command of cultural capital—in short, to switch scripts. By Charlotte's account, the white principal interprets the art teacher's behavior as gauche, even though there may be no basis for this conclusion aside from the teacher's refusal to engage a white script. "He kinda doesn't make the—he's an artist, and he's eccentric, and he's just *him,* and he doesn't do the bullshit." Charlotte added parenthetically, "And see . . . they want that, they want him to do that."

As Charlotte's narrative makes clear, some middle-class blacks believe that social acceptance in the public sphere is contingent upon their ability to script-switch. They believe that they are less likely to be hassled in white settings if they are willing to script-switch. Blacks who refuse to do so or are uncomfortable doing so may be penalized, just as the teacher at Charlotte's school was targeted.

Asserting Public Identities

Undoubtedly, all persons attempting to cross class boundaries have to spend time thinking about clothing, language, tastes, and mannerisms; they risk being identified as a member of a lower class if they make a mistake. Concerns among the socially mobile about needing to properly appropriate the general skills and cultural styles of the middle class in a convincing way are well-documented in the sociological literature and in fictional accounts. However, middle-class blacks' ambiguous position in the racial hierarchy means that they have to spend more time thinking about what they will wear in public, work harder at pulling off a middle-class presentation of self, and be more demonstrative at it than white middle-class people who are also exhibiting their status and negotiating for deference. Moreover, while the fault line for upwardly mobile whites today is strictly class, middle-class blacks must negotiate class boundaries as well as the stereotypes associated with their racial group. In this section, I demonstrate how public identities are put to work in three public spaces, each with its own distinct pattern of black-white interaction. While shopping and in the workplace, these middle-class blacks employ public identities to establish their distinctiveness from the black poor and from subordinate workers. In the context of house-hunting, middle-class blacks perceive that class bias operates; therefore, public identities are constructed to establish their overlap with the white middle class.

Exclusionary Boundary-Work

Shopping

An obvious way for middle-class blacks to signal their class position is through physical appearance.

In real terms, this would mean selecting clothing that contrasts sharply with the attire associated with black popular culture. Philip, who wears a suit to his job as a corporate executive, observed: "Being black is a negative, particularly if you're not lookin' a certain way. You . . . go in an elevator dressed in what I have on now [he is wearing a blue polo shirt and white shorts], white women start holding their pocketbooks. But if I'm dressed like I normally go to work, then it's fine."

Philip implies that he can control the extent to which he will be evaluated on the basis of whites' stereotypes about poor blacks by the type of clothing he decides to wear to the store. If Philip decides to assert his public identity, he will shop in his suit. Once he begins to make purchases, additional signifiers of his social status such as credit cards and zip code assure the store clerk that he is a legitimate member of the middle class. Through his performance, Philip believes that he annuls a stigmatized racial identity. He believes that when he is dressed as a professional, whites see his class status first and respond to him as a member of that social group.

The assurance of these additional middle-class signifiers allows middle-class blacks to occasionally engage in subversive expressions of their class identity, much to their delight. Terry complained that store clerks react negatively to blacks who are "dressed down" under the assumption that they cannot afford to buy anything. "Going into a store, somebody follows you around the store. But [store clerks] don't help you [at all] if you go into a specialty store. They just refuse to walk up to you. Then you see a white person walk in and they immediately run to help them." However, as a member of the middle class, Terry is pleased that she has the leisure time and the requisite skills to voice a complaint:

> Now lately I will write a complaint. I will find out who owns the store and write a complaint. Before I used to just tell the [salesperson], "I guess you didn't know who walked into your store. It's a shame that you treat people like this because you don't know how much money I have." I love going to expensive stores in jeans and a T-shirt. Because they don't know how much money you have. And, you know, I may have a thousand dollars to give away that day. [She laughs.] They just don't know. And the way people treat you, I think it's a shame, based on your appearance.

Convinced that a store clerk ignored her because her clothing belied her actual class status, Terry went on to test her suspicion by varying the style of clothing that she sports while shopping. Terry enjoys "dressing down," but this subversive presentation of self appears to be enjoyable precisely because she can shed this role at a moment's notice, reassuming her actual middle-class identity. She then drew on her resources as a member of the middle class to file a formal complaint.

Because their performance as members of the middle class is perceived as legitimate when they are clothed in a way that signifies their social status, these blacks believe that using this strategy helps them to avoid the discrimination that blacks of a lower-class status experience. This perception is illustrated by Michael, a stylish corporate manager who suggested that his appearance, coupled with his Sherwood Park zip code and his assets, lead others to draw the conclusion that he is middle-class. He boasted, "When I apply for anything [that requires using] credit, I just give my name, address.... You have to fill out the credit application ... you put your address down there, then you put down your collateral, IRA, all that stuff. So I don't know if I've been discriminated against that way. I mean, I can go to the store and buy what I want to buy."

In cases of racial discrimination, blacks are typically precluded from achieving a desired goal, such as obtaining a desired product or entering a particular establishment. Since middle-class blacks in this study enter stores and "buy what they want to buy," (when they are dressed in a manner that reflects their social status), they conclude that they have not experienced racial discrimination. This suggests that when interviewees "buy what they want to buy" without interference from whites, they have successfully conveyed their class position to store clerks.

Others also suggested that when middle-class blacks are dressed down, that is, not engaging public identities, their shopping experience is often extremely unpleasant. Michelle attempted to shop while dressed down, and was dismissed by the store clerk: "I went somewhere and they tried to tell me how I couldn't afford something . . . I was in the mood to buy. They were saying, 'Well, it might cost this or that.' I mean I went there seriously looking to shop. But I wasn't dressed that way." In response to my question, "Was it clothing or race?" Michelle looked slightly puzzled, as if she hadn't consider this possibility, then waffled as to the explanation for the store clerk's behavior.

> Probably a mixture of both, I don't know. See, I don't know what it's like to be white and dressed poorly and [to] try and buy something. I've always had in my mind where someone told me that you could wear holey jeans as long as you have on two-hundred-dollar shoes. People know that you got money. That's when you're worried about what people think about you. But, you know, on a relaxed day, I don't care what they [white people] think. You either have the money or you don't.

Sorting out the store clerk's motivation is difficult for Michelle in part because whites' stereotypes of the face of poverty are conflated with race. When most whites think abstractly about the middle class, they see a white family, not a black one. This same image leads whites to associate poverty with blacks. In order for whites to believe that the blacks appearing before them are middle-class, they would have to erase the indelible image linking the concept "middle class" exclusively to whites. Middle-class blacks in the Washington, D.C., area convey this status by engaging their public identity, expressed through clothing that signals their middle-class status to others. In the workplace, middle-class blacks focus on a different form of cultural capital—professional title and credentials—to minimize racial tension in the workplace and to underscore their position as managers or supervisors.

The Workplace

Perhaps no public setting better reflects the cultural styles and preferences of the American mainstream than the corporate world. As Feagin and Sikes observe in *Living with Racism,* blacks "in corporate America are under constant pressure to adapt . . . to the values and ways of the white word" (Feagin and Sikes 1994: 135).

White colleagues and clients still register surprise when they encounter corporate blacks who speak intelligently about the topic at hand, black managers still confront "glass ceilings," and black managers still endure subjective critiques assessing their "fit" with the corporate culture. I focus here on two such problems faced by black managers today: managing white subordinates and negotiating racial disputes. Like shopping sites, workplace settings are characterized by a low regard for black cultural styles. According to Mary Jackman, many whites perceive black cultural styles as "inappropriate for occupational tasks involving responsibility or authority." (Jackman 1994: 130). This means that black managers' credibility resides in their ability to switch to the script associated with white cultural styles. Therefore, in the workplace, black managers assert public identities by demonstrating their command of the cultural capital appropriate for their title or position. Indeed, they must, since the workplace experiences of middle-class blacks are characterized by frequent episodes of discriminatory treatment.

Michael, a corporate manager, has a dry sense of humor and enjoys putting people in their place. He established his role as an authority figure at the outset by highlighting impermeable boundaries between himself and his receptionist. One such boundary is the telephone. Clearly annoyed, Michael explained, "The receptionist always bitches about answering telephones, but that's her job. She's the receptionist. I ain't never gonna answer the telephone." Michael does not answer his own telephone because his conception of a manager means that subordinates handle mundane details such as phone messages. Answering the phone would reduce his social status to that of a subordinate.

In his position as corporate manager, Michael says he has never experienced any racial discrimination. He attributes this feat to the weight of his title and his ability to utilize it. "On *this* job . . . I always came in with some authority. Hey, you know, like, 'I'm corporate manager. You all can do whatever you want to do, but remember, I'm the one that signs [off]. I'm the one that signs.' And, when you're the one that signs, you got the power. So even if they don't like you, they got to smile, which is okay by me." Greg used a similar strategy with white employees who resented having to work under him after his company was awarded a lucrative contract. He begins to smile as he remembers how he handled the conflict:

> There are two folks that I know of that are stone redneck. I mean they're the biggest rednecks you ever did see. They now came over to work for me. They couldn't accept that. So we had several briefings and I said, "Okay, here's how we're gonna do this and here's how we're gonna work." Well, the people they work for . . . were also big rednecks, so they just sort of go along together. Well, they refused—not openly refused, but just subtly. They wouldn't come to meetings. . . . I ended up basically saying to them, "Look, y'all can do what you want to do. But when it comes time for bonuses, and it comes time for yearly wages and all that kind of stuff, now you can go to Jim [white supervisor who reports to Greg], and he can tell you what to do and y'all can go do it. But if he

> doesn't tell me that you did it, I won't know. So when it comes time for your annual evaluation, I'll just say, 'Didn't do nothing.' So, it's y'all's fault." Well, then they sort of opened up.

In contrast to previous studies of middle-class blacks in the workplace, those surveyed here feel empowered to negotiate workplace discrimination. Situated in positions of power, middle-class blacks rely on public identities—for example, their role as supervisor or manager—to solidify their identities as persons of considerable social status. Once their status is established, these middle-class blacks are in a position to extinguish racial conflicts in the workplace. In other instances, middle-class blacks decide that such effort is "not worth it," and juxtapose the pleasant aspects of their high-status occupations against such racial incidents as they arise. Though racial discrimination in the workplace has hardly disappeared, black professionals have become more adept at using class-based resources to resolve these kinds of conflicts.

Inclusionary Boundary-Work

House-Hunting

In *American Apartheid,* Douglas Massey and Nancy Denton argue that middle-class blacks have not had the opportunity to live wherever they want, to live "where people of their means and resources usually locate" (Massey and Denton 1993:138). They conclude that a major factor in blacks' exclusion is racial discrimination by real estate agents, who serve as the "gatekeepers" of predominantly white neighborhoods to which blacks, even those with the requisite resources, seldom gain entry.

Yet middle-class blacks interviewed for this study insist that one of the benefits of being middle-class is the option of living in any neighborhood one desires. Their housing decisions are no longer restricted by the behavior of real estate agents. John, who chose the majority-black but upper-middle-class Sherwood Park community, explained, "We could have lived anywhere we wanted to. We could have afforded to live a lot of different places, but we chose here." He and most of the middle-class blacks in this study minimize the likelihood that they have experienced racial discrimination while house-hunting because, in so many other aspects of their lives, they use class-based resources to secure a desired good. How do blacks use their public identity while house-hunting? To manage their interactions with white real estate agents, these middle-class blacks place a good deal of emphasis on displays of cultural capital—particularly appropriate clothing, apt language, and knowledge of the housing market. Yet house-hunting is a more complicated site for the construction and use of public identity because in house-hunting, unlike shopping and the workplace, respondents are unsure as to whether real estate agents are responding to their race or their class.

The preoccupation with presenting a middle-class appearance is evident in Lydia's description of her experience while viewing a model home in a predominantly white suburban subdivision located in the same greater metropolitan area where she and her husband eventually bought a home.

In response to the question "Have you ever experienced racial discrimination while house-hunting?" Lydia replied:

> I guess I never really thought about that in terms of racial, but economically, I think I have been. I tend not to be a person that dresses up. [She chuckles.] A couple times I've gone looking for houses and I'll just wear sweatpants. And you go out looking for a house that's in expensive neighborhoods, I don't know what they expect me to drive up in. That has nothing to do with how much money I have in the bank. And I've had that happen . . . a couple times. . . . I went to a house. . . . I don't know what we were driving, probably an old beat-up car. So I pull into the driveway, and I had on sweatpants, my [baseball] hat, I go in and see the house. I'd asked [the real estate agent] about the house, asked her for the information, and I said I wanted to take a tour. She immediately said to me, "Is this your price range?" [Dramatic pause.]

> I asked her how much was the house. She told me, and it was my price range, no big deal.

I asked, "Was this a black real estate agent?" Lydia answered, "White. She wanted to discuss my income before she would show me the house. Basically, I told her I'll take a look at the house, and I'll let her know when I'm finished."

Lydia felt that the real estate agent was attempting to discourage her from viewing the house. However, she believed that her choice of clothing, her baseball hat, and her old car all signaled the wrong social class status to the agent—not that the agent objected to black home-seekers. I attempted to clarify the kind of discrimination Lydia felt she had experienced by posing a follow-up question: "Is that standard procedure? Do real estate agents normally ask you how much you make before they show you the house?" Lydia responded, shaking her head slightly from side to side, "No, no. I had been looking at lots and lots of houses. And I *knew* what she was doing. It was her way of saying, 'Oh, *God*, who is this person coming in here?' Because when I was there, a white couple came in, and I stopped to listen to what [the agent] would say to them. None of that, none of that."

I asked, "Did they have on sweatpants too?"

"No, they were dressed up," she laughed.

I persisted, "But you think it was because you had on sweatpants, not because you were a black person?"

"I think she probably, maybe looked at me and felt maybe I didn't make enough money to afford the house. That was part of it. I'm not sure, looking at houses, that we ever experienced any kind of *racial* discrimination. Because the real estate agents we had . . . they all took us to predominantly white neighborhoods. It wasn't that they were trying to steer us toward any type of neighborhood. They were willing to take our money anywhere." She burst into laughter.

I asked, "So do you think you could have actually bought one of those houses if you'd wanted to?"

"Oh yeah, oh yeah."

Lydia had an opportunity to test her suspicion that the real estate agent associated her with the poor when a white couple arrived to view the same home, even though the white couple had been well-dressed, not like her, and no comparison that controlled on clothing had been possible. Lydia concluded that the agent assessed the couple more favorably based on the quality of their clothing, and to support her position, Lydia identified occasions when white real estate agents had accepted her middle-class performance as a legitimate expression of who she is, showing her expensive homes in white neighborhoods.

Lydia's account illustrates the difficulty in pinpointing racial discrimination in the housing market. She did go on to view the model home. Though she was dressed in an overly casual way, Lydia used strong language to inform the real estate agent that she intended to tour the home. And the agent did not move to prevent her from walking through the model home. Consequently, in Lydia's view, the encounter did not qualify as a "racial" one. So long as they are permitted to view the homes of their choice, the middle-class blacks in this study do not perceive racial discrimination in the housing market as affecting their own housing choices.

For instance, Audrey, now sixty-four, remembers her disheartening experience with a real estate agent over twenty years ago, when she and her husband moved into the area from another city:

> The agent took us down south [of the city] mostly, to . . . where more of the blacks lived . . . where they seemed to be feeding the black people that came into the area. . . . And when she started out showing us property [south of the city] . . . we told her we wanted to be closer. The things she showed us [that were closer] . . . it was gettin' worse. The properties . . . weren't as nice. So when she showed us the properties [in a black and Hispanic low-income section], we kind of like almost accepted the fact that this was what you're going to be getting.

Audrey and her family moved into the undesirable housing, but "from that day on," she said, "we never stopped looking at houses. We took it upon ourselves to continue just to look, to explore different areas." A year later, they moved to a more attractive neighborhood. Now distrustful of real estate agents, Audrey drew on cultural knowledge she'd acquired on her own—about desirable neighborhoods, schools, and the housing market in general—to locate a home in a neighborhood more suited to her family's tastes. Acquiring this kind of detailed information takes leisure time and research skills.

Greg and his wife found that their real estate agent also directed them to undesirable housing when they returned to the United States from a work assignment in Taiwan. Greg remembers:

> The agent kept showing us older homes. . . . They were ten-year-old homes, twelve-year old homes, and they just weren't our style. . . . Some of the homes, they had beautician shops in the basement, and she thought that was a great deal, you know, you could wash, you could style your hair. And I'm thinking, "I don't need that." So Andrea [his wife] just said, "I'm not interested in those." So we came here [to Sherwood Park] just on a whim, I guess. And they had a girl [real estate agent] named [Liz], she said, "Let me show you these," and we looked at 'em. We went, "Oh," and, "Ah, yeah, okay." Then Andrea just said, "Hey, that's what I want."

In each of these examples, middle-class blacks confront discrimination from real estate agents. Audrey and her husband were steered to a section of the city where many blacks already lived. Greg and his wife were shown older, less attractive homes within their general area of choice. But the fact that these families were able to successfully find a home that did appeal to them leads them to the conclusion that widespread discrimination against blacks no longer effectively bars blacks of their social status from entering the neighborhood of their choice. Recall John's comment: "We could have lived anywhere we wanted to. . . . We chose here." When real estate agents fail them, middle-class blacks simply find an attractive home by driving around on their own. As Audrey made clear, they "never stop looking," or they happen to find a desirable home "on a whim," as Greg and his wife did.

In cases where respondents do recognize discriminatory practices, they rely on two strategies to secure desirable housing, both of which require middle-class blacks to assert public identities. Some blacks confront real estate agents directly, the option Lydia chose when she advised the agent that she would "take a look at the house, and . . . let her know when [she had] finished." Though she was "dressed down," Lydia used unmistakable language to articulate her middle-class identity to the agent. In short, Lydia attempted to show that she belonged there, viewing the model home, just as much as the well-dressed white couple. In doing so, she relied on the class conviction that her access to cultural capital effectively challenged the real estate agent's potential roadblock.

Other middle-class blacks forgo the agent-client relationship completely, locating homes on their own as they drive through potential neighborhoods. Many already have friends living in the neighborhoods where they find their homes. Through these social networks, they are made aware of homes coming up for sale. This strategy can be likened to the self-reliant script that middle-class blacks make use of in the workplace. In the workplace, these middle-class blacks place little faith in the EEOC to resolve racial conflicts; instead, they resolve them on their own. While house-hunting, they dispense with real estate agents who are unwilling to help them find adequate housing. To locate a home on one's own requires skill and resources: a car, a working knowledge of the area's neighborhoods, leisure time to search, and so on. Thus, the middle-class blacks in this study realize that racial discrimination persists in the housing market, but they do not feel that their housing options are severely limited by it. After all, in the end, Lakeview and Riverton residents do locate a home that pleases them, and because they have no way to systematically assess whether their housing search compares unfavorably to that of their white counterparts, they tend to wave off the practices of prejudicial real estate agents as inconsequential in their housing decisions. Relying on middle-class resources and networks to negotiate these public interactions and to secure their dream home is a reasonably satisfying solution.

Conclusion

Although what makes the evening news is corporate discrimination scandals—multimillion-dollar lawsuits filed against companies accused of engaging in various forms of modern racism—the everyday instances of racial discrimination experienced by middle-class blacks warrant additional attention from scholars and the public. Feagin's racial stigma theory suggests that a middle-class standing does not protect blacks from racial discrimination. However, I have shown that this conclusion may not be invariantly true. Middle-class blacks in the Washington, D.C., area use public identities to reduce the probability that racial discrimination will determine important outcomes in their lives. By examining how public identities are employed in various public settings, we gain insight into the informal strategies blacks develop as a result of their experiences in a racialized society. These informal strategies are far more common than the occasional discrimination suits filed by blacks and profiled in the media.

Public identities constitute a form of cultural capital in which blacks with the knowledge and skills valorized by the American mainstream are in a position to manipulate public interactions to their advantage. Previous studies have not examined how high-status minority group members come to possess cultural capital. I introduced two conceptual devices to explain this process and to connect the acquisition of cultural capital to the construction and assertion of public identities in adulthood: improvisational processes and script-switching. To assert public identities, middle-class blacks first acquire cultural capital through their childhood introduction to integrated settings and through their ongoing interactions in the American mainstream, where white cultural styles rule the day. These improvisational and script-switching socialization processes allow middle-class blacks to demonstrate their familiarity with the cultural codes and practices associated with the white middle class. I also show that the cultural capital so critical to doing well in school is influential beyond the school setting as well: in shopping malls, the workplace, and to some extent, with real estate agents.

Among these middle class blacks, projecting public identities is an opportunity to shore up their status as a group that is not merely black, but distinctly black and *middle-class.* Interviewees noted that "the world is not fair" and that "people will look at [them] in special ways because [they] are black." But these middle-class blacks also tend to associate persistent racial discrimination in public spaces with lower-class blacks, not their class grouping. As members of the middle class, they firmly believe in their ability to engage in strategies that minimize the amount and severity of discrimination directed toward their group. On the rare occasions when they believe that they do experience discrimination—from sales clerks, for example—middle-class blacks associate the incidents with an inability on their part to effectively signal their class position to store employees.

The findings presented in this [reading] do not negate the racial stigma paradigm. Rather, these findings call attention to a neglected aspect of the model, namely, the mobilization of class-related strategies as a bulwark against racial discrimination. Indeed, the data suggest that social class may figure more centrally in middle-class blacks' subjective understanding of their public interactions than previous studies allow.

REFERENCES

Aschaffenburg, Karen, and Ineke Maas. 1997. "Cultural and Educational Careers: The Dynamics of Social Reproduction." *American Sociological Review* 62:573–87.

Bourdieu, Pierre, and Jean-Claude Passeron. 1977. *Reproduction in Education, Society, and Culture.* Beverly Hills, Calif.: Sage.

Feagin, Joe. 1991. "The Continuing Significance of Race: Antiblack Discrimination in Public Places." *American Sociological Review* 56: 101–16.

Feagin, Joe and Melvin Sikes. 1994. *Living with Racism: The Black Middle-Class Experience.* Boston: Beacon Press.

Jackman, Mary. 1994. *The Velvet Glove.* Berkeley: University of California Press.

Kochman, Thomas. 1981. *Black and White Styles in Conflict.* Chicago: University of Chicago Press.

Massey, Douglas, and Nancy Denton. 1993. *American Apartheid.* Cambridge, Mass.: Harvard University Press.

THINKING ABOUT THE READING

According to Lacy, what was the key distinction between middle-class blacks and middle-class whites? What are public identities, and how did the study participants use these in their public interactions with whites? What is cultural capital, and how was it used in the creation of these public identities? What do the findings in the study mean for racial stigma theory? How might other minority groups use public identities in the architecture of their social environments?

The Girl Hunt

Urban Nightlife and the Performance of Masculinity as Collective Activity

David Grazian

(2007)

Young urbanites identify downtown clusters of nightclubs as *direct sexual marketplaces*, or markets for singles seeking casual encounters with potential sex partners (Laumann et al. 2004).

In this article I examine girl hunting—a practice whereby adolescent heterosexual men aggressively seek out female sexual partners in nightclubs, bars, and other public arenas of commercialized entertainment. In this article I wish to emphasize the performative nature of contemporary flirtation rituals by examining how male-initiated games of heterosexual pursuit function as strategies of impression management in which young men sexually objectify women to heighten their own performance of masculinity. While we typically see public sexual behavior as an interaction between *individuals,* I illustrate how these rituals operate as collective and homosocial group activities conducted in the company of men.

The Performance of Masculinity as Collective Activity

Girl hunting in nightclubs would not seem to serve as an especially efficacious strategy for locating sexual partners, particularly when compared with other methods (such as meeting through mutual friends, colleagues, classmates, or other trusted third parties; common participation in an educational or recreational activity; or shared membership in a civic or religious organization). In fact, the statistical rareness of the one-night stand may help explain why successful lotharios are granted such glorified status and prestige among their peers in the first place (Connell and Messerschmidt 2005:851). But if this is the case, then why do adolescent men persist in hassling women in public through aggressive sexual advances and pickup attempts (Duneier and Molotch 1999; Snow et al. 1991; Whyte 1988), particularly when their chances of meeting sex partners in this manner are so slim?

I argue that framing the question in this manner misrepresents the actual sociological behavior represented by the girl hunt, particularly since adolescent males do not necessarily engage in girl hunting to generate sexual relationships, even on a drunken short-term basis. Instead, three counterintuitive attributes characterize the girl hunt. First, the girl hunt is as much *ritualistic* and *performative* as it is utilitarian—it is a social drama through which young men perform their interpretations of manhood. Second, as demonstrated by prior studies (Martin and Hummer 1989; Polk 1994; Sanday 1990; Thorne and Luria 1986), girl hunting is not always a purely heterosexual pursuit but can also take the form of an inherently *homosocial* activity. Here, one's male peers are the intended audience for competitive games of sexual reputation and peer status, public displays of situational dominance and rule transgression, and in-group rituals of solidarity and loyalty. Finally, the emotional effort and logistical deftness required by rituals of sexual pursuit (and by extension the public performance of masculinity itself) encourage some young men to seek out safety in numbers by participating in the girl hunt as a kind of *collective* activity, in which they enjoy the social and psychological resources generated by group cohesion and dramaturgical teamwork (Goffman 1959). Although tales of sexual adventure traditionally feature a single male hero, such as Casanova, the performance of heterosexual conquest more often resembles the exploits of the dashing Christian de Neuvillette and his better-spoken coconspirator Cyrano de Bergerac (Rostand 1897). By aligning themselves with similarly oriented accomplices, many young men convince themselves of the importance and efficacy of the girl hunt (despite its poor track record), summon the courage to pursue their female targets (however clumsily), and assist one another in "mobilizing masculinity" (Martin 2001) through a collective performance of gender and heterosexuality.

Methods and Data

I draw on firsthand narrative accounts provided by 243 heterosexual male college students attending the University of Pennsylvania, an Ivy League research university situated in Philadelphia. These data represent part of a larger study involving approximately 600 college students (both men and women).

Because young people are likely to self-consciously experiment with styles of public behavior (Arnett 1994, 2000), observing undergraduates can help researchers understand how young heterosexual men socially construct masculinity through gendered interaction rituals in the context of everyday life. But just as there is not one single mode of masculinity but many *masculinities* available to young men, respondents exhibited a variety of socially recognizable masculine roles in their accounts, including the doting boyfriend, dutiful son, responsible escort, and perfect gentleman. In the interests of exploring the girl hunt as *one among many types* of social orientation toward the city at night, the findings discussed here represent only the accounts of those heterosexual young men whose accounts revealed commonalities relevant to the girl hunt, as outlined above.

The Girl Hunt and the Myth of the Pickup

It is statistically uncommon for men to successfully attract and "pick up" female sexual partners in bars and nightclubs. However, as suggested by a wide selection of mass media—from erotic films to hardcore pornography—heterosexual young men nevertheless sustain fantasies of successfully negotiating chance sexual encounters with anonymous strangers in urban public spaces (Bech 1998), especially dance clubs, music venues, singles bars, cocktail lounges, and other nightlife settings. According to Aaron, a twenty-one-year-old mixed-race junior:

> I am currently in a very awkward, sticky, complicated and bizarre relationship with a young lady here at Penn, where things are pretty open right now, hopefully to be sorted out during the summer when we both have more time. So my mentality right now is to go to the club with my best bud and seek out the ladies for a night of great music, adventure and female company off of the grounds of campus.

Young men reproduce these normative expectations of masculine sexual prowess—what I call *the myth of the pickup*—collectively through homosocial group interaction. According to Brian, a nineteen-year-old Cuban sophomore:

> Whether I would get any girl's phone number or not, the main purpose for going out was to try to get with hot girls. That was our goal every night we went out to frat parties on campus, and we all knew it, even though we seldom mention that aspect of going out. *It was implicitly known that tonight, and every night out, was a girl hunt.* Tonight, we were taking that goal to Philadelphia's nightlife. In the meanwhile, we would have fun drinking, dancing, and joking around. (emphasis added)

For Brian and his friends, the "girl hunt" articulates a shared orientation toward public interaction in which the group collectively negotiates the city at night. The heterosexual desire among men for a plurality of women (hot *girls*, as it were) operates at the individual and group level. As in game hunting, young men frequently evaluate their erotic prestige in terms of their raw number of sexual conquests, like so many notches on a belt. Whereas traditional norms of feminine desire privilege the search for a singular and specified romantic interest (Prince Charming, Mr. Right, or his less attractive cousin, Mr. Right Now), heterosexual male fantasies idealize the pleasures of an endless abundance and variety of anonymous yet willing female sex partners (Kimmel and Plante 2005).

Despite convincing evidence to the contrary (Laumann et al. 2004), these sexual fantasies seem deceptively realizable in the context of urban nightlife. To many urban denizens, the city and its never-ending flow of anonymous visitors suggests a sexualized marketplace governed by transactional relations and expectations of personal noncommitment (Bech 1998), particularly in downtown entertainment zones where nightclubs, bars, and cocktail lounges are concentrated. The density of urban nightlife districts and their tightly packed venues only intensifies the pervasive yet improbable male fantasy of successfully attracting an imaginary surplus of amorous single women.

Adolescent men strengthen their belief in this fantasy of the sexual availability of women in the city—the myth of the pickup—through collective reinforcement in their conversations in the hours leading up to the girl hunt. While hyping their sexual prowess to the group, male peers collectively legitimize the myth of the pickup and increase its power as a model for normative masculine behavior. According to Dipak, an eighteen-year-old Indian freshman:

> I finished up laboratory work at 5:00 pm and walked to my dormitory, eagerly waiting to "hit up a club" that night. . . . I went to eat with my three closest friends at [a campus dining hall]. We acted like high school freshmen about to go to our first mixer. We kept hyping up the night and saying we were going to meet and dance with many girls. Two of my friends even bet with each other over who can procure the most phone numbers from girls that night. Essentially, the main topic of discussion during dinner was the night yet to come.

Competitive sex talk is common in male homosocial environments (Bird 1996) and often acts as a catalyst for sexual pursuit among groups of adolescent and young adult males. For example, in his ethnographic work on Philadelphia's black inner-city neighborhoods, Anderson (1999) documents how sex codes among youth evolve in a context of peer pressure in which young black males "run their game" by women as a means of pursuing in-group status. Moreover, this type of one-upmanship heightens existing heterosexual fantasies and the myth of the pickup while creating a largely unrealistic set of sexual and gender expectations for young men seeking in-group status among their peers. In doing so, competitive sexual boasting may have the effect of momentarily energizing group participants. However, in the long run it is eventually likely to deflate the confidence of those who inevitably continue to fall short of such exaggerated expectations and who consequently experience the shame of a spoiled masculine identity (Goffman 1963).

Preparing for the Girl Hunt Through Collective Ritual

Armed with their inflated expectations of the nightlife of the city and its opportunities for sexual conquest, young men at Penn prepare for the girl hunt by crafting a specifically gendered and class-conscious nocturnal self (Grazian 2003)—a presentation of masculinity that relies on prevailing fashion cues and upper-class taste emulation. According to Edward, a twenty-year-old white sophomore, these decisions are made strategically:

> I hadn't hooked up with a girl in a couple weeks and I needed to break my slump (the next girl you hook up with is commonly referred to as a "slump-bust" in my social circle). So I was willing to dress in whatever manner would facilitate in hooking up.

Among young college men, especially those living in communal residential settings (i.e., campus dormitories and fraternities), these preparations for public interaction serve as *collective rituals of confidence building*—shared activities that generate group solidarity and cohesion while elevating the personal resolve and self-assuredness of individual participants mobilizing for the girl hunt. Frank, a nineteen-year-old white sophomore, describes the first of these rituals:

> As I began observing both myself and my friends tonight, I noticed that there is a distinct pre-going-out ritual that takes place. I began the night by blasting my collection of rap music as loud as possible, as I tried to overcome the similar sounds resonating from my roommate's room. Martin seemed to play his music in order to build his confidence. It appears that the entire ritual is simply there to build up one's confidence, to make one more adept at picking up the opposite sex.

Frank explains this preparatory ritual in terms of its collective nature, as friends recount tall tales that celebrate character traits commonly associated with traditional conceptions of masculinity, such as boldness and aggression. Against a soundtrack of rap music—a genre known for its misogynistic lyrics and male-specific themes, including heterosexual boasting, emotional detachment, and masculine superiority (McLeod 1999)—these shared ritual moments of homosociality are a means of generating group resolve and bolstering the self-confidence of each participant. Again, according to Frank:

> Everyone erupted into stories explaining their "high-roller status." Martin recounted how he spent nine hundred dollars in Miami one weekend, while Lance brought up his cousins who spent twenty-five hundred dollars with ease one night at a Las Vegas bachelor party. Again, all of these stories acted as a confidence booster for the night ahead.

Perhaps unsurprisingly, this constant competitive jockeying and one-upmanship so common in male-dominated settings (Martin 2001) often extends to the sexual objectification of women. While getting dressed among friends in preparation for a trip to a local strip club, Gregory, a twenty-year-old white sophomore, reports on the banter: "We should all dress rich and stuff, so we can get us some hookers" Like aggressive locker-room boasting, young male peers bond over competitive sex talk by laughing about real and make-believe sexual exploits and misadventures (Bird 1996). This joking strengthens male group intimacy and collective heterosexual identity and normalizes gender differences by reinforcing dominant myths about the social roles of men and women (Lyman 1987).

After engaging in private talk among roommates and close friends, young men (as well as women) commonly participate in a more public collective ritual known among American college students as "pregaming." As Harry, an eighteen-year-old white freshman, explains,

> Pregaming consists of drinking with your "boys" so that you don't have to purchase as many drinks while you are out to feel the desired buzz. On top of being cost efficient, the actual event of pregaming can get any group ready and excited to go out.

The ritualistic use of alcohol is normative on college campuses, particularly for men (Martin and Hummer 1989), and students largely describe pregaming as an economical and efficient way to get drunk before going out into the city. This is especially the case for underage students who may be denied access to downtown nightspots. However, it also seems clear that pregaming is a bonding ritual that fosters social cohesion and builds confidence among young men in anticipation of the challenges that accompany the girl hunt. According to Joey, an eighteen-year-old white freshman:

> My thoughts turn to this girl, Jessica. . . . I was thinking about whether or not we might hook up tonight. . . . As I turn to face the door to 301, I feel the handle, and it is shaking from the music and dancing going on in the room. I open the door and see all my best friends just dancing together. . . . I quickly rush into the center of the circle and start doing my "J-walk," which I have perfected over the years. My friends love it and begin to chant, "Go Joey—it's your birthday." I'm feeling connected with my friends and just know that we're about to have a great night. . . . Girls keep coming in and out of the door, but no one really pays close attention to them. Just as the "pregame" was getting to its ultimate height, each boy had his arms around each other jumping in unison, to a great hip-hop song by Biggie Smalls. One of the girls went over to the stereo and turned the power off. We yelled at her to turn it back on, but the mood was already lost and we decided it was time to head out.

In this example, Joey's confidence is boosted by the camaraderie he experiences in a male-bonding ritual in which women—supposedly the agreed-upon raison d'être for the evening—are ignored or, when they make their presence known, scolded. As these young men dance arm-in-arm with one another, they generate the collective effervescence and sense of social connectedness necessary to plunge into the nightlife of the city. As such, pregaming fulfills the same function as the last-minute huddle (with all hands in the middle) does for an athletic team (Messner 2002). It is perhaps ironic that Joey's ritual of "having fun with my boys" prepares him for the girl hunt (or more specifically in his case, an opportunity to "hook up" with Jessica) even as it requires those boys to exclude their female classmates. At the same time, this men-only dance serves the same function as the girl hunt: it allows its participants to expressively perform hegemonic masculinity through an aggressive display of collective identification.

During similar collective rituals leading up to the girl hunt, young men boost each other's confidence in their abilities of sexual persuasion by watching films about male heterosexual exploits in urban nightlife, such as Doug Liman's *Swingers* (1996), which chronicles the storied escapades of two best friends, Mike and Trent. According to Kevin, an eighteen-year-old white freshman:

> I knew that [my friend] Darryl needed to calm down if he wanted any chance of a second date. At about 8:15 pm, I sat him down and showed him (in my opinion), the movie that every man should see at least once—I've seen it six times—*Swingers*. . . . Darryl immediately related to Mike's character, the self-conscious but funny gentleman who is still on the rebound from a long-term relationship. At the same time, he took Trent's words for scripture (as I planned): "There's nothing wrong with showing the beautiful babies that you're money and that you want to party." His mind was clearly eased at the thought of his being considered "money." Instead of being too concerned with not screwing up and seeming "weird or desperate," Darryl now felt like he was in control. The three of us each went to our own rooms to get ready.

This collective attention to popular cultural texts helps peer groups generate common cultural references, private jokes, and speech norms as well as build in-group cohesion (Eliasoph and Lichterman 2003; Fine 1977; Swidler 2001).

Girl Hunting and the Collective Performance of Masculinity

Finally, once the locus of action moves to a more public venue such as a bar or nightclub, the much-anticipated "girl hunt" itself proceeds as a strategic display of masculinity best performed with a suitable game partner. According to Christopher, a twenty-two-year-old white senior, he and his cousin Darren "go out together a lot. We enjoy each other's company and we seem to work well together when trying to meet women." Reporting on his evening at a local dance club, Lawrence, a twenty-one-year-old white junior, illustrates how the girl hunt itself operates as collective activity:

> We walk around the bar area as we finish [our drinks]. After we are done, we walk down to the regular part of the club. We make the rounds around the dance floor checking out the girls. . . . We walk up to the glassed dance room and go in, but leave shortly because it is really hot and there weren't many prospects.

Lawrence and his friends display their elaborated performance of masculinity by making their rounds together as a pack in search of a suitable feminine target. Perhaps it is not surprising that the collective nature of their pursuit should also continue *after* such a prize has been located:

> This is where the night gets really interesting. We walk back down to the main dance floor and stand on the outside looking at what's going on and I see a really good-looking girl behind us standing on the other side of the wall with three friends. After pointing her out to my friends, I decide that I'm going to make the big move and talk to her. So I turn around and ask her to dance. She accepts and walks over. My friends are loving this, so they go off to the side and watch. . . .
>
> After dancing for a little while she brings me over to her friends and introduces me. They tell me that they are all freshman [*sic*] at [a local college], and we go through the whole small talk thing again. I bring her over to my two boys who are still getting a kick out of the whole situation . . . My boys tell me about some of the girls they have seen and talked to, and they inform me that they recognized some girls from Penn walking around the club.

Why do Lawrence and his dance partner both introduce each other to their friends? Lawrence seems to gain almost as much pleasure from his *friends'* excitement as from his own exploits, just as they are "loving" the vicarious thrill of watching their comrade succeed in commanding the young woman's attention, as if their own masculinity is validated by his success.

In this instance, arousal is not merely individual but represents a collectively shared experience as well (Thorne and Luria 1986:181). For these young men the performance of masculinity does not necessarily require successfully meeting a potential sex partner as long as one enthusiastically participates in the ritual *motions* of the girl hunt in the company of men. When Lawrence brings over his new female friend, he does so to celebrate his victory with his buddies, and in return, they appear gratified by their *own* small victory by association. (And while Lawrence celebrates with them, perhaps he alleviates some of the pressure of actually conversing with her.)

As Christopher remarked above on his relationship with his cousin, the collective aspects of the girl hunt also highlight the efficacy of conspiring with peers to meet women: "We go out together a lot. We enjoy each other's company and we seem to work well together when trying to meet women." In the language of the confidence game, men eagerly serve as each other's shills (Goffman 1959; Grazian 2004; Maurer 1940) and sometimes get roped into the role unwittingly with varying degrees of success.

Among young people, the role of the passive accomplice is commonly referred to in contemporary parlance as a *wingman*. In public rituals of courtship, the wingman serves multiple purposes: he provides validation of a leading man's trustworthiness, eases the interaction between a single male friend and a larger group of women, serves as a source of distraction for the friend or friends of a more desirable target of affection, can be called on to confirm the wild (and frequently misleading) claims of his partner, and, perhaps most important, helps motivate his friends by building up their confidence. Indeed, men describe the role of the wingman in terms of loyalty, personal responsibility, and dependability, traits commonly associated with masculinity (Martin and Hummer 1989; Mishkind et al. 1986). According to Nicholas, an eighteen-year-old white freshman:

> As we were beginning to mobilize ourselves and move towards the dance floor, James noticed Rachel, a girl he knew from Penn who he often told me about as a potential girlfriend. Considering James was seemingly into this girl, Dan and I decided to be good wingmen and entertain Rachel's friend, Sarah.

Hegemonic masculinity is not only expressed by competitiveness but camaraderie as well, and many young men will take their role as a wingman quite seriously and at a personal cost to their relationships with female friends. According to Peter, a twenty-year-old white sophomore:

> "It sounds like a fun evening," I said to Kyle, "but I promised Elizabeth I would go to her date party." I don't like to break commitments. On the other hand, I didn't want to leave Kyle to fend for himself at this club . . . Kyle is the type of person who likes to pick girls up at clubs. If I were to come see him, I would want to meet other people as well. Having Elizabeth around would not only prevent me from meeting (or even just talking to) other girls, but it would also force Kyle into a situation of having no "wing man."

In the end, Peter takes Elizabeth to a nightclub where, although he *himself* will not be able to meet available women, he will at least be able to assist Kyle in meeting them:

> Behind Kyle, a very attractive girl smiles at me. Yes! Oh, wait. Damnit, Elizabeth's here. . . ." Hey, Kyle," I whisper to him. "That girl behind you just smiled at you. Go talk to her." Perhaps Kyle will have some luck with her. He turns around, takes her by the hand, and begins dancing with her. She looks over at me and smiles again, and I smile back. I don't think Elizabeth noticed. I would have rather been in Kyle's position, but I was happy for him, and I was dancing with Elizabeth, so I was satisfied for the moment.

By the end of the night, as he and Kyle chat in a taxi on the way back to campus, Peter learns that he was instrumental in securing his friend's success in an additional way:

> "So what ever happened with you and that girl?" I ask. "I hooked up with her. Apparently she's a senior." I ask if she knew he was a freshman. "Oh, yeah. She asked how old you were, though. I said you were a junior. I had to make one of us look older."

Peter's willingness to serve as a wingman demonstrates his complicity in sustaining the ideals of hegemonic masculinity, which therefore allows him to benefit from the resulting "patriarchal dividends"—acceptance as a member of his male homosocial friendship network and its attendant prestige—even when he himself does not personally seek out the sexual rewards of the girl hunt.

In addition, the peer group provides a readily available audience that can provide emotional comfort to all group members, as well as bear witness to any individual successes that might occur. As demonstrated by the preceding examples, young men deeply value the erotic prestige they receive from their conspiratorial peers upon succeeding in the girl hunt. According to Zach, a twenty-year-old white sophomore:

> About ten minutes later, probably around 2:15 am, we split up into cabs again, with the guys in one and the girls in another. . . . This time in the cab, all the guys want to talk about is me hooking up on the dance floor. It turns out that they saw the whole thing. I am not embarrassed; in fact I am proud of myself.

As an audience, the group can collectively validate the experience of any of its members and can also internalize an individual's success as a shared victory. Since, in a certain sense, a successful sexual interaction must be recognized by one's peers to gain status as an in-group "social fact," the group can transform a private moment into a celebrated public event—thereby making it "count" for the male participant and his cohorts.

A participant's botched attempt at an ill-conceived pickup can solidify the male group's bonds as much as a successful one. According to Brian, the aforementioned nineteen-year-old Cuban sophomore:

> We had been in the club for a little more than half an hour, when the four of us were standing at the perimeter of the main crowd in the dancing room. It was then when Marvin finished his second Corona and by his body gestures, he let it be known that he was drunk enough and was pumped up to start dancing. He started dancing behind a girl who was dancing in a circle with a few other girls. Then the girl turned around and said "Excuse me!" Henry and I saw what happened. We laughed so hard and made so much fun of him for the rest of the night. I do not think any of us has ever been turned away so directly and harshly as that time.

In this instance, Marvin's abruptly concluded encounter with an unwilling female participant turns into a humorous episode for the rest of his peer group, leaving his performance of masculinity bruised yet intact. Indeed, in his gracelessness Marvin displays an enthusiastic male heterosexuality as emphasized by his drunken attempts to court an unsuspecting target before a complicit audience of his male peers. And as witnesses to his awkward sexual advance, Brian and Henry take pleasure in the incident, as it not only raises *their* relative standing within the group in comparison with Marvin but can also serve as a narrative focus for future "signifying" episodes (or ceremonial exchanges of insults) and other rituals of solidarity characteristic of joking relationships among male adolescents (Lyman 1987:155). Meanwhile, these young men can bask in their collective failure to attract a woman without ever actually challenging the basis of the girl hunt itself: the performance of adolescent masculinity.

In the end, young men may enjoy this performance of masculinity—the hunt itself—even more than the potential romantic or sexual rewards they hope to gain by its successful execution. In his reflections on a missed opportunity to procure the phone number of a law student, Christopher, the aforementioned twenty-two-year-old senior, admits as much: "There's something about the chase that I really like. Maybe I subconsciously neglected to get her number. I am tempted to think that I like the idea of being on the look out for her better than the idea of calling her to go out for coffee." While Christopher's excuse may certainly function as a compensatory face-saving strategy employed in the aftermath of another lonely night (Berk 1977), it might also indicate a possible acceptance of the limits of the girl hunt despite its potential opportunities for male bonding and the public display of adolescent masculinity.

REFERENCES

Anderson, Elijah. 1999. *Code of the Street: Decency, Violence, and the Moral Life of the Inner City.* New York: Norton.

Arnett, Jeffrey Jensen. 1994. "Are College Students Adults? Their Conceptions of the Transition to Adulthood." *Journal of Adult Development* *1*(4):213–24.

———. 2000. "Emerging Adulthood: A Theory of Development from the Late Teens through the Twenties." *American Psychologist 55*(5): 469–80.

Bech, Henning. 1998. "Citysex: Representing Lust in Public." *Theory, Culture & Society 15*(3–4): 215–41.

Berk, Bernard. 1977. "Face-Saving at the Singles Dance." *Social Problems 24*(5):530–44.

Bird, Sharon R. 1996. "Welcome to the Men's Club: Homosociality and the Maintenance of Hegemonic Masculinity." *Gender & Society 10*(2):120–32.

Connell, R. W. and James W. Messerschmidt. 2005. "Hegemonic Masculinity: Rethinking the Concept." *Gender & Society 19*(6):829–59.

Duneier, Mitchell and Harvey Molotch. 1999. "Talking City Trouble: Interactional Vandalism, Social Inequality, and the 'Urban Interaction Problem.'" *American Journal of Sociology 104*(5):1263–95.

Eliasoph, Nina and Paul Lichterman. 2003. "Culture in Interaction." *American Journal of Sociology 108*(4):735–94.

Fine, Gary Alan. 1977. "Popular Culture and Social Interaction: Production, Consumption, and Usage." *Journal of Popular Culture 11*(2): 453–56.

Goffman, Erving. 1959. *The Presentation of Self in Everyday Life.* Garden City, NY: Anchor Books.

———. 1963. *Stigma: Notes on the Management of Spoiled Identity.* New York: Simon & Schuster.

Grazian, David. 2003. *Blue Chicago: The Search for Authenticity in Urban Blues Clubs.* Chicago: University of Chicago Press.

———. 2004. "The Production of Popular Music as a Confidence Game: The Case of the Chicago Blues." *Qualitative Sociology 27*(2):137–58.

Kimmel, Michael S. and Rebecca F. Plante. 2005. "The Gender of Desire: The Sexual Fantasies of Women and Men." In *The Gender of Desire: Essays on Male Sexuality,* edited by M. S. Kimmel. Albany: State University of New York Press.

Laumann, Edward O., Stephen Ellingson, Jenna Mahay, Anthony Paik, and Yoosik Youm, eds. 2004. *The Sexual Organization of the City.* Chicago: University of Chicago Press.

Lyman, Peter. 1987. "The Fraternal Bond as a Joking Relationship: A Case Study of the Role of Sexist Jokes in Male Group Bonding." In *Changing Men: New Directions in Research on Men and Masculinity,* edited by M. S. Kimmel. Newbury Park, CA: Sage.

Martin, Patricia Yancey. 2001. "'Mobilizing Masculinities': Women's Experiences of Men at Work." *Organization 8*(4):587–618.

Martin, Patricia Yancey and Robert A. Hummer. 1989. "Fraternities and Rape on Campus." *Gender & Society 3*(4):457–73.

Maurer, David W. 1940. *The Big Con: The Story of the Confidence Man.* New York: Bobbs-Merrill.

McLeod, Kembrew. 1999. "Authenticity within Hip-Hop and Other Cultures Threatened with Assimilation." *Journal of Communication 49*(4): 134–50.

Messner, Michael A. 2002. *Taking the Field: Women, Men, and Sports.* Minneapolis: University of Minnesota Press.

Mishkind, Marc, Judith Rodin, Lisa R. Silberstein, and Ruth H. Striegel-Moore. 1986. "The Embodiment of Masculinity." *American Behavioral Scientist 29*(5):545–62.

Polk, Kenneth. 1994. "Masculinity, Honor, and Confrontational Homicide." In *Just Boys Doing Business? Men, Masculinities, and Crime,* edited by T. Newburn and E. A. Stanko. London: Routledge.

Rostand, Edmond. 1897. *Cyrano de Bergerac.*

Sanday, Peggy Reeves. 1990. *Fraternity Gang Rape: Sex, Brotherhood, and Privilege on Campus.* New York: New York University Press.

Snow, David A., Cherylon Robinson, and Patricia L. McCall. 1991. "'Cooling Out' Men in Singles Bars and Nightclubs: Observations on the Interpersonal Survival Strategies of Women in Public Places." *Journal of Contemporary Ethnography 19*(4):423–19.

Swidler, Ann. 2001. *Talk of Love: How Culture Matters.* Chicago: University of Chicago Press.

Thorne, Barrie and Zella Luria. 1986. "Sexuality and Gender in Children's Daily Worlds." *Social Problems 33*(3):176–90.

Whyte, William H. 1988. *City: Rediscovering the Center.* New York: Doubleday.

THINKING ABOUT THE READING

What is the "girl hunt"? According to the author, the purpose of the girl hunt is male social bonding. What does he mean by this? The author uses Goffman's framework as a way to understand why men engage in these activities. What are some of the interaction rituals and performances these young men engage in when they go out to clubs? What are the implications of this kind of research for understanding social issues such as gender, dating, and sexual violence?

Building Social Relationships

7

Intimacy and Family

In U.S. society, close, personal relationships are one standard by which we judge the quality and happiness of our everyday lives. Yet in a complex, individualistic society like ours, these relationships are becoming more difficult to establish and sustain. Although we like to think that the things we do in our relationships are completely private experiences, they are continually influenced by large-scale political interests and economic pressures. Like every other aspect of our lives, close relationships are best understood within the broader social context. Laws, customs, and social institutions often regulate the form relationships can take, our behavior in them, and even the ways in which we can exit them. At a more fundamental level, societies determine which relationships can be considered "legitimate" and therefore entitled to cultural and institutional recognition. Relationships that lack societal validation are often scorned and stigmatized.

If you were to ask couples applying for a marriage license why they were getting married, most, if not all, would no doubt mention the love they feel for one another. But as Stephanie Coontz discusses in "The Radical Idea of Marrying for Love," love hasn't always been a prerequisite or even a justification for marriage. Until relatively recently, marriage was principally an economic arrangement, and love, if it existed at all, was a sometimes irrational emotion that was of secondary importance. In fact, in some past societies, falling in love before marriage was considered disruptive, even threatening, to the extended family. Today, however, it's hard to imagine a marriage that begins without love.

Speaking of love, gay fatherhood has presented a challenge to the socially constructed and legitimate images of paternity and masculinity in society. Even within the gay community, gay fatherhood has brought up questions around the sexual norms of gay culture. Judith Stacey interviews gay fathers in Los Angeles to explore the growing social character of paternity and the deliberate as well as difficult terrain these men navigate as they become fathers. In the same way that Coontz shows how marriage has entered contested terrain, paths to "planned parenthood" for gay men are full of tension as they negotiate both cultural and institutional recognition.

The covenant marriage movement developed as a response to what some U.S. religious leaders and organizations saw as the breakdown of the family, in large part attributed to the fracturing of heterosexual marriages. This cultural phenomenon became institutional as various states in the U.S. passed covenant marriage laws that were designed to make marriages stronger and deeper in their commitment. In "Covenant Marriage: Reflexivity and Retrenchment in the Politics of Intimacy," Dwight Fee uses the notion of reflexivity to understand the burgeoning covenant marriage movement. Although members of the movement steadfastly believe that they are revitalizing the belief in the sanctity of marriage, Fee points out that they might just be bucking tradition and custom and opening up marriage to the possibility of transformation.

Something to Consider as You Read

Each of these selections emphasizes the significance of external or structural components in shaping family experiences. As you read, keep track of factors such as income level and job opportunities and consider how these factors affect the choices families make. For example, consider what choices a family with a high income might have regarding how best to assist an ailing grandparent or how to deal with an unexpected teen pregnancy or in providing children with extracurricular activities. Consider how these choices are related to the appearance of "traditional family values." How does legal marriage support families? For instance, what kinds of benefits and social assistance do married couples receive that assist them in raising children? How might a redefinition of marriage impact different groups in society? How do different routes to parenthood affect the cultural and institutional recognition of families?

The Radical Idea of Marrying for Love

Stephanie Coontz

(2005)

The Real Traditional Marriage

To understand why the love-based marriage system was so unstable and how we ended up where we are today, we have to recognize that for most of history, marriage was not primarily about the individual needs and desires of a man and woman and the children they produced. Marriage had as much to do with getting good in-laws and increasing one's family labor force as it did with finding a lifetime companion and raising a beloved child.

Marriage, a History

Reviewing the role of marriage in different societies in the past and the theories of anthropologists and archaeologists about its origins, I came to reject two widespread, though diametrically opposed, theories about how marriage came into existence among our Stone Age ancestors: the idea that marriage was invented so men would protect women and the opposing idea that it was invented so men could exploit women. Instead, marriage spoke to the needs of the larger group. It converted strangers into relatives and extended cooperative relations beyond the immediate family or small band by creating far-flung networks of in-laws. . . .

Certainly, people fell in love during those thousands of years, sometimes even with their own spouses. But marriage was not fundamentally about love. It was too vital an economic and political institution to be entered into solely on the basis of something as irrational as love. For thousands of years the theme song for most weddings could have been "What's Love Got to Do with It?" . . .

For centuries, marriage did much of the work that markets and governments do today. It organized the production and distribution of goods and people. It set up political, economic, and military alliances. It coordinated the division of labor by gender and age. It orchestrated people's personal rights and obligations in everything from sexual relations to the inheritance of property. Most societies had very specific rules about how people should arrange their marriages to accomplish these tasks.

Of course there was always more to marriage than its institutional functions. At the end of the day—or at least in the middle of the night—marriage is also a face-to-face relationship between individuals. The actual experience of marriage for individuals or for particular couples seldom conforms exactly to the model of marriage codified in law, custom, and philosophy in any given period. But institutions do structure people's expectations, hopes, and constraints. For thousands of years, husbands had the right to beat their wives. Few men probably meted out anything more severe than a slap. But the law upheld the authority of husbands to punish their wives physically and to exercise forcibly their "marital right" to sex, and that structured the relations between men and women in *all* marriages, even loving ones.

The Radical Idea of Marrying for Love

George Bernard Shaw described marriage as an institution that brings together two people "under the influence of the most violent, most insane, most delusive, and most transient of passions. They are required to swear that they will remain in that excited, abnormal, and exhausting condition continuously until death do them part."[1]

Shaw's comment was amusing when he wrote it at the beginning of the twentieth century, and it still makes us smile today, because it pokes fun at the unrealistic expectations that spring from a dearly held cultural ideal—that marriage should be based on intense, profound love and a couple should maintain their ardor until death do them part. But for thousands of years the joke would have fallen flat.

For most of history it was inconceivable that people would choose their mates on the basis of something as fragile and irrational as love and then focus all their sexual, intimate, and altruistic desires on the resulting marriage. In fact, many historians, sociologists, and anthropologists used to think romantic love was a recent Western invention. This is not true. People have always fallen in love, and throughout the ages many couples have loved each other deeply.[2]

But only rarely in history has love been seen as the main reason for getting married. When someone did advocate such a strange belief, it was no laughing matter. Instead, it was considered a serious threat to social order.

In some cultures and times, true love was actually thought to be incompatible with marriage. Plato believed love was a wonderful emotion that led men to behave honorably. But the Greek philosopher was referring not to the love of women, "such as the meaner men feel," but to the love of one man for another.[3]

Other societies considered it good if love developed after marriage or thought love should be factored in along with the more serious considerations involved in choosing a mate. But even when past societies did welcome or encourage married love, they kept it on a short leash. Couples were not to put their feelings for each other above more important commitments, such as their ties to parents, siblings, cousins, neighbors, or God.

In ancient India, falling in love before marriage was seen as a disruptive, almost antisocial act. The Greeks thought lovesickness was a type of insanity, a view that was adopted by medieval commentators in Europe. In the Middle Ages the French defined love as a "derangement of the mind" that could be cured by sexual intercourse, either with the loved one or with a different partner.[4] This cure assumed, as Oscar Wilde once put it, that the quickest way to conquer yearning and temptation was to yield immediately and move on to more important matters.

In China, excessive love between husband and wife was seen as a threat to the solidarity of the extended family. Parents could force a son to divorce his wife if her behavior or work habits didn't please them, whether or not he loved her. They could also require him take a concubine if his wife did not produce a son. If a son's romantic attachment to his wife rivaled his parents' claims on the couple's time and labor, the parents might even send her back to her parents. In the Chinese language the term *love* did not traditionally apply to feelings between husband and wife. It was used to describe an illicit, socially disapproved relationship. In the 1920s a group of intellectuals invented a new word for love between spouses because they thought such a radical new idea required its own special label.[5]

In Europe, during the twelfth and thirteenth centuries, adultery became idealized as the highest form of love among the aristocracy. According to the Countess of Champagne, it was impossible for true love to "exert its powers between two people who are married to each other."[6]

In twelfth-century France, Andreas Capellanus, chaplain to Countess Marie of Troyes, wrote a treatise on the principles of courtly love. The first rule was that "marriage is no real excuse for not loving." But he meant loving someone outside the marriage. As late as the eighteenth century the French essayist Montaigne wrote that any man who was in love with his wife was a man so dull that no one else could love him.[7]

Courtly love probably loomed larger in literature than in real life. But for centuries, noblemen and kings fell in love with courtesans rather than the wives they married for political reasons. Queens and noblewomen had to be more discreet than their husbands, but they too looked beyond marriage for love and intimacy.

This sharp distinction between love and marriage was common among the lower and middle classes as well. Many of the songs and stories popular among peasants in medieval Europe mocked married love.

The most famous love affair of the Middle Ages was that of Peter Abelard, a well-known theologian in France, and Héloïse, the brilliant niece of a fellow churchman at Notre Dame. The two eloped without marrying, and she bore him a child. In an attempt to save his career but still placate Héloïse's furious uncle, Abelard proposed they marry in secret. This would mean that Héloïse would not be living in sin, while Abelard could still pursue his church ambitions. But Héloïse resisted the idea, arguing that marriage would not only harm his career but also undermine their love.[8] . . .

"Happily Ever After"

Through most of the past, individuals hoped to find love, or at least "tranquil affection," in marriage.[9] But nowhere did they have the same recipe for marital happiness that prevails in most contemporary Western countries. Today there is general agreement on what it takes for a couple to live "happily ever after." First, they must love each other deeply and choose each other unswayed by outside pressure. From then on, each must make the partner the top priority in life, putting that relationship above any and all competing ties. A husband and wife, we believe, owe their highest obligations and deepest loyalties to each other and the children they raise. Parents and in-laws should not be allowed to interfere in the marriage. Married couples should be best friends, sharing their most intimate feelings and secrets. They should express affection openly but also talk candidly about problems. And of course they should be sexually faithful to each other.

This package of expectations about love, marriage, and sex, however, is extremely rare. When we look at the historical record around the world, the customs of modern America and Western Europe appear exotic and exceptional. . . .

About two centuries ago Western Europe and North America developed a whole set of new values about the way to organize marriage and sexuality, and many of these values are now spreading across the globe. In this Western model, people expect marriage to satisfy more of their psychological and social needs than ever before. Marriage is supposed to be free of the coercion, violence, and gender inequalities that were tolerated in the past. Individuals want marriage to meet most of their needs for intimacy and affection and all their needs for sex.

Never before in history had societies thought that such a set of high expectations about marriage was either realistic or desirable. Although many Europeans and Americans found tremendous joy in building their relationships around these values, the adoption of these unprecedented goals for marriage had unanticipated and revolutionary consequences that have since come to threaten the stability of the entire institution.

The Era of Ozzie and Harriet: The Long Decade of "Traditional" Marriage

The long decade of the 1950s, stretching from 1947 to the early 1960s in the United States and from 1952 to the late 1960s in Western Europe, was a unique moment in the history of marriage. Never before had so many people shared the experience of courting their own mates, getting married at will, and setting up their own households. Never had married couples been so independent of extended family ties and community groups. And never before had so many people agreed that only one kind of family was "normal."

The cultural consensus that everyone should marry and form a male breadwinner family was like a steamroller that crushed every alternative view. By the end of the 1950s even people who had grown up in completely different family systems had come to believe that universal marriage at a young age into a male breadwinner family was the traditional and permanent form of marriage.

In Canada, says historian Doug Owram, "every magazine, every marriage manual, every advertisement . . . assumed the family was based on the . . . male wage-earner and the child-rearing, home-managing housewife." In the United States, marriage was seen as the only culturally acceptable route to adulthood and independence. Men who chose to remain bachelors were branded "narcissistic," "deviant," "infantile," or "pathological." Family advice expert Pat Landes argued that practically everyone, "except for the sick, the badly crippled, the deformed, the emotionally warped and the mentally defective," ought to marry. French anthropologist Martine Segalen writes that in Europe the postwar period was characterized by the overwhelming "weight of a single family model." Any departure from this model—whether it was late marriage, nonmarriage, divorce, single motherhood, or even delayed childbearing—was considered deviant. Everywhere psychiatrists agreed and the mass media affirmed that if a woman did not find her ultimate fulfillment in homemaking, it was a sign of serious psychological problems.[10]

A 1957 survey in the United States reported that four out of five people believed that anyone who preferred to remain single was "sick," "neurotic" or "immoral." Even larger majorities agreed that once married, the husband should be the breadwinner and the wife should stay home. As late as 1962 one survey of young women found that almost all expected to be married by age twenty-two, most hoped to have four children, and all expected to quit work permanently when the first child was born.[11]

During the 1950s even women who had once been political activists, labor radicals, or feminists—people like my own mother, still proud of her work to free the Scottsboro Boys from legal lynching in the 1930s and her job in the shipyards during the 1940s—threw themselves into homemaking. It's hard for anyone under the age of sixty to realize how profoundly people's hunger for marriage and domesticity during the 1950s was shaped by their huge relief that two decades of depression and war were finally over and by their amazed delight at the benefits of the first real mass consumer economy in history. "It was like a miracle," my mother once told me, to see so many improvements, so quickly, in the quality of everyday life. . . .

This was the first chance many people had to try to live out the romanticized dream of a private family, happily ensconced in its own nest. They studied how the cheery husbands and wives on their favorite television programs organized their families (and where the crabby ones went wrong). They devoured articles and books on how to get the most out of marriage and their sex lives. They were even interested in advertisements that showed them how to use home appliances to make their family lives better. . . .

Today strong materialist aspirations often corrode family bonds. But in the 1950s, consumer aspirations were an integral part of constructing the postwar family. In its April 1954 issue, *McCall's* magazine heralded the era of "togetherness," in which men and women were constructing a "new and warmer way of life . . . as a family sharing a common experience." In women's magazines that togetherness was always pictured in a setting filled with modern appliances and other new consumer products. The essence of modern life, their women readers learned, was "abundance, emancipation, social progress, airy house, healthy children, the refrigerator, pasteurised milk, the washing-machine, comfort, quality and accessibility."[12] And of course marriage.

Television also equated consumer goods with family happiness. Ozzie and Harriet hugged each other in front of their Hotpoint appliances. A man who had been a young father in the 1950s told a student of mine that he had no clue how to cultivate the family "togetherness" that his wife kept talking about until he saw an episode of the sitcom *Leave It to Beaver*, which gave him the idea of washing the car with his son to get in some "father-son" time.

When people could not make their lives conform to those of the "normal" families they saw on TV, they blamed themselves—or their parents. . . . "Why didn't she clean the house in high heels and shirtwaist dresses like they did on television?"[13]

At this early stage of the consumer revolution, people saw marriage as the gateway to the good life. Americans married with the idea of quickly buying their first home, with the wife working for a few years to help accumulate the down payment or furnish it with the conveniences she would use once she became a full-time housewife. People's newfound spending money went to outfit their homes and families. In the five years after World War II, spending on food in the United States rose by a modest 33 percent and clothing expenditures by only 20 percent, but purchases of household furnishings and appliances jumped by 240 percent. In 1961, Phyllis Rosenteur, the author of an American advice book for single women, proclaimed: "Merchandise plus Marriage equals our economy."[14]

In retrospect, it's astonishing how confident most marriage and family experts of the 1950s were that they were witnessing a new stabilization of family life and marriage. The idea that marriage should provide both partners with sexual gratification, personal intimacy, and self-fulfillment was taken to new heights in that decade. Marriage was the place not only where people expected to find the deepest meaning in their lives but also where they would have the most fun. Sociologists noted that a new "fun morality," very different "from the older 'goodness morality,'" pervaded society. "Instead of feeling guilty for having too much fun, one is inclined to feel ashamed if one does not have enough." A leading motivational researcher of the day argued that the challenge for a consumer society was "to demonstrate that the hedonistic approach to life is a moral, not an immoral, one."[15]

But these trends did not cause social commentators the same worries about the neglect of societal duties that milder ideas about the pleasure principle had triggered in the 1920s. Most 1950s sociologists weren't even troubled by the fact that divorce rates were *higher* than they had been in the 1920s, when such rates had been said to threaten the very existence of marriage. The influential sociologists Ernest Burgess and Harvey Locke wrote matter-of-factly that "the companionship family relies upon divorce as a means of rectifying a mistake in mate selection." They expressed none of the panic that earlier social scientists had felt when they first realized divorce was a permanent feature of the love-based marital landscape. Burgess and Locke saw a small amount of divorce as a safety valve for the "companionate" marriage and expected divorce rates to stabilize or decrease in the coming decades as "the services of family-life education and marriage counseling" became more widely available.[16]

The marriage counseling industry was happy to step up to the plate. By the 1950s Paul Popenoe's American Institute of Family Relations employed thirty-seven counselors and claimed to have helped twenty thousand people become "happily adjusted" in their marriages. "It doesn't require supermen or superwomen to succeed in marriage," wrote Popenoe in a 1960 book on saving marriages. "Success can be attained by almost anyone."[17]

There were a few dissenting voices. American sociologist Robert Nisbet warned in 1953 that people were loading too many "psychological and symbolic functions" on the nuclear family, an institution too fragile to bear such weight. In the same year, Mirra Komarovsky decried the overspecialization of gender roles in American marriage and its corrosive effects on women's self confidence.[18]

But even when marriage and family experts acknowledged that the male breadwinner family created stresses for women, they seldom supported any change in its division of labor. The world-renowned American sociologist Talcott Parsons recognized that because most women were not able to forge careers, they might feel a need to attain status in other ways. He suggested that they had two alternatives. The first was to be a "glamour girl" and exert sexual sway over men. The second was to develop special expertise in "humanistic" fields, such as the arts or community volunteer work. The latter, Parsons thought, was socially preferable, posing less of a threat to society's moral standards and to a woman's own self-image as she aged. He never considered the third alternative: that women might actually win access to careers. Even Komarovsky advocated nothing more radical than expanding part-time occupations to give women work that didn't interfere with their primary role as wives and mothers.[19]

Marriage counselors took a different tack in dealing with housewives' unhappiness. Popenoe wrote dozens of marital advice books, pamphlets, and syndicated newspaper columns, and he pioneered the *Ladies' Home Journal* feature "Can This Marriage Be Saved?," which was based on case histories from his Institute of Family Relations. The answer was almost always yes, so long as the natural division of labor between husbands and wives was maintained or restored. . . .

In retrospect, the confidence these experts expressed in the stability of 1950s marriage and gender roles seems hopelessly myopic. Not only did divorce rates during the 1950s never drop below the highs reached in 1929, but as early as 1947 the number of women entering the labor force in the United States had begun to surpass the number of women leaving it.[20] Why were the experts so optimistic about the future of marriage and the demise of feminism?

Some were probably unconsciously soothed into complacency by the mass media, especially the new television shows that delivered nightly images of happy female homemakers in stable male breadwinner families. . . .

When divorce did occur, it was seen as a failure of individuals rather than of marriage. One reason people didn't find fault with the 1950s model of marriage and gender roles was that it was still so new that they weren't sure they were doing it right. Millions of people in Europe and America were looking for a crash course on how to attain the modern marriage. Confident that "science" could solve their problems, couples turned not just to popular culture and the mass media but also to marriage experts and advice columnists for help. If the advice didn't work, they blamed their own inadequacy.[21] . . .

At every turn, popular culture and intellectual elites alike discouraged women from seeing themselves as productive members of society. In 1956 a *Life* magazine article commented that women "have minds and should use them . . . so long as their primary interest is in the home." . . . Adlai Stevenson, the two-time Democratic Party candidate for president of the United States, told the all-female graduating class of Smith College that "most of you" are going to assume "the humble role of housewife," and "whether you like the idea or not just now," later on "you'll like it."[22]

Under these circumstances, women tried their best to "like it." By the mid-1950s American advertisers reported that wives were using housework as a way to express their individuality. It appeared that Talcott Parsons was right: Women were compensating for their lack of occupational status by expanding their role as consumer experts and arbiters of taste and style. First Lady Jackie Kennedy was the supreme exemplar of this role in the early 1960s.[23]

Youth in the 1950s saw nothing to rebel against in the dismissal of female aspirations for independence. The number of American high school students agreeing that it would be good "if girls could be as free as boys in asking for dates" fell from 37 percent in 1950 to 26 percent in 1961, while the percentage of those who thought it would be good for girls to share the expenses of dates declined from 25 percent to 18 percent. The popular image was that only hopeless losers would engage in such egalitarian behavior. A 1954 Philip Morris ad in the *Massachusetts Collegian* made fun of poor Finster, a boy who finally found a girl who shared his belief in "the equity of Dutch treat." As a result, the punch line ran, "today Finster goes everywhere and shares expenses fifty-fifty with Mary Alice Hematoma, a lovely three-legged girl with side-burns."[24]

No wonder so many social scientists and marriage counselors in the 1950s thought that the instabilities associated with the love-based "near-equality" revolution in gender roles and marriage had been successfully contained. Married women were working outside the home more often than in the past, but they still identified themselves primarily as housewives. Men seemed willing to support women financially even in the absence of their older patriarchal rights, as long as their meals were on the table and their wives kept themselves attractive. Moreover, although men and women aspired to personal fulfillment in marriage, most were willing to stay together even if they did not get it. Sociologist Mirra Komarovsky interviewed working-class couples at the end of the 1950s and found that "slightly less than one-third [were] happily or very happily married." In 1957, a study of a cross section of all social classes found that only 47 percent of U.S. married couples described themselves as "very happy." Although the proportion of "very happy" marriages was lower in 1957 than it was to be in 1976, the divorce rate was also lower.[25]

What the experts failed to notice was that this stability was the result of a unique moment of equilibrium in the expansion of economic, political, and personal options. Ironically, this one twenty-year period in the history of the love-based "near-equality" marriage when people stopped predicting disaster turned out to be the final lull before the long-predicted storm.

The seeming stability of marriage in the 1950s was due in part to the thrill of exploring the new possibilities of married life and the size of the rewards that men and women received for playing by the rules of the postwar economic boom. But it was also due to the incomplete development of the "fun morality" and the consumer revolution. There were still many ways of penalizing nonconformity, tamping down aspirations, and containing discontent in the 1950s.

One source of containment was the economic and legal dependence of women. Postwar societies continued the century-long trend toward increasing women's legal and political rights outside the home and restraining husbands from exercising heavy-handed patriarchal power, but they stopped short of giving wives equal authority with their husbands. Legal scholar Mary Ann Glendon points out that right up until the 1960s, "nearly every legislative attempt to regulate the family decision-making process gave the husband and father the dominant role."[26]

Most American states retained their "head and master" laws, giving husbands the final say over questions like whether or not the family should move. Married women couldn't take out loans or credit cards in their own names. Everywhere in Europe and North America it was perfectly legal to pay women less than men for the same work. Nowhere was it illegal for a man to force his wife to have sex. One legal scholar argues that marriage law in the 1950s had more in common with the legal codes of the 1890s than the 1990s.[27]

Writers in the 1950s generally believed that the old-style husband and father was disappearing and that this was a good thing. The new-style husband, said one American commentator, was now "partner in the family firm, part-time man, part-time mother and part-time maid." Family experts and marital advice columnists advocated a "fifty-fifty design for living," emphasizing that a husband should "help out" with child rearing and make sure that sex with his wife was "mutually satisfying."[28]

But the 1950s definition of fifty-fifty would satisfy few modern couples. Dr. Benjamin Spock, the famous parenting advice expert, called for men to get more involved in parenting but added that he wasn't suggesting equal involvement. "Of course I don't mean that the father has to give just as many bottles, or change just as many diapers as the mother," he explained in a 1950s edition of his perennial bestseller *Baby and Child Care.* "But it's fine for him to do these things occasionally. He might make the formula on Sunday."[29]

The family therapist Paul Popenoe was equally cautious in his definition of what modern marriage required from the wife. A wife should be "sympathetic with her husband's work and a good listener," he wrote. But she must never consider herself "enough of an expert to criticize him."[30] . . .

Many 1950s men did not view male breadwinning as a source of power but as a burdensome responsibility made worthwhile by their love for their families. A man who worked three jobs to support his family told interviewers, "Although I am somewhat tired at the moment, I get pleasure out of thinking the family is dependent on me for their income." Another described how anxious he had been to finish college and "get to . . . acting as a husband and father should, namely, supporting my family." Men also remarked on how wonderful it felt to be able to give their children things their families had been unable to afford when they were young.[31]

A constant theme of men and women looking back on the 1950s was how much better their family lives were in that decade than during the Depression and World War II. But in assessing their situation against a backdrop of such turmoil and privation, they had modest expectations of comfort and happiness, so they were more inclined to count their blessings than to measure the distance between their dreams and their real lives.

Modest expectations are not necessarily a bad thing. Anyone who expects that marriage will always be joyous, that the division of labor will always be fair, and that the earth will move whenever you have sex is going to be often disappointed. Yet it is clear that in many 1950s marriages, low expectations could lead people to put up with truly terrible family lives.

Historian Elaine Tyler May comments that in the 1950s "the idea of 'working marriage' was one that often included constant day-to-day misery for one or both partners." Jessica Weiss recounts interviews conducted over many years in the Berkeley study with a woman whose husband beat her and their children. The wife often threw her body between her husband and the young ones, taking the brunt of the violence on herself because "I can take it much easier than the kids can." Her assessment of the marriage strikes the modern observer as a masterpiece of understatement: "We're really not as happy as we should be." She was not even indignant that her neighbors rebuffed her children when they fled the house to summon help. "I can't say I blame the neighbors," she commented. "They didn't want to get involved." Despite two decades of such violence, this woman did not divorce until the late 1960s.[32]

A 1950s family that looked well functioning to the outside world could hide terrible secrets. Both movie star Sandra Dee and Miss America of 1958, Marilyn Van Derbur, kept silent about their fathers' incestuous abuse until many years had passed. If they had gone public in the 1950s or early 1960s, they might not even have been believed. Family "experts" of the day described incest as a "one-in-a-million occurrence," and many psychiatrists claimed that women who reported incest were simply expressing their own oedipal fantasies.[33]

In many states and countries a nonvirgin could not bring a charge of rape, and everywhere the idea that a man could rape his own wife was still considered absurd. Wife beating was hardly ever treated seriously. The trivialization of family violence was epitomized in a 1954 report of a Scotland Yard commander that "there are only about twenty murders a year in London and not all are serious—some are just husbands killing their wives."[34] . . .

Still, these signs of unhappiness did not ripple the placid waters of 1950s complacency. The male breadwinner marriage seemed so pervasive and popular that social scientists decided it was a necessary and inevitable result of modernization. Industrial societies, they argued, needed the division of labor embodied in the male breadwinner nuclear family to compensate for their personal demands of the modern workplace. The ideal family—or what Talcott Parsons called "the normal" family—consisted of a man who specialized in the practical, individualistic activities needed for subsistence and a woman who took care of the emotional needs of her husband and children.[35]

The close fit that most social scientists saw between the love-based male breadwinner family and the needs of industrial society led them to anticipate that this form of marriage would accompany the spread of industrialization across the globe and replace the wide array of other marriage and family systems in traditional societies. This view was articulated in a vastly influential 1963 book titled *World Revolution and Family Patterns,* by American sociologist William F. Goode. Goode's work became the basis for almost all high school and college classes on family life in the 1960s, and his ideas were popularized by journalists throughout the industrial world.[36]

Goode surveyed the most up-to-date family data in Europe and the United States, the Middle East, sub-Saharan Africa, India, China, and Japan and concluded that countries everywhere were evolving toward a conjugal family system characterized by the "love pattern" in mate selection. The new international marriage system, he said, focused people's material and psychic investments on the nuclear family and increased the "emotional demands which each spouse can legitimately make upon each other," elevating loyalty to spouse above obligations to parents. Goode argued that such ideals would inevitably eclipse other forms of marriage, such as polygamy. Monogamous marriage would become the norm all around the world.

The ideology of the love-based marriage, according to Goode, "is a radical one, destructive of the older traditions in almost every society." It "proclaims the right of the individual to choose his or her own spouse. . . . It asserts the worth of the *individual* as against the inherited elements of wealth or ethnic group." As such, it especially appealed "to intellectuals, young people, women, and the disadvantaged." . . .

Despite women's legal gains and the "radical" appeal of the love ideology to women and youth, Goode concluded that a destabilizing "full equality" was not in the cards. Women had not become more "career-minded" between 1900 and the early 1960s, he said. In his 380-page survey of world trends, Goode did not record even one piece of evidence to suggest that women might become more career-minded in the future.

Most social scientists agreed with Goode that the 1950s family represented the wave of the future. They thought that the history of marriage had in effect reached its culmination in Europe and North America and that the rest of the world would soon catch up. As late as 1963 nothing seemed more obvious to most family experts and to the general public than the preeminence of marriage in people's lives and the permanence of the male breadwinner family.

But clouds were already gathering on the horizon.

When sustained prosperity turned people's attention from gratitude for survival to a desire for greater personal satisfaction . . .

When the expanding economy of the 1960s needed women enough to offer them a living wage . . .

When the prepared foods and drip-dry shirts that had eased the work of homemakers also made it possible for men to live comfortable, if sloppy bachelor lives . . .

When the invention of the birth control pill allowed the sexualization of love to spill over the walls of marriage . . .

When the inflation of the 1970s made it harder for a man to be the sole breadwinner for a family . . .

When all these currents converged, the love-based male-provider marriage would find itself buffeted from all sides.

Notes

1. Quoted in John Jacobs, *All You Need Is Love and Other Lies About Marriage* (New York: HarperCollins, 2004), p. 9.

2. William Jankowiak and Edward Fischer, "A Cross-Cultural Perspective on Romantic Love," *Ethnology* 31 (1992).

3. Ira Reiss and Gary Lee, *Family Systems in America* (New York: Holt, Rinehart and Winston, 1988), pp. 91–93.

4. Karen Dion and Kenneth Dion, "Cultural Perspectives on Romantic Love," *Personal Relationships* 3 (1996); Vern Bullough, "On Being a Male in the Middle Ages," in Clare Less, ed., *Medieval*

Masculinities (Minneapolis: University of Minnesota Press, 1994); Hans-Werner Goetz, *Life in the Middle Ages, from the Seventh to the Thirteenth Century* (Notre Dame, Ind.: University of Notre Dame Press, 1993).

5. Francis Hsu, "Kinship and Ways of Life," in Hsu, ed., *Psychological Anthropology* (Cambridge, U.K.: Schenkman, 1972), and *Americans and Chinese: Passage to Differences* (Honolulu: University Press of Hawaii, 1981); G. Robina Quale, *A History of Marriage Systems* (Westport, Conn.: Greenwood Press, 1988); Marilyn Yalom, "Biblical Models," in Yalom and Laura Carstensen, eds., *Inside the American Couple* (Berkeley: University of California Press, 2002).

6. Andreas Capellanus, *The Art of Courtly Love* (New York: W. W. Norton, 1969), pp. 106–07.

7. Ibid., pp.106–07, 184. On the social context of courtly love, see Theodore Evergates, ed., *Aristocratic Women in Medieval France* (Philadelphia: University of Pennsylvania Press, 1999); Montaigne, quoted in Olwen Hufton, *The Prospect Before Her: A History of Women in Western Europe, 1500–1800* (New York: Alfred A. Knopf, 1996), p. 148.

8. Betty Radice, trans., *Letters of Abelard and Heloise* (Harmondsworth, U.K.: Penguin, 1974).

9. The phrase is from Chiara Saraceno, who argues that until the end of the nineteenth century, Italian families defined love as the development of such feelings over the course of a marriage. Saraceno, "The Italian Family," in Antoine Prost and Gerard Vincent, eds., *A History of Private Life: Riddles of Identity in Modern Times* (Cambridge, Mass.: Belknap Press, 1991), p. 487.

10. Owram, *Born at the Right Time*, p. 22 (see chap. 13, n. 20); Elaine Tyler May, *Homeward Bound: American Families in the Cold War Era* (New York: Basic Books, 1988); Barbara Ehrenreich, *The Hearts of Men: American Dreams and the Flight from Commitment* (Garden City, N.Y.: Anchor Press, 1983), pp. 14–28; Douglas Miller and Marson Nowak, *The Fifties: The Way We Really Were* (Garden City, N.Y.: Doubleday, 1977), p. 154; Duchen, *Women's Rights* (see chap. 13, n. 28); Marjorie Ferguson, *Forever Feminine: Women's Magazines and the Cult of Femininity* (London: Heinemann, 1983); Moeller, *Protecting Motherhood* (see chap. 13, n. 22); Martine Segalen, "The Family in the Industrial Revolution," in Burguière et al., p. 401 (see chap. 8, n. 2).

11. Daniel Yankelovich, *New Rules: Searching for Self-Fulfillment in a World Turned Upside Down* (New York: Random House, 1981); Lois Gordon and Alan Gordon, *American Chronicle: Seven Decades in American Life, 1920–1989* (New York: Crown, 1990).

12. Alan Ehrenhalt, *The Lost City: Discovering the Forgotten Virtues of Community in the Chicago of the 1950s* (New York: Basic Books, 1995), p. 233; modernity quote from the French woman's magazine *Marie-Claire*, in Duchen, *Women's Rights and Women's Lives*, p. 73 (see chap. 13, n. 28).

13. Quoted in Ruth Rosen, *The World Split Open: How the Modern Women's Movement Changed America* (New York: Viking, 2000), p. 44.

14. Coontz, *The Way We Never Were*, p. 25; Rosenteur, quoted in Bailey, *From Front Porch to Back Seat*, p. 76 (see chap. 12, n. 11).

15. Martha Wolfenstein, "Fun Morality" [1955], in Warren Susman, ed., *Culture and Commitment, 1929–1945* (New York: George Braziller, 1973), pp. 84, 90; Coontz, *The Way We Never Were*, p. 171.

16. Ernest Burgess and Harvey Locke, *The Family: From Institution to Companionship* (New York: American Book Company, 1960), pp. 479, 985, 538.

17. Molly Ladd-Taylor, "Eugenics, Sterilisation and Modern Marriage in the USA," *Gender & History* 13 (2001), pp. 312, 318.

18. Nisbet, quoted in John Scanzoni, "From the Normal Family to Alternate Families to the Quest for Diversity with Interdependence," *Journal of Family Issues* 22 (2001); Mirra Komarovsky, *Women in the Modern World: Their Education and Their Dilemmas* (Boston: Little, Brown, 1953).

19. Talcott Parsons, "The Kinship System of the United States" in Parsons, *Essays in Sociological Theory* (Glencoe, Ill.: Free Press, 1954); Parsons and Robert Bales, *Family, Socialization, and Interaction Processes* (Glencoe, Ill.: Free Press, 1955).

20. *Historical Statistics of the United States: Colonial Times to the Present* (Washington, D.C.: U.S. Department of Commerce, Bureau of the Census, 1975); Sheila Tobias and Lisa Anderson, "What Really Happened to Rosie the Riveter," *Mss Modular Publications* 9 (1973).

21. Beth Bailey, "Scientific Truth . . . and Love: The Marriage Education Movement in the United States," *Journal of Social History* 20 (1987).

22. Miller and Nowak, *The Fifties*, pp. 164–65; Weiss, *To Have and to Hold*, p. 19 (see chap. 13, n. 29); Rosen, *World Split Open*, p. 41.

23. Glenna Mathews, "*Just a Housewife*": *The Rise and Fall of Domesticity in America* (New York:

Oxford University Press, 1987); Betty Friedan, *The Feminine Mystique* (New York: Dell, 1963).

24. Bailey, *From Front Porch to Back Seat,* p. 111.

25. Mirra Komarovsky, *Blue-Collar Marriage* (New Haven: Vintage, 1962), p. 331. Mintz and Kellogg, *Domestic Revolutions,* p. 194; Norval Glenn, "Marital Quality," in David Levinson, ed., *Encyclopedia of Marriage and the Family* (New York: Macmillan, 1995), vol. 2, p. 449.

26. Mary Ann Glendon, *The Transformation of Family Law* (Chicago: University of Chicago Press, 1989), p. 88. On Europe, Gisela Bock, *Women in European History* (Oxford, U.K.: Blackwell Publishers, 2002), p. 248; Bonnie Smith, *Changing Lives: Women in European History Since 1700* (Lexington, Mass.: D. C. Heath, 1989), p. 492.

27. Sara Evans, *Tidal Wave: How Women Changed America at Century's End* (New York: Free Press, 2003), pp. 1–20; John Ekelaar, "The End of an Era?," *Journal of Family History* 28 (2003), p. 109. See also Lenore Weitzman, *The Marriage Contract* (New York: Free Press, 1981).

28. Ehrenhalt, *Lost City,* p. 233.

29. Quoted in Michael Kimmell, *Manhood in America: A Cultural History* (New York: Free Press, 1996), p. 246.

30. Ladd-Taylor, "Eugenics," p. 319.

31. Ibid., p. 32; Robert Rutherdale, "Fatherhood, Masculinity, and the Good Life During Canada's Baby Boom," *Journal of Family History* 24 (1999), p. 367.

32. May, *Homeward Bound,* p. 202; Weiss, *To Have and to Hold,* pp. 136–38.

33. Marilyn Van Derbur Atler, "The Darkest Secret," *People* (June 10, 1991); Dodd Darin, *The Magnificent Shattered Life of Bobby Darin and Sandra Dee* (New York: Warner Books, 1995); Elizabeth Pleck, *Domestic Tyranny* (New York: Oxford University Press, 1987); Linda Gordon, *Heroes of Their Own Lives: The Politics and History of Family Violence, 1880–1960* (New York: Viking, 1988).

34. Coontz, *The Way We Never Were,* p. 35; Leonore Davidoff et al., *The Family Story* (London: Longmans, 1999), p. 215.

35. Parsons, "The Kinship System of the United States"; Parsons, "The Normal American Family," in Seymour Farber, Piero Mustacchi, and Roger Wilson, eds., *Man and Civilization: The Family's Search for Survival* (New York: McGraw-Hill, 1965); Parsons and Bales, *Family, Socialization, and Interaction Processes.* For similar theories in British sociology, see Michael Young and Peter Willmott's *The Symmetrical Family* (London: Pelican, 1973), pp. 28–30; *Family and Kinship in East London* (Glencoe, Ill.: The Free Press, 1957); and *Family and Class in a London Suburb.*

36. The quotations and figures in this and the following paragraphs are from Goode, *World Revolution.*

THINKING ABOUT THE READING

According to Coontz, if not for romance, what are some of the common reasons throughout history that people marry? What are the characteristics of the "male breadwinner, love-based marriage"? What social conditions are necessary for this kind of family arrangement to prevail? Does Coontz think this family form is viable in the long-term future? Do you?

Gay Parenthood and the End of Paternity as We Knew It

Judith Stacey

(2011)

Because let's face it, if men weren't always hungry for it, nothing would ever happen. There would be no sex, and our species would perish.

—Sean Elder, "Why My Wife Won't Sleep With Me," 2004

Because homosexuals are rarely monogamous, often having as many as three hundred or more partners in a lifetime—some studies say it is typically more than one thousand—children in those polyamorous situations are caught in a perpetual coming and going. It is devastating to kids, who by their nature are enormously conservative creatures.

—James Dobson, "Same-Sex Marriage Talking Points"

Unlucky in love and ready for a family, [Christie] Malcomson tried for 4½ years to get pregnant, eventually giving birth to the twins when she was 38. Four years later, again without a mate, she had Sarah. "I've always known that I was meant to be a mother," Malcomson, 44, said. "I tell people, I didn't choose to be a single parent. I choose to be a parent."

—Lornet Turnbull, "Family Is . . . Being Redefined All the Time," 2004

Gay fathers were once as unthinkable as they were invisible. Now they are an undeniable part of the contemporary family landscape. During the same time that the marriage promotion campaign in the United States was busy convincing politicians and the public to regard rising rates of fatherlessness as a national emergency (Stacey 1998), growing numbers of gay men were embracing fatherhood. Over the past two decades, they have built a cornucopia of family forms and supportive communities where they are raising children outside of the conventional family. Examining the experiences of gay men who have openly pursued parenthood against the odds can help us to understand forces that underlie the decline of paternity as we knew it. Contrary to the fears of many in the marriage-promotion movement, however, gay parenting is not a new symptom of the demise of fatherhood, but of its creative, if controversial, reinvention. When I paid close attention to gay men's parenting desires, efforts, challenges, and achievements, I unearthed crucial features of contemporary paternity and parenthood more generally. I also came upon some inspirational models of family that challenge widely held beliefs about parenthood and child welfare.

The Uncertainty of Paternity

Access to effective contraception, safe abortions, and assisted reproductive technologies (ART) unhitches traditional links between heterosexual love, marriage, and baby carriages. Parenthood, like intimacy more generally, is now contingent. Paths to parenthood no longer appear so natural, obligatory, or uniform as they used to but have become voluntary, plural, and politically embattled. Now that children impose immense economic and social responsibilities on their parents, rather than promising to become a reliable source of family labor or social security, the pursuit of parenthood depends on an emotional rather than an economic calculus. "The men and women who decide to have children today," German sociologists Ulrich Beck and Elisabeth Beck-Gernsheim correctly point out, "certainly do not do so because they expect any material advantages. Other motives closely linked with the emotional needs of the parents play a significant role; our children mainly have 'a psychological utility'" (Beck and Beck-Gernsheim 1995:105). Amid the threatening upheavals, insecurities, and dislocations of life under global market and military forces, children can rekindle opportunities for hope, meaning, and connection. Adults who wish to become parents today typically seek the intimate bonds that children seem to promise. More reliably than a lover or spouse, parenthood beckons to many who hunger for lasting love, intimacy, and kinship—for that elusive "haven in a heartless world" (Lasch 1995).

Gay men confront these features of postmodern parenthood in a magnified mode. They operate from cultural premises antithetical to what U.S. historian Nicholas Townsend termed "the package deal" of (now eroding) modern masculinity—marriage, work, and fatherhood (Townsend 2002). Gay men who choose to become primary parents challenge conventional definitions of masculinity and paternity and even dominant sexual norms of gay culture itself.

Gay fatherhood represents "planned parenthood" in extremis. Always deliberate and often difficult, it offers fertile ground for understanding why and how people do and do not choose to become parents today. Unlike most heterosexuals or even lesbians, gay men have to struggle for access to "the means of reproduction" without benefit of default scripts for achieving or practicing parenthood. They encounter a range of challenging, risky, uncertain options—foster care, public and private forms of domestic and international adoption, hired or volunteered forms of "traditional" or gestational surrogacy, contributing sperm to women friends, relatives, or strangers who agree to co-parent with them, or even resorting to an instrumental approach to old-fashioned heterosexual copulation.

Compared with maternity, the social character of paternity has always been more visible than its biological status. Indeed, that's why prior to DNA testing, most modern societies mandated a marital presumption of paternity. Whenever a married woman gave birth, her husband was the 'presumed and legal' father. Gay male paternity intensifies this emphasis on social rather than biological definitions of parenthood. Because the available routes to genetic parenthood for gay men are formidably expensive, very difficult to negotiate, or both, most prospective gay male parents pursue the purely social paths of adoption or foster care (Brodzinsky, Patterson, and Vaziri 2002).

Stark racial, economic, and sexual asymmetries characterize the adoption marketplace. Prospective parents are primarily white, middle-class, and relatively affluent, but the available children are disproportionately from poorer and darker races and nations. Public and private adoption agencies, as well as birth mothers and fathers, generally consider married heterosexual couples to be the most desirable adoptive parents (Human Rights Campaign 2009). These favored straight married couples, for their part, typically seek healthy infants, preferably from their own race or ethnic background. Because there are not enough of these to meet the demand, most states and counties allow single adults, including gay men, to shop for parenthood in their overstocked warehouse of "hard to place" children. This is an index of expediency more than tolerance. The state's stockpiled children have been removed from parents who were judged to be negligent, abusive, or incompetent. Disproportionate numbers are children of color, and the very hardest of these to place are older boys with "special needs," such as physical, emotional, and cognitive disabilities.

The gross disjuncture between the market value of society's adoptable children and the supply of prospective adoptive parents allows gay men to parent a hefty share of them. Impressive numbers of gay men willingly rescue such children from failing or devastated families. Just as in their intimate adult relationships, gay men more readily accept children across boundaries of race, ethnicity, class, and even health.

The multi-racial membership of so many of gay men's families visually signals the social character of most gay fatherhood. In addition, as we will see, some gay men willingly unhitch their sexual and romantic desires from their domestic ones in order to become parents. For all of these reasons, gay men provide frontier terrain for exploring noteworthy changes in the meanings and motives for paternity and parenthood.

Finding Pop Luck in the City of Angels

Gay paternity is especially developed and prominent in L.A.—again, not the environment where most people would expect to find it, but which, for many reasons, became a multiethnic mecca for gay parenthood. According to data reported in Census 2000, both the greatest number of same-sex couple households in the United States and of such couples who were raising children were residing in Los Angeles County (Sears and Badgett 2004). It is likely, therefore, that the numbers there exceeded those of any metropolis in the world.

Local conditions in Los Angeles have been particularly favorable for gay and lesbian parenthood. L.A. County was among the first in the United States to openly allow gay men to foster or adopt children under its custody, and numerous local, private adoption agencies, lawyers, and services emerged that specialized in facilitating domestic and international adoptions for a gay clientele. In 2001 California enacted a domestic-partnership law that authorized second-parent adoptions, and several family-court judges in California pioneered the still-rare practice of granting prebirth custody rights to same-sex couples who planned to co-parent. The City of Angels became the surrogacy capital of the gay globe, thanks especially to Growing Generations, the world's first gay- and lesbian-owned professional surrogacy agency founded to serve an international clientele of prospective gay parents (Strah and Margolis 2003).

The gay men I studied were among the first cohort of gay men young enough to even imagine parenthood outside heterosexuality and mature enough to be in a position to choose or reject it. I intentionally over-sampled for gay fathers. Nationally 22 percent of male same-sex-couple households recorded in Census 2000 included children under the age of eighteen (Simmons and O'Connell 2003:10). However, fathers composed half of my sample overall and more than 60 percent of the men who were then in same-sex couples. Depending on which definition of fatherhood one uses, between twenty-four and twenty-nine of my fifty primary interviewees were fathers of thirty-five children, and four men who were not yet parents declared their firm intention to become so.[1] Only sixteen men in contrast, depicted themselves as childless more or less by choice. Also by design, I sampled to include the full gamut of contemporary paths to gay paternity. Although most children with gay fathers in the United States were born within heterosexual marriages before their fathers came out, this was true for only six of the thirty-four children that the men in my study were raising. All of the others were among the pioneer generation of children with gay dads who chose to parent after they had come out of the closet. Fifteen of the children had been adopted (or were in the process of becoming so) through county and private agencies or via independent, open adoption agreements with birth mothers; four were foster-care children; five children had been conceived through surrogacy contracts, both gestational and "traditional";

and four children had been born to lesbians who conceived with sperm from gay men with whom they were co-parenting. In addition, five of the gay men in my study had served as foster parents to numerous teenagers, and several expected to continue to accept foster placements. Two men, however, were biological but not social parents, one by intention, the other unwittingly.[2]

The fathers and children in my study were racially and socially diverse, and their families, like gay-parent families generally, were much more likely to be multi-racial and multi-cultural than are other families in the United States, or perhaps anywhere in the world. Two-thirds of the gay-father families in my study were multi-racial. The majority (fifteen) of the twenty-four gay men who were parenting during the time of my study were white, but most (twenty-one) of their thirty-four children were not.[3] Even more striking, only two of the fifteen children they had adopted by 2003 were white, both of these through open adoption arrangements with birth mothers; seven adoptees were black or mixed race, and six were Latino. In contrast, nine of the twelve adoptive parents were white, and one each was black, Latino, and Asian American.

It is difficult to assess how racially representative this is of gay men, gay parents, and their families in the city, the state, or the nation. Although the dominant cultural stereotype of gay men and gay fathers is white and middle class, U.S. Census 2000 data surprisingly report that racial minorities represented a higher proportion of same-sex-couple-parent households in California than of heterosexual married couples (Sears and Badgett 2004). The vast majority of the children in these families, however, were born within their gay parents' former heterosexual relationship (Gates 2005). Contemporary gay paths to paternity are far more varied and complex.

Predestined Progenitors

Of the men I interviewed, eighteen who had become dads and four who planned to do so portrayed their passion for parenthood in terms so ardent that I classify them as predestined parents. The following two stories illustrate typical challenges and triumphs of different paths to predestined parenthood. The first depicts another blessedly compatible and privileged couple, and the second is about a courageous, much less affluent gay man who was "single by chance, parent by choice."

Predestined Pairing

Eddie Leary and Charles Tillery, a well-heeled, white, Catholic couple, had three children born through gestational surrogacy. Their firstborn was a genetic half-sibling to a younger set of twins. The same egg donor and the same gestational surrogate conceived and bore the three children, but Charles is the genetic father of the first child, and Eddie's sperm conceived the twins. At the time I first interviewed them in 2002, their first child was three years old, the twins were infants, and the couple had been together for eighteen years. Eddie told me that they had discussed their shared desire to parent on their very first date. In fact, by then Eddie had already entered a heterosexual marriage primarily for that purpose, but he came out to his wife and left the marriage before achieving it. Eddie claimed that he always knew that he "was meant to be a parent." He recalled that during his childhood whenever adults had posed the clichéd question to him, "What do you want to be when you grow up?" his ready answer was "a daddy."

Charles and Eddie met and spent their first ten years together on the East Coast, where they built successful careers in corporate law and were gliding through the glamorous DINC (double income, no children) fast lane of life. By their mid-thirties, however, they were bored and began to ask themselves the existential question, "Is this all there is?" They had already buried more friends than their parents had by their sixties, which, Eddie believed, "gives you a sense of gravitas." In addition, he reported, "My biological clock was definitely ticking." In the mid-1990s, the couple migrated to L.A., lured by the kind of gay family life style and the ample job opportunities it seemed to offer. They spent the next five years riding an emotional roller coaster attempting to become parents. At first Eddie and Charles considered adoption, but they became discouraged when they learned that then-governor Pete Wilson's administration was preventing joint adoptions by same-sex couples. Blessed with ample financial and social resources, they decided to shift their eggs, so to speak, into the surrogacy basket. One of Charles's many cousins put the couple in touch with her college roommate, Sally, a married mother of two in her mid-thirties who lived in Idaho. Sally was a woman who loved both bearing and rearing children, and Charles's cousin knew that she had been fantasizing about bestowing the gift of parenthood on a childless couple. Although Sally's imaginary couple had not been gay, she agreed to meet them. Eddie and Sally both reported that they bonded instantly, and she agreed to serve as the men's gestational surrogate.

To secure an egg donor and manage the complex medical and legal processes that surrogacy requires at a moment just before Growing Generations had opened shop, Eddie and Charles became among the first gay clients of a surrogate parenthood agency that mainly served infertile heterosexual couples. Shopping for DNA in the agency's catalog of egg donors, they had selected Marya, a Dutch graduate student who had twice before served as an anonymous donor for married couples in order to subsidize her education. Marya had begun to long for maternity herself, however, and she was loathe to subject her body and soul yet again to the grueling and hormonally disruptive process that donating ova entails. Yet when she learned that the new candidates for her genes were gay men, she found herself taken with the prospect of openly aiding such a quest. Like Sally, she felt an immediate affinity with Eddie and agreed to enter a collaborative egg-donor relationship with him and Charles. When she had served as egg donor for infertile married couples, Marya explained, "the mother there can get a little jealous and a little threatened, because she's already feeling insecure about being infertile, and having another woman having that process and threatening the mother's role, I think is a big concern." With a gay couple, in contrast, "you get to be—there's no exclusion, and there's no threatened feelings."

Because Eddie is a few years older than Charles, he wanted to be the first to provide the sperm, and all four parties were thrilled when Sally became pregnant on the second in-vitro fertilization (IVF) attempt. Elation turned to despair, however, when the pregnancy miscarried in the thirteenth week. Eddie described himself as devastated, saying, "I grieved and mourned the loss of my child, just as if I'd been the one carrying it." In fact, Sally recovered from the trauma and was willing to try again before Eddie, who said, "I couldn't bear the risk of losing another of my children." Instead, Charles wound up supplying the sperm for what became the couple's firstborn child, Heather. Two years later, eager for a second child, the couple had persuaded both reluctant women to subject their bodies to one more IVF surrogacy, this time with Eddie's sperm. A pair of healthy twin boys arrived one year later, with all four procreative collaborators, as well as Sally's husband, present at the delivery to welcome the boys into what was to become a remarkable, surrogacy-extended family.

Occasionally Marya, the egg donor, continued to visit her genetic daughter, but Eddie and Sally quickly developed an extraordinary, deep, familial bond. They developed the habit of daily, long-distance phone calls that were often lengthy and intimate. "Mama Sally," as Heather started to call her, began to make regular use of the Leary-Tillery guest room, accompanied sometimes by her husband and their two children. Often she came to co-parent with Eddie as a substitute for Charles, who had to make frequent business trips. The two families began taking joint vacations skiing or camping together in the Rockies, and once Marya had come along. Sally's then ten-year-old daughter and eight-year-old son began to refer to Heather as their "surrogate sister."

Eddie and Charles jointly secured shared legal custody of all three children through some of the earliest pre-birth decrees granted in California. From the start, the couple had agreed that Eddie, a gourmet cook who had designed the family's state-of-the-culinary-art kitchen, would stay home as full-time parent, and Charles would be the family's sole breadwinner. After the twins arrived, they hired a daytime nanny to assist Eddie while Charles was out earning their sustenance, and she sometimes minded the twins when Eddie and Heather joined the weekly playgroup of the Pop Luck Club (PLC), composed of at-home dads and tots. Charles, for his part, blessed with Herculean energy and scant need for sleep, would plunge into his full-scale second shift of baby feedings, diapers, baths, and bedtime storytelling the moment he returned from the office. Although Eddie admitted to some nagging concerns that he "may have committed career suicide by joining the mom's club in the neighborhood," he also believed he'd met his calling: "I feel like this is who I was meant to be."

Parent Seeking Partner

Armando Hidalgo, a Mexican immigrant, was thirty-four years old when I interviewed him in 2001. At that point, he was in the final stages of adopting his four-year-old black foster son, Ramon. Armando had been a teenage sexual migrant to Los Angeles almost twenty years earlier. He had run away from home when he was only fifteen in order to conceal his unacceptable sexual desires from his large, commercially successful, urban Mexican family. The youthful Armando had paid a coyote to help him cross the border. He had survived a harrowing illegal immigration experience which culminated in a Hollywood-style footrace across the California desert to escape an INS patrol in hot pursuit. By working at a Taco Bell in a coastal town, Armando put himself through high school. Drawing upon keen intelligence, linguistic facility, and a prodigious work ethic and drive, he had built a stable career managing a designer furniture showroom and he had managed to secure U.S. citizenship as well.

Four years after Armando's sudden disappearance from Mexico, he had returned there to come out to his family, cope with their painful reactions to his homosexuality and exile, and begin to restore his ruptured kinship bonds. He had made annual visits to his family ever since, and on one of these he fell in love with Juan, a Mexican language teacher. Armando said that he told Juan about his desire to parent right at the outset, and his new lover had seemed enthusiastic: "So, I thought we were the perfect match." Armando brought his boyfriend back to Los Angeles, and they lived together for five years.

However, when Armando began to pursue his lifelong goal of parenthood, things fell apart. To initiate the adoption process, Armando had enrolled the couple in the county's mandatory foster-care class. However, Juan kept skipping class and neglecting the homework, and so he failed to qualify for foster-parent status. This behavior jeopardized Armando's eligibility to adopt children as well as Juan's. The county then presented Armando with a "Sophie's choice." They would not place a child in his home unless Juan moved out. Despite Armando's primal passion for parenthood, "at the time," he self-critically explained to me, "I made the choice of staying with him, a choice that I regret. I chose him over continuing with my adoption." This decision ultimately exacted a fatal toll on the relationship. In Armando's eyes, Juan was preventing him from fulfilling his lifelong dream of having children. His resentment grew, but it took another couple of years before his passion for parenthood surpassed his diminishing passion for his partner. That is when Armando moved out and renewed the adoption application as a single parent.

Ramón was the first of three children that Armando told me he had "definitely decided" to adopt, whether or not he found another partner. His goal was to adopt two more children, preferably a daughter and another son, in that order. Removed at birth from crack-addicted parents, Ramón had lived in three foster homes in his first three years of life, before the county placed him with Armando through its fost-adopt program. Ramón had suffered from food allergies, anxiety, and hyperactivity when he arrived, and the social worker warned Armando to anticipate learning disabilities as well. Instead, after nine months under Armando's steady, patient, firm, and loving care, Ramón was learning rapidly and appeared to be thriving. And so was Armando. He felt so lucky to have Ramón, whom he no longer perceived as racially different from himself: "To me he's like my natural son. I love him a lot, and he loves me too much. Maybe I never felt so much unconditional love."

In fact, looking back, Armando attributed part of the pain of the years he spent struggling to accept his own homosexuality to his discomfort with gay male sexual culture and its emphasis on youth and beauty. "I think it made me fear that I was going to grow old alone," he reflected. "Now I don't have to worry that I'm gay and I'll be alone." For in addition to the intimacy that Armando savored with Ramón, his son proved to be a vehicle for building much closer bonds with most of his natal family. Several of Armando's eleven siblings had also migrated to Los Angeles. Among these were a married brother, his wife, and their children, who provided indispensable back-up support to the single working father. Ramón adored his cousins, and he and his father spent almost every weekend and holiday with them.

Ramón had acquired a devoted, long-distance *abuela* (grandmother) as well. Armando's mother had begun to travel regularly from Mexico to visit her dispersed brood, and, after years of disapproval and disappointment, she had grown to admire and appreciate her gay son above all her other children. Armando reported with sheepish pride that during a recent phone call his mother had stunned and thrilled him when she said, "You know what? I wish that all your brothers were like you. I mean that they liked guys." Astonished, Armando had asked her, "Why do you say that?" She replied, "I don't know. I just feel that you're really good to me, you're really kind. And you're such a good father." Then she apologized for how badly she had reacted when Armando told the family that he was gay, and she told him that now she was really proud of him. "'Now I don't have to accept it,'" Armando quoted her, "'because there's nothing to accept. You're natural, you're normal. You're my son, I don't have to accept you.' And she went on and on. It was so nice, it just came out of her. And now she talks about gay things, and she takes a cooking class from a gay guy and tells me how badly her gay cooking teacher was treated by his family when they found out and how unfair it is and all."

Although Armando had begun to create the family he always wanted, he still dreamt of sharing parenthood with a mate who would be more compatible than Juan: "I would really love to meet someone, to fall in love." Of course, the man of his dreams was someone family-oriented: "Now that's really important, family-oriented, because I am very close to my family. I always do family things, like my nephews' birthday parties, going to the movies with them, family dinners, etcetera. But these are things that many gay men don't like to do. If they go to a straight family party, they get bored." Consequently, Armando was pessimistic about finding a love match. Being a parent, moreover, severely constrained his romantic pursuits. He didn't want to subject Ramón, who had suffered so much loss and instability in his life, to the risk of becoming attached to yet another new parental figure who might leave him. In addition, he didn't want Ramón "to think that gay men only have casual relationships, that there's no commitment." "But," he observed, with disappointment, "I haven't seen a lot of commitment among gay men." Armando took enormous comfort, however, in knowing that even if he never found another boyfriend, he will "never really be alone": "And I guess that's one of the joys that a family brings." Disappointingly, I may never learn whether Armando found a co-parent and adopted a sister and brother for Ramón, because I was unable to locate him again in 2008.

Adopting Diversity

While Eddie, Charles, and Armando all experienced irrepressible parental yearnings, they pursued very different routes to realizing this common "destiny." Gestational surrogacy, perhaps the newest, the most high-tech, and certainly the most expensive path to gay parenthood, is available primarily to affluent couples, the overwhelming majority of whom are white men who want to have genetic progeny. Adoption, on the other hand, is one of the oldest forms of "alternative" parenthood. It involves bureaucratic and social rather than medical technologies, and the county fost-adopt program which Armando and six other men in my study employed is generally the least expensive, most accessible route to gay paternity. Like Armando, most single, gay prospective parents pursue this avenue and adopt "hard-to-place" children who, like Ramón, are often boys of color with "special needs."

The demographics of contrasting routes to gay parenthood starkly expose the race and class disparities in the market value of children. Affluent, mainly white couples, like Charles and Eddie, can purchase the means to reproduce white infants in their own image, or even an enhanced, eugenic one, by selecting egg donors who have traits they desire with whom to mate their own DNA. In contrast, for gay men who are single, less privileged, or both, public agencies offer a grab bag of displaced children who are generally older, darker, and less healthy (U.S. Department of Health and Human Services 2003; Kapp, McDonald, and Diamond 2001). Somewhere in between these two routes to gay paternity are forms of "gray market," open domestic or international adoptions, or privately negotiated sperm-donor agreements with women, especially lesbians, who want to co-parent with men. Independent adoption agencies and the Internet enable middle-class gay men, again typically white couples, to adopt newborns in a variety of hues.

Price does not always determine the route to parenthood that gay men choose, or the race, age, health, or pedigree of the children they agree to adopt. During the period of my initial research, only one white, middle-class couple in my study had chosen to adopt healthy white infants. Some affluent white men enthusiastically adopted children of color, even when they knew that the children had been exposed to drugs prenatally. Drew Greenwald, a very successful architect who could easily have afforded assisted reproductive technology (ART), was the most dramatic example of this. He claimed, "It never would have occurred to me to do surrogacy. I think it's outrageous because there are all these children who need good homes. And people have surrogacy, they say, in part it's because they want to avoid the complications of adoption, but in candor they are really in love with their own genes. . . . I just think there is a bit of narcissism about it."

Drew had opted for independent, open, transracial adoption instead. When I first interviewed him in 2002, he had just adopted his second of two multi-racial babies born to two different women who both had acknowledged using drugs during their pregnancies. Soon after adopting his first infant, Drew reunited with James, a former lover who had fallen "wildly in love" with Drew's new baby. James moved in while Drew was in the process of adopting a second child, and they have co-parented together ever since. Indeed, parenthood is the "glue" that cemented a relationship between the couple that Drew believed might otherwise have failed. Shared parenting provided them with a "joint project, a focus, and a source of commitment."

I was indulging in my guilty pleasure of reading the Style section of the Sunday *New York Times* one morning in the fall of 2008, when I stumbled across a wedding photo and announcement that Drew and James, "the parents of five adopted children," had just married. Several weeks later, on a conference trip to Los Angeles, I visited the bustling, expanded family household. I learned that the white birth mother of their second child had since had two more unwanted pregnancies, one with the same African American man as before and one with a black Latino. She had successfully appealed to Drew and James to add both of these mixed-race siblings to their family. After the first of these two new brothers had joined their brood, Drew and James began to worry that because only one of their children was a girl, she would find it difficult to grow up in a family with two dads and only brothers. And so they turned to the Internet, where they found a mixed-race sister for their first daughter. Three of the five children suffered from learning or attention-deficit difficulties, but Drew took this in stride. He was well aware, he said, that he and James had signed on "for all sorts of trauma, challenge, heartache" in the years ahead. He was both determined and financially able to secure the best help available for his children. Nonetheless, Drew acknowledged, "I fully expect that the kids will break my heart at some point in various ways, but it's so worth it" It was sufficiently worth it, apparently, that the year after my 2008 visit, I received an email from Drew announcing that their child head count had climbed to six, because their "jackpot birth mom" had given birth yet again. "We're up to four boys and two girls," Drew elaborated. "It's a lot, as you can imagine, but wonderful."

Situational Parents

Despite the fact that I over-sampled for gay parents, the majority of men in my study fell into the intermediate range on the passion-to-parent continuum. I would classify twenty-six of my fifty primary research subjects as having been situationally with or without children. Nine men whose personal desire to parent had ranged from reluctant, unenthusiastic, or indifferent to ambivalent, hesitant, or even mildly interested became situational parents after they succumbed to the persuasive entreaties of a fervently motivated mate, or if they fell in love with a man who was already a parent. Sixteen men who had remained childless expressed a similar range of sentiments, and in one case even a portion of regret. These men would have agreed to co-parent with a predestined partner or, in some cases, with even just a willing one. They had remained childless, however, either because they were single or because their partners were refuseniks or other situationists. None of them had a passion for parenthood that was potent enough to overcome the resistance of a reluctant mate or to confront alone the formidable challenges that prospective parents, and especially gay men, must meet.

Persuasive Partner

Glenn Miya, a Japanese American who was thirty-six years old when we first met, liked children enough to spend his workday life as a pediatrician. Nonetheless, he had not felt an independent desire to fill his home life with them as well. His long-term partner, Steven Llanusa, a Cuban-Italian elementary school teacher, however, was a predestined parent who, eight years into their relationship, had given Glenn an ultimatum to co-parent with him or part. Glenn's initial misgivings had been serious enough to rupture the couple's relationship for several months. Looking backward on this period, Glenn thought that he had been "suffering a bit of pre-parental panic," while Steven felt that he "was being forced to make a choice between his partner or being a parent," just the way Armando had felt. Although Steve had not wanted to face this choice, he had been determined that he "was not going to renege" on his commitment to parenthood. Fortunately for both men and, as it turns out, for the three Latino brothers whom they later adopted, couples counseling helped Glenn to work through his reservations and to reunite the couple.

Their co-parenting career began, Glenn said, by "parenting backwards." First they had signed up with a foster-care-parent program and taken in several teenagers, including one who was gay. Both the positive and negative aspects of their experiences as foster parents convinced them that they were ready to make a more permanent commitment to children. The couple's combined income was clearly sufficient to cover the expense of independent adoption, and perhaps even surrogacy, had they wished to pursue these options. Instead, however, they had enrolled in the county's fost-adopt program, choosing "very consciously to adopt elementary-school-age kids," because they believed that they could not afford to stay home as full-time parents and did not want to hire a nanny to take care of infants or toddlers. They chose, in other words, to undertake what most authorities consider to be the most difficult form of adoptive parenthood. Nor had they chosen to start, or to stop, with one "difficult-to-place" child. Rather, they had accepted first a set of seven-year-old Mexican American twin boys and their five-year-old brother soon afterward. The county had removed the three boys from drug-addicted parents. Both twins had acquired learning disabilities from fetal alcohol syndrome, and one had a prosthetic leg. All three boys had suffered parental neglect and been physically abused by their father, who was serving a prison sentence for extensive and repeated domestic violence.

Despite the formidable challenges of transracially adopting three school-age abused and neglected children with cognitive, physical, and emotional disabilities, or perhaps partly because of these facts, the Miya-Llanusa family had become a literal California poster family for gay fatherhood. Both parents and their three sons played active leadership roles in the Pop Luck Club; they all participated in public education and outreach within the gay community and beyond; they spoke frequently to the popular media; they hosted massive community and holiday parties; and they served as general goodwill ambassadors for gay and multi-cultural family values in the boys' schools, sports teams, and dance classes and in their Catholic parish and their white, upper-middle-class suburban neighborhood.

Although Steve had been the predestined parent, and Glenn initially had been a reluctant, situational one, Glenn was the one who told me that he wouldn't mind emulating Eddie Leary's pattern of staying home to parent full-time, if his family had been able to afford forgoing the ample income that his pediatric practice earned.

The Miya-Llanusa clan was still going strong and still going public with their enduring love and family story when I caught up with them again in October 2008. Love certainly had come first for this family, but it had taken twenty-two years before the state of California briefly allowed marriage to follow. In August 2008, Steve and Glenn had seized the moment and held a glorious, almost-traditional, religious and legal wedding ceremony, with all three, now teenage sons as ushers, and more than one hundred of their beaming family and friends in attendance. By then, Proposition 8 was on the California ballot, and Glenn and Steve had contributed their time, resources, and a photo-album slide show portraying the history of their love, marriage, and family to that unsuccessful political campaign to keep marriage legal for other California families like theirs.

Poly-Parent Families

Independent adoption often generates complex family ties. Many pregnant women choose this option so that they can select adoptive parents whom they like for their babies and who will maintain contact with them after the adoption has been finalized. That is one of the reasons for the steady growth in the number of children Drew and James were raising. Although there are no reliable data on this, gay men seem to have an advantage over lesbian or single straight women who seek gray-market babies, because some birth mothers find it easier to relinquish their babies to men than to women, just as Marya had felt about donating her eggs. A pregnant woman who chooses gay men to adopt her offspring can hold on to her maternal status and avoid competitive, jealous feelings with infertile, adopting mothers.

It is true that most of the men in my study who adopted children through the gray market wanted their children to stay in touch with their birth mothers, and sometimes with their birth fathers as well. Drew and James even chose to operate "on a first-name basis" with their six (so far!) adopted children in order to reserve the terms *Mommy and Daddy* for their children's various genetic parents. Poly-parenting families do not always spring from such contingencies, however. Pursuing parenthood outside the box inspires some people to create intentional multi-parent families.

Front House/Back House

After thirteen years of close friendship, Paul (a white gay man) and Nancy (a white lesbian) decided to try to start a family together through alternative insemination. The two self-employed professionals spent the next two years carefully discussing their familial visions, values, expectations, anxieties, and limits. In October 1999, when Nancy began attempting to conceive their first child, they composed and signed a co-parenting agreement. They understood that the document would lack legal force but believed that going through the process of devising it would lay a crucial foundation for co-parenting. This agreement could serve as a model of ethical, sensitive planning for egalitarian, responsible co-parenting. In fact, it has already done so for several lesbian and gay friends of Paul's and Nancy's, and for two of mine. I do not know of any heterosexual couples who have approached the decision to parent together so thoughtfully. Perhaps this agreement can inspire some of them to do so too. Nancy and Paul were delighted, devoted biological and legal co-parents of a preschool-age son and an infant daughter when I interviewed them in 2001. They were not, however, the children's only parents. Before Nancy became pregnant with their first child, Cupid tested Paul's ability to live up to the sixth of the pair's prenatal pledges. Nancy had met and entered a romantic relationship with Liza, a woman who long had wanted to have children. Paul had risen to the challenge of supporting and incorporating Liza into his parenting alliance with Nancy, and so their son and daughter were born into a three-parent family. Nancy and Paul more than honored all of the pertinent terms in their shared parenting plan. Jointly they had purchased a duplex residential property. During the period of my study, Nancy and Liza lived together in the front house, Paul inhabited the back house, their toddler was sleeping alternate nights in each, and the breastfed infant still was sharing her two mothers' bedroom every night. Paul and Nancy, the two primary parents, were fully sharing the major responsibilities and expenses along with the joys of parenthood. Both had reduced their weekly work schedules to three days so that each could devote two days weekly to full-time parenting. A hired nanny cared for the children on the fifth day. Liza, who was employed full-time, did early evening child care on the days that Nancy and Paul worked late, and she fully co-parented with them on weekends and holidays.

This three-parent family enjoyed the support of a thick community of kith and kin. One of Paul's former lovers was godfather to the children, and he visited frequently. The three-parent family celebrated holidays with extended formal and chosen kin, including another gay-parent family.

The family was still intact when I contacted Paul and Nancy again in October 2008. Nancy and Liza had just celebrated their tenth anniversary as a couple, and Paul was still single.

Careful Fourplay

A second successfully planned poly-parent family included two moms, two dads, and two homes. Lisa and Kat, a monogamous, white lesbian couple, had initiated this family when after fifteen years together, they had asked their dear friend and former housemate, Michael Harwood, to serve as the sperm donor and an acknowledged father to the children they wished to rear. It had taken Michael, a white gay man who was single at that time, five years of serious reflection and discussions before he finally agreed to do so. "There is really no way to express the complexity of my journey," Michael related in an account he wrote for a gay magazine, "or to impart the richness of the experience. Given the rare opportunity to truly think about whether or not I wanted to be a parent (as opposed to having it sprung upon me), I left no rock unturned—no hiking trail was untread."[4]

Gradually Michael had realized that he did not wish to become a parent unless he too had a committed mate: "I told them that I could not do it alone (without a partner). I thought about what it would be like going through parenthood without a significant partner with whom to discuss and share things. It seemed too isolating."[5] Fortuitously, just when his lesbian friends were reaching the end of their patience, Michael met and fell in love with Joaquin, a Chicano, gay predestined parent who had always wanted children. The new lovers asked Lisa and Kat to give them a year to solidify their union before embarking on co-parenthood. Both couples reported that they spent that year in a four-way parental courtship:

> Joaquin and I had many talks and all four of us were, quite frankly, falling in love with each other in a way that can only be described as romantic love. There were flowers, there were candlelight dinners, and there were many beach walks and much laughter. There were many brave conversations about our needs and our fears and our excitement. There was nothing that could prepare us for the first night when Joaquin and I went to Lisa and Kat's home to make love and leave a specimen. . . . By the way, it is not a turkey baster but a syringe that is used. Love was the main ingredient, though, and Joaquin and I experienced a transcendent epiphany as we walked along the beach after the exchange. We knew that our lives and our relationship to Lisa and Kat would never be the same even if the conception did not happen. We shared, perhaps, the most intimate of experiences with Lisa and Kat.[6]

Since that magical night, the two couples also had shared many of the intimate joys and burdens of parenting two children. Unlike Nancy and Paul, however, they did not try to equalize parental rights and responsibilities. Lisa and Michael are the children's biological and legal parents, with both of their names on both of the birth certificates. The children resided, however, with Lisa and Kat, who are their primary, daily caretakers and their chief providers. Lisa, who gave birth to and breastfed both children, also spent the most time with them, primarily because Kat's employment demanded more time outside the home. Although Michael and Joaquin lived and worked more than seventy-five miles away, they had visited their children every single weekend of the children's lives as well as on occasional weeknights. They also conferred with the co-moms and spoke, sang, read, or sent emails to their preschooler almost daily. In addition, the adults consciously sustained, monitored, and nurtured their co-parenting alliance and friendship by scheduling periodic "parent time" for the four adults to spend together while the children slept.

This four-parent family, like the three-parent front-house/back-house family and like the surrogacy-extended family that Eddie and Mama Sally nurtured, regularly shared holidays and social occasions with a wide array of legal and chosen kin. They too were immersed in a large local community of lesbian- and gay-parent families, a community which Lisa had taken the initiative to organize. Three proud sets of doting grandparents were constantly vying for visits, photos, and contact with their grandchildren. In painful contrast, Kat's parents had rejected her when she came out, and they refused to incorporate, or even to recognize, their grandchildren or any of their lesbian daughter's family members within their more rigid, ideological understanding of family.

The Contingency of Contemporary Parenthood

This colorful quilt of lucky, and less lucky, gay pop stories from my research opens a window onto the vagaries of contemporary paths to parenthood generally and to paternity specifically. Because I intentionally over-sampled for fathers when I was recruiting participants for my study, I wound up including a disproportionate number of predestined parents. Their stories help us to understand some complex connections between romantic partnership and parenthood today. Most, if not all, of the fervently motivated dads strongly wished to combine the two forms of intimacy. Some even had made parenthood a pivotal courtship criterion, and the luckiest of these, like Eddie and Charles, found compatible predestined partners. However, if push comes to shove for a predestined parent, children will trump coupledom and can even thwart it, as we have seen. Although Armando deeply desired and attempted to combine partnership with parenthood, he was ultimately unwilling to sacrifice the latter on the pyre of adult intimacy. On the other hand, parenthood can prove a pathway to coupling for a fortunate few who, like Drew and Bernardo, find that their parental status enhances their appeal to other predestined parents.

There are numerous reasons to believe that fewer straight men than gay men feel a predestined urge to parent. For one thing, by definition, if not by disposition, gay men are already gender dissidents. Living without wives or girlfriends, they have to participate in caretaking and domestic chores more than straight men do and are less likely to find these activities threatening to their masculine identities. Second, gay men are more likely to be single than are straight men or than are women of whatever sexual orientation (Bell and Weinberg 1978). That translates into a higher percentage of men like Armando, who are apt to feel drawn to seek compensatory intimacy through parenthood. On the carrot side of the ledger, gay dads enjoy easier access than most straight dads do to primary parenting status and its rewards and to support networks for their families.

Gay men also face less pressure to conform to gender scripts for parenting or to defer to women's biological and cultural advantages for nurturing young children. Gay fatherhood, that is to say, occupies terrain more akin to conventional motherhood than to dominant forms of paternity.

The unmooring of masculinity from paternity exposes the situational character of contemporary fatherhood and fatherlessness. No longer a mandatory route to masculine adult social status, paternity today is increasingly contingent on the fate of men's romantic attachments. In fact, to attain any form of parenthood today requires either the unequivocal yearning of at least one adult or a more or less accidental pregnancy, like egg donor Marya's. In other words, contemporary maternity has also become increasingly situational, a fact that is reflected in declining fertility rates.

Nonetheless, the majority of women still skew toward the predestined pole of the desire-to-parent continuum. Men, in contrast, regardless of their sexual inclinations, generally cluster along the situational bandwidth. Heterosexual "situations" lead most straight men into paternity (and straight women to maternity). Homosexual situations, on the other hand, lead most gay men to forgo parenthood (lesbian situations likely are somewhere in between) (Simmons and O'Connell 2003). If this contrast seems obvious, even tautological, it was not always the case. Instead, most contemporary gay fathers became parents while they were enmeshed in closeted homosexual "situations." The past few decades of hard-won gains in gay struggles for social acceptance have diminished the need for men with homoerotic desires to resort to this ruse.

Paradoxically, the same shift from closeted to open homosexuality which has made gay fatherhood so visible might also reduce its incidence. Beyond the closet, far fewer gay men than before will become situational parents because they entered heterosexual marriages to pass as straight. Openly gay paternity, by definition, is never accidental. It requires the determined efforts of at least one gay man, like Armando or Eddie, whose passion for parenthood feels predestined—a man, that is, whose parental desires more conventionally might be labeled maternal rather than paternal.

The gay dads I studied did not feel that parenting made them less, or more, of a man. Instead, most felt free to express a full palette of gender options. As Drew put it, "I feel that I have a wider emotional range available to me than maybe most of the straight men I know. And I feel comfortable being mother, father, silly gay man, silly queen, tough negotiator in business. I feel like I'm not bound by rules." Rather than a bid for legitimate masculine status, or a rejection of it, intentional gay parenthood represents a search for enduring love and intimacy in a world of contingency and flux.

Of course, there is nothing distinctively gay or masculine about this quest. Heterosexual masculinity also no longer depends upon marriage or parenthood. Indirectly, therefore, gay male paths to planned parenthood highlight the waning of traditional incentives for pursuing the status of fatherhood as we knew it. Parenthood, like marriage and mating practices, has entered contingent terrain.

The fact that gay men now pursue parenthood outside social conventions of gender, marriage, and procreation catapults them into the vanguard of contemporary parenting. Just as gay men are at once freer and more obliged than most of the rest of us to craft the basic terms of their romantic and domestic unions, so too they have to make more self-conscious decisions about whether to parent, with whom, and how. I hope that the thoughtful, magnanimous, child-centered co-parenting agreement that Paul and Nancy devised will inspire throngs of prospective parents to undertake similar discussions before deciding whether baby should make three, or four or more, for that matter.

Notes

1. Twenty-four men were actively parenting children. In addition, two men were step-fathers to a partner's non-residential children; one man with his mother formerly co-foster-parented teenagers; four of the adoptive fathers had also formerly fostered teenagers, and two of these intended to resume this practice in the future; one man served as a known sperm donor for lesbian-couple friends; and one man was a genetic father who does not parent his offspring.

2. One man, a sperm dad who nicknamed himself a "spad," had facilitated a lesbian friend's desire to conceive a child with a donor willing to be an avuncular presence in her child's life. The other unwittingly impregnated a former girlfriend who chose to keep the child and agreed not to reveal its paternity.

3. Of the gay parents, five are Latino, three are black or Caribbean, and one is Asian American. Thirteen of the thirty-four children are white; nine are Latino; eight are black, Caribbean, or mixed race; and four are multi-racial Asian.

4. "Love Makes a Family," unpublished speech to a gay community group, on file with author. Additional information about this speech is withheld to protect the anonymity of my informant.

5. Ibid.

6. Ibid.

REFERENCES

Beck, Ulrich, and Elizabeth Beck-Gernsheim. 1995. *The Normal Chaos of Love.* Cambridge, UK: Polity.

Bell, Alan P., and Martin S. Weinberg. 1978. *Homosexualities: A Study of Diversity among Men and Women.* New York: Simon and Schuster.

Brodzinsky, David, Charlotte J. Patterson, and Mahnoush Vaziri. 2002. "Adoption Agency Perspectives on Lesbian and Gay Prospective Parents: A National Study." *Adoption Quarterly* 5(3): 5–23.

Gates, Gary. Distinguished Scholar at the Williams Institute, UCLA Law School, personal communication, May 17, 2005.

Human Rights Campaign. 2009. "Equality from State to State 2009." http://www.hrc.org/documents/HRC_States_Report_09.pdf.

Kapp, Stephen, Thomas P. McDonald, and Kandi L. Diamond. 2001. "The Path to Adoption for Children of Color." *Child Abuse and Neglect* 25(2): 215–229.

Lasch, Christopher. 1995. *Haven in a Heartless World: The Family Besieged.* New York: Norton.

Sears, R. Bradley, and M.V. Lee Badgett. 2004. "Same-Sex Couples and Same-Sex Couples Raising Children in California: Data from Census 2000." Williams Project on Sexual Orientation and the Law, UCLA Law School.

Simmons, Tavia, and Martin O'Connell. 2003. "Married-Couple and Unmarried-Partner Households: 2000." U.S. Census Bureau, February.

Stacey, Judith. 1998. "Dada-ism in the Nineties: Getting Past Baby Talk about Fatherlessness." In *Lost Fathers: The Politics of Fatherlessness*, ed. Cynthia Daniels. New York: St. Martin's.

Strah, David, and Susanna Margolis. 2003. *Gay Dads.* New York: J.T. Tacher/Putnam.

Townsend, Nicholas. 2002. *The Package Deal: Marriage, Work, and Fatherhood in Men's Lives.* Philadelphia: Temple University Press.

U.S. Department of Health and Human Services, Administration for Children and Families, Administration on Children, Youth, and Families, Children's Bureau. 2003. *The AFCARS Report.* http://www.acf.hhs.gov/programs/cb/publications/afcars/report8.pdf.

THINKING ABOUT THE READING

According to Stacey, how do gay men who choose to parent challenge conventional definitions of masculinity and paternity? How have these men made the "social character of paternity" more visible? How do the stories of Stacey's interviewees compare with images of gay male families you have observed in the media? Stacey's interviewees all lived in Los Angeles, a city that she notes is one of the most favorable cities for the architectures of gay and lesbian parenthood. Due to different legal and social restrictions on gay parenthood and families in other areas of the United States, how might the experiences of gay men who wish to parent and form families be different in these places?

Covenant Marriage

Reflexivity and Retrenchment in the Politics of Intimacy

Dwight Fee

(2011)

In recent years, sociologists have pointed to many transformations in personal life. We have heard quite a bit about the "questioning of tradition," the "redefinition of gender," the "reworking of relationships," or the "transformation of intimacy" and so on. Some sociologists have understood changes in private life in terms of an increase in "reflexivity" (see Giddens 1991, 1992; Beck and Beck-Gernsheim 1995; Swidler 2001; Weeks 1995; Weeks, Heaphy and Donovan 2001). Generally speaking, reflexivity means that, in a time of change and heightened social diversity, people no longer are able unconsciously to rely on traditions and customs to determine how they live. Applied to intimacy and sexuality, people are thrown back upon themselves to define their relationships and their identities within them. Crudely put, we must make decisions for ourselves once ingrained institutions and traditions are questioned, or once it becomes harder to say, "That's just the way the world is."

Therefore, once traditions are questioned, conventional intimate arrangements assume the status of mere *choices* that exist among many other competing ones. Not everyone has the same choices or can act on them as easily as others, but nevertheless, most of the time choice rules. Of course, tradition "hangs around" among all the options—but that hardly sounds like a tradition.

The Travails of Reflexivity

Being thrown back upon oneself when figuring out relationships and sexuality is surely challenging. For example, it would stand to reason that "commitment" itself would have to be debated and defined within each relationship, rather than simply assumed across all of them. And because we can't assume much cultural uniformity about such things, how do we establish trust in our relationships? Perhaps more than anything else, *risk* comes to paint the entire landscape of intimate life.

Despite all of the problems and ambiguities, however, most of those researching the growing uncertainty surrounding intimacy are encouraged. After all, people have to talk more, figure things out together, "be open." Consider Beck and Beck-Gernsheim's (1995:5) view of the situation:

> [I]t is no longer possible to pronounce in some binding way what family, marriage, parenthood, sexuality, or love mean, or what they should or could be; rather, these vary in substance, exceptions, norms and morality from individual to individual and from relationship to relationship. The answers to the questions above must be worked out, negotiated, arranged and justified in all the details of how, what, why or why not, even if this might unleash the conflicts and devils that lie slumbering among the details and were assumed to be tamed. Increasingly, the individuals who want to live together are, or more precisely [are] becoming, the legislators of their own way of life, the judges of their own transgressions, the priests who absolve their own sins, and the therapists who loosen the bonds of their own past. . . . Love is becoming a blank that the lovers must fill in themselves, across the widening trenches of biography. . . .

If these authors are right, even when we pick up the pieces of the old system we are patterning new relational forms, if only subtly. It may be that in many cases this reflexive work is opening up new avenues for autonomy in relationships, making our lives more "our own" and authentic, and, perhaps most crucially, making equality in relationships more possible.

Giddens (1992) calls this mode of relationality the "pure relationship." By calling it "pure" Giddens is suggesting that the viability of this type of relationship depends only on the people involved. The participants are the ones in charge; in this way it is "internally self-referential" through mutual disclosure. Reflexivity "disarms" those forming and moving through relationships. All that there is, is that other person and you—"free floating" as Giddens (1992) puts it. For some, it sounds a lot less romantic; for others, it is the beginning of possibility. For still others, as we will soon see, it reflects a moral decline, as relationships are seen as increasingly whimsical and self-serving.

Covenant Marriage: "Super-Sizing" Matrimony?

On Valentine's Day 2005, Governor Mike Huckabee (Republican-Arkansas) and his wife entered into a covenant marriage in front of about 6,400 onlookers. Already married for thirty years, the Huckabees took a new kind of plunge, one that was established to "inspire confidence" in marriage, and one usually discussed by proponents as important counter-strategies to the high divorce rate and to the "changing social values" that "threaten marriage." According to an Associated Press article in the *New York Times* (February 15, 2005), the governor announced to the crowd: "There is a crisis in America. The crisis is divorce. It is easier to get out of a marriage than [to get out of a] contract to buy a used car." After the Huckabees renewed their vows, the governor instructed the couples in the audience to do the same—to face each other and to repeat the vows of the Governor and First Lady. Many couples followed suit, crying, and then kissing after their spontaneous recitations.

Originally emerging from conservative Protestant churches in the late 1980s and early 1990s, the covenant marriage movement began as a response to a declared "divorce culture" and a "crisis of the family" in the U.S. Religious leaders and organizations quickly targeted legislative change so as to make the marriage bond a weightier, more durable (and, if only indirectly, religiously-based) commitment. The Covenant Marriage Law was first established in Louisiana in 1997, and similar laws were passed soon after in Arkansas and Arizona. While mainly in Southern states, there is now some kind of covenant marriage legislation afoot in some twenty states, including Minnesota, Iowa, Indiana, and Maryland, which is part of other widespread "divorce reform" legislative activity.

While there has been an increasing amount of public and media-based attention paid to covenant marriage since the Huckabee ceremony, it has so far fallen short of some proponents' early predictions that covenant marriage would "boom" and "could soon sweep the nation." Studies are scant, but the consensus seems to be that numbers are down, and were never really up. Of about 35,000 marriages in Arkansas in 2004, only 164 were of the covenant variety—most being conversions of existing marriages. According to Gilgoff (2005), rates are similar in Arizona and Louisiana—with no more than 2 percent of marriages being covenant. Still, it's worth considering what's going on here, now that covenant marriage has at least some salience within the broadening array of marriage debates. (Many proponents attribute the low numbers of covenant marriages to people simply not knowing about the option.)

Covenant marriage, of course, is more than a declaration of traditional marriage; it has very specific, legal dimensions. Advocates for covenant marriage want to offer an alternative to what they see as a blasé, or self-serving, or "test-drive" approach to marriage, since "no fault" divorce was ushered in during the 1970s. In the three states that have actually passed and instituted covenant marriage laws—Louisiana, Arkansas, and Arizona—couples are given a choice between standard marriage and the "CM" option. It's as easy as checking the appropriate box for the court clerk—but, according to Nock, Wright, and Sanchez (1999), here are the differences for the CM couple:

- the couple will seek premarital counseling—which must include discussions of the seriousness of marriage—and have a signed affidavit (signed by the counselor and the couple) to prove their participation;
- likewise, divorce is only possible if the couple goes to counseling, and after a two year waiting or cooling-off period.
- dissolving a covenant marriage in less than two years requires that one person prove fault on the part of the other. Acceptable faults are felony convictions, abuse, abandonment or adultery. Irreconcilable differences ("we just don't get along") are not acceptable grounds for divorce before two years (2.5 years if you have kids);
- and, couples can "upgrade" to a CM, like Governor Huckabee and his wife.

At the root of CM is the hope of revitalizing a belief in marriage and its sanctity through critiquing the supposed "contract mentality" of recent years. As Gary Chapman argues in *Covenant Marriage: Building Communication and Intimacy* (2003), the legalistic side of marriage is surely important, but the contract mentality has replaced "as long as we both shall live" with "we are committed to each other so long as this relationship is mutually beneficial for us." By contrast, covenant marriage offers deep spirituality and (ideally) a life-long commitment to the other's well being that is "above one's self." As one Louisiana woman put it, "we know that if we have problems, we can't just say I'm leaving" (Loconte 1998).

Covenant Marriage: Political Statement or Personal Choice?

There are many debates around covenant marriage, and some center on the specific problems that exist, or potentially exist, inside of them. Obviously, the fact that it becomes harder to get out of this form of marriage is a major concern in cases of marital violence and abuse. The CM laws state that divorce can be granted in such situations, but many are skeptical that these instances will be "verified" by those charged with that responsibility, which we would presume are mostly pastors and other church-based counselors. (Remember that abuse must be "proved.") Whether women get trapped in CMs remains unclear. Given the recent instigation of covenant marriage, I have not seen any systematic research to argue the situation either way. Predictably, the little research that has been done on CM has unambiguously shown that the large majority of supporters hold highly traditional attitudes about gender and the roles of men and women within marriage (Nock, Wright, and Sanchez 1999), which could itself worry some critics when it comes to issues of abuse.

However, advocates are quick to argue that most marriages fall apart because of "low-level" conflict, where the couple drifts apart, often without confronting their problems openly. In this sense, covenant marriage proponents say, "we're not erecting a barricade . . . we're just putting in some speed bumps" (Loconte 1998). They might also point out that the requirement to seek counseling before marriage—and subsequently, if problems arise—is not something men are often willing to do. As a progressive reform, CM could help men transcend "traditional" codes of masculinity by prompting them to develop effective communication and coping skills.

But then there is the larger issue of its cultural and political significance. On the one hand, proponents are right about the challenges of marriage; however, the supposed moral vacuum or "collapse" that they see behind it—as if statistics reflect ethical stances—has an obvious reactionary subtext. While covenant marriages are hardly widespread, it may not be going too far to say that we are witnessing the latest attempt to redefine marriage along religious and otherwise conservative lines.

According to the website ReligiousTolerance .org, some states are considering abolishing conventional marriage and offering only the covenant version—and obviously this is just when debates about gay marriage are particularly salient. For many, it makes sense that covenant marriage would emerge in the wake of gay marriage initiatives and the passage of the Defense of Marriage Act. Even though we have to make a focused effort to find much in the rhetoric about gay marriage, it is easy to assume that the CM movement is only a knee-jerk political expression. Proponents, though, might say it is simply a way to exemplify God's vision of marriage: "one man, one woman, forever—above their own shifting desires." As far as I have been able to determine, the part about "one man, one woman" is written into the actual legislation that is on the books in Louisiana, Arkansas and Arizona. This wording, we must assume, reaches out beyond covenant marriage itself. From this perspective, then, it is no accident that the Huckabee ceremony and all of the subsequent journalistic coverage comes at a time when gay marriage has arguably become the most salient social issue thus far in twenty-first-century America. Gay groups, in fact, were in attendance at the Huckabee event, fundraising and raising awareness about how marriage—any kind of marriage—is not available to same-sex couples.

In this sense, covenant marriage is at least an *implicit* socio-political statement about a "return" to most traditional forms of heterosexual relations. Put another way, *personal understandings and choices about marriage are intersecting with (or becoming articulated within) discourses of social and political reform.* This is tricky because we are not always dealing with, on the one hand, people's solely "personal" concerns about their relationship choices, or on the other hand, an explicit and intentional political backlash. Covenant marriage, in the broadest sense, is a place where a multitude of personal and political strategies are at work—so much so that the two realms are often indistinguishable. Of course, this predicament is nothing new; it is what many theorists and researchers have discussed in terms of the displacement of the private onto the public within the "politics of intimacy," or in debates about "sexual citizenship" and so on. Virtually all intimate choices now intertwine with various "culture wars" about sexuality, morality, and, if only indirectly, marriage itself. Even if covenant marriage only bears a kin-relationship to other more obvious political appropriations of marriage by conservatives, *covenant marriage is implicitly political, whether or not its supporters see themselves in such a light.*

Reflexivity and Retrenchment

Whatever the politics of CM supporters, there is something highly *performative* about covenant marriage from a sociological perspective: the willingness to step apart from the crowd, to make one's choice visible and different, to say (and to do so in an almost public way) "this particular alternative is the best way to go." In a strict sociological sense, this development is "anti-traditional," as it makes reflexivity and innovation central to decision-making about marriage. The centrality of therapy in covenant marriage makes it even more so—couples must deliberate, disclose their fears, and ostensibly work together. We might say it is "*doing* intimacy" in a world where virtually no one can simply blend into the background and not give voice to their choices (Seidman 2002). Covenant marriage is presumably about creating options, new possibilities, and, we would assume, the creation of more satisfying relationships. The equality piece is more ambiguous, but the innovation is there, whether or not one approves of the particular vision. As one advocate put it, "[covenant marriage] has everything to do with giving people more choices" (Nock, Wright, and Sanchez 1999). The difference here, however, is that reflexive processes are paradoxically moving, or hoping to move, in the direction of "tradition," or at least the way that tradition is being defined by the covenant marriage instigators.

This irony of providing more and more choices is not lost on some conservatives. We need only take note of the reactionary discourse about gay marriage to get the gist of the "slippery slope" argument: "so after gay marriage, what's next, marrying your cat?" If covenant marriage proponents take this view, we could easily grant covenant marriage the official status of *moral panic.* But it goes further: this slippery slope viewpoint is one reason why traditionalists themselves are part of the heretofore modest cultural impact of covenant marriage. When given the option of tinkering or not tinkering with marriage, many invested in orthodoxy and traditionalism will invariably side on the latter approach of sticking with the status quo. If something is so sacred and natural, there is something irreverent and contradictory about breaking it into differing levels and subcategories. It is here that the ironies of tradition/de-tradition come full circle: can reflexivity in intimate life be effectivcly used to reaffirm heteronormativity, which has historically thrived on the very *absence* of it? Can choice be used to fend off other choices seen as threatening or dangerous? In sum, how can the covenant marriage movement advocate a reflexive program when it comes very close to saying that reflexivity itself is the problem with marriage today?

REFERENCES

Associated Press (2005) "Thousands Renew Vows in Arkansas," *New York Times*, February 15.

Beck, U. and E. Beck-Gernsheim (1995) *The normal chaos of love.* Cambridge: Polity.

Chapman, G. (2003) *Covenant marriage: building communication and intimacy.* Nashville, TN: Broadman and Holman.

Giddens, A. (1991) *Modernity and self-identity: self and society in the late-modem world.* Stanford, CA: Stanford University Press.

——(1992) *The transformation of intimacy: sexuality, love and eroticism in modem societies.* Stanford, CA: Stanford University Press.

Gilgoff, D. (2005) "Tying a right knot," *US News and World Report*, February 28.

Loconte, J. (1998) "I'll Stand Bayou: Louisiana couples choose a more muscular marriage contract," *Policy Review* 89 (5).

Nardi, P. (1999) *Gay men's friendships: invincible communities.* Chicago, IL and London: University of Chicago Press.

Nock, S., J. Wright, and L. Sanchez (1999) "America's Divorce Problem," *Society* (May–June) (36); 4.

Plummer, K. (2003) *Intimate citizenship: private decisions and public dialogues.* Seattle, WA and London: University of Washington Press.

Seidman, S. (2002) *Beyond the closet.* London and New York: Routledge.

Swidler, A. (2001) *Talk of love.* Chicago, IL: University of Chicago Press.

Weeks, J. (1995) *Invented moralities: sexual values in the age of uncertainty.* New York: Columbia University Press.

Weeks, J., B. Heaphy, and C. Donovan (2001) *Same-sex intimacies.* London and New York: Routledge.

THINKING ABOUT THE READING

Drawing on Fee's definition of reflexivity that refers to how we consciously think about something that we might otherwise take for granted, how, if at all, have you employed reflexivity in your relationships and identities within them (e.g., romantic relationships, friends, family members)? Do you rely on long-standing traditions and customs, or do you make other choices that challenge these traditions and customs? Fee suggests that it is important to note that not everyone has the opportunity to be as reflexive as the next person. In what situations might an individual feel restricted from making choices that challenge tradition and custom? Give some examples. Finally, taking the central example of covenant marriage, do you agree or disagree with Fee that the covenant marriage movement is advocating a reflexive program while simultaneous suggesting that reflexivity is the problem with marriage today? What, if any, concerns would you have with the equality issue in covenant marriages? Considering the steady and ongoing change in marriage and family forms in recent decades, is the covenant marriage movement a viable option for couples in contemporary society?

Constructing Difference

8

Social Deviance

According to most sociologists, deviance is not an inherent feature or personality trait. Instead, it is a consequence of a definitional process. Like beauty, it is in the eye of the beholder. Deviant labels can impede everyday social life by forming expectations in the minds of others. Some sociologists argue that the definition of deviance is a form of social control exerted by more powerful people and groups over less powerful ones.

At the structural level, the treatment of people defined as deviant is often more a function of *who* they are than of *what* they did. In particular, sex, age, class, ethnic, and racial stereotypes often combine to influence social reactions to individuals who have broken the law. In "Watching the Canary," Lani Guinier and Gerald Torres provide several explanations for the disproportionate number of black and brown young men in U.S. prisons. They examine the intersection of racial profiling tactics, the war on drugs, and our mass incarceration policies to illustrate why these men are at greater risk for arrest. On the basis of race, these men are already defined as deviant and often expected to be engaged in criminal activity.

Similarly, our perceptions of deviant social problems can also be influenced by the identities of people most closely associated with the behavior in question. The use of marijuana is frequently associated with the stereotype of the "pothead." Who uses marijuana for medical purposes, and how do these individuals negotiate the deviant identity and politics associated with its use? It may surprise readers to learn that this group consists of children and older people as well as individuals who might normally be seen as possible pot users. In "Patients, 'Potheads,' and Dying to Get High," Wendy Chapkis describes the various strategies that providers and users of medical marijuana engage in to offset the impression of the deviant "pothead."

The definitional process that results in the labeling of some people as deviant can occur at the institutional as well as the individual level. Powerful institutions are capable of creating a definition of deviance that the public comes to accept as truth. One such institution is the field of medicine. We usually think of medicine as a benevolent institution whose primary purpose is to help sick people get better. But in "Healing (Disorderly) Desire: Medical-Therapeutic Regulation of Sexuality," P. J. McGann shows how the medical institution shapes dominant images and expectations of gender and sexuality. She points to the way in which contemporary sexual difficulties have been defined as violations of culturally approved sexual rules. The medical-therapeutic profession made up of physicians, psychiatrists, psychologists, counselors, and other specialists serves as an agent of social control by enforcing these definitions at the cultural and individual levels.

Something to Consider as You Read

In reading and comparing these selections, consider who has the power to define others as *deviant*. Think about the role of social institutions in establishing definitions of deviance. For example, how does medicine or religion or law participate in describing certain behaviors as abnormal and/or immoral and/or illegal? Does it make a difference which social institution defines certain behaviors as deviant? Why do you think certain deviant behaviors fall under the domain of medicine and others fall under the domain of the law? For instance, over time, alcohol use has moved from being an illegal activity to being a medical condition. Who makes the decisions to define certain behaviors not only as deviant but as deviant within a particular social domain?

Watching the Canary

Lani Guinier and Gerald Torres

(2002)

"To my friends, I look like a black boy. To white people who don't know me, I look like a wanna-be punk. To the cops I look like a criminal." Niko, now fourteen years old, is reflecting on the larger implications of his daily journey, trudging alone down Pearl Street, backpack heavy with books, on his way home from school. As his upper lip darkens with the first signs of a moustache, he is still a sweet, sometimes kind, unfailingly polite upper-middle-class black boy. To his mom and dad he looks innocent, even boyish. Yet his race, his gender, and his baggy pants shout out a different, more alarming message to those who do not know him. At thirteen, Niko was aware that many white people crossed the street as he approached. Now at fourteen, he is more worried about how he looks to the police. After all, he is walking while black.

One week after Niko made these comments to his mom, the subject of racial profiling was raised by a group of Cambridge eighth graders who were invited to speak in a seminar at Harvard Law School. Accompanied by their parents, teachers, and the school principal, the students read essays they had written in reaction to a statement of a black Harvard Law School student whose own arrest the year before in New York City had prompted him to write about racial profiling.[1] One student drew upon theories of John Locke to argue that "the same mindset as slavery provokes police officers to control black people today." Another explained a picture he had drawn showing a black police officer hassling a black woman because the officer assumed she was a prostitute. Black cops harass black people too, he said aloud. "It just seems like all the police are angry and have a lot of aggression coming out." A third boy concluded that when the cops see a black person they see "the image of a thug." Proud that he knew the *American Heritage Dictionary*'s definition of a thug—a "cut-throat or ruffian"—he concluded that the cops are not the key to understanding racial profiling. Nor did he blame the white people who routinely crossed the street as he approached. If what these white people see is a thug, "they would normally want to pull their purse away." He blamed the media for this "psychological enslavement," as well as those blacks who allowed themselves to be used to "taint our image."

One boy spoke for fifteen minutes in a detached voice, showing little emotion; but he often strayed from his prepared text to describe in great detail the story of relatives who had been stopped by the police or to editorialize about what he had written. Only after all the students left did the professor discover why the boy had talked so long—and why so many adults had shown up for this impromptu class.

Several of the boys, including the one who had spoken at length, had already had personal encounters with the police. Just the week before, two of the boys had been arrested and had spent six hours locked in separate cells. . . .

Watching the Canary

Rashid and Jonathan (not their real names) are the sons of a lawyer and a transit employee, respectively. "Why don't you arrest *them?*" one of the boys asked the officer, referring to the white kids walking in the same area. "We only have two sets of cuffs," the officer replied. These cops knew whom to take in: the white kids were innocent; the black boys were guilty.

In the words of one of their classmates, black boys like Rashid and Jonathan are viewed as thugs, despite their class status. Aided by the dictionary and the media, our eighth-grade informant says this is racial profiling. Racial profiling, he believes, is a form of "psychological enslavement." . . .

But these black boys are not merely victims of racial profiling. They are canaries. And our political-race project asks people to pay attention to the canary. The canary is a source of information for all who care about the atmosphere in the mines—and a source of motivation for changing the mines to make them safer. The canary serves both a diagnostic and an innovative function. It offers us more than a critique of the way social goods are distributed. What the canary lets us see are the hierarchical arrangements of power and privilege that have naturalized this unequal distribution.

We have urged those committed to progressive social change to watch the canary—and to assure the most vulnerable among us a space to experiment with democratic practice and discover their own power. Even though the canary is in a cage, it continues to have agency and voice. If the miners were watching the canary, they would not wait for it to fall off its perch, legs up. They would notice that it is talking to them. "I can't breathe, but you know what? You are being poisoned too. If you save me, you will save yourself. Why is that mine owner sending all of us down here to be poisoned anyway?" The miners might then realize that they cannot escape this life-threatening social arrangement without a strategy that disrupts the way things are.

What would we learn if we watched these particular two black boys? First, we would discover that from the moment they were born, each had a 30 percent chance of spending some portion of his life in prison or jail or under the supervision of the criminal justice system. . . . Among black men between the ages of 18 and 30 who drop out of high school, more become incarcerated than either go on to attend college or hold a job.[2]. . . .

In the United States, if young men are not tracked to college and they are black or brown, we wait for their boredom, desperation, or sense of uselessness to catch up with them. We wait, in other words, for them to give us an excuse to send them to prison. The criminal justice system has thus become our major instrument of urban social policy.

David Garland explains that imprisonment has ceased to be the incarceration of individual offenders and has instead become "the systematic imprisonment of whole groups of the population"—in this case, young black and Latino males in large urban municipalities. Or as the political scientist Mary Katzenstein observes, "Policies of incarceration in this country are fundamentally about poverty, about race, about addiction, about mental illness, about norms of masculinity and female accommodation among men and women who have been economically, socially, and politically demeaned and denied."[3] . . .

But how does this "race to incarcerate" happen disproportionately to young black and Latino boys? Why is it that increasingly the nation's prisons and jails have become temporary or permanent cages for our canaries? One reason is that white working-class youth enjoy greater opportunities in the labor market than do black and Latino boys, owing in part to lingering prejudice. . . .

A second reason for the disproportionate impact of incarceration on the black and brown communities is the increased discretion given to prosecutors and police officers and the decreased discretion given to judges, whose decisions are exposed to public scrutiny in open court, unlike the deals made by prosecutors and police. Media sensationalism and political manipulation around several high profile cases (notably Willie Horton and Polly Klaas) led to mandatory minimum sentences in many states. Meanwhile, laws such as "three strikes and you're out" channeled unreviewable discretion to prosecutors, who decide which strikes to call and which to ignore. . . .

A third and, according to some commentators, the most important explanation for the disproportionate incarceration of black and Latino young men is the war on drugs. In this federal campaign—one of the most volatile issues in contemporary politics—drug users and dealers are routinely painted as black or Latino, deviant and criminal. This war metaphorically names drugs as the enemy, but it is carried out in practice as a massive incarceration policy focused on black, Latino, and poor white young men. It has also swept increasing numbers of black and Latina women into prison. . . .

Presidents Ronald Reagan and George Bush had a distinct agenda, according to Marc Mauer: to "reduce the powers of the federal government," to "scale back the rights of those accused of crime," and to "diminish privacy rights."[4] Their goal was to shrink one branch of government (support for education and job training), while enlarging another (administration of criminal justice). Mauer concludes that the political and fiscal agendas of both the Reagan and first Bush administrations were quite successful. They reduced the social safety net and government's role in helping the least well off. Their success stemmed, in part, from their willingness to "polarize the debate" on a variety of issues, including drugs and prison.

Racial targeting by police (racial profiling) works in conjunction with the drug war to criminalize black and Latino men. Looking for drug couriers, state highway patrols use a profile, developed ostensibly at the behest of federal drug officials, that suggests black and Latinos are more likely to be carrying drugs. The disproportionate stops of cars driven by blacks or Latinos as well as the street sweeps of pedestrians certainly helps account for some of the racial disparity in sentencing and conviction rates. And because much of the drug activity in the black and Latino communities takes place in public, it is easier to target. . . .

A fourth explanation for the high rates of incarceration of black and brown young men is the economic boon that prison-building has brought to depressed rural areas. Prison construction has become—next to the military—our society's major public works program. And as prison construction has increased, money spent on higher education has declined, in direct proportion. Moreover, federal funds that used to go to economic or job training programs now go exclusively to building prisons. . . .

A fifth explanation is the need for a public enemy after the Cold War. Illegal drugs conveniently fit that role. President Nixon started this effort, calling drugs "public enemy number one." George Bush continued to escalate the rhetoric, declaring that drugs are "the greatest domestic threat facing our nation" and are turning our cities "into battlegrounds." By contrast, the use and abuse of alcohol and prescription drugs, which are legal, rarely result in incarceration. . . .

When drunk drivers do serve jail time, they are typically treated with a one- or two-day sentence for a first offense. For a second offense they may face a mandatory sentence of two to ten days. Compare that with a person arrested and convicted for *possession* of illegal drugs. Typical state penalties for a first-time offender are up to five years in prison and one to ten years for a second offense. . . .

We do not, by any means, claim to have exhaustively researched the criminal justice implications of racial profiling, the war on drugs, or our nation's mass incarceration policies. What we do claim is that canary watchers should pay attention to these issues if they want to understand what is happening in the United States. The cost of these policies is being subsidized by all taxpayers; one immediate result is that government support for other social programs has become an increasingly scarce resource.

Notes

1. Bryonn Bain, "Walking While Black," *The Village Voice*, April 26, 2000, at 1, 42. Bain and his brother and cousin were arrested, held overnight and then released, with all charges eventually dropped, after the police in New York City, looking for young men who were throwing bottles on the Upper West Side, happened upon Bain et al. as they exited a Bodega. Bain, at the time, had his laptop and law books in his backpack, because he was en route to the bus station where he intended to catch a bus back to Cambridge. Bain's essay in *The Voice* generated 90,000 responses.

2. Bruce Western and Becky Pettit, "Incarceration and Racial Inequality in Men's Employment," 54 *Industrial and Labor Relations Review* 3 (2000).

3. "Remarks on Women and Leadership: Innovations for Social Change," sponsored by Radcliffe Association, Cambridge, Massachusetts, June 8, 2001. In her talk, Katzenstein cites David Garland. "Introduction: The Meaning of Mass Imprisonment," 3(1) *Punishment and Society* 5–9 (2001).

4. Marc Mauer. (1999) *Race to Incarcerate,* New York: New Press.

THINKING ABOUT THE READING

Make a list of the social factors that Guinier and Torres link to the high incarceration rate of African American and Latino men. Discuss why these factors may affect these men more than white men. Do you think economic opportunity is related to these factors? In other words, are all African American and Latino men equally at risk for incarceration? What other factors do you think might be part of this equation? Groups who oppose the death penalty often argue that it is applied unevenly and discriminates among certain groups of people. Discuss this argument in light of what you have just read. As you think about this, consider each of the phases of the judicial process: the processes of arrest, the decision to charge with a crime, the availability of legal defense, the jury selection, and the sentencing guidelines. Who or what is making the decisions in each of these instances? Do you think the different people and agencies involved in each step of the process are all in agreement, or might there be disagreement between, say, the police, judges, and lawmakers? How might these relationships affect the likelihood of a defendant being treated justly?

Healing (Disorderly) Desire

Medical-Therapeutic Regulation of Sexuality[1]

P. J. McGann

(2011)

Sex matters—to individuals, to be sure, but also to social groups. Consequently all societies define and enforce norms of how to "do it," with whom, when, where, how often, and why. Yet how such sexual norms are enforced, indeed which acts are even considered to *be* sex ("it"), varies tremendously. In some cases sexual regulation is informal, as when girls or women admonish one another to control their sexual appetites lest one gain a "reputation." In others, regulation is more formal, as when a female prostitute is arrested and sentenced for her sexual misconduct. Of course, the legal system is not the only institution that formally regulates acceptable and unacceptable sexual practices. Religion also helps construct and enforce ideals of normal sex, defining some acts as sinful, others as righteous. In both cases, the moral language of sin and crime renders the social control aspects of legal and religious sexual regulation apparent.

But what of therapeutic approaches, as when a girl viewed as having "too much" sex is referred to juvenile court for "correction" of her incorrigibility? Is not intervention then for the girl's own good? Might it, for example, derail her developing delinquency, perhaps even prevent her subsequent involvement in prostitution? And what of the prostitute herself? What if rather than sending her to jail we instead direct her to therapy—based on the belief that a woman who sells her sexuality to others must, *obviously*, be sick? Do these therapeutic approaches also count as sexual regulation?

Here it is helpful to speak of social control, a broad concept that refers to any acts or practices that encourage conformity to and/or discourage deviations from norms (Conrad and Schneider 1992). From this perspective a medical-therapeutic response to violations of sexual rules *is* a form of sexual social control. However, in contrast to the transparently moral language of law and religion—good/bad, righteous/sinful, right/wrong—therapeutic regulation of sex relies on more opaque dichotomies of health and illness, normality and abnormality. Although such terms may camouflage the moral evaluation being made, the result is the same; whether the means are legal, religious, or therapeutic, a negative social judgment is made and a sexual hierarchy is produced (Rubin 1993). A dichotomy of good versus bad sexual practices, good versus bad "sexual citizens" (Seidman 2002), is thus created and enforced:

> Individuals whose behavior stands high in this hierarchy are rewarded with certified mental health, respectability, legality, social and physical mobility, institutional support, and material benefits. As sexual behaviors or occupations fall lower on the scale, the individuals who practice them are subjected to a presumption of mental illness, disreputability, criminality, restricted social and physical mobility, loss of institutional support, and economic sanctions. (Rubin 1993:12)

Medical-therapeutic approaches—medicine, psychiatry, psychology, social work, and juvenile justice—are part of a web of practices that help define and enforce a society's sexual hierarchy and sexual norms. The "'helping" ethos of therapeutic approaches, however, disguises their regulatory dynamics and effects.

An illness diagnosis provides a seemingly positive rationale for restricting or changing sexual behaviors found to be disturbing; intervention is, after all, *for our own good*. Even so, what is considered a sexual disorder may have disciplinary consequences. Whether or not a sexual activity is "really" a dysfunction or even causes distress for those diagnosed, individual sexual choices deemed non-typical are curtailed, and sexual culture is restricted in the name of health.

This [reading] explores some of the politics of "healing" disorderly desire. Using three contemporary sexual difficulties—erectile dysfunction, gender identity disorder, and sexual addiction/compulsion—I show how medical-therapeutic approaches shape and direct sexual expression. Some forms of regulation are directly repressive; they limit or deny sexual options construed as unnatural, abnormal, or unhealthy. Other forms of medical-therapeutic regulation are more subtle; their "normalizing" dynamics work by producing cultural ideals of natural and healthy sexuality. The ostensibly objective medical model of sex is especially important in this regard. It provides both the taken-for-granted understanding of what "sex" is (and is for) and the reference point from which sexual abnormality and sexual disorders are defined. As we shall see, this intertwining of the individual and cultural levels, and of repressive and normalizing forms of power, is a central dynamic in medical-therapeutic sexual regulation. Moreover, given that some individuals who "have" sexual disorders suffer neither distress nor impairment, it seems that diagnostic categories are not purely scientific entities, but social constructs that reflect social and political dynamics and concerns.

Medicalized Sex and Medical Social Control

When something is "medicalized" it is conceptually placed in a medical framework. The "problem" is then understood using medical language, typically as a disorder, dysfunction, disease, or syndrome, and is approached or solved via medical means (Conrad and Schneider 1992). A "sex offender," for example, might be sentenced to rehabilitative therapy rather than prison, or a man concerned about his homosexual desire might consult a psychiatrist rather than a priest. Although such medical-therapeutic regulation may be less punitive than criminal or religious sexual intervention, medicalizing sex produces positive and negative results.

On the plus side, defining sexual difficulties as medical problems may make it easier for people to talk more openly about sex and thus seek information and advice. Accordingly a medical approach to sex may enhance individual sexual pleasure. Yet medicalization also raises the possibility of "medical social control" (Conrad and Schneider 1992)—in this case, the use of medical means to increase conformity to sexual norms and/or to decrease sexual deviance. Prescription drugs, talk therapy, behavioral modification, negative or aversive conditioning, and/or confinement in a juvenile or mental health facility, can be enlisted to ensure adherence to sexual norms. Even without such direct medical intervention, viewing sex with a "medical gaze" often leads to a limited, biologically reductionist understanding. Stripped from its social context, sexual *difference* may become sexual *pathology*.

Medical social control sometimes has a slippery, elusive character. When individuals consult therapeutic professionals regarding sexual matters, they typically anticipate alleviation of sexual distress rather than restriction of sexual freedom. For their part psychiatrists, doctors, therapists, and social workers may neither intend nor understand their therapeutic practice as tools of sexual repression. Despite this mismatch of intent and effect, therapeutic intervention has regulatory consequences. The "promiscuous" girl can be held against her will in a mental or juvenile justice institution. The man who desires multiple sex partners might be forced to remain monogamous and to refrain from masturbation lest his sexual "addiction" overtake him.

Sexology and Its Legacies

Although some of our categories of sexual disease are new, medicine and sex have long been entangled in North America and Europe. Sexology, the science of sex, originated in mid-nineteenth-century Europe when physicians such as Magnus Hirschfeld, Richard von Krafft-Ebing, and Havelock Ellis turned their attention to sexual behavior. At the time Europe was caught up in a cultural mood of scientific rationality, evolutionism, and fantasies of white racial superiority. These currents inspired detailed scientific description of sexual diversity, and the delineation of sexual practices into normal and abnormal types. The latter were dubbed "perversions" and seen as sickness rather than sin. With the emergence of sexology, formal regulation of sexuality shifted from predominantly religious to secular modes of social control. Regulation of deviant sexuality thus became the province of medical authority (Foucault 1990).

Commonsense understandings of categories of disease view them as morally neutral descriptions of states of un-health. Yet even cursory consideration of the malleability of sexological categories shows that sexual disorders reflect more than just the accumulated sexual knowledge of the time. Forms of sexual behavior once considered abnormal and diseased are now "known" to be a normal part of sexual health. Some sexual illnesses reflect the normative standards of more powerful groups at the expense of those with differing sexual tastes and less power. And some sexual disorders seem more like reflections of prevailing cultural currents rather than actual sexual dysfunctions.

Masturbation, for example, was once the disease of "Onanism" (Conrad and Schneider 1992). A dangerous illness on its own, "self-abuse" was also a sort of gateway disease—a disorder that could so weaken the afflicted that he might fall prey to other perverse "infections" such as homosexuality or sadism. Now, though, masturbation is considered a "natural" (even if private) part of healthy sexuality; in fact, masturbation is prescribed as a therapeutic treatment for some sexual disorders, such as "premature ejaculation" (ejaculation that occurs before coitus) or "anorgasmia" (inability to orgasm). Healthy and normal sexual practices may also become disordered or unsavory over time. Visits to female prostitutes, for example, were once part of the prescribed treatment for male (but not female) "lovesickness." Massage of female external genitalia by a doctor or midwife was once the preferred treatment for "hysteria" (Maines 1999). However, in most locales today the former is illegal and the latter might be considered sexual misconduct or abuse. The female "psychopathic hypersexual" illustrates how disease categories reflect cultural concerns. Female sexual psychopaths "suffered" from an excessive amount of sexual desire at a time when it was "known" that girls and women were naturally modest and chaste, or at least sexually passive. Interestingly, the hypersexual female diagnosis emerged at the end of the Victorian era—a time of changing gender relations and rising anxiety over the increasing independence and agency of women. The psychopathic hypersexual diagnosis reflects these concerns and codifies the violation of normative gender standards as disease. Finally, the declassification of homosexuality as a mental disorder in response to social and political developments outside psychiatry is the example *par excellence* that disease categories rest on more than scientific facts (Conrad and Schneider 1992). One wonders: if historical categories of sexual disease are so obviously shaped by non-scientific factors, might the same be true of contemporary constructions of normal and abnormal, healthy and diseased sex?

The Medical Model of Sex

Although many concepts from classic sexology are no longer accepted, there are continuities between nineteenth- and twenty-first-century medical approaches to sex. The thrust to describe and delineate the diversity of sexual practices and types persists. So, too, does the "medical model" of sex. This view posits sex as an innate, natural essence or drive contained in and released from the body. Bodies, in turn, are understood as machine-like composites of parts. When the parts are in proper working order, bodies are able to achieve their functional purposes. Sexual organs become engorged as blood and other bodily fluids accumulate in anticipation of sexual activity. These changes, as well as sexual drives, patterns of sexual behavior, and even sexual types (bi, homo, hetero), are understood as universal properties of individuals independent of society. Cultural variation in sexual practice is seen as relatively superficial; changes in surface social details do not alter the deeper biological reality of sex (Tieffer 1995). Because reproduction is considered the natural function of sex, the medical model depicts heterosexuality as natural and neutral, not in need of explanation or scrutiny—unless, that is, something goes awry with the hydraulic sexual machine. Thus, nineteenth-century sexology and modem sexual science have mostly observed, described, and catalogued deviations from or problems with sex oriented toward reproduction.

The contemporary Human Sexual Response Cycle (HSRC) is the iconic embodiment of this approach. First conceived by Masters and Johnson in the 1960s, the HSRC describes a presumably universal pattern of physiological changes that occur during "sex": excitement, plateau, orgasm, and resolution. In the "excitement" stage, for example, penises become engorged with blood and vaginas lubricate in preparation for "sex." Most medical-therapeutic professionals concerned with sexual disorders now rely on a three-stage derivative model of desire, arousal, and orgasm. Despite its supposed scientific neutrality—the HSRC model was based on seemingly disinterested laboratory observation of heterosexual genital intercourse—the HSRC has been critiqued as heteronormative and androcentric, and for reifying a limited understanding of "sex" as the cultural sexual ideal (Tieffer 1995).

Although the array of potentially normal human sexual activity is vast, HSRC constructs only a narrow range of acts relating to coitus (penile-vaginal intercourse) as constituting "sex." Other forms of sexual activity are relegated to "foreplay"—preparatory, albeit pleasurable, preparations for the "real thing." Forms of sexual activity that do not culminate in coitus are viewed as perverse substitutions for, or distractions from, the real thing (Tieffer 1995). Oral sex followed by heterosexual coitus may be normal foreplay, for example, whereas oral sex in the absence of coitus is considered abnormal or dysfunctional. The HSRC thus constructs both what *should* and *should not* be done during normal sex. In so doing the HSRC helps constitute "sex" itself.

What, for example, comes to mind when one person says to another, "We had sex"? Despite the nearly endless possibilities (given the number of bodies and body parts that may or may not be involved, variations in sequence, pace, position, sexual aids or toys, and the like) the meaning of "We had sex" is typically unproblematic in everyday life. In fact, a common response might be a titillated "Really! How many times?" The answer to this question is also typically unproblematic, given that we know both what "sex" is and what "counts" as a time. Now, though, let's make it explicit that the two people who "had sex" are of the same sex. Did the meaning of "having sex" change in your mind? What counts as a "time" now? Will your answer change if our partners are male, female, or transgendered? What if we add a third or fourth participant? Is the sex that was had still the *normal* kind? Or does it now appear abnormal, maybe even *sick,* despite being consensual and mutually pleasurable?

The HSRC is also critiqued as androcentric (male-centered), given that the orgasm in question is the man's. Clinically, male ejaculation/orgasm marks the transition to the "resolution" stage. As such, male orgasm is the basis of counting how many "times" sex occurs, or even if sex is "had" at all. Moreover, although the HSRC codifies "foreplay" as an official part of sex, there is no category of "afterplay"—such as sexual activity focused on female orgasm after the man ejaculates. Even adding this concept, though, leaves coitus intact as the defining sexual moment. The (heterosexual) male's orgasmic experience thus defines "sex," whereas female orgasm is not considered. Indeed, female orgasmic pleasure is not a necessary part of real sex. Female orgasm does not count, at least not from the perspective of the supposedly universal and natural HSRC—unless, that is, the counting concerns "abnormal" female sex response. Since at least 70 percent of women do not achieve orgasm from penile-vaginal penetration alone, the HSRC focus on coitus and male ejaculation as the goal and purpose of "sex" renders most women sexually unhealthy, defective, disordered, or dysfunctional (Tieffer 1995). Despite these shortcomings, sexual activities that diverge from the HSRC are defined as abnormal, pathological, deviant, unnatural, dysfunctional, and disordered.

Sexual Disorder in DSM

"DSM" is short for *Diagnostic and Statistical Manual of Mental Disorder.* Published by the American Psychiatric Association, DSM is a professionally approved listing of diagnostic categories and criteria. It is the central text for those working in the mental and sexual health fields in the USA, and the key to second-party reimbursement for medical-therapeutic services. DSM has undergone four revisions since its initial publication in 1952. A roman numeral in the title denotes placement in the revision sequence: DSM-II in 1968, DSM-III in 1980, DSM-III-R in 1987, DSM-IV in 1994, and DSM-IV-TR in 2000. Categories of disease are refined and reconceptualized over the course of these editions. Sometimes this results in a disorder being relocated within DSM's typological system, as when homosexuality shifted from psychopathic personality disturbance (DSM-I) to type of sexual deviance (DSM-II); sometimes it leads to the complete removal of a disorder, as when homosexuality was left out of DSM-III

The most recent DSM lists three major classifications of Sexual Disorder: Paraphilias, Sexual Dysfunctions, and Gender Identity Disorders. "Paraphilias" include exhibitionism, fetishism, frotteurism, pedophilia, sexual masochism, sexual sadism, transvestic fetishism, and voyeurism. (Many of these types were the original "perversions" described by nineteenth-century sexology). "Sexual Dysfunctions" concern impairments or disturbances related to coitus. Three subtypes directly mirror the derivative HSRC model: disorders of desire, disorders of arousal, and disorders of orgasm. The fourth subtype, pain during coitus, also reflects the centrality of coitus in constructing the sexual dysfunctions. The last major classification of sexual disorder in DSM is "Gender Identity Disorders." With subtypes for adults and children, these disorders represent deviations from "normal" gender embodiment.

DSM facilitates communication among an array of sexual helping professionals with diverse training, specialization, and institutional placements and practices. Using DSM categories, individuals as varied as a social worker with one year of postgraduate academic training, a psychiatrist with over ten years of medical and clinical training, or a college intern working in a residential juvenile treatment program, can communicate with one another. This is helpful, to be sure. But the shared language of DSM may also make the sexual disorders seem less politically contested and more objectively real than they really are. This may, in turn, make it more difficult for practitioners to recognize the biases built into DSM diagnostic categories. Far from being neutral classifications, the major DSM categories of sexual disorder encode normative assumptions of sexuality. Paraphilias delineate that which we should do sexually; sexual dysfunctions reflect incapacities in what we *should* do; and gender identity disorders concern how we should appear and who we should be while doing it.

One way to avoid uncritically replicating these biases is to rename the major types of sexual disorder based on their ideological effects rather than naming the types based on their relationship to coitus. "Sexual dysfunctions," for example, seems a neutral and comprehensive term; in reality it is a specific reference to *hetero*sexual dysfunctions of penile-vaginal intercourse. Consider instead "disorders of prowess"—disorders based on one's compromised ability to engage in coitus. We could similarly disrupt the heteronormativity of HSRC and speak of "disorders of appetite"—disorders of too much or too little desire to engage in coitus, and/or having desires that do not include or that extend beyond coitus. Finally, "gender violations" seems an apt tag for forms of gender expression and embodiment that violate the traditional gender styles underpinning normative heterosexuality. Having linguistically interrupted diagnostic business as usual, I now turn to some specific disorders and consider how each reflects and reproduces dominant North American sexual norms and the sexual hierarchy built on them.

Sexual Disorders and the Maintenance of Ideal Sexuality

Some people do experience distress in relation to sexual matters. Treatment of sexual disorders may alleviate such distress and thereby enhance sexual pleasure. This does not mean, however, that medical-therapeutic intervention does not also produce negative consequences or operate in repressive fashion. Moreover, since some of the DSM's sexual disorders are not necessarily *dysfunctions* but violations of dominant sexual norms, therapeutic intervention may enforce conformity to the dominant sexual ideal. Sexual codes, though, are political, ethical, moral, and existential matters—not medical ones. While bringing one's sexual practices into line with prevailing sexual norms may alleviate individual distress, it also obscures the social, cultural, and ideological sources of sexual difficulties.

The Western sexual ideal has become less oriented to procreation and more pleasure-based over time. This new "relational" sexual code (Levine and Troiden 1988) understands sexual activity as creating, expressing, and enhancing a couple's intimacy, and mutual sexual pleasure is accordingly thought of as a normal part of healthy sex. This does not mean, of course, that anything goes. Ideally sexual pleasure occurs within an on-going committed monogamous relationship between two conventionally gendered people of "opposite" sexes (Rubin 1993). In some locales homosexuality may be approaching this sexual ideal—provided, that is, the same-sex couple and their relationship is otherwise normal: the partners are committed to one another, their sex is an expression of love and caring, and they are, like heterosexuals, either masculine men or feminine women (Seidman 2002). Ideal sex occurs in private, is genitally centered (as per the HSRC construction of coitus as "sex"), and is caring rather than aggressive or violent (Rubin 1993). Individuals who engage in such normal sex are good sexual citizens, while those who engage in non-normative sex are thought of as bad. The latter are perceived as immoral, abnormal, unhealthy, diseased, perverted and socially dangerous (Seidman 2002).

Erectile Dysfunction: A Prowess Disorder

At first glance it may be hard to grasp how improving a man's erection may be a form of sexual regulation and repression. Certainly treatment of Erectile Dysfunction (ED) holds the promise of increased sexual pleasure! It also, however, channels pleasure toward particular sexual acts and body parts in a manner that reflects and reinforces the limited HSRC construction of coitus as the be-all and end-all of "sex." The focus on erections also helps reproduce traditional masculinity and associated stereotypes of "natural" male and female sexuality.

Previously known as the psychological and interpersonal disorder of "impotence," ED is now understood as a physiological impairment of arousal. ED has recently risen to prominence alongside the increased visibility of drug-based treatments; indeed, the discovery of a pharmacological treatment is intertwined with the discovery that impotence is "really" a hydraulic and mechanical disorder. In this, the "Viagra age" (Loe 2004) is emblematic of the biological reductionism that often results when sex is medicalized. The individualized focus removes sexual problems from their interpersonal, social, and cultural contexts. ED thus seems a purely medical rather than a political matter.

The proliferation of penile fixes—Viagra, Cialis, Levitra, and the like—has made it possible for many men to achieve the full, long-lasting erections they desire. The penile fix has also raised expectations and created new norms of male sexual performance. In the Viagra age it is easy enough to rebuild him, make him bigger, harder, and get him that way faster regardless of his age, fatigue, or emotional state. With the little blue pill and some physical stimulation a real man can get the job done, whether or not his heart is in it. That the perfect penis is now but a swallow away reinforces the cultural understanding of male sexuality as machine-like, uncomplicated, straightforward, and readily available. This reinforces the commonsense view that men are "about" sex whereas women are "about" relationships, while constructing female sexuality as complicated and mysterious (Loe 2004). These contrasting images of male and female sexuality reflect the notion that men and women are "opposite" sexes. This gendered assumption in turn bolsters the seeming neutrality of heterosexuality as complementary opposites that "naturally" attract the other. In this way ED reflects and reinforces heteronormative cultural ideals of sexuality *and* gender.

The ED treatment focus on producing erections "sufficient" for penetration also disciplines the sexuality of individual men. For starters, it directs attention to preparing the man for "sex"—understood, of course, as coitus. While this constructs sex in an image of male orgasm, it also constrains the realm of pleasurable and culturally valuable sexual activity. The phallic focus comes at the expense of the man's other body parts and their pleasures, and construes other sexual activities, including other types of intercourse (anal or oral), as less than the real thing. The phallic focus even deflects attention from the wider possibilities of pleasure linked to other penile states (the soft penis, for example, or movements between soft and hard). Preoccupation with the size and "quality" of erections also reinforces a sexual "work ethic" that emphasizes active male performance rather than sexual enjoyment and/or receptivity. The man's sexuality is thus restricted and restrained, his potential pleasures lessened. Pharmaceutical ads disseminate these messages widely in their depictions of ED treatments as being for caring, committed heterosexual couples rather than, say, homosexual couples, single men, men who masturbate or use pornography, or men who engage in multiple-partner sex (Loe 2004).

Gender Identity Disorder: A Gender Violation

The gendered nature of ED and its treatment suggests the close coupling of traditional gender and normative sexuality. As Seidman (2002) points out, good sexual citizens are gender-normal citizens: their gender identities and expressions fit traditional gender images and understandings. Thus, normal sex occurs between individuals whose gender styles and gender identities are seen as appropriate for their sex category—men are masculine and see themselves as male; women are feminine and see themselves as female. The gender identity disorders reflect and reinforce these essentialist understandings of the "natural" relationship between sex category, gender identity, and gender embodiment, by pathologizing alternative configurations of sex and gender (McGann 1999). Non-normative ways of doing gender—a feminine man, for example—and atypical gender identities—such as a female-bodied person who identifies as male—are examples of clinical "gender dysphoria." Whether or not their ego functioning is impaired, they experience distress related to their condition, or other psychopathology exists, gender dysphoric individuals, including children, may be diagnosed with and treated for Gender Identity Disorder (GID). Although a GID diagnosis can have the positive result of facilitating access to medical technologies of bodily transformation, it does so by constructing gender difference as disease. GID thus also provides a rationale for medical social control of gender difference—an especially troubling possibility for gender-different children.

When GID first appeared in 1980, it included two types of diagnostic criteria. One concerned impairments in cognitive functioning centered on sexual anatomy, such as a boy thinking his penis ugly or wishing he did not have a penis, or a girl insisting that she could one day grow a penis. The other diagnostic criteria were thought to indicate the child's desire to "be" the other sex. In actuality, however, these criteria focused on cultural violations of gendered appearance or activity norms—boys who look and "act like" girls, for example. Although a child need not demonstrate distress regarding the condition—in fact, DSM notes that most children deny distress—a child had to demonstrate *both* the cognitive functioning and cultural criteria to be diagnosed. That is, a child could not be diagnosed with GID on the basis of cultural gender role violations alone. This two-tiered diagnostic requirement has weakened over subsequent DSM editions. Since 1994 (DSM-IV) it has been possible to diagnose a child as gender disordered based *only* on cultural criteria—that is, based only on the child's violations of social standards of traditional masculinity and femininity in the absence of demonstrated impairment of cognitive function. Thus, a girl with short hair, whose friends are boys, and who refuses to wear dresses, may now "have" GID. In effect, the diagnostic net has widened; a tomboy considered normal under DSM-III-R became abnormal in DSM-IV.

Interestingly, this expansion of GID has occurred alongside the removal of homosexuality from DSM-III and the increasing "normalization" of homosexuality in everyday life (Seidman 2002). Nonetheless, organizations such as Focus on the Family publicize and support therapeutic treatment of gender-different children in order to stave off their future homosexuality. Because children can be and are diagnosed and treated solely for gendered appearance and role violations, GID enforces our cultural gender dichotomy and our understanding of heterosexuality as the natural attraction of gendered opposites (McGann 1999).

One need not be gender-dysphoric oneself to suffer GID's disciplinary effects. As noted earlier, medical judgments of health and illness influence everyday life understandings. In this case the construction of atypical gender as illness discourages gender openness and fluidity for all. GID also regulates sexual expression directly by limiting normal sex to that which occurs between traditionally gendered people; cross-dressing sex play by individuals who are otherwise gender-normal, for example, is "known" to be abnormal or perverse. The sexuality of gender-atypical but non-dysphoric people is also distorted by GID. The wholly normative heterosexual desire of a "tomboyish" woman may be invisible to her potential partner, for example. Alternatively, her erotic draw to males may be dismissed as unbelievable or insincere since she is, *obviously,* a lesbian, based on her appearance (McGann 1999).

Sexual Addiction/Compulsion: A Disorder of Appetite

At times the terms "sexual addiction" and "sexual compulsion" are used interchangeably; at times they refer to different disorders. Neither is currently listed as an official mental disorder in the DSM. Nonetheless, patients are treated for sexual addiction/compulsion, books and articles are published on the disorder, practitioners are trained in its treatment modalities, therapeutic institutions specialize in it, and TV documentaries such as Discovery Health's *Sex Mania!* present it as a valid diagnostic category. Sexual addiction/compulsion is thought to be similar to other chemical or behavioral dependencies, such as those on alcohol or food. In practice, the diagnosis can refer to nearly any sexual behavior deemed "excessive" in the therapist's or clinician's professional judgment. In this, sexual addiction is a near-perfect obverse of the prevailing sexual ideal, the dark shadow of the good sexual citizen. As with GID, many individuals diagnosed with sex addiction deny that their disorder causes them distress or harm, and helping professionals often have to work long and hard to convince their "patients" that they are in fact "sick." For this reason sexual addiction aptly illustrates how disease categories crystallize political differences regarding sexual norms. It also shows how disease categories reflect the social currents and concerns of their origin.

Sexual addiction was "unthinkable" in the relatively sexually permissive, sex positive 1970s (Levine and Troiden 1988). At the time sex was seen in a more recreational, pleasure-based light and therapeutic concern consequently focused on "Inhibited Sexual Disorder" (Irvine 2005). But in the 1980s and the early days of the AIDS epidemic, fears of sexual chaos came to the fore and therapeutic attention turned instead to excessive—thus dangerous sexual desire. As the dominant sexual ideal shifted from a recreational to a relational code, forms of sexual expression that had been normalized in the 1970s were pathologized as addictive and compulsive (Levine and Troiden 1988).

Gay men were at first thought to be especially prone to sex addiction. Indeed, the gay press worried that the diagnosis was a medical form of homophobia, a way to pathologize behavior construed as deviant by the hetero majority, but that was normative within some gay communities (Levine and Troiden 1988). While gay male acceptance of anonymous and/or public sex may have initially swelled the sex addict rank, heterosexuals also "suffer" from the disorder. Straight women who deviate from relationally-oriented monogamous sex may be considered addicts, for example. A broad range of sexual activity outside of coital monogamy is considered indicative of addiction/compulsion, including multiple partners (at the same time or as successive couplings), "frequent" masturbation, the use of pornography, "recreational" sex (sex solely for pleasure), anonymous or public sex. Sex addiction also manifests within the otherwise sexually normal hetero couple, as when one partner desires coitus more frequently, and/or wants to engage in activities in addition to or instead of coitus. In these cases a therapist may tip the balance in favor of the more traditional partner by elevating one personal preference as "normal" while deeming the other compulsive or addictive. Here, use of the term "lust" rather than "desire" in the sex addiction literature reveals a moral evaluation masquerading as neutral medical description. Other morally charged "retro-purity" terms are also common in the sexual addiction literature, such as promiscuity, nymphomania, and womanizing (Irvine 2005).

Although the activities presumed indicative of sexual addiction or compulsion may be atypical, they are not inherently pathological. It seems, then, that medical ideology has retained the theme of morality but has done so in seemingly apolitical terms. The sex addict diagnosis codifies prevailing erotic values as health (Levine and Troiden 1988). This move privileges a certain style of sexual expression while marginalizing others. Indeed, the sex addict diagnostic guidelines read like a description of a dangerous sexual citizen. The construction of sexual activities not oriented toward coitus, polyamorous relationships, and non-relational, pleasure-based sex as illness ends political debate on these matters before it begins.

Sexual Disorders or Disorderly Sex?

Medicalized sex is not necessarily the enemy of pleasure. As Rachel Maines (1999:3) documents regarding the preferred treatment of hysteria: "Massage to orgasm of female patients was a staple of medical practice among some (but certainly not all) Western physicians from the time of Hippocrates until the 1920s." This example is titillating, of course. It is also instructive: it points to the necessity of separating the therapeutic professional's *intent* from the potentially repressive *effects* of therapeutic intervention, and to the importance of viewing both in the context of cultural understandings of sex and eroticism. The clinical phenomenon of "hysterical paroxysm" certainly looks now to be orgasm. But in a cultural moment that understood vaginal penetration as necessary for "sex" to occur it was seen instead as the climax of illness. Just as the physician treating the hysterical patient did not necessarily intend to incite his patient's pleasure, contemporary helping professionals may not intend to restrict the sexual freedom of their patient-clients.

Medical-therapeutic sexual regulation works at the cultural level via a normalizing dynamic that constructs a limited range of sexual activity as healthy, natural, normal sex. Medical-therapeutic regulation also works repressively directly on individuals, limiting their sexual choices and/or serving as justification for coercive "therapeutic" responses to non-normative sexual variation. Both dynamics and more are apparent in the sexual disorders just discussed. Erectile dysfunction illustrates the strait-jacket that is the medical model of sex. It also shows the chameleon-like nature of medical social control; therapeutic response to a sexual problem can simultaneously enhance and reduce pleasure. GID demonstrates how medical constructions of "normality" at the cultural level intertwine with the individual level; the enforcement of normal gender on individuals reinforces the cultural concepts of gender that the naturalness of heterosexuality is built on. Together, ED and GID show that one need not be diagnosed to have one's sexuality regulated by medical-therapeutic approaches. Finally, much like the female hypersexual diagnosis, sexual addiction/compulsion demonstrates the political danger that arises when illness categories embody prevailing erotic ideals. At such times sexual disorder can be wielded as a "baton" to force erotically unconventional individuals to adhere to sexual norms (Levine and Troiden 1988). Perhaps then, rather than speaking of sexual *disorders*—a term that suggests objective disease and dysfunction—we could more accurately speak of *disorderly* sex; sex that is socially disruptive, sex that disturbs the dominant cultural sexual ideal.

Culturally-defined sexual ideals regarding valid forms of sexual activity and relationship are institutionally supported (defining marriage as the union of two rather than three persons, for example); such institutionalization confers legitimacy, value, and power such that questioning or challenging the norm seems to threaten disorder.

Medical diagnoses may be preferable when other definitional options include depravity (you sinner! you freak!) or personal moral failing (how *could* you?). But medical neutrality is false neutrality given the negative social judgment that is illness (Irvine 2005; Conrad and Schneider 1992). The helping ethos and humanitarian ideal of medicine may obfuscate but does not negate the reality that therapeutic intervention in sexual matters has disciplinary effects. Disease categories are forms of power that enshrine and enforce prevailing sexual standards in the name of sexual health. Disorders of desire are thus as much about the *social* body as they are about the corporeal one. Medical-therapeutic discourse, though, disguises the ways in which the personal has always been political when it comes to sex.

Note

1. This work was supported in part by NIMH Grant T32MH19996 and benefited from many stimulating discussions with Kim Greenwell.

REFERENCES

Conrad, Peter, and Joseph Schneider. 1992. *Deviance and Medicalization: From Badness to Sickness.* Philadelphia, PA: Temple University Press.

Foucault, Michel. 1990 (1978). *The History of Sexuality* New York: Vintage.

Irvine, Janice. 2005. *Disorders of Desire* (2nd ed.). Philadelphia. PA: Temple University Press.

Levine, Martin P. and Richard R. Troiden. 1988. "The myth of sexual compulsivity" *Journal of Sex Research* 25, 3:347–63.

Loe, Meika. 2004. *The Rise of Viagra: How the Little Blue Pill Changed Sex in America.* New York: NYU Press.

Maines, Rachel P. 1999. *The Technology of Orgasm "Hysteria," the Vibrator, and Women's Sexual Satisfaction.* Baltimore, MD: Johns Hopkins University Press.

McGann. P. J. 1999. "Skirting the gender normal divide: a tomboy life story." In Mary Romero and Abigail J. Stewart (eds.), *Women's Untold Stories; Breaking Silence, Talking Back, Voicing Complexity.* New York: Routledge.

Rubin, Gayle S. 1993. "Thinking sex." In Henry Abelove, Michele Aina Barale, and David M. Halperin (eds.), *The Lesbian and Gay Studies Reader.* New York: Routledge.

Seidman, Steven. 2002. *Beyond the Closet: The Transformation of Gay and Lesbian Life.* New York: Routledge.

Tieffer, Leonore. 1995. *Sex is Not a Natural Act and Other Essays.* Oxford: Westview Press.

THINKING ABOUT THE READING

According to McGann, how do the medical institution and other institutions (legal, religious, etc.) help create and enforce "normal" sexual activity and relationships? Taking one of the three examples McGann analyzes, explain how these definitions work at both the cultural and individual levels. Taking McGann's definition of "disorderly sex," think of some examples of this concept beyond the ones she provides in this article.

Patients, "Potheads," and Dying to Get High

Wendy Chapkis

(2006)

The Wo/Men's Alliance for Medical Marijuana (WAMM) is an organization that is not easily classified. WAMM is not, as the federal Drug Enforcement Administration (DEA) might suggest, a cover for illicit drug dealing to recreational users, but neither is it properly characterized as a pharmacy dispensing physician-recommended medicine. In this article, I describe briefly the origins of WAMM and then discuss the problem of trying to divide medical marijuana users into "real patients" and "potheads." Such classification is complicated further by the relationship between medicinal cannabis use and the experience of getting "high."

What Is WAMM?

WAMM was founded in California in 1993 by medical marijuana patient Valerie Corral and her husband, Michael Corral, a master gardener. In April 2004, WAMM drew national and international attention when it successfully won a temporary injunction against the U.S. Justice Department; as a result, the alliance now operates the only nongovernmental legal medical marijuana garden in the country. The organization is unique in other respects as well. It is organized as a *cooperative.* Marijuana is grown and distributed collectively and *without charge* to the 250 patient participants. Instead of paying for their marijuana, members are expected, as their health permits, to contribute volunteer hours to the organization by working in the garden; assisting with fundraising; making cannabis tinctures, milk, capsules, and muffins; or volunteering in the office.

Over a five-year period (from 1999 to 2004), I conducted more than three dozen interviews with WAMM members about their involvement with the organization and their therapeutic use of marijuana. The WAMM members interviewed for this article reported use of marijuana with a physician's recommendation for a range of conditions including nausea related to chemotherapy (for cancer and AIDS), spasticity (multiple sclerosis), seizures (epilepsy), and chronic and acute pain. In order to become members of WAMM, each of the patient participants had to discuss with his or her doctor the possible therapeutic value of marijuana and to have been told explicitly (and in writing) that cannabis might prove useful in managing the specific symptoms associated with his or her illness, disability, or course of treatment. The number of members the program can accommodate is limited by the amount of marijuana the organization is able to grow. There is an extensive waiting list to join the organization; with more than 80% of its members living with a life-threatening illness, the standing joke is that "people are literally dying to get into WAMM."

Financial support for the organization comes largely from external donations. In 1998, however, the federal government revoked WAMM's nonprofit status on the grounds that it was involved in supplying a federally prohibited substance. WAMM's struggle for survival further intensified in the fall of 2002 when the federal DEA raided the WAMM garden and arrested the two cofounders (to date, no charges have been filed against them). Despite these challenges, WAMM has continued to operate with the full support of California's elected officials and in close cooperation with local law enforcement. In April 2004, Judge Jeremy Fogel of the federal district court in San

Jose (citing a recent Ninth Circuit Court of Appeals decision, *Raich v. Ashcroft,* soon to be reviewed by the U.S. Supreme Court) barred the Justice Department from interfering with the Corrals, WAMM patients, or the collective's garden. The federal injunction has provided at least temporary respite in the ongoing battle with the federal government.

Who Are the WAMM Members? "Worthy Patients" or "Unworthy Potheads"?

Dorothy Gibbs is the sort of patient voters are encouraged to imagine when the question of medical marijuana is before them. At ninety-four and confined to a bed in a Santa Cruz nursing home, this WAMM member is hardly the stereotypical "pothead" many critics believe to be hiding behind the medical marijuana movement. Cannabis, for Dorothy Gibbs, has never been anything but a medicine, a particularly effective analgesic that relieves severe pain associated with her post-polio syndrome:

> I never smoked marijuana before; I had no reason to. But the relief I got was wonderful and long lasting and pretty immediate, too. I didn't really have any misgivings about using marijuana; I figured it had to be better than what I'd got. They had me on lots of other medications, but I couldn't stand them; they made me so sick.

Most Americans (80% according to a recent CNN/Time poll; Stein, 2002) support the right of seriously ill patients like Dorothy Gibbs to access and use medical marijuana. This broad support is coupled, however, with lingering concerns that medical marijuana may be, as described in *Time* magazine, largely "a kind of ruse" (Stein). From this perspective, medical marijuana campaigns are seen as a cover for drug legalization and most, if not all, "medicinal use" as nothing more than a recreational habit dressed up in a doctor's recommendation.

Tensions between medical and social uses of marijuana are unavoidable in a political context in which nonmedicinal use is at once widespread, formally prohibited, and often severely punished. Because of the social and legal penalties associated with recreational use, it is reasonable that some consumers would attempt to acquire a measure of legitimacy and protection by identifying a medical need for marijuana. Medical marijuana users, then, become divided in the public arena between patients, like Dorothy Gibbs, who have never used marijuana except as a medicine, and "pretenders" who have a social relationship to the drug. As with other discreditable identities (like the prostitute, the poor person, or the single mother), a line is then drawn between a small class of deserving "victims" and a much larger group of the willfully bad who are unworthy of protection or support.

Such divisions are both illusory and dangerous. In the case of marijuana use, the identities of medical and social users are not neatly dichotomous. Some medical users have had prior experience with marijuana as a means of enhancing pleasure before they had occasion to become familiar with its potential in relieving pain. Other patients discovered the reasons for the plant's popularity as a recreational drug only after being introduced to it for a more narrowly therapeutic purpose.

With the majority of WAMM members living with life-threatening conditions and many of the chronically ill confined to wheelchairs, this is an organization that presents the legitimate face of medical marijuana, the sick and dying who are widely seen as deserving of the drug. Yet even within this population, neat divisions between medical and social users are unworkable. "Di," for example, a WAMM member in her midforties living with AIDS, acknowledges,

> I'm just going to be totally honest—it wasn't AIDS that introduced me to pot. I had smoked marijuana as a kid and I liked it even then. When I tested positive in 1991, I felt that it was kind of a benefit that I got to use the term "medical marijuana," but I didn't quite own it as medicine because it had just been my lifestyle. But then, a few years ago, I traveled out of state [without access to marijuana]. I

> spent a week traveling and then went to Florida with my mom. By the time we got there, I was in so much pain from the neuropathy, I couldn't get up. We went to Urgent Care and they gave me morphine. The pain just wouldn't go away. I took the morphine for a week until I got back to California. When I got home, I started smoking pot again as normal and it took about three or four days and I stopped taking the morphine. I realized I had probably kept myself from having this really severe nerve pain for a long time by smoking every day. It was like this big validation that I really was using good medicine.

"Maria," a fifty-two-year-old single mother living with metastatic ovarian cancer, had no current relationship to marijuana when she fell ill, but she did associate the drug with the recreational use of her youth. This past association made it difficult for her to accept that cannabis might have therapeutic value:

> I don't even know if I would have believed [that marijuana was medicine] if I hadn't tried it for medical purposes myself. I hadn't smoked for many years since I had my daughter. But a good friend said that they had heard it was really good for the nausea [related to chemotherapy] and turned me on to WAMM. . . . What an incredible difference; the pharmaceuticals don't hold a candle [to marijuana] in terms of immediate relief . . . I don't think I would have believed it because it had always been recreational to me.

A Cover for Drug Dealing or an Alternative Pharmacy?

Because of confusions about the legitimacy of marijuana as medicine and of users as patients, provider organizations such as WAMM are often misunderstood as well. Even within communities largely tolerant of marijuana use, such as Santa Cruz, suspicions remain about the role of a provider organization. "Betty," now a WAMM volunteer, initially assumed WAMM was little more than a cover for recreational users to obtain their drug of choice:

> A friend of the family developed stomach cancer, and when he got the prescription for marijuana he said to me, "I got into this organization; it's called WAMM." I had heard of WAMM but had never been involved with it or anything. Anyway, he says, "Everybody wants to be my caregiver but they all smoke and they're going to steal my pot. I know you don't do marijuana, so would you do this for me?" At first I said, "No, I don't think so." A couple of weeks later, he came back and asked again. So I said, "I'll tell you what—I'll go and check this out, but I'm not going to be sitting around with a bunch of potheads. I'm really not into that, I might as well be honest. But I'll go with you and check it out." So I went and I was really surprised at what I found. These people aren't potheads. These people aren't drug addicts. They're not derelicts. It's nothing like I had envisioned in my mind. I was very surprised . . . these people are really sick. And it's not like they all sit around and get stoned. I was amazed.

Similarly, "Hal," a seventy-year-old with severe neurological pain from failed back surgery, remembers that when a friend suggested he consider marijuana to manage the pain and WAMM as a way to access that marijuana, he was suspicious:

> Right, "medical" marijuana, sure. But [after trying it] I couldn't deny I felt better. I didn't know anything about WAMM; I'd never even heard of a cannabis buying club. I just wasn't in that world. I immediately jumped to the wrong conclusion. I thought "you're a bunch of potheads who are scamming the system." Right? So I'll be a pothead and scam the system. I don't care because I need it. I need it.

Hal's suspicion that WAMM was largely a cover for drug dealing to recreational users was shattered only when he attended his first weekly membership meeting:

> The first time I went to WAMM, with all these misconceptions in my mind, I looked around the room and thought, "My god, these people are really not well." I went home and said to my wife, "I'm going to have to rethink this whole thing. I'm going to have to stop jumping to conclusions here because this was an incredible experience."

Attending a WAMM meeting is indeed consciousness altering; new patients and guests enter expecting a room thick with marijuana smoke and instead find a room filled with human suffering and a collectively organized attempt to alleviate it. In fact, no marijuana is smoked at WAMM membership meetings. Rather, the hour-and-a-half gathering is spent building community: sharing news about the needs of the organization and the needs of the membership. Announcements are made not only about volunteer "opportunities" to work in the garden or the office, but also about members needing hospital visits, meals, or informal hospice support. Memorials are planned for those who have recently died, and holiday parties are organized for those with a desire to socialize and to celebrate. Information is exchanged about the practical dimensions of living with chronic or terminal illness and about coping with the often cascading challenges of pain, poverty, and social isolation. The meetings conclude with members picking up a week's supply of medical marijuana.

Neither Drug Dealing nor a Pharmacy: A New Model of Community Health Care

Although WAMM is not, then, a cover for recreational drug dealing, neither is it simply a pharmacy dispensing physician-recommended medicine. Indeed, by its very design, the organization does not "dispense" marijuana at all. Rather, members collectively grow, harvest, clean, and store the plants; transform them into tinctures, baked goods, and other products; and draw their share throughout the year according to medical need. This is made all the more remarkable by the fact that WAMM never charges for that medicine. The paradigm-breaking phenomenon of patients collectively producing their own medicines not only challenges the "pharmaceuticalization" of healing, it also creates a therapeutic setting that effectively disrupts the atomized experience of illness and treatment characteristic of conventional medical practice. And, because WAMM members are on the front lines of legal and political battles around medical marijuana, the organization also necessarily facilitates civic engagement.

In short, WAMM approaches "health care" in the most expansive terms, addressing "afflictions" of the body, mind, and spirit, as well as those of the body politic. In this way, the organization more closely resembles women's health care cooperatives (originating in the feminist health care movement of the 1970s) and AIDS self-help and community support organizations (organized through the gay community in the 1980s and 1990s) than it does a pharmacy.

Like these earlier manifestations of community-based health initiatives, WAMM, too, deliberately challenges the monopoly of medical professionals, the pharmaceutical industry, and the state to determine the conditions of treatment, access to drugs, and even the terms of life and death. The objective of WAMM activists is not simply to add another drug to a patient's medicine cabinet but rather to create community. WAMM cofounder Valerie Corral observes:

> We came together around the marijuana, but it's not just the marijuana, it's the community. If the government has its way, and we have to go to a pharmacy to get our prescriptions filled, then we do it all alone. We would lack that coming together, and that is as important as anything else. Totally important. There is magic in joining together with other beings in suffering. That's the "joyful participation in the sorrows of the world" that the Buddhists talk about. It's how you recognize something is bigger than you and it's a paradigm breaker.

Patient participants initially may join WAMM for no other reason than to access doctor-recommended medication. But it is difficult for members to relate to the organization as nothing more than a dispensary. Weekly attendance at a ninety-minute participants' meeting is required for pickup of marijuana. At a minimum, this means that members must become familiar with each other's faces, witness each other's suffering, and confront repeated requests for assistance by both the organization and by individual participants. In other words, just because the marijuana in WAMM is free and organic doesn't make it without cost, at least in terms of emotional investment. For some, like thirty-seven-year-old "John," living with HIV, the price feels very high indeed.

> I've had a strange relationship to WAMM because a dispensary is really what I would have rather had it be. I'm a matter-of-fact kind of person, and if I have to have this condition, and I have to use a substance, I want to be able to get it and go and not be a part of anything. I don't want to know who my pharmacist is. That's exactly how I feel about WAMM. I go to the meetings because it is a requirement, but it's not necessarily what I would opt to do. I bet everybody who goes just wants to pick up their medicine and leave. Basically, what we want is our medicine and to get on with our lives.

Indeed, those "with a life" and, perhaps more important, an income may prefer a dispensary or buyers' club over a demandingly intimate self-help collective. But for many who remain members, marijuana becomes only one of a number of threads tying them to the organization. "Joe," a forty-year-old man with a severe seizure disorder, explains:

> The medicine is actually turning into a secondary or tertiary part of what WAMM is all about for me now. It's more about the group itself, the fellowship that goes on, the ways we help each other. Actually that's the biggest thing I want to rave about: that de-isolation that takes place. Isolation that accompanies illness gets to everybody eventually. Suddenly you are removed from any kind of social matrix, like being in school or at work so you don't have the day-to-day contacts with people that make all the difference in your life. WAMM takes you out of that isolation by putting you in contact with other people, like it or not. That's what I really like about the requirement that you come every week to get what you need. You have to be there. That's the only rule actually, and that's what makes it work. People get there whether they'd rather stay at home and then they start finding things in themselves that relate to other people. It's a way for patients to get a hold of their own lives and feel whole, feel human.

One of the most distinctive features of belonging to the WAMM community is, in the words of one participant, the possibility of "dying in the embrace of friends." Because the majority of members are living with life-threatening illness, death is a close companion. For the most active members, this is both the source of great social cohesion and, simultaneously, an almost unbearably painful aspect of collective life. "Kurt," a forty-two-year-old living with AIDS, explains:

> At first I came because I heard that this Mother Teresa was giving out the best medicinal marijuana in the fucking world. And I wanted to know what this was about. What I found was a collective. WAMM has become the most unique group I've belonged to in my whole life. Sometimes it's hard for me, though, because it's a place for the sick and dying. And I'm sick but I'm not dying. I've pulled away from WAMM these past years because every time I become close to someone, they've died. And I was like "fuck this. I'm not going to go through this every time." But what WAMM has done for me—and for everybody they've supplied—is what nobody could do. Whether it's the marijuana or the tincture, or Valerie just coming and sitting by your side. So many people have died, but at least they had somebody sitting by their side.

Dying to Get High

Given decades of condemnation in this country of "reefer madness" and the supposedly dangerously intoxicating high produced by marijuana, it is not surprising that medical marijuana advocates have steered away from any discussion of the consciousness-altering properties of the substance. In an effort to distinguish medical from recreational use, the medical marijuana movement has focused almost entirely on the utility of medical marijuana in physical symptom management (that is, on its effects on nausea, pain, appetite, muscle spasms, ocular pressure, and seizure disorders). It is as if the "high" that inspires recreational use either disappears with medicinal use or, at best, is an unintended and unfortunate side effect. The medical marijuana movement, in other words, seems to have decided that talking about the psychoactive properties of cannabis will serve only to further discredit the drug, working against efforts to transform it into a "medicine."

WAMM cofounder Valerie Corral has resisted this impulse. In addition to gathering data on how marijuana affects members' physical symptom management, Valerie has been encouraging members to reflect on how marijuana might be affecting their psychospiritual well-being. Valerie observes:

> I've gotten criticism for even talking about "consciousness"—I think people are afraid it will be used against us. But I really think it's interesting that the government is so determined to take the "high" out of marijuana before they legalize it as a prescription medicine. You never hear them talking about taking the "low" out of opiates; we allow medicine that relieves pain and is addictive but puts a veil over consciousness. So why is it so important to remove access to a drug that relieves pain and allows for an opening of consciousness?

My interviews suggest that living with severe and chronic pain and with an enhanced awareness of death is, in itself, profoundly consciousness altering. Medical interventions that ignore this dimension are increasingly recognized as inadequate by those involved in palliative care. "Healing," in such situations, is necessarily distinct from "curing" and involves interventions in body, mind, and spirit. Dr. Bal Mount, the founder and director of the Palliative Care Unit at the Royal Victorian Hospital in Montreal, argues, "Healing doesn't necessarily have to do with just the physical body. If one has a broader idea of what healing and wellness are, all kinds of people die as well people" (quoted in Webb, 1999, p. 317).

"Bill," a fifty-three-year-old gay man who has been a caregiver to several WAMM members living with and dying of AIDS, was a cofounder of the local AIDS project in the 1980s and has extensive experience with the medical use of marijuana. He reports:

> The gay community was well aware of marijuana, and early on in the AIDS epidemic we realized that it solved several major things: it solved problems of appetite when somebody wouldn't eat anymore; and it solved nausea, which I didn't think it would, but it did. And it seemed to really help somebody get past the stuck spot they were in of being sick and not being able to be helped, being in that all-alone space. Get them stoned and they got past that.

Deborah Silverknight, a fifty-one-year-old African American/Native American woman with chronic pain from a broken back, notes that this association between marijuana and a sense of enhanced well-being is an old one:

> My great-grandmother referred to marijuana as the "mother plant"—the one you smoke that helps you medicinally and spiritually. Marijuana is a meditative thing for me. It's not only about the physical pain, but relief of pain of the spirit. If I'm having a terrible back spasm, then it's mostly about the need for physical pain relief. But even then, it has that other dimension as well.

The psychospiritual effects of marijuana were frequently remarked upon by WAMM members who reported that the consciousness-altering properties of marijuana were a necessary component of its therapeutic value. Altered consciousness enhances physical symptom relief by helping them to deal with situational depression associated with chronic pain and illness.

"Barb," a forty-five-year-old white woman with post-polio syndrome, observes:

> When I get a pain flair, I smoke and it helps to relieve the pain and relieve the spasms but it also means I don't get as depressed. My attitude is more like, "oh, okay, I'm going to be in pain today, but I'm going to enjoy what I can enjoy and get through the day." Marijuana never fails to lift my mood. I smoke and think "okay, I'm just going to have to go with the pain today. It's beautiful outside and I'm going to go tool around the garden in my chair." It takes you to another level mentally of acceptance about being in this kind of pain. So when I'm in that "I can't handle this another minute stage," it produces a positive shift and I can go on to something else. The other drugs I'm prescribed have such major side effects, but if I smoke a joint, the biggest side effect is a mental lift. And that's a side effect I can live with.

"Hal," the seventy-year-old living with severe neuropathy, notes:

> When you are in constant pain, your focus is 100% on yourself: I can't move this way, I can't twist this way, I can't put my foot down that way. That kind of thing. It's terrible.
>
> I'd never been one who was self-absorbed to the extent that I would forget about other things. I became that way [because of the pain]. I don't know how my wife could stand it. But, I find at this point [with the marijuana], I am in a sense witnessing my pain; what's happened as a result is that I am no longer so absorbed in my pain. I can rise above it, and then I'm able to do whatever I have to do.

One common objection in antidrug literature to the "high" associated with psychoactive drug use is that it offers only a "distortion" or "escape" from reality. The implication is that escape is somehow unworthy or undesirable, and the "alternative reality" accessed through drugs is illegitimate. But, in the context of chronic pain or terminal illness, one might question whether, in the words of Lily Tomlin, reality isn't "greatly overrated."

Pamela Cutler, a thirty-eight-year-old white woman in the final stages of living with metastatic breast cancer, observes:

> Marijuana kind of helps dull the reality of this situation. And anyone who says it's not a tough reality . . . I mean your mind will barely even take it in. It does dull it, and I don't think there is anything wrong with that if I want to dull it. That's fine . . . was diagnosed with breast cancer and had a radical mastectomy . . . the whole thing was a big shock. I mean I was thirty-six. I was like, "What?"And ever since then it's been like a roller coaster: okay, it's spread to your bones, and then it's spread to my lungs, and then my liver . . . [Marijuana] makes it easier to take for me. It doesn't really take it away. It just dulls the sharpness of it—like "oh my god, I'm going to be dead." I just think that's just incredible.

Although most of the individuals I interviewed commented on the therapeutic value of the psychoactive properties of marijuana, a number were careful to make a distinction between the psychoactive effect of marijuana when used medicinally and the effects when used recreationally. For many of them, the contrast between "getting loaded for fun" and medicinal use was described as profound. In part, this difference may be a question of, as Norman Zinberg (1984) has phrased it, the effects of "set" (the user's mindset) and "setting" (the context in which the drug is taken). A substance taken in expectation of pleasurable intoxication by a healthy individual may produce a substantially different effect from that experienced by an individual living with a life-threatening illness or in chronic pain.

"Kurt," the forty-two-year-old white man living with AIDS, notes:

> I used to smoke recreationally, but it became a whole different thing when I became HIV positive and needed it as a medicinal thing. I had never had a life-threatening disease, and now I was watching everyone die in front of me. It wasn't just getting high anymore; it let me think about why I am still here after they are all gone.

Some individuals suggested that the different experience of marijuana when used medicinally was less about changed context and more about simple drug habituation. "Cher," a fifty-two-year-old white woman with chronic pain and seizure disorder, reports:

> I have to use marijuana every day to deal with pain . . . I feel like I'm so habituated that it really doesn't do that much to my mood anymore or at least it's hard to tell . . . It was more fun being a big pothead than it is being a medical user, for me. You get higher when you are a recreational user. Anytime you use something every day, your system almost naturalizes it. And frankly, I do not really feel stoned anymore unless I smoke because I am so habituated to eating it. And when I smoke, I remember how wonderful it was when marijuana actually got you high. It's like any drug; your body gets habituated.

Hal too reports a diminishing "high" as he became more accustomed to the drug:

> After a couple of months, I found that I wasn't getting high as much as I was getting calm. It took a couple of months though. At first, I'd smoke and get really high and have a wonderful afternoon or evening or whatever. But then it started to change and I just got calm.

Hal's shift from "high" to "calm" may be the result not only of biochemical tolerance but also the effect of increased familiarity with the altered state so that it becomes the quotidian reality rather than the "alter."

Interview subjects who had a history of drug abuse and recovery were especially insistent on making a clear distinction between a recreational "high" and the effects of medicinal marijuana use. By drawing a clear line between "getting loaded" and "taking medicine," these individuals were able to maintain their sense of sobriety while using cannabis therapeutically. Inocencio Manjon-McFaline, a fifty-four-year-old African American/Latino man with cancer and a former cocaine addict who got sober in 1998, described the difference like this:

> What's strange now is that I don't feel the effects [from marijuana] that I remember from when I would smoke it before, smoking to get loaded. It's a different time for me. I'm not smoking it looking for a high. Maybe it's just psychologically different knowing I'm smoking it for medicinal use. But I haven't felt loaded. I smoke only for the pain. It gets me out of there, out of that frame of mind. I'll smoke and I'll tend to focus on what I want to focus on. Generally, that's my breathing and my heartbeat. And I'll get really into plants. I just really get into that and forget about pain. . . . Every breath I take is a blessing. I don't fear death, but I don't look forward to it. I really treasure life.

Inocencio's comments raise an important question about what it means to get "high." Clearly he is no longer looking to "get loaded" and argues that he no longer gets "high." Yet, his description of his medicinal use suggests that the marijuana assists him not only in dealing with pain and nausea but also in "focusing" on his breath, on his heartbeat, on the blessings of being alive. This state may seem more "altered" when there are more conventional demands on one's time than when one is in a state of dying. The present-tense focus, which is an aspect of getting "high" that is often commented on, matches the needs of the end-of-life process and therefore may not feel "altering" but rather "confirming" or "enhancing."

Qualities associated with the psychoactive effects of cannabis—such as present-tense focus, mood elevation, and a deepened appreciation of the "minor miracles" of life—may be especially usefully enhanced in the face of anxiety over death or chronic pain. Given anti-drug rhetoric in our culture, it is not surprising that some medical marijuana patients may downplay the psychoactive effects of cannabis use. And certainly some medical marijuana users may become habituated to the effects of the THC and to the altered or enhanced state to which it provides access. But "habituation," "tolerance," or "familiarity" are not synonymous with "no effect." For those living with chronic pain, terminal illness, or both, the psychotherapeutic and metaphysical effects of marijuana may complement the mindset and setting in which the substance is used. As these accounts suggest, although medical marijuana use is most certainly not just about getting "loaded" in any conventional recreational sense, its therapeutic value may be strongly tied to the psychoactive properties of *the plant* (rather than "of cannabis"). This suggestion has significant implications for both the practice of medicine and the transformation of public policy.

REFERENCES

Stein, J. (2002, November 4). The new politics of pot. *Time.* Retrieved March 4, 2005, from http://www.time.com/time/archive/preview/0,10987,1101021104–384830,00.html

Webb, M. (1999). *The good death.* New York: Bantam.

Zinberg, N. (1984). *Drug, set, and setting.* New Haven, CT: Yale University Press.

THINKING ABOUT THE READING

People who routinely use marijuana are often considered socially deviant. How is this social conception of deviance applied to people who use marijuana for medical purposes? How do common conceptions of deviance affect the availability of medical marijuana? What are some of the strategies that health care providers and users of medical marijuana use to reduce the stigma? What are the factors that determine why some drugs are considered medicinal (and therefore legal) and others considered recreational (and therefore illegal)? Can you think of other types of activities that may be considered beneficial in one social situation and deviant in another?

PART III

Social Structure, Institutions, and Everyday Life

The Structure of Society

Organizations and Social Institutions

One of the great sociological paradoxes is that we live in a society that so fiercely extols the virtues of rugged individualism and personal accomplishment, yet we spend most of our lives responding to the influence of larger organizations and social institutions. These include both nurturing organizations, such as churches and schools, and larger, more impersonal bureaucratic institutions.

No matter how powerful and influential they are, organizations are more than structures, rules, policies, goals, job descriptions, and standard operating procedures. Each organization, and each division within an organization, develops its own norms, values, and language. This is usually referred to as *organizational culture.* Organizational cultures are usually pervasive and entrenched, yet, individuals often find ways to exert some control over their lives within the confines of these organizations. Accordingly, organizations are dynamic entities in which individuals struggle for personal freedom and expression while also existing under the rules and procedures that make up the organization. Given this dynamic activity, an organization is rarely what it appears to be on the surface.

For example, many people are unaware of and unconcerned with the harsh conditions under which our most coveted products are made. William Greider, in "These Dark Satanic Mills," discusses the exploitative potential of relying on "third world" factories. He uses a particular tragedy, the 1993 industrial fire at the Kader Industrial Toy Company in Thailand, to illustrate how global economics create and sustain international inequality. Greider shows us the complex paradox of the global marketplace: While foreign manufacturing facilities free factory workers from certain poverty, they also ensnare the workers in new and sometimes lethal forms of domination.

In "The Smile Factory," John Van Maanen examines the organizational culture of one of U.S. society's most enduring icons: Disney theme parks. Disneyland and Disney World have a highly codified and strict set of conduct standards. Variations from tightly defined employee norms are not tolerated. You'd expect in such a place that employees would be a rather homogeneous group. However, Van Maanen discovers that beneath the surface of this self-proclaimed "Happiest Place on Earth" lies a mosaic of distinct groups that have created their own status system and that work hard to maintain the status boundaries between them.

Murry Milner, in "Creating Consumers: Freaks, Geeks, and Cool Kids," explores how the structure of secondary education provides an opportunity for teenagers to create their own informal social world. He argues that since teenagers lack legitimate economic and political powers within the educational sphere, they resort to creating their own status system where they define the norms, values, and rules of teenage life. Indeed, the organizational culture of this teenage world determines who fulfils the prerequisites to be conferred with status power in it. What's more, aspects of this teenage status system are clearly reflected in the commodities of consumer capitalism.

Something to Consider as You Read

As you read these selections, think about a job you've had and the new procedures you had to learn when you started. Was the job just about the procedures or did you also have to learn new (and perhaps informal) cultural norms? Think about some of the ways in which the organizational environment induces you to behave in ways that are very specific to that situation. As you read, compare some of these organizational environments to the ones discussed in this section—work and education. How might other social institutions such as sport and religion shape behavior and beliefs?

These Dark Satanic Mills

William Greider

(1997)

. . . If the question were put now to everyone, everywhere—do you wish to become a citizen of the world?—it is safe to assume that most people in most places would answer, no, they wish to remain who they are. With very few exceptions, people think of themselves as belonging to a place, a citizen of France or Malaysia, of Boston or Tokyo or Warsaw, loyally bound to native culture, sovereign nation. The Chinese who aspire to get gloriously rich, as Deng instructed, do not intend to become Japanese or Americans. Americans may like to think of themselves as the world's leader, but not as citizens of "one world."

The deepest social meaning of the global industrial revolution is that people no longer have free choice in this matter of identity. Ready or not, they are already of the world. As producers or consumers, as workers or merchants or investors, they are now bound to distant others through the complex strands of commerce and finance reorganizing the globe as a unified marketplace. The prosperity of South Carolina or Scotland is deeply linked to Stuttgart's or Kuala Lumpur's. The true social values of Californians or Swedes will be determined by what is tolerated in the factories of Thailand or Bangladesh. The energies and brutalities of China will influence community anxieties in Seattle or Toulouse or Nagoya.

. . . Unless one intends to withdraw from modern industrial life, there is no place to hide from the others. Major portions of the earth, to be sure, remain on the periphery of the system, impoverished bystanders still waiting to be included in the action. But the patterns of global interconnectedness are already the dominant reality. Commerce has leapt beyond social consciousness and, in doing so, opened up challenging new vistas for the human potential. Most people, it seems fair to say, are not yet prepared to face the implications. . . .

Two centuries ago, when the English industrial revolution dawned with its fantastic invention and productive energies, the prophetic poet William Blake drew back in moral revulsion. Amid the explosion of new wealth, human destruction was spread over England—peasant families displaced from their lands, paupers and poorhouses crowded into London slums, children sent to labor at the belching ironworks or textile looms. Blake delivered a thunderous rebuke to the pious Christians of the English aristocracy with these immortal lines:

> And was Jerusalem builded here
> Among these dark Satanic mills?

Blake's "dark Satanic mills" have returned now and are flourishing again, accompanied by the same question.[1]

On May 10, 1993, the worst industrial fire in the history of capitalism occurred at a toy factory on the outskirts of Bangkok and was reported on page 25 of the *Washington Post.* The *Financial Times* of London, which styles itself as the daily newspaper of the global economy, ran a brief item on page 6. The *Wall Street Journal* followed a day late with an account on page 11. The *New York Times* also put the story inside, but printed a dramatic photo on its front page: rows of small shrouded bodies on bamboo pallets—dozens of them—lined along the damp pavement, while dazed rescue workers stood awkwardly among the corpses. In the background, one could see the collapsed, smoldering structure of a mammoth factory where the Kader Industrial Toy Company of Thailand had employed three thousand workers manufacturing stuffed toys and plastic dolls, playthings destined for American children.[2]

The official count was 188 dead, 469 injured, but the actual toll was undoubtedly higher since the four-story buildings had collapsed swiftly in the intense heat and many bodies were incinerated. Some of the missing were never found; others fled home to their villages. All but fourteen of the dead were women, most of them young, some as young as thirteen years old. Hundreds of the workers had been trapped on upper floors of the burning building, forced to jump from third- or fourth-floor windows, since the main exit doors were kept locked by the managers, and the narrow stairways became clotted with trampled bodies or collapsed.

When I visited Bangkok about nine months later, physical evidence of the disaster was gone—the site scraped clean by bulldozers—and Kader was already resuming production at a new toy factory, built far from the city in a rural province of northeastern Thailand. When I talked with Thai labor leaders and civic activists, people who had rallied to the cause of the fire victims, some of them were under the impression that a worldwide boycott of Kader products was under way, organized by conscience-stricken Americans and Europeans. I had to inform them that the civilized world had barely noticed their tragedy.

As news accounts pointed out, the Kader fire surpassed what was previously the worst industrial fire in history—the Triangle Shirtwaist Company fire of 1911—when 146 young immigrant women died in similar circumstances at a garment factory on the Lower East Side of Manhattan. The Triangle Shirtwaist fire became a pivotal event in American politics, a public scandal that provoked citizen reform movements and energized the labor organizing that built the International Ladies Garment Workers Union and other unions. The fire in Thailand did not produce meaningful political responses or even shame among consumers. The indifference of the leading newspapers merely reflected the tastes of their readers, who might be moved by human suffering in their own communities but were inured to news of recurring calamities in distant places. A fire in Bangkok was like a typhoon in Bangladesh, an earthquake in Turkey.

The Kader fire might have been more meaningful for Americans if they could have seen the thousands of soot-stained dolls that spilled from the wreckage, macabre litter scattered among the dead. Bugs Bunny, Bart Simpson and the Muppets. Big Bird and other *Sesame Street* dolls. Playskool "Water Pets." Santa Claus. What the initial news accounts did not mention was that Kader's Thai factory produced most of its toys for American companies—Toys "R" Us, Fisher-Price, Hasbro, Tyco, Arco, Kenner, Gund and J. C. Penney—as well as stuffed dolls, slippers and souvenirs for Europe.[3]

Globalized civilization has uncovered an odd parochialism in the American character: Americans worried obsessively over the everyday safety of their children, and the U.S. government's regulators diligently policed the design of toys to avoid injury to young innocents. Yet neither citizens nor government took any interest in the brutal and dangerous conditions imposed on the people who manufactured those same toys, many of whom were mere adolescent children themselves. Indeed, the government position, both in Washington and Bangkok, assumed that there was no social obligation connecting consumers with workers, at least none that governments could enforce without disrupting free trade or invading the sovereignty of other nations.

The toy industry, not surprisingly, felt the same. Hasbro Industries, maker of Playskool, subsequently told the *Boston Globe* that it would no longer do business with Kader, but, in general, the U.S. companies shrugged off responsibility. Kader, a major toy manufacturer based in Hong Kong, "is extremely reputable, not sleaze bags," David Miller, president of the Toy Manufacturers of America, assured *USA Today*. "The responsibility for those factories,"

Miller told ABC News, "is in the hands of those who are there and managing the factory."[4]

The grisly details of what occurred revealed the casual irresponsibility of both companies and governments. The Kader factory compound consisted of four interconnected, four-story industrial barns on a three-acre lot on Buddhamondhol VI Road in the Sampran district west of Bangkok. It was one among Thailand's thriving new industrial zones for garments, textiles, electronics and toys. More than 50,000 people, most of them migrants from the Thai countryside, worked in the district at 7,500 large and small firms. Thailand's economic boom was based on places such as this, and Bangkok was almost choking on its own fantastic growth, dizzily erecting luxury hotels and office towers.

The fire started late on a Monday afternoon on the ground floor in the first building and spread rapidly upward, jumping to two adjoining buildings, all three of which swiftly collapsed. Investigators noted afterwards that the structures had been cheaply built, without concrete reinforcement, so steel girders and stairways crumpled easily in the heat. Thai law required that in such a large factory, fire-escape stairways must be sixteen to thirty-three feet wide, but Kader's were a mere four and a half feet. Main doors were locked and many windows barred to prevent pilfering by the employees. Flammable raw materials—fabric, stuffing, animal fibers—were stacked everywhere, on walkways and next to electrical boxes. Neither safety drills nor fire alarms and sprinkler systems had been provided.

Let some of the survivors describe what happened.

A young woman named Lampan Taptim: "There was the sound of yelling about a fire. I tried to leave the section but my supervisor told me to get back to work. My sister who worked on the fourth floor with me pulled me away and insisted we try to get out. We tried to go down the stairs and got to the second floor; we found that the stairs had already caved in. There was a lot of yelling and confusion. . . . In desperation, I went back up to the windows and went back and forth, looking down below. The smoke was thick and I picked the best place to jump in a pile of boxes. My sister jumped, too. She died."

A young woman named Cheng: "There is no way out [people were shouting], the security guard has locked the main door out! It was horrifying. I thought I would die. I took off my gold ring and kept it in my pocket and put on my name tag so that my body could be identifiable. I had to decide to die in the fire or from jumping down from a three stories' height." As the walls collapsed around her, Cheng clung to a pipe and fell downward with it, landing on a pile of dead bodies, injured but alive.

An older woman named La-iad Nadsnguen: "Four or five pregnant women jumped before me. They died before my eyes." Her own daughter jumped from the top floor and broke both hips.

Chauweewan Mekpan, who was five months pregnant: "I thought that if I jumped, at least my parents would see my remains, but if I stayed, nothing would be left of me." Though her back was severely injured, she and her unborn child miraculously survived.

An older textile worker named Vilaiwa Satieti, who sewed shirts and pants at a neighboring factory, described to me the carnage she encountered: "I got off work about five and passed by Kader and saw many dead bodies lying around, uncovered. Some of them I knew. I tried to help the workers who had jumped from the factory. They had broken legs and broken arms and broken heads. We tried to keep them alive until they got to the hospital, that's all you could do. Oh, they were teenagers, fifteen to twenty years, no more than that, and so many of them, so many."

This was not the first serious fire at Kader's factory, but the third or fourth. "I heard somebody yelling 'fire, fire,'" Tumthong Podhirun testified, " . . . but I did not take it seriously because it has happened before. Soon I smelled smoke and very quickly it billowed inside the place. I headed for the back door but it was locked. . . . Finally, I had no choice but to join the others and jumped out of the window. I saw many of my friends lying dead on the ground beside me."[5]

In the aftermath of the tragedy, some Bangkok activists circulated an old snapshot of two smiling peasant girls standing arm in arm beside a thicket of palm trees. One of them, Praphai Prayonghorm, died in the 1993 fire at Kader. Her friend, Kammoin Konmanee, had died in the 1989 fire. Some of the Kader workers insisted afterwards that their factory had been haunted by ghosts, that it was built on the site of an old graveyard, disturbing the dead. The folklore expressed raw poetic truth: the fire in Bangkok eerily resembled the now-forgotten details of the Triangle Shirtwaist disaster eighty years before. Perhaps the "ghosts" that some workers felt present were young women from New York who had died in 1911.

Similar tragedies, large and small, were now commonplace across developing Asia and elsewhere. Two months after Kader, another fire at a Bangkok shirt factory killed ten women. Three months after Kader, a six-story hotel collapsed and killed 133 people, injuring 351. The embarrassed minister of industry ordered special inspections of 244 large factories in the Bangkok region and found that 60 percent of them had basic violations similar to Kader's. Thai industry was growing explosively—12 to 15 percent a year—but workplace injuries and illnesses were growing even faster, from 37,000 victims in 1987 to more than 150,000 by 1992 and an estimated 200,000 by 1994.

In China, six months after Kader, eighty-four women died and dozens of others were severely burned at another toy factory fire in the burgeoning industrial zone at Shenzhen. At Dongguan, a Hong Kong-owned raincoat factory burned in 1991, killing more than eighty people (Kader Industries also had a factory at Dongguan where two fires have been reported since 1990). In late 1993, some sixty women died at the Taiwanese-owned Gaofu textile plant in Fuzhou Province, many of them smothered in their dormitory beds by toxic fumes from burning textiles. In 1994, a shoe factory fire killed ten persons at Jiangmen; a textile factory fire killed thirty-eight and injured 160 at the Qianshan industrial zone.[6]

"Why must these tragedies repeat themselves again and again?" the *People's Daily* in Beijing asked. The official *Economic Daily* complained: "The way some of these foreign investors ignore international practice, ignore our own national rules, act completely lawlessly and immorally and lust after wealth is enough to make one's hair stand on end."[7]

America was itself no longer insulated from such brutalities. When a chicken-processing factory at Hamlet, North Carolina, caught fire in 1991, the exit doors there were also locked and twenty-five people died. A garment factory discovered by labor investigators in El Monte, California, held seventy-two Thai immigrants in virtual peonage, working eighteen hours a day in "sub-human conditions." One could not lament the deaths, harsh working conditions, child labor and subminimum wages in Thailand or across Asia and Central America without also recognizing that similar conditions have reappeared in the United States for roughly the same reasons.

Sweatshops, mainly in the garment industry, scandalized Los Angeles, New York and Dallas. The grim, foul assembly lines of the poultry-processing industry were spread across the rural South; the *Wall Street Journal*'s Tony Horwitz won a Pulitzer Prize for his harrowing description of this low-wage work. "In general," the U.S. Government Accounting Office reported in 1994, "the description of today's sweatshops differs little from that at the turn of the century."[8]

That was the real mystery: Why did global commerce, with all of its supposed modernity and wondrous technologies, restore the old barbarisms that had long ago been forbidden by law? If the information age has enabled multinational corporations to manage production and marketing spread across continents, why were their managers unable—or unwilling—to organize such mundane matters as fire prevention?

The short answer, of course, was profits, but the deeper answer was about power: Firms behaved this way because they could, because nobody would stop them. When law and social values retreated before the power of markets, then capitalism's natural drive to maximize returns had no internal governor to check its social behavior. When one enterprise took the low road to gain advantage, others would follow.

The toy fire in Bangkok provided a dramatic illustration for the much broader, less visible forms of human exploitation that were flourishing in the global system, including the widespread use of children in manufacturing, even forced labor camps in China or Burma. These matters were not a buried secret. Indeed, American television has aggressively exposed the "dark Satanic mills" with dramatic reports. ABC's *20/20* broadcast correspondent Lynn Sherr's devastating account of the Kader fire; CNN ran disturbing footage. Mike Wallace of CBS's *60 Minutes* exposed the prison labor exploited in China. NBC's *Dateline* did a piece on Wal-Mart's grim production in Bangladesh. CBS's *Street Stories* toured the shoe factories of Indonesia.

The baffling quality about modern communications was that its images could take us to people in remote corners of the world vividly and instantly, but these images have not as yet created genuine community with them. In terms of human consciousness, the "global village" was still only a picture on the TV screen.

Public opinion, moreover, absorbed contradictory messages about the global reality that were difficult to sort out. The opening stages of industrialization presented, as always, a great paradox: the process was profoundly liberating for millions, freeing them from material scarcity and limited life choices, while it also ensnared other millions in brutal new forms of domination. Both aspects were true, but there was no scale on which these opposing consequences could be easily balanced, since the good and ill effects were not usually apportioned among the same people. Some human beings were set free, while other lives were turned into cheap and expendable commodities.

Workers at Kader, for instance, earned about 100 baht a day for sewing and assembling dolls, the official minimum wage of $4, but the constant stream of new entrants meant that many at the factory actually worked for much less—only $2 or $3 a day—during a required "probationary" period of three to six months that was often extended much longer by the managers. Only one hundred of the three thousand workers at Kader were legally designated employees; the rest were "contract workers" without permanent rights and benefits, the same employment system now popularized in the United States.

"Lint, fabric, dust and animal hair filled the air on the production floor," the International Confederation of Free Trade Unions based in Brussels observed in its investigative report. "Noise, heat, congestion and fumes from various sources were reported by many. Dust control was nonexistent; protective equipment inadequate. Inhaling the dust created respiratory problems and contact with it caused skin diseases." A factory clinic dispensed antihistamines or other drugs and referred the more serious symptoms to outside hospitals. Workers paid for the medication themselves and were reimbursed, up to $6, only if they had contributed 10 baht a month to the company's health fund.

A common response to such facts, even from many sensitive people, was: yes, that was terrible, but wouldn't those workers be even worse off if civil standards were imposed on their employers since they might lose their jobs as a result? This was the same economic rationale offered by American manufacturers a century before to explain why American children must work in the coal mines and textile mills. U.S. industry had survived somehow (and, in fact, flourished) when child labor and the other malpractices were eventually prohibited by social reforms. Furthermore, it was not coincidence that industry always assigned the harshest conditions and lowest pay to the weakest members of a society—women, children, uprooted migrants. Whether the factory was in Thailand or the United States or Mexico's *maquiladora* zone, people who were already quite powerless were less likely to resist, less able to demand decency from their employers. . . .

After the fire Thai union members, intellectuals and middle-class activists from social rights organizations (the groups known in developing countries as nongovernmental organizations, or NGOs) formed the Committee to Support Kader Workers and began demanding justice from the employer. They sent a delegation to Hong Kong to confront Kader officials and investigate the complex corporate linkages of the enterprise. What they discovered was that Kader's partner in the Bangkok toy factory was actually a fabulously wealthy Thai family, the Chearavanonts, ethnic Chinese merchants who own the Charoen Pokphand Group, Thailand's own leading multinational corporation.

The CP Group owns farms, feed mills, real estate, air-conditioning and motorcycle factories, food-franchise chains—two hundred companies worldwide, several of them listed on the New York Stock Exchange. The patriarch and chairman, Dhanin Chearavanont, was said by *Fortune* magazine to be the seventy-fifth richest man in the world, with personal assets of $2.6 billion (or 65 billion baht, as the *Bangkok Post* put it). Like the other emerging "Chinese multinationals," the Pokphand Group operates through the informal networks of kinfolk and ethnic contacts spread around the world by the Chinese diaspora, while it also participates in the more rigorous accounting systems of Western economies. . . .

In the larger context, this tragedy was not explained by the arrogant power of one wealthy family or the elusive complexities of interlocking corporations. The Kader fire was ordained and organized by the free market itself. The toy industry—much like textiles and garments, shoes, electronics assembly and other low-wage sectors—existed (and thrived) by exploiting a crude ladder of desperate competition among the poorest nations. Its factories regularly hopped to new locations where wages were even lower, where the governments would be even more tolerant of abusive practices. The contract work assigned to foreign firms, including thousands of small sweatshops, fitted neatly into the systems of far-flung production of major brand names and distanced the capital owners from personal responsibility. The "virtual corporation" celebrated by some business futurists already existed in these sectors and, indeed, was now being emulated in some ways by advanced manufacturing—cars, aircraft, computers.

Over the last generation, toy manufacturers and others have moved around the Asian rim in search of the bottom-rung conditions: from Hong Kong, Korea and Taiwan to Thailand and Indonesia, from there to China, Vietnam and Bangladesh, perhaps on next to Burma, Nepal or Cambodia. Since the world had a nearly inexhaustible supply of poor people and supplicant governments, the market would keep driving in search of lower rungs; no one could say where the bottom was located. Industrial conditions were not getting better, as conventional theory assured the innocent consumers, but in many sectors were getting much worse. In America, the U.S. diplomatic opening to Vietnam was celebrated as progressive politics. In Southeast Asia, it merely opened another trapdoor beneath wages and working conditions.

A country like Thailand was caught in the middle: if it conscientiously tried to improve, it would pay a huge price. When Thai unions lobbied to win improvements in minimum-wage standards, textile plants began leaving for Vietnam and elsewhere or even importing cheaper "guest workers" from Burma. When China opened its fast-growing industrial zones in Shenzhen, Dongguan and other locations, the new competition had direct consequences on the factory floors of Bangkok.

Kader, according to the ICFTU [International Confederation of Free Trade Unions], opened two new factories in Shekou and Dongguan where young people were working fourteen-hour days, seven days a week, to fill the U.S. Christmas orders for Mickey Mouse and other American dolls. Why should a company worry about sprinkler systems or fire escapes for a dusty factory in Bangkok when it could hire brand-new workers in China for only $20 a month, one fifth of the labor cost in Thailand?

The ICFTU report described the market forces: "The lower cost of production of toys in China changes the investment climate for countries like Thailand. Thailand competes with China to attract investment capital for local toy production. With this development, Thailand has become sadly lax in enforcing its own legislation. It turns a blind eye to health violations, thus allowing factory owners to ignore safety standards. Since China entered the picture, accidents in Thailand have nearly tripled."

The Thai minister of industry, Sanan Kachornprasart, described the market reality more succinctly: "If we punish them, who will want to invest here?" Thai authorities subsequently filed charges against three Kader factory managers, but none against the company itself nor, of course, the Chearavanont family.[9]

. . . The fire in Bangkok reflected the amorality of the marketplace when it has been freed of social obligations. But the tragedy also mocked the moral claims of three great religions, whose adherents were all implicated. Thais built splendid golden temples exalting Buddha, who taught them to put spiritual being before material wealth. Chinese claimed to have acquired superior social values, reverence for family and community, derived from the teachings of Confucius. Americans bought the toys from Asia to celebrate the birth of Jesus Christ. Their shared complicity was another of the strange convergences made possible by global commerce. . . .

In the modern industrial world, only the ignorant can pretend to self-righteousness since only the primitive are truly innocent. No advanced society has reached that lofty stage without enduring barbaric consequences and despoliation along the way; no one who enjoys the uses of electricity or the internal combustion engine may claim to oppose industrialization for others without indulging in imperious hypocrisy.

Americans, one may recall, built their early national infrastructure and organized large-scale agriculture with slave labor. The developing American nation swept native populations from their ancient lands and drained the swampy prairies to grow grain. It burned forests to make farmland, decimated wildlife, dammed the wild rivers and displaced people who were in the way. It assigned the dirtiest, most dangerous work to immigrants and children. It eventually granted political rights to all, but grudgingly and only after great conflicts, including a terrible civil war.

The actual history of nations is useful to remember when trying to form judgments about the new world. Asian leaders regularly remind Americans and Europeans of exactly how the richest nation-states became wealthy and observe further that, despite their great wealth, those countries have not perfected social relations among rich and poor, weak and powerful. The maldistribution of incomes is worsening in America, too, not yet as extreme as Thailand's, but worse than many less fortunate nations. . . .

Coming to terms with one's own history ought not only to induce a degree of humility toward others and their struggles, but also to clarify what one really believes about human society. No one can undo the past, but that does not relieve people of the burden of making judgments about the living present or facing up to its moral implications. If the global system has truly created a unified marketplace, then every worker, every consumer, every society is already connected to the other. The responsibility exists and invoking history is not an excuse to hide from the new social questions.

Just as Americans cannot claim a higher morality while benefiting from inhumane exploitation, neither can developing countries pretend to become modern "one world" producers and expect exemption from the world's social values. Neither can the global enterprises. The future asks: Can capitalism itself be altered and reformed? Or is the world doomed to keep renewing these inhumanities in the name of economic progress?

The proposition that human dignity is indivisible does not suppose that everyone will become equal or alike or perfectly content in his or her circumstances. It does insist that certain well-understood social principles exist internationally which are enforceable and ought to be the price of admission in the global system. The idea is very simple: every person—man, woman and child—regardless of where he or she exists in time and place or on the chain of economic development, is entitled to respect as an individual being.

For many in the world, life itself is all that they possess; an economic program that deprives them of life's precious possibilities is not only unjust, but also utterly unnecessary. Peasants may not become kings, but they are entitled to be treated with decent regard for their sentient and moral beings, not as cheap commodities. Newly industrialized nations cannot change social patterns overnight, any more than the advanced economies did before them, but they can demonstrate that they are changing.

This proposition is invasive, no question, and will disturb the economic and political arrangements within many societies. But every nation has a sovereign choice in this matter, the sort of choice made in the marketplace every day. If Thailand or China resents the intrusion of global social standards, it does not have to sell its toys to America. And Americans do not have to buy them. If Singapore rejects the idea of basic rights for women, then women in America or Europe may reject Singapore—and multinational firms that profit from the subordination of women. If people do not assert these values in global trade, then their own convictions will be steadily coarsened.

In Bangkok, when I asked Professor Voravidh to step back from Thailand's problems and suggest a broader remedy, he thought for a long time and then said: "We need cooperation among nations because the multinational corporations can shift from one country to another. If they don't like Thailand, they move to Vietnam or China. Right now, we are all competing and the world is getting worse. We need a GATT [General Agreement on Tariffs and Trade] on labor conditions and on the minimum wage, we need a standard on the minimum conditions for work and a higher standard for children."

The most direct approach, as Voravidh suggested, is an international agreement to incorporate such standards in the terms of trade, with penalties and incentives, even temporary embargoes, that will impose social obligations on the global system, the firms and countries. Most of the leading governments, including the United States, have long claimed to support this idea—a so-called social clause for GATT—but the practical reality is that they do not. Aside from rhetoric, when their negotiators are at the table, they always yield readily to objections from the multinational corporations and developing nations. Both the firms and the governing elites of poor countries have a strong incentive to block the proposition since both profit from a free-running system that exploits the weak. A countering force has to come from concerned citizens. Governments refuse to act, but voters and consumers are not impotent, and, in the meantime, they can begin the political campaign by purposefully targeting the producers—boycotting especially the well-known brand names that depend upon lovable images for their sales. Americans will not stop buying toys at Christmas, but they might single out one or two American toy companies for Yuletide boycotts, based on their scandalous relations with Kader and other manufacturers. Boycotts are difficult to organize and sustain, but every one of the consumer-goods companies is exquisitely vulnerable.

In India, the South Asian Coalition on Child Servitude, led by Kailash Satyarthi, has created a promising model for how to connect the social obligations of consumers and workers. Indian carpet makers are notorious for using small children at their looms—bonded children like Thailand's bonded prostitutes—and have always claimed economic necessity. India is a poor nation and the work gives wage income to extremely poor families, they insist. But these children will never escape poverty if they are deprived of schooling, the compulsory education promised by law.

The reformers created a "no child labor" label that certifies the rugs were made under honorable conditions and they persuaded major importers in Germany to insist upon the label. The exporters in India, in turn, have to allow regular citizen inspections of their workplaces to win the label for their rugs. Since this consumer-led certification system began, the carpet industry's use of children has fallen dramatically. A Textile Ministry official in New Delhi said: "The government is now contemplating the total eradication of child labor in the next few years."[10]

Toys, shoes, electronics, garments—many consumer sectors are vulnerable to similar approaches, though obviously the scope of manufacturing is too diverse and complex for consumers to police it. Governments have to act collectively. If a worldwide agreement is impossible to achieve, then groups of governments can form their own preferential trading systems, introducing social standards that reverse the incentives for developing countries and for capital choosing new locations for production.

The crucial point illustrated by Thailand's predicament is that global social standards will help the poorer countries escape their economic trap. Until a floor is built beneath the market's social behavior, there is no way that a small developing country like Thailand can hope to overcome the downward pull of competition from other, poorer nations. It must debase its citizens to hold on to what it has achieved. The path to improvement is blocked by the economics of an irresponsible marketplace.

Setting standards will undoubtedly slow down the easy movement of capital—and close down the most scandalous operations—but that is not a harmful consequence for people in struggling nations that aspire to industrial prosperity or for a global economy burdened with surpluses and inadequate consumption. When global capital makes a commitment to a developing economy, it ought not to acquire the power to blackmail that nation in perpetuity. Supported by global rules, those nations can begin to improve conditions and stabilize their own social development. At least they would have a chance to avoid the great class conflicts that others have experienced.

In the meantime, the very least that citizens can demand of their own government is that it no longer use public money to finance the brutal upheavals or environmental despoliation that has flowed from large-scale projects of the World Bank and other lending agencies. The social distress in the cities begins in the countryside, and the wealthy nations have often financed it in the name of aiding development. The World Bank repeatedly proclaims its new commitment to strategies that address the development ideas of indigenous peoples and halt the destruction of natural systems. But social critics and the people I encountered in Thailand and elsewhere have not seen much evidence of real change.

The terms of trade are usually thought of as commercial agreements, but they are also an implicit statement of moral values. In its present terms, the global system values property over human life. When a nation like China steals the property of capital, pirating copyrights, films or technology, other governments will take action to stop it and be willing to impose sanctions and penalty tariffs on the offending nation's trade. When human lives are stolen in the "dark Satanic mills," nothing happens to the offenders since, according to the free market's sense of conscience, there is no crime.

Notes

1. William Blake's immortal lines are from "Milton," one of his "prophetic books" written between 1804 and 1808. *The Portable Blake,* Alfred Kazin, editor (New York: Penguin Books, 1976).

2. *Washington Post, Financial Times* and *New York Times,* May 12, 1993, and *Wall Street Journal,* May 13, 1993.

3. The U.S. contract clients for Kader's Bangkok factory were cited by the International Confederation of Free Trade Unions headquartered in Brussels in its investigatory report, "From the Ashes: A Toy Factory Fire in Thailand," December 1994. In the aftermath, the ICFTU and some

nongovernmental organizations attempted to mount an "international toy campaign" and a few sporadic demonstrations occurred in Hong Kong and London, but there never was a general boycott of the industry or any of its individual companies. The labor federation met with associations of British and American toy manufacturers and urged them to adopt a "code of conduct" that might discourage the abuses. The proposed codes were inadequate, the ICFTU acknowledged, but it was optimistic about their general adoption by the international industry.

4. Mitchell Zuckoff of the *Boston Globe* produced a powerful series of stories on labor conditions in developing Asia and reported Hasbro's reaction to the Kader fire, July 10, 1994. David Miller was quoted in *USA Today,* May 13, 1993, and on ABC News *20/20,* July 30, 1993.

5. The first-person descriptions of the Kader fire are but a small sampling from survivors' horrifying accounts, collected by investigators and reporters at the scene. My account of the disaster is especially indebted to the investigative report by the International Confederation of Free Trade Unions; Bangkok's English-language newspapers, the *Post* and *The Nation;* the Asia Monitor Resource Center of Hong Kong; and Lynn Sherr's devastating report on ABC's *20/20,* July 30, 1993. Lampan Taptim and Tumthong Podhirun, "From the Ashes," ICFTU, December 1994; Cheng: *Asian Labour Update,* Asia Monitor Resource Center, Hong Kong, July 1993; La-iad Nads-nguen: *The Nation,* Bangkok, May 12, 1993; and Chaweewan Mekpan: *20/20.*

6. Details on Thailand's worker injuries and the litany of fires in China are from the ICFTU report and other labor bulletins, as well as interviews in Bangkok.

7. The *People's Daily* and *Economic Daily* were quoted by Andrew Quinn of Reuters in *The Daily Citizen* of Washington, DC, January 18, 1994.

8. Tony Horwitz described chicken-processing employment as the second fastest growing manufacturing job in America: *Wall Street Journal,* December 1, 1994. U.S. sweatshops were reviewed in "Garment Industry: Efforts to Address the Prevalence and Conditions of Sweatshops," U.S. Government Accounting Office, November 1994.

9. Sanan was quoted in the *Bangkok Post,* May 29, 1993.

10. The New Delhi-based campaign against child labor in the carpet industry is admittedly limited to a narrow market and expensive product, but its essential value is demonstrating how retailers and their customers can be connected to a distant factory floor. See, for instance, Hugh Williamson, "Stamp of Approval," *Far Eastern Economic Review,* February 2, 1995, and N. Vasuk Rao in the *Journal of Commerce,* March 1, 1995.

THINKING ABOUT THE READING

Greider argues that the tragedy of the Kader industrial fire cannot be explained simply by focusing on greedy families and multinational corporations. Instead, he blames global economics and the organization of the international toy industry. He writes, "The Kader fire was ordained and organized by the free market itself." What do you suppose he means by this? Given the enormous economic pressures that this and other multinational industries operate under, are such tragedies inevitable? Why have attempts to improve the working conditions in "third world" factories been so ineffective?

The Smile Factory

Work at Disneyland

John Van Maanen

(1991)

Part of Walt Disney Enterprises includes the theme park Disneyland. In its pioneering form in Anaheim, California, this amusement center has been a consistent money maker since the gates were first opened in 1955. Apart from its sociological charm, it has, of late, become something of an exemplar for culture vultures and has been held up for public acclaim in several best-selling publications as one of America's top companies. . . . To outsiders, the cheerful demeanor of its employees, the seemingly inexhaustible repeat business it generates from its customers, the immaculate condition of park grounds, and, more generally, the intricate physical and social order of the business itself appear wondrous.

Disneyland as the self-proclaimed "Happiest Place on Earth" certainly occupies an enviable position in the amusement and entertainment worlds as well as the commercial world in general. Its product, it seems, is emotion—"laughter and well-being." Insiders are not bashful about promoting the product.Bill Ross, a Disneyland executive, summarizes the corporate position nicely by noting that "although we focus our attention on profit and loss, day-in and day-out we cannot lose sight of the fact that this is a feeling business and we make our profits from that."

The "feeling business" does not operate, however, by management decree alone. Whatever services Disneyland executives believe they are providing to the 60 to 70 thousand visitors per day that flow through the park during its peak summer season, employees at the bottom of the organization are the ones who most provide them. The work-a-day practices that employees adopt to amplify or dampen customer spirits are therefore a core concern of this feeling business. The happiness trade is an interactional one. It rests partly on the symbolic resources put into place by history and park design but it also rests on an animated workforce that is more or less eager to greet the guests, pack the trams, push the buttons, deliver the food, dump the garbage, clean the streets, and, in general, marshal the will to meet and perhaps exceed customer expectations. False moves, rude words, careless disregard, detected insincerity, or a sleepy and bored presence can all undermine the enterprise and ruin a sale. The smile factory has its rules.

It's a Small World

. . . This rendition is of course abbreviated and selective. I focus primarily on such matters as the stock appearance (vanilla), status order (rigid), and social life (full), and swiftly learned codes of conduct (formal and informal) that are associated with Disneyland ride operators. These employees comprise the largest category of hourly workers on the payroll. During the summer months, they number close to four thousand and run the 60-odd rides and attractions in the park.

They are also a well-screened bunch. There is—among insiders and outsiders alike—a rather fixed view about the social attributes carried by the standard-make Disneyland ride operator. Single, white males and females in their early twenties, without facial blemish, of above average height and below average weight, with straight teeth, conservative grooming standards, and a chin-up, shoulder-back posture radiating the sort of good health suggestive of a recent history in sports are typical of these social identifiers. There are representative minorities on the payroll but because ethnic displays are sternly discouraged by management, minority employees are rather close copies of the standard model Disneylander, albeit in different colors.

This Disneyland look is often a source of some amusement to employees who delight in pointing out that even the patron saint, Walt himself, could not be hired today without shaving off his trademark pencil-thin mustache. But, to get a job in Disneyland and keep it means conforming to a rather exacting set of appearance rules. These rules are put forth in a handbook on the Disney image in which readers learn, for example, that facial hair or long hair is banned for men as are aviator glasses and earrings and that women must not tease their hair, wear fancy jewelry, or apply more than a modest dab of makeup. Both men and women are to look neat and prim, keep their uniforms fresh, polish their shoes, and maintain an upbeat countenance and light dignity to complement their appearance—no low spirits or cornball raffishness at Disneyland.

The legendary "people skills" of park employees, so often mentioned in Disneyland publicity and training materials, do not amount to very much according to ride operators. Most tasks require little interaction with customers and are physically designed to practically insure that is the case. The contact that does occur typically is fleeting and swift, a matter usually of only a few seconds. In the rare event sustained interaction with customers might be required, employees are taught to deflect potential exchanges to area supervisors or security. A Training Manual offers the proper procedure: "On misunderstandings, guests should be told to call City Hall.... In everything from damaged cameras to physical injuries, don't discuss anything with guests... there will always be one of us nearby." Employees learn quickly that security is hidden but everywhere. On Main Street security cops are Keystone Kops; in Frontierland, they are Town Marshalls; on Tom Sawyer's Island, they are Cavalry Officers, and so on.

Occasionally, what employees call "line talk" or "crowd control" is required of them to explain delays, answer direct questions, or provide directions that go beyond the endless stream of recorded messages coming from virtually every nook and cranny of the park. Because such tasks are so simple, consisting of little more than keeping the crowd informed and moving, it is perhaps obvious why management considers the sharp appearance and wide smile of employees so vital to park operations. There is little more they could ask of ride operators whose main interactive tasks with visitors consist of being, in their own terms, "information booths," "line signs," "pretty props," "shepherds," and "talking statues."

A few employees do go out of their way to initiate contact with Disneyland customers but, as a rule, most do not and consider those who do to be a bit odd. In general, one need do little more than exercise common courtesy while looking reasonably alert and pleasant. Interactive skills that are advanced by the job have less to do with making customers feel warm and welcome than they do with keeping each other amused and happy. This is, of course, a more complex matter.

Employees bring to the job personal badges of status that are of more than passing interest to peers. In rough order, these include: good looks, college affiliation, career aspirations, past achievements, age (directly related to status up to about age 23 or 24 and inversely related thereafter), and assorted other idiosyncratic matters. Nested closely alongside these imported status badges are organizational ones that are also of concern and value to employees.

Where one works in the park carries much social weight. Postings are consequential because the ride and area a person is assigned provide rewards and benefits beyond those of wages. In-the-park stature for ride operators turns partly on whether or not unique skills are required. Disneyland neatly complements labor market theorizing on this dimension because employees with the most differentiated skills find themselves at the top of the internal status ladder, thus making their loyalties to the organization more predictable.

Ride operators, as a large but distinctly middle-class group of hourly employees on the floor of the organization, compete for status not only with each other but also with other employee groupings whose members are hired for the season from the same applicant pool. A loose approximation of the rank ordering among these groups can be constructed as follows:

1. The upper-class prestigious Disneyland Ambassadors and Tour Guides (bilingual young women in charge of ushering—some say rushing—little bands of tourists through the park);
2. Ride operators performing coveted "skilled work" such as live narrations or tricky transportation tasks like those who symbolically control customer access to the park and drive the costly entry vehicles (such as the antique trains, horse-drawn carriages, and Monorail);
3. All other ride operators;
4. The proletarian Sweepers (keepers of the concrete grounds);
5. The sub-prole or peasant status Food and Concession workers (whose park sobriquets reflect their lowly social worth—"pancake ladies," "peanut pushers," "coke blokes," "suds divers," and the seemingly irreplaceable "soda jerks").

Pay differentials are slight among these employee groups. The collective status adheres, as it does internally for ride operators, to assignment or functional distinctions. As the rank order suggests, most employee status goes to those who work jobs that require higher degrees of special skill, [offer] relative freedom from constant and direct supervision, and provide the opportunity to organize and direct customer desires and behavior rather than to merely respond to them as spontaneously expressed.

The basis for sorting individuals into these various broad bands of job categories is often unknown to employees—a sort of deep, dark secret of the casting directors in personnel. When prospective employees are interviewed, they interview for "a job at Disneyland," not a specific one. Personnel decides what particular job they will eventually occupy. Personal contacts are considered by employees as crucial in this job-assignment process as they are in the hiring decision. Some employees, especially those who wind up in the lower ranking jobs, are quite disappointed with their assignments as is the case when, for example, a would-be Adventureland guide is posted to a New Orleans Square restaurant as a pot scrubber. Although many of the outside acquaintances of our pot scrubber may know only that he works at Disneyland, rest assured, insiders will know immediately where he works and judge him accordingly.

Uniforms are crucial in this regard for they provide instant communication about the social merits or demerits of the wearer within the little world of Disneyland workers. Uniforms also correspond to a wider status ranking that casts a significant shadow on employees of all types. Male ride operators on the Autopia wear, for example, untailored jump-suits similar to pit mechanics and consequently generate about as much respect from peers as the grease-stained outfits worn by pump jockeys generate from real motorists in gas stations. The ill-fitting and homogeneous "whites" worn by Sweepers signify lowly institutional work tinged, perhaps, with a reminder of hospital orderlies rather than street cleanup crews. On the other hand, for males, the crisp, officer-like Monorail operator stands alongside the swashbuckling Pirate of the Caribbean, the casual cowpoke of Big Thunder Mountain, or the smartly vested Riverboat pilot as carriers of valued symbols in and outside the park. Employees lust for these higher status positions and the rights to small advantages such uniforms provide. A lively internal labor market exists wherein there is much scheming for the more prestigious assignments.

For women, a similar market exists although the perceived "sexiness" of uniforms, rather than social rank, seems to play a larger role. To wit, the rather heated antagonisms that developed years ago when the ride "It's a Small World" first opened and began outfitting the ride operators with what were felt to be the shortest skirts and most revealing blouses in the park. Tour Guides, who traditionally headed the fashion vanguard at Disneyland in their above-the-knee kilts, knee socks, tailored vests, black English hats, and smart riding crops were apparently appalled at being upstaged by their social inferiors and lobbied actively (and, judging by the results, successfully) to lower the skirts, raise the necklines, and generally remake their Small World rivals. . . .

Movement across jobs is not encouraged by park management, but some does occur (mostly within an area and job category). Employees claim that a sort of "once a sweeper, always a sweeper" rule obtains but all know of at least a few exceptions to prove the rule. The exceptions offer some (not much) hope for those working at the social margins of the park and perhaps keep them on the job longer than might otherwise be expected. Dishwashers can dream of becoming Pirates, and with persistence and a little help from their friends, such dreams just might come true next season (or the next).

These examples are precious, perhaps, but they are also important. There is an intricate pecking order among very similar categories of employees. Attributes of reward and status tend to cluster, and there is intense concern about the cluster to which one belongs (or would like to belong). To a degree, form follows function in Disneyland because the jobs requiring the most abilities and offering the most interest also offer the most status and social reward. Interaction patterns reflect and sustain this order. Few Ambassadors or Tour Guides, for instance, will stoop to speak at length with Sweepers who speak mostly among themselves or to Food workers. Ride operators, between the poles, line up in ways referred to above with only ride proximity (i.e., sharing a break area) representing a potentially significant intervening variable in the interaction calculation. . . .

Paid employment at Disneyland begins with the much renowned University of Disneyland whose faculty runs a day-long orientation program (Traditions I) as part of a 40-hour apprenticeship program, most of which takes place on the rides. In the classroom, however, newly hired ride operators are given a very thorough introduction to matters of managerial concern and are tested on their absorption of famous Disneyland fact, lore, and procedure. Employee demeanor is governed, for example, by three rules:

> First, we practice the friendly smile.
>
> Second, we use only friendly and courteous phrases.
>
> Third, we are not stuffy—the only Misters in Disneyland are Mr. Toad and Mr. Smee.

Employees learn too that the Disneyland culture is officially defined. The employee handbook put it in this format:

> Dis-ney Cor-po-rate Cul-ture (diz'ne kor'pr'it kul'cher) *n* 1. Of or pertaining to the Disney organization, as *a:* the philosophy underlying all business decisions; *b:* the commitment of top leadership and management to that philosophy; *c:* the actions taken by individual cast members that reinforce the image.

Language is also a central feature of university life, and new employees are schooled in its proper use. Customers at Disneyland are, for instance, never referred to as such, they are "guests." There are no rides at Disneyland, only "attractions." Disneyland itself is a "Park," not an amusement center, and it is divided into "back-stage," "on-stage," and "staging" regions. Law enforcement personnel hired by the park are not policemen, but "security hosts." Employees do not wear uniforms but check out fresh "costumes" each working day from "wardrobe." And, of course, there are no accidents at Disneyland, only "incidents." . . .

Classes are organized and designed by professional Disneyland trainers who also instruct a well-screened group of representative hourly employees straight from park operations on the approved newcomer training methods and materials. New-hires seldom see professional trainers in class but are brought on board by enthusiastic peers who concentrate on those aspects of park procedure thought highly general matters to be learned by all employees. Particular skill training (and "reality shock") is reserved for the second wave of socialization occurring on the rides themselves as operators are taught, for example, how and when to send a mock bobsled caroming down the track or, more delicately, the proper ways to stuff an obese adult customer into the midst of children riding the Monkey car on the Casey Jones Circus Train or, most problematically, what exactly to tell an irate customer standing in the rain who, in no uncertain terms, wants his or her money back and wants it back now.

During orientation, considerable concern is placed on particular values the Disney organization considers central to its operations. These values range from the "customer is king" verities to the more or less unique kind, of which "everyone is a child at heart when at Disneyland" is a decent example. This latter piety is one few employees fail to recognize as also attaching to everyone's mind as well after a few months of work experience. Elaborate checklists of appearance standards are learned and gone over in the classroom and great efforts are spent trying to bring employee emotional responses in line with such standards. Employees are told repeatedly that if they are happy and cheerful at work, so, too, will the guests be at play. Inspirational films, hearty pep talks, family imagery, and exemplars of corporate performance are all representative of the strong symbolic stuff of these training rites. . . .

Yet, like employees everywhere, there is a limit to which such overt company propaganda can be effective. Students and trainers both seem to agree on where the line is drawn, for there is much satirical banter, mischievous winking, and playful exaggeration in the classroom. As young seasonal employees note, it is difficult to take seriously an organization that provides its retirees "Golden Ears" instead of gold watches after 20 or more years of service. All newcomers are aware that the label "Disneyland" has both an unserious and artificial connotation and that a full embrace of the Disneyland role would be as deviant as its full rejection. It does seem, however, because of the corporate imagery, the recruiting and selection devices, the goodwill trainees hold toward the organization at entry, the peer-based employment context, and the smooth fit with real student calendars, the job is considered by most ride operators to be a good one. The University of Disneyland, it appears, graduates students with a modest amount of pride and a considerable amount of fact and faith firmly ingrained as important things to know (if not always accept). . . .

Employees learn quickly that supervisors and, to a lesser degree, foremen are not only on the premises to help them, but also to catch them when they slip over or brazenly violate set procedures or park policies. Because most rides are tightly designed to eliminate human judgment and minimize operational disasters, much of the supervisory monitoring is directed at activities ride operators consider trivial: taking too long a break; not wearing parts of one's official uniform such as a hat, standard-issue belt, or correct shoes; rushing the ride (although more frequent violations seem to be detected for the provision of longer-than-usual rides for lucky customers); fraternizing with guests beyond the call of duty; talking back to quarrelsome or sometimes merely querisome customers; and so forth. All are matters covered quite explicitly in the codebooks ride operators are to be familiar with, and violations of such codes are often subject to instant and harsh discipline. The firing of what to supervisors are "malcontents," "trouble-makers," "bumblers," "attitude problems," or simply "jerks" is a frequent occasion at Disneyland, and among part-timers, who are most subject to degradation and being fired, the threat is omnipresent. There are few workers who have not witnessed firsthand the rapid disappearance of a co-worker for offenses they would regard as "Mickey Mouse." Moreover, there are few employees who themselves have not violated a good number of operational and demeanor standards and anticipate, with just cause, the violation of more in the future. . . .

Employees are also subject to what might be regarded as remote controls. These stem not from supervisors or peers but from thousands of paying guests who parade daily through the park. The public, for the most part, wants Disneyland employees to play only the roles for which they are hired and costumed. If, for instance, Judy of the Jets is feeling tired, grouchy, or bored, few customers want to know about it. Disneyland employees are expected to be sunny and helpful; and the job, with its limited opportunities for sustained interaction, is designed to support such a stance. Thus, if a ride operator's behavior drifts noticeably away from the norm, customers are sure to point it out—"Why aren't you smiling?" "What's wrong with you?" "Having a bad day?" "Did Goofy step on your foot?" Ride operators learn swiftly from the constant hints, glances, glares, and tactful (and tactless) cues sent by their audience what their role in the park is to be, and as long as they keep to it, there will be no objections from those passing by.

> I can remember being out on the river looking at the people on the Mark Twain looking down on the people in the Keel Boats who are looking up at them. I'd come by on my raft and they'd all turn and stare at me. If I gave them a little wave and a grin, they'd all wave back and smile; all ten thousand of them. I always wondered what would happen if I gave them the finger? (Ex-ride operator, 1988)

Ride operators also learn how different categories of customers respond to them and the parts they are playing on-stage. For example, infants and small children are generally timid, if not frightened, in their presence. School-age children are somewhat curious, aware that the operator is at work playing a role but sometimes in awe of the role itself. Nonetheless, these children can be quite critical of any flaw in the operator's performance. Teenagers, especially males in groups, present problems because they sometimes go to great lengths to embarrass, challenge, ridicule, or outwit an operator. Adults are generally appreciative and approving of an operator's conduct provided it meets their rather minimal standards, but they sometimes overreact to the part an operator is playing (positively) if accompanied by small children. . . .

The point here is that ride operators learn what the public (or, at least, their idealized version of the public) expects of their role and find it easier to conform to such expectations than not. Moreover, they discover that when they are bright and lively others respond to them in like ways. This . . . balancing of the emotional exchange is such that ride operators come to expect good treatment. They assume, with good cause, that most people will react to their little waves and smiles with some affection and perhaps joy. When they do not, it can ruin a ride operator's day. . . .

By and large, however, the people-processing tasks of ride operators pass good naturedly and smoothly, with operators hardly noticing much more than the bodies passing in front of view (special bodies, however, merit special attention as when crew members on the subs gather to assist a young lady in a revealing outfit on board and then linger over the hatch to admire the view as she descends the steep steps to take her seat on the boat). Yet, sometimes, more than a body becomes visible, as happens when customers overstep their roles and challenge employee authority, insult an operator, or otherwise disrupt the routines of the job. In the process, guests become "dufusses," "ducks," and "assholes" (just three of many derisive terms used by ride operators to label those customers they believe to have gone beyond the pale). Normally, these characters are brought to the attention of park security officers, ride foremen, or area supervisors who, in turn, decide how they are to be disciplined (usually expulsion from the park).

Occasionally, however, the alleged slight is too personal or simply too extraordinary for a ride operator to let it pass unnoticed or merely inform others and allow them to decide what, if anything, is to be done. Restoration of one's respect is called for, and routine practices have been developed for these circumstances. For example, common remedies include: the "seatbelt squeeze," a small token of appreciation given to a deviant customer consisting of the rapid cinching-up of a required seatbelt such that the passenger is doubled-over at the point of departure and left gasping for the duration of the trip; the "break-toss," an acrobatic gesture of the Autopia trade whereby operators jump on the outside of a norm violator's car, stealthily unhitching the safety belt, then slamming on the brakes, bringing the car to an almost instant stop while the driver flies on the hood of the car (or beyond); the "seatbelt slap," an equally distinguished (if primitive) gesture by which an offending customer receives a sharp, quick snap of a hard plastic belt across the face (or other parts of the body) when entering or exiting a seat-belted ride; the "break-up-the-party" gambit, a queuing device put to use in officious fashion whereby bothersome pairs are separated at the last minute into different units, thus forcing on them the pain of strange companions for the duration of a ride through the Haunted Mansion or a ramble on Mr. Toad's Wild Ride; the "hatch-cover ploy," a much beloved practice of Submarine pilots who, in collusion with mates on the loading dock, are able to drench offensive guests with water as their units pass under a waterfall; and, lastly, the rather ignoble variants of the "Sorry-I-didn't-see-your-hand" tactic, a savage move designed to crunch a particularly irksome customer's hand (foot, finger, arm, leg, etc.) by bringing a piece of Disneyland property to bear on the appendage, such as the door of a Thunder Mountain railroad car or the starboard side of a Jungle Cruise boat. This latter remedy is, most often, a "near miss" designed to startle the little criminals of Disneyland.

All of these unofficial procedures (and many more) are learned on the job. Although they are used sparingly, they are used. Occasions of use provide a continual stream of sweet revenge talk to enliven and enrich colleague conversation at break time or after work. Too much, of course, can be made of these subversive practices and the rhetoric that surrounds their use. Ride operators are quite aware that there are limits beyond which they dare not pass. If they are caught, they know that restoration of corporate pride will be swift and clean.

In general, Disneyland employees are remarkable for their forbearance and polite good manners even under trying conditions. They are taught, and some come to believe, for a while at least, that they are really "on-stage" at work. And, as noted, surveillance by supervisory personnel certainly fades in light of the unceasing glances an employee receives from the paying guests who tromp daily through the park in the summer. Disneyland employees know well that they are part of the product being sold and learn to check their more discriminating manners in favor of the generalized countenance of a cheerful lad or lassie whose enthusiasm and dedication is obvious to all.

At times, the emotional resources of employees appear awesome. When the going gets tough and the park is jammed, the nerves of all employees are frayed and sorely tested by the crowd, din, sweltering sun, and eyeburning smog. Customers wait in what employees call "bullpens" (and park officials call "reception areas") for up to several hours for a 3½ minute ride that operators are sometimes hell-bent on cutting to 2½ minutes. Surely a monument to the human ability to suppress feelings has been created when both users and providers alike can maintain their composure and seeming regard for one another when in such a fix.

It is in this domain where corporate culture and the order it helps to sustain must be given its due. Perhaps the depth of a culture is visible only when its members are under the gun. The orderliness—a good part of the Disney formula for financial success—is an accomplishment based not only on physical design and elaborate procedures, but also on the low-level, part-time employees who, in the final analysis, must be willing, even eager, to keep the show afloat. The ease with which employees glide into their kindly and smiling roles is, in large measure, a feat of social engineering. Disneyland does not pay well; its supervision is arbitrary and skin-close; its working conditions are chaotic; its jobs require minimal amounts of intelligence or judgment; and asks a kind of sacrifice and loyalty of its employees that is almost fanatical. Yet, it attracts a particularly able workforce whose personal backgrounds suggest abilities far exceeding those required of a Disneyland traffic cop, people stuffer, queue or line manager, and button pusher. As I have suggested, not all of Disneyland is covered by the culture put forth by management. There are numerous pockets of resistance and various degrees of autonomy maintained by employees. Nonetheless, adherence and support for the organization are remarkable. And, like swallows returning to Capistrano, many part-timers look forward to their migration back to the park for several seasons.

The Disney Way

Four features alluded to in this unofficial guide to Disneyland seem to account for a good deal of the social order that obtains within the park. First, socialization, although costly, is of a most selective, collective, intensive, serial, sequential, and closed sort. These tactics are notable for their penetration into the private spheres of individual thought and feeling. . . . Incoming identities are not so much dismantled as they are set aside as employees are schooled in the use of new identities of the situational sort. Many of these are symbolically powerful and, for some, laden with social approval. It is hardly surprising that some of the more problematic positions in terms of turnover during the summer occur in the food and concession domains where employees apparently find little to identify with on the job. Cowpokes on Big Thunder Mountain, Jet Pilots, Storybook Princesses, Tour Guides, Space Cadets, Jungle Boat Skippers, or Southern Belles of New Orleans Square have less difficulty on this score. Disneyland, by design, bestows identity through a process carefully set up to strip away the job relevance of other sources of identity and learned response and replace them with others of organizational relevance. It works.

Second, this is a work culture whose designers have left little room for individual experimentation. Supervisors, as apparent in their focused wandering and attentive looks, keep very close tabs on what is going on at any moment in all the lands. Every bush, rock, and tree in Disneyland is numbered and checked continually as to the part it is playing in the park. So too are employees. Discretion of a personal sort is quite limited while employees are "on-stage." Even "back-stage" and certain "off-stage" domains have their corporate monitors.

Employees are indeed aware that their "off-stage" life beyond the picnics, parties, and softball games is subject to some scrutiny, for police checks are made on potential and current employees. Nor do all employees discount the rumors that park officials make periodic inquiries on their own as to a person's habits concerning sex and drugs. Moreover, the sheer number of rules and regulations is striking, thus making the grounds for dismissal a matter of multiple choice for supervisors who discover a target for the use of such grounds. The feeling of being watched is, unsurprisingly, a rather prevalent complaint among Disneyland people, and it is one that employees must live with if they are to remain at Disneyland.

Third, emotional management occurs in the park in a number of quite distinct ways. From the instructors at the university who beseech recruits to "wish every guest a pleasant good day," to the foremen who plead with their charges to, "say thank you when you herd them through the gate," to the impish customer who seductively licks her lips and asks, "what does Tom Sawyer want for Christmas?" appearance, demeanor, and etiquette have special meanings at Disneyland. Because these are prized personal attributes over which we normally feel in control, making them commodities can be unnerving. Much self-monitoring is involved, of course, but even here self-management has an organizational side. Consider ride operators who may complain of being "too tired to smile" but, at the same time, feel a little guilty for uttering such a confession. Ride operators who have worked an early morning shift on the Matterhorn (or other popular rides) tell of a queasy feeling they get when the park is opened for business and they suddenly feel the ground begin to shake under their feet and hear the low thunder of the hordes of customers coming at them, oblivious of civil restraint and the small children who might be among them. Consider, too, the discomforting pressures of being "on-stage" all day and the cumulative annoyance of having adults ask permission to leave a line to go to the bathroom, whether the water in the lagoon is real, where the well-marked entrances might be, where Walt Disney's cryogenic tomb is to be found, or—the real clincher—whether or not one is "really real."

The mere fact that so much operator discourse concerns the handling of bothersome guests suggests that these little emotional disturbances have costs. There are, for instance, times in all employee careers when they put themselves on "automatic pilot," "go robot," "can't feel a thing," "lapse into a dream," "go into a trance," or otherwise "check out" while still on duty. Despite a crafty supervisor's (or curious visitor's) attempt to measure the glimmer in an employee's eye, this sort of willed emotional numbness is common to many of the "on-stage" Disneyland personnel. Much of this numbness is, of course, beyond the knowledge of supervisors and guests because most employees have little trouble appearing as if they are present even when they are not. It is, in a sense, a passive form of resistance that suggests there still is a sacred preserve of individuality left among employees in the park.

Finally, taking these three points together, it seems that even when people are trained, paid, and told to be nice, it is hard for them to do so all of the time. But, when efforts to be nice have succeeded to the degree that is true of Disneyland, it appears as a rather towering (if not always admirable) achievement. It works at the collective level by virtue of elaborate direction. Employees—at all ranks—are stage-managed by higher ranking employees who, having come through themselves, hire, train, and closely supervise those who have replaced them below. Expression rules are laid out in corporate manuals. Employee time-outs intensify work experience. Social exchanges are forced into narrow bands of interacting groups. Training and retraining programs are continual. Hiding places are few. Although little sore spots and irritations remain for each individual, it is difficult to imagine work roles being more defined (and accepted) than those at Disneyland. Here, it seems, is a work culture worthy of the name.

THINKING ABOUT THE READING

What is the significance of the title "The Smile Factory"? What, exactly, is the factory-made product that Disney sells in its theme parks? How does the Disney organizational culture shape the lives of employees? Disney has been criticized for its strict—some would say oppressive—employee rules and regulations. Would it be possible to run a "smile factory" with a more relaxed code of conduct? Disney theme parks in countries such as France and Japan have not been nearly as successful as Disneyland and Disney World. What are some of the reasons why the "feeling business" doesn't export as well to other countries? Consider also the social rankings that employees create. Describe examples of social rankings in your own experience. What are the criteria for these rankings?

Creating Consumers

Freaks, Geeks, and Cool Kids

Murry Milner

(2006)

Teenagers have long been preoccupied with the clothes they wear, how they fix their hair, the cars they drive, the latest music, and what constitutes being "cool." Adults have been complaining about teenagers in general and their materialism and consumerism in particular for more than fifty years. Why has little been done to change these patterns of adolescent behavior? Certainly in other societies and in other historical periods in our own society adults have exercised more control and authority over young people. If so many adults are critical of teenage behavior, why have they hesitated to exercise more control and authority? Why have the numerous attempts at curriculum revision and school reform had so little impact on these patterns? Why do adults both bemoan the consumerism of teenagers and yet do many things to encourage it? The answer to this set of questions has relatively little to do with family values, liberalism, or progressive education. Rather, it has to do with the benefits that adults, especially parents and businesses, gain from the present way of organizing young people's lives.

A teenager's status in the eyes of his or her peers is extremely important to most adolescents. Why this near obsession with status? It is because they have so little real economic or political power. They must attend school for most of the day and they have only very limited influence on what happens there. They are pressured to learn complex and esoteric knowledge like algebra, chemistry, and European history, which rarely has immediate relevance to their day-to-day lives. They do, however, have one crucial kind of power: the power to create an informal social world in which they evaluate one another. That is, they can and do create their own status systems—usually based on criteria that are quite different from those promoted by parents or teachers. In short, the main kind of power teenagers have is status power. Predictably, their status in the eyes of their peers becomes very important in their day-to-day lives.

The usual ways of trying to change the basic patterns of adolescent behavior are not promising. Better parenting styles, compensatory education for those from disadvantaged backgrounds, and reform schools are likely to have only modest effects on teenage behavior. This conclusion is congruent with sociologist Randall Collins' more general analysis of power. In his *The Credential Society,* Collins argued that significant social power is primarily rooted in the extensiveness of the informal social network one is able to maintain. In his terms, power is rooted in political labor rather than productive labor (He is using "political" and "productive" in a very broad sense and not restricting politics to the government or productivity to the economy.). Stated in more concrete terms, who you know and how you relate to them is often more important than your technical skill or any objective measure of productivity. Adolescent status groups are often the starting point for learning "political" skills and building social networks. Many teenagers seem to have an intuitive sense of this and accordingly they care more about their friendships than they care about their grades.

Are Peers Important?

If parents, social class, and schools all have less impact than is usually assumed, what shapes teenagers' behavior? Steinberg and Harris both see peers—other teenagers—as playing a very important role. Studies of younger children have also found that peer cultures develop quite early and are an important feature of children's lives. William Corsaro, one of the foremost sociologists of children, emphasizes that we should not study childhood as simply a period of socialization and training—a prelude to adulthood. Rather the social lives of children and adolescents need to be studied in their own right, and their role in the wider society analyzed. Much of children's socialization results from interaction with peers. The peer cultures that emerge are not iron scripts that are simply inherited and determine young people's behavior. Rather, according to Corsaro, children engage in "interpretive reproduction." This involves innovation and individual adaptation, as well as reproduction of what has been socially inherited.

I think the emphasis that Harris, Steinberg, and Corsaro place on young peoples' peer relationships and cultures is especially warranted in the study of adolescents. Adolescents spend more time with each other and less time with adults than younger children do and hence the importance of peer cultures almost certainly increases. In some situations the impact of peers seems to outweigh the impact of family influences, in shaping economic and occupational goals and opportunities. Teenagers themselves clearly think peer relations are crucial. When teenagers phone hotline peer counseling centers, what they most commonly discuss is not family problems, a lack of money, academic stress—or even sex and drugs—but their relationships with other students.

There is room for argument about *how* influential peers are, but the argument seems to range from quite important to extremely important. Given that peers are very significant, *my focus is on why and how teenagers organize themselves when they are left more or less to their own devices.* Some of this happens outside of school hours; much of adolescent social life, however, takes place within school facilities, even if it is not part of the formal educational program.

Why Is Status so Important to Teenagers?

If peers are so influential, how do they exercise this influence? A primary means is through creating status differences. Of course, all social systems create some kinds of status differences. So why are these so important for teenagers? In all societies, as individuals mature they develop some level of independence from their parents; their autonomy increases. School children in modern societies, in effect, "go off to work" and spend most of their day away from their parents. Teachers and administrators supervise them, but the scope of these adults' authority is much narrower than that of parents. The ratio of supervisors to subordinates also decreases significantly. This new autonomy and reduced control by adults usually means that the influence of peers is amplified dramatically. All of these processes are further intensified when students reach high school. They move to larger, more complex schools, gain increased mobility (often by driving cars), and greater communication facilities (via the telephone, e-mail, and the Internet). Not only school time, but most leisure time is spent in the presence of peers or in communication with them. "In a typical week, even discounting time spent in classroom instruction (23% of an average student's waking hours), high school students spend twice as much of their time with peers (29%) as with parents or other adults (15%)." The separation of school and peers from family, the increased mobility, and the independent communication networks mean that the actions of students are less visible to adults, and hence less subject to supervision and control.

Adolescents have more autonomy, but little economic or political power. They cannot change the curriculum, hire or fire the teachers, decide who will be admitted to their school, or move to another school without the permission of adults. At the time of life when the biological sources of sexuality are probably strongest, in a social environment saturated with sexual imagery and language, they are exhorted to avoid sex. In many situations they are treated as inferior citizens who are looked upon as at best a nuisance. They are denied the right to buy alcohol or see "adult" movies and are subject to the control not only of parents, teachers, and police, but numerous petty clerks in stores, movies, and nightclubs who "check their IDs."

In one realm, however, their power is supreme; they control their evaluations of one another. That is, the kind of power they do have is status power: the power to create their own status systems based on their own criteria. Predictably, the creation and distribution of this kind of power is often central to their lives.

The Economy and the Status System: Profiting from Teenagers

Teenagers play a much broader and more important economic role than is usually assumed, but let us begin with the obvious ways they affect consumer capitalism—as a market for businesses.

Teenagers often seek to maintain or enhance their status by the acquisition of fashionable status symbols. Children and especially teenagers are important consumers not only because they are potential customers in the present but also for two other reasons. According to Professor James U. McNeal in *Kids as Customers: A Handbook of Marketing to Children,* there are three markets: children as a primary market spending their own money; as a significant influence on their parents' spending; and as a future market when they become adults. All of these become more important as children develop into adolescents.

Taking Aim at Teenagers

Because of the three markets outlined above, teenagers have become the target of massive marketing and advertising campaigns. These campaigns shape their selection and use of a broad range of commodities. Advertising is also a core feature of the symbolic milieu within which young people grow up—and hence their very being and sense of the world are shaped by such ads. By the time American children graduate from high school they will have spent 18,000–22,000 hours in front of the television compared to only 13,000 hours in the classroom. Children and adolescents spend more time watching TV and videos than any other activity besides sleeping.

Teenagers are willing to experiment with new products and so companies see them as customers whose loyalty is up for grabs. As an article in the traditionally conservative *Economist* reported, the strategy is "Hook them on a brand today, and with any luck they will still be using it in the next century." These companies not only advertise the nature of their products, but also attempt to associate their products with cultural images that appeal to children, including the Flintstones, Batman, and a whole array of Disney characters and various "superheroes." Most sport shoe and apparel companies have one or more celebrity athletes in their ads, which are aimed at teenagers. The Nike advertising campaign featuring Michael Jordan and his special line of athletic shoes is a classic example. Rock and rap music stars are also featured in ads aimed specifically at teenagers.

Very few ad-free zones exist for adolescents, as marketing permeates TV, the Internet, magazines, and even public schools. Others have written about the commercialization of schools, but I will give a brief overview of some of these trends. One of the most noted efforts at marketing in schools is Channel One. This is a TV program beamed by satellite to participating schools. It features ten minutes of youth-oriented news reporting and two minutes of advertising daily. Twelve thousand schools nationwide had opted to receive Channel One in 1993, serving over eight million teenagers, about forty percent of 12- to 17-year-olds enrolled in school. Advertisers on

Channel One pay $200,000 for a thirty-second spot. This by no means exhausts the advertising in schools. According to Consumers Union, in some schools ads adorn bathroom stalls and wall space above urinals. Nutritional posters depicting Rice Krispies cereal, compliments of Kellogg, fill hallways. Ronald McDonald speaks at no charge to schools about self-esteem and other supposedly pertinent topics. School districts now sell exclusive contracts to Coke and Pepsi for vending machine rights. Even Eli Lily, the pharmaceutical company that produces the antidepressant Prozac, spoke free of charge at a high school in Maryland, then handed out school supplies depicting Prozac insignia. School buses emblazoned with advertisements now serve New York City and Colorado Springs school districts. Some educators are concerned that the results of advertising in an educational environment could have far deeper effects than other advertising. Advertising products in an "educational context may further heighten their credibility and impact." In the terms of the theory of status relations, the legitimacy of advertisements is enhanced by their association with educational institutions.

There is little doubt that marketers work hard to shape and influence the choices of teenagers, but their success is by no means automatic. Teenagers are not dolts totally manipulated by marketers and the media. Styles and products frequently fall flat; the "kids" simply do not think they are cool. In his excellent journalistic account of those who study what teenagers want, Malcolm Gladwell refers to these marketing research efforts as "the coolhunt." This uncertainty about what will be cool is the primary reason that firms spend considerable amounts of money and time on such research.

To the degree that adults are concerned about the consumerism of teenagers, they usually blame the "usual suspects": the businesses that make products aimed at teenagers and the marketers and advertisers that attempt to persuade adolescents to buy these. But to understand both the reasons for teenage consumerism, and more generally the patterns of behavior characteristic of teenagers, we need to look beyond the "usual suspects" and consider the role of some "un-indicted co-conspirators": teachers, public officials, and parents.

Teachers, Officials, and Parents: Profiting in Other Ways

Despite a half century of adult complaints, and sometimes outrage, few changes have occurred in the way adults organize the lives of adolescents: They are sent off to schools for five days a week, sorted by age, and supervised by a few teachers. As school attendance and years of schooling increased, more and more young people have been kept in a state of postponed maturity for longer and longer. Clearly, this is likely to increase the importance of peers and decrease the significance of adults—resulting in behaviors that adults have long grumbled about. When those with real power complain loud and long about the behavior of subordinates—without really doing anything—we need to ask why. Often those in authority have a vested interest in the very patterns they bemoan. More accurately, they are wed to patterns of social organization that make the forms of behavior they lament likely, if not inevitable.

* * *

I will focus on three categories of adults that benefit from the existing patterns: parents, teachers, and public officials.

Teachers and public officials: Teachers, school administrators, and public officials are rarely in favor of shrinking or even radically changing the education system. Politicians find supporting education attractive because it is much easier to promise a better future than to actually improve the present. Even supporters of educational vouchers usually assume most young people will remain in school and that taxes will pay for this. In general, politicians, educational administrators, and teachers gain in power and influence when the scope of the school system is increased. Nearly everyone wants schools to be more effective, and most teachers work hard to accomplish this. Better teaching, revised curricula, and improved facilities may help. Such measures do not, however, address the issue of youth cultures, which frequently resist adult visions of what should happen in schools. Radically reducing adolescent autonomy, which makes youth cultures and resistance possible, would require a level of coercion and exclusion that is not available to teachers—even if they wanted to use it. One or both of two things would be required. Schools would need to become more like prisons. Or, many students would have to be excluded from schools. Both the financial and the political costs would be enormous. The deinstitutionalization of the mentally ill has visibly impacted American cities by creating thousands of homeless people. Expelling all of the high school students who were not serious about their studies would have a much more dramatic and negative impact on both the job market and "law and order." In short, while our present schools often frustrate politicians, educational officials, and teachers, they are seldom in favor of a fundamental reorganization of education. Instead politicians, officials, educators and citizens engage in their favorite activity: revising the curriculum. Some of these revisions may actually change students' behaviors in modest ways, but they have little impact on the basic ways young people spend their lives. More commonly such reforms are taken in stride or ignored as young people go about the more important business of spending time with their friends.

Parents: First, let me affirm a legitimate truism: most parents love their children, want the best for them, and often make significant sacrifices on their behalf. But this does not mean that parents do not indirectly benefit from the social arrangements that cause the kind of adolescent behavior they criticize. One aspect of this might be called "*leaving children.*" Raising children is very hard work and most parents—good, loving parents—are more than happy to see them go off to school for the better part of the day. Adults then have time to do other things with their lives. Many children and adolescents—good kids who love their parents—will still engage in risky behaviors if they can do so undetected. I am *not* suggesting that parents should stay home, but only that not doing so has significant consequences, not simply at the level of individual children and families, but at the level of the neighborhood, the community, and the society.

A second form of adult behavior might be called "*indulging children.*" Parents' absence from their children's lives may be a source of guilt for a parent. One way to appease this guilt is by spending money on them. Joan Chiaramonte, vice president of a market research firm, states that because parents have such a limited amount of time with their children, they do not "want to spend it arguing over whether to go to McDonald's or Burger King," and so give in quickly to their children's demands.

"*Using children*" to display the family's status is another way parents reinforce teenage behavior. Some parents attempt to demonstrate their wealth and status via their children. Until the middle of the twentieth century, men earned most of the income for the family, while women, as "traditional housewives," were responsible for providing services to family members and converting money into status. This was seen most clearly in the value placed on nearness, cleanliness, decorations, and entertaining as lavishly as budgets would allow. These responsibilities spawned a whole industry of magazines, such as *Good Housekeeping, Ladies Home Journal, Better Homes and Gardens,* and *McCalls.* In general, today's adults are too busy earning income to attend to many of the subtle fine points of showing it off within local neighborhoods and communities. Increasingly children are their status symbols. Of course, parents in many societies and historical periods have spent considerable time and resources to improve their family status by enhancing their children's attractiveness to others—as terms like dowry, bride price, finishing school, and debutante ball suggest.

What is new is that the children are increasingly the key decision makers; they choose their peers and the commodities—including major purchases like automobiles—needed to enhance status among their peers and the family's status in the community. For parents who have less time and opportunity to display their wealth in their neighborhood and local community, teenagers eagerly step in to take on this "responsibility." In this respect they are often the contemporary "housewives" of the professional and managerial classes.

Last, but not least, parents contribute to the behavior they object to in adolescents by "*copying teenagers.*" Parents not only raise their own status by having successful, attractive teenagers; they raise their own status by being like their teenagers.

If, however, parents gain status by being more like teenagers, it is difficult for them to exercise authority to change or shape the behaviors of those teenagers. Most parents try to responsibly cope with the dilemma they face: maintaining their own status through displays of youthfulness and sexuality, and limiting and guiding the sexuality of their immature adolescents. This is, however, a fine and difficult line to walk. Many parents do this thoughtfully and skillfully. Some, however, have attempted to resolve this dilemma by helping their children to take on the characteristics of adults. Through dress, cosmetics, language, and body movements the children take on the symbols of sexuality at an early age. The result is the sexy sixth grader keenly attuned to the latest fashions and status symbols, and the juvenile parent preoccupied with the same concerns.

My intention is not to scold teachers, public officials, and parents, or to call into question their good intentions, sincerity, and hard work, but to suggest that understanding adolescents requires paying serious attention to the things adults do to create and sustain the teenage behaviors they lament.

My argument is that the structure of American secondary education keeping teenagers in their own isolated world with little economic and political power or few non-school responsibilities results in the status preoccupations of teenagers. These status concerns, in turn, play a significant contributing role in the development and maintenance of consumer capitalism.

Economist Juliet Schor has emphasized that people's base of comparison has been changed because of the mass media and the decline of social ties in neighborhoods. They no longer simply try to "keep up with the Joneses" who live down the street; they don't know who lives down the street. Now they aspire to the lifestyles portrayed on television—images that disproportionately focus on the "rich and famous" or greatly exaggerate the standard of living of "typical" people. This increase in the level of aspiration has been further accentuated by growing income inequality and an extensive system of consumer credit.

There is a well-established line of theory and research that claims the influence of the mass media and other impersonal institutions is often mediated through concrete networks of interpersonal ties. Early research on voting, marketing, and the diffusion of ideas found that "opinion leaders" were crucial. Most people were not directly influenced by impersonal messages, but by the opinions of those they knew and interacted with.

My argument is that the high school status system typically serves this role in the lives of teenagers. The "opinion leaders" who mediate the influence of the mass media are largely the peers in high school crowds and cliques. Their role is more than simply being the forerunners in the adoption of images from the mass media. They also create new styles. As we have seen, the producers for the adolescent market spend considerable time and effort observing and attempting to anticipate what high school opinion leaders will embrace as the next hot item, the thing that must be purchased in order to be cool. Hence, without wanting to deny the importance of other influences on consumer markets, attention needs to be drawn to the crucial role that these adolescent status systems play in the broader economy and society.

Our educational system plays a central role, not just in giving people technical skills, but also in molding their desires and ambitions. Life with one's peers, in and out of the classroom, powerfully shapes people's worldviews and personalities. The peer status system is central to this process. Currently that status system is an integral part of consumer capitalism; learning to consume is one of the most important lessons taught in our high schools. The question we need to face is whether this is the kind of education we want to give our children and the kind of people we want them to become. The answer to that question will be determined, in large measure, by how adults choose to organize their own lives.

Let me recapitulate the arguments of this [reading]. Maintaining high levels of consumption has become crucial to the economic prosperity of advanced capitalist societies, which can legitimately be characterized as consumer capitalism. Changes in fashion and more generally the desire to acquire status symbols have become central to maintaining high levels of consumption and economic demand. High school status systems play an important role in socializing people to be concerned about their status and more specifically the way this status is displayed through the acquisition of consumer commodities. "Learning to consume," not "learning to labor," is the central lesson taught in American high schools.

While most adults complain about teenagers' preoccupation with status and status symbols, adults support the basic institutions that encourage these adolescent behaviors because in certain important respects the grown-ups benefit from the existing social arrangements. Many businesses have become very-self-conscious of how their economic interests are linked to the status preoccupations of adolescents. They have developed extensive marketing campaigns to encourage the preoccupations of both young people and adults with status, consumption, and the associated ideals of youthfulness, self-indulgence, and hedonism. Consumption now consumes much of life.

THINKING ABOUT THE READING

Milner states that "'learning to consume,' not 'learning to labor,' is the central lesson taught in American high schools." Indeed, teenagers hold little power in the economic and political spheres, but they do have one crucial type of power within their teenage networks. According to Milner, what is that supreme power and how is it linked to consumption? Apply Milner's discussion of this type of power to your experience as a teenager or your observation of teenagers. How, if at all, does technology and social networking play a role in the creation and maintenance of this power? Milner suggests that "un-indicted co-conspirators" like teachers, public officials, and parents complain about how this aforementioned power organizes the world of teenagers. Nonetheless, discuss the ways he *also* suggests that these adults benefit and even profit from this power held by teenagers.

The Architecture of Stratification

10

Social Class and Inequality

Inequality is woven into the fabric of all societies through a structured system of *social stratification.* Social stratification is a ranking of entire groups of people that perpetuates unequal rewards and life chances in society. The structural-functionalist explanation of stratification is that the stability of society depends on all social positions being filled—that is, there are people around to do all the jobs that need to be done. Higher rewards, such as prestige and large salaries, are afforded to the most important positions, thereby ensuring that the most qualified individuals will occupy the highest positions. In contrast, conflict theory argues that stratification reflects an unequal distribution of power in society and is a primary source of conflict and tension.

Social class is the primary means of stratification in American society. Contemporary sociologists are likely to define a person's class standing as a combination of income, wealth, occupational prestige, and educational attainment. It is tempting to see class differences as simply the result of an economic stratification system that exists at a level above the individual. Although inequality is created and maintained by larger social institutions, it is often felt most forcefully and is reinforced most effectively in the chain of interactions that take place in our day-to-day lives.

The media play a significant role in shaping people's perceptions of class. But instead of providing accurate descriptive information about different classes, the media—especially the news media—give the impression that the United States is largely a classless society. According to Gregory Mantsios in "Making Class Invisible," when different classes are depicted in the media, the images tend to hover around stereotypes that reinforce the cultural belief that people's position in society is largely a function of their own effort and achievement or, in the case of "the poor," lack of effort and achievement.

The face of American poverty has changed somewhat over the past several decades. The economic status of single mothers and their children has deteriorated while that of people older than age 65 has improved somewhat. What hasn't changed is the ever-widening gap between the rich and the poor. Poverty persists because in a free market and competitive society, it serves economic and social functions. In addition, poverty receives institutional "support" in the form of segmented labor markets and inadequate educational systems. The ideology of competitive individualism—that to succeed in life, all one has to do is work hard and win in competition with others—creates a belief that poor people are to blame for their own suffering. So although the problem of poverty remains serious, public attitudes toward poverty and poor people are frequently indifferent or even hostile. Fred Block and his colleagues call this attitude "the compassion gap." In this reading, the authors discuss the tendency of intolerance toward the poor in U.S. society. According to their research, this cultural attitude of indifference or disdain is rooted in individualism and a lack of understanding of economic conditions over time (e.g., the relative difficulty of owning a home today as compared with the period just after World War II, when much government assistance was available). The authors see the "compassion gap" as an attitude that gets in the way of establishing more workable social policies for the poor.

Social class affects nearly every aspect of our daily lives. In "Branded with Infamy: Inscriptions of Poverty and Class in America," Vivyan Adair describes the various ways in which poor women's bodies are marked as "unclean" or "unacceptable." These markings are the result of a life of poverty: lack of access to adequate health care, lack of proper nutrition and shelter, and a consequence of difficult and demanding physical and emotional labor. But instead of seeing the impact of economic circumstances on the lives of these women, more affluent people tend to view them (and their children) as members of an undesirable social class who should be discipline, controlled, and punished.

Something to Consider as You Read

In reading these selections, pay careful attention to the small ways in which economic resources affect everyday choices and behavior. For instance, how might poverty, including the lack of access to nice clothing, affect one's ability to portray the best possible image at a job interview? Consider further the connection between media portrayals and self-image. Where do people get their ideas about their own self-worth, their sense of entitlement, and how they fit into society generally? How do these ideals differ across social class, and how are they similar? Some observers have suggested that people in the United States don't know how to talk about class, except in stereotypical terms. How might this lack of "class discourse" perpetuate stereotypes and the myth that the poor deserve their fate? Consider examples of the "compassion gap" in your own life and as reflected in recent news and policy decisions.

Making Class Invisible

Gregory Mantsios

(1998)

Of the various social and cultural forces in our society, the mass media is arguably the most influential in molding public consciousness. Americans spend an average twenty-eight hours per week watching television. They also spend an undetermined number of hours reading periodicals, listening to the radio, and going to the movies. Unlike other cultural and socializing institutions, ownership and control of the mass media is highly concentrated. Twenty-three corporations own more than one-half of all the daily newspapers, magazines, movie studios, and radio and television outlets in the United States. The number of media companies is shrinking and their control of the industry is expanding. And a relatively small number of media outlets is producing and packaging the majority of news and entertainment programs. For the most part, our media is national in nature and single-minded (profit-oriented) in purpose. This media plays a key role in defining our cultural tastes, helping us locate ourselves in history, establishing our national identity, and ascertaining the range of national and social possibilities. In this essay, we will examine the way the mass media shapes how people think about each other and about the nature of our society.

The United States is the most highly stratified society in the industrialized world. Class distinctions operate in virtually every aspect of our lives, determining the nature of our work, the quality of our schooling, and the health and safety of our loved ones. Yet remarkably, we, as a nation, retain illusions about living in an egalitarian society. We maintain these illusions, in large part, because the media hides gross inequities from public view. In those instances when inequities are revealed, we are provided with messages that obscure the nature of class realities and blame the victims of class-dominated society for their own plight. Let's briefly examine what the news media, in particular, tells us about class.

About the Poor

The news media provides meager coverage of poor people and poverty. The coverage it does provide is often distorted and misleading.

The Poor Do Not Exist

For the most part, the news media ignores the poor. Unnoticed are forty million poor people in the nation—a number that equals the entire population of Maine, Vermont, New Hampshire, Connecticut, Rhode Island, New Jersey, and New York combined. Perhaps even more alarming is that the rate of poverty is increasing twice as fast as the population growth in the United States. Ordinarily, even a calamity of much smaller proportion (e.g., flooding in the Midwest) would garner a great deal of coverage and hype from a media usually eager to declare a crisis, yet less than one in five hundred articles in the *New York Times* and one in one thousand articles listed in the *Readers Guide to Periodic Literature* are on poverty. With remarkably little attention to them, the poor and their problems are hidden from most Americans.

When the media does turn its attention to the poor, it offers a series of contradictory messages and portrayals.

The Poor Are Faceless

Each year the Census Bureau releases a new report on poverty in our society and its results are duly reported in the media. At best, however, this coverage emphasizes annual fluctuations (showing how the numbers differ from previous years) and ongoing debates over the validity of the numbers (some argue the number should be lower, most that the number should be higher). Coverage like this desensitizes us to the poor by reducing poverty to a number. It ignores the human tragedy of poverty—the suffering, indignities, and misery endured by millions of children and adults. Instead, the poor become statistics rather than people.

The Poor Are Undeserving

When the media does put a face on the poor, it is not likely to be a pretty one. The media will provide us with sensational stories about welfare cheats, drug addicts, and greedy panhandlers (almost always urban and Black). Compare these images and the emotions evoked by them with the media's treatment of middle-class (usually white) "tax evaders," celebrities who have a "chemical dependency," or wealthy businesspeople who use unscrupulous means to "make a profit." While the behavior of the more affluent offenders is considered an "impropriety" and a deviation from the norm, the behavior of the poor is considered repugnant, indicative of the poor in general, and worthy of our indignation and resentment.

The Poor Are an Eyesore

When the media does cover the poor, they are often presented through the eyes of the middle class. For example, sometimes the media includes a story with panhandlers. Rather than focusing on the plight of the poor, these stories are about middle-class opposition to the poor. Such stories tell us that the poor are an inconvenience and an irritation.

The Poor Have Only Themselves to Blame

In another example of media coverage, we are told that the poor live in a personal and cultural cycle of poverty that hopelessly imprisons them. They routinely center on the Black urban population and focus on perceived personality or cultural traits that doom the poor. While the women in these stories typically exhibit an "attitude" that leads to trouble or a promiscuity that leads to single motherhood, the men possess a need for immediate gratification that leads to drug abuse or an unquenchable greed that leads to the pursuit of fast money. The images that are seared into our mind are sexist, racist, and classist. Census figures reveal that most of the poor are white not Black or Hispanic, that they live in rural or suburban areas not urban centers, and hold jobs at least part of the year. Yet, in a fashion that is often framed in an understanding and sympathetic tone, we are told that the poor have inflicted poverty on themselves.

The Poor Are Down on Their Luck

During the Christmas season, the news media sometimes provides us with accounts of poor individuals or families (usually white) who are down on their luck. These stories are often linked to stories about soup kitchens or other charitable activities and sometimes call for charitable contributions. These "Yule time" stories are as much about the affluent as they are about the poor: they tell us that the affluent in our society are a kind, understanding, giving people—which we are not.[1] The series of unfortunate circumstances that have led to impoverishment are presumed to be a temporary condition that will improve with time and a change in luck.

Despite appearances, the messages provided by the media are not entirely disparate. With each variation, the media informs us what poverty is not (i.e., systemic and indicative of American society) by informing us what it is. The media tells us that poverty is either an aberration of the American way of life (it doesn't exist, it's just another number, it's unfortunate but temporary) or an end product of the poor themselves (they are a nuisance, do not deserve better, and have brought their predicament upon themselves).

By suggesting that the poor have brought poverty upon themselves, the media is engaging in what William Ryan has called "blaming the victim." The media identifies in what ways the poor are different as a consequence of deprivation, then defines those differences as the cause of poverty itself. Whether blatantly hostile or cloaked in sympathy, the message is that there is something fundamentally wrong with the victims—their hormones, psychological makeup, family environment, community, race, or some combination of these—that accounts for their plight and their failure to lift themselves out of poverty.

But poverty in the United States is systemic. It is a direct result of economic and political policies that deprive people of jobs, adequate wages, or legitimate support. It is neither natural nor inevitable: there is enough wealth in our nation to eliminate poverty if we chose to redistribute existing wealth or income. The plight of the poor is reason enough to make the elimination of poverty the nation's first priority. But poverty also impacts dramatically on the nonpoor. It has a dampening effect on wages in general (by maintaining a reserve army of unemployed and underemployed anxious for any job at any wage) and breeds crime and violence (by maintaining conditions that invite private gain by illegal means and rebellion-like behavior, not entirely unlike the urban riots of the 1960s). Given the extent of poverty in the nation and the impact it has on us all, the media must spin considerable magic to keep the poor and the issue of poverty and its root causes out of the public consciousness.

About Everyone Else

Both the broadcast and the print news media strive to develop a strong sense of "we-ness" in their audience. They seek to speak to and for an audience that is both affluent and like-minded. The media's solidarity with affluence, that is, with the middle and upper class, varies little from one medium to another. Benjamin DeMott points out, for example, that the *New York Times* understands affluence to be intelligence, taste, public spirit, responsibility, and a readiness to rule and "conceives itself as spokesperson for a readership awash in these qualities." Of course, the flip side to creating a sense of "we," or "us," is establishing a perception of the "other." The other relates back to the faceless, amoral, undeserving, and inferior "underclass." Thus, the world according to the news media is divided between the "underclass" and everyone else. Again the messages are often contradictory.

The Wealthy Are Us

Much of the information provided to us by the news media focuses attention on the concerns of a very wealthy and privileged class of people. Although the concerns of a small fraction of the populace, they are presented as though they were the concerns of everyone. For example, while relatively few people actually own stock, the news media devotes an inordinate amount of broadcast time and print space to business news and stock market quotations. Not only do business reports cater to a particular narrow clientele, so do the fashion pages (with $2,000 dresses), wedding announcements, and the obituaries. Even weather and sports news often have a class bias. An all news radio station in New York City, for example, provides regular national ski reports. International news, trade agreements, and domestic policies issues are also reported in terms of their impact on business climate and the business community. Besides being of practical value to the wealthy, such coverage has considerable ideological value. Its message: the concerns of the wealthy are the concerns of us all.

The Wealthy (as a Class) Do Not Exist

While preoccupied with the concerns of the wealthy, the media fails to notice the way in which the rich as a class of people create and shape domestic and foreign policy. Presented as an aggregate of individuals, the wealthy appear without special interests, interconnections, or unity in purpose. Out of public view are the class interests of the wealthy, the interlocking business links, the concerted actions to preserve their class privileges and business interests (by running for public office, supporting political candidates, lobbying, etc.). Corporate lobbying is ignored, taken for granted, or assumed to be in the public interest. (Compare this with the media's portrayal of the "strong arm of labor" in attempting to defeat trade legislation that is harmful to the interests of working people.) It is estimated that two-thirds of the U.S. Senate is composed of millionaires. Having such a preponderance of millionaires in the Senate, however, is perceived to be neither unusual nor antidemocratic; these millionaire senators are assumed to be serving "our" collective interests in governing.

The Wealthy Are Fascinating and Benevolent

The broadcast and print media regularly provide hype for individuals who have achieved "super" success. These stories are usually about celebrities and superstars from the sports and entertainment world. Society pages and gossip columns serve to keep the social elite informed of each other's doings, allow the rest of us to gawk at their excesses, and help to keep the American dream alive. The print media is also fond of feature stories on corporate empire builders. These stories provide an occasional "insider's" view of the private and corporate life of industrialists by suggesting a rags to riches account of corporate success. These stories tell us that corporate success is a series of smart moves, shrewd acquisitions, timely mergers, and well thought out executive suite shuffles. By painting the upper class in a positive light, innocent of any wrongdoing (labor leaders and union organizations usually get the opposite treatment), the media assures us that wealth and power are benevolent. One person's capital accumulation is presumed to be good for all. The elite, then, are portrayed as investment wizards, people of special talent and skill, who even their victims (workers and consumers) can admire.

The Wealthy Include a Few Bad Apples

On rare occasions, the media will mock selected individuals for their personality flaws. Real estate investor Donald Trump and New York Yankees owner George Steinbrenner, for example, are admonished by the media for deliberately seeking publicity (a very un-upper class thing to do); hotel owner Leona Helmsley was caricatured for her personal cruelties; and junk bond broker Michael Milkin was condemned because he had the audacity to rob the rich. Michael Parenti points out that by treating business wrongdoing as isolated deviations from the socially beneficial system of "responsible capitalism," the media overlooks the features of the system that produce such abuses and the regularity with which they occur. Rather than portraying them as predictable and frequent outcomes of corporate power and the business system, the media treats abuses as if they were isolated and atypical. Presented as an occasional aberration, these incidents serve not to challenge, but to legitimate, the system.

The Middle Class Is Us

By ignoring the poor and blurring the lines between the working people and the upper class, the news media creates a universal middle class. From this perspective, the size of one's income becomes largely irrelevant: what matters is that most of "us" share an intellectual and moral superiority over the disadvantaged. As *Time* magazine once concluded, "Middle America is a state of mind." "We are all middle class," we are told, "and we all share the same concerns": job security, inflation, tax burdens, world peace, the cost of food and housing, health care, clean air and water, and the safety of our streets. While the concerns of the wealthy are quite distinct from those of the middle class (e.g., the wealthy worry about investments, not jobs), the media convinces us that "we [the affluent] are all in this together."

The Middle Class Is a Victim

For the media, "we" the affluent not only stand apart from the "other"—the poor, the working class, the minorities, and their problems—"we" are also victimized by the poor (who drive up the costs of maintaining the welfare roles), minorities (who commit crimes against us), and by workers (who are greedy and drive companies out and prices up). Ignored are the subsidies to the rich, the crimes of corporate America, and the policies that wreak havoc on the economic well-being of middle America. Media magic convinces us to fear, more than anything else, being victimized by those less affluent than ourselves.

The Middle Class Is Not a Working Class

The news media clearly distinguishes the middle class (employees) from the working class (i.e., blue collar workers) who are portrayed, at best, as irrelevant, outmoded, and a dying breed. Furthermore, the media will tell us that the hardships faced by blue collar workers are inevitable (due to progress), a result of bad luck (chance circumstances in a particular industry), or a product of their own doing (they priced themselves out of a job). Given the media's presentation of reality, it is hard to believe that manual, supervised, unskilled, and semiskilled workers actually represent more than 50 percent of the adult working population. The working class, instead, is relegated by the media to "the other."

In short, the news media either lionizes the wealthy or treats their interests and those of the middle class as one and the same. But the upper class and the middle class do not share the same interests or worries. Members of the upper class worry about stock dividends (not employment), they profit from inflation and global militarism, their children attend exclusive private schools, they eat and live in a royal fashion, they call on (or are called upon by) personal physicians, they have few consumer problems, they can escape whenever they want from environmental pollution, and they live on streets and travel to other areas under the protection of private police forces.[2]

The wealthy are not only a class with distinct life-styles and interests, they are a ruling class. They receive a disproportionate share of the country's yearly income, own a disproportionate amount of the country's wealth, and contribute a disproportionate number of their members to governmental bodies and decision-making groups—all traits that William Domhoff, in his classic work *Who Rules America,* defined as characteristic of a governing class.

This governing class maintains and manages our political and economic structures in such a way that these structures continue to yield an amazing proportion of our wealth to a minuscule upper class. While the media is not above referring to ruling classes in other countries (we hear, for example, references to Japan's ruling elite), its treatment of the news proceeds as though there were no such ruling class in the United States.

Furthermore, the news media inverts reality so that those who are working class and middle class learn to fear, resent, and blame those below, rather than those above them in the class structure. We learn to resent welfare, which accounts for only two cents out of every dollar in the federal budget (approximately $10 billion) and provides financial relief for the needy,[3] but learn little about the $11 billion the federal government spends on individuals with incomes in excess of $100,000 (not needy), or the $17 billion in farm subsidies, or the $214 billion (twenty times the cost of welfare) in interest payments to financial institutions.

Middle-class whites learn to fear African Americans and Latinos, but most violent crime occurs within poor and minority communities and is neither interracial[4] nor interclass. As horrid as such crime is, it should not mask the destruction and violence perpetrated by corporate America. In spite of the fact that 14,000 innocent people are killed on the job each year, 100,000 die prematurely, 400,000 become seriously ill, and 6 million are injured from work-related accidents and diseases, most Americans fear government regulation more than they do unsafe working conditions.

Through the media, middle-class—and even working-class—Americans learn to blame blue collar workers and their unions for declining purchasing power and economic security. But while workers who managed to keep their jobs and their unions struggled to keep up with inflation, the top 1 percent of American families saw their average incomes soar 80 percent in the last decade. Much of the wealth at the top was accumulated as stockholders and corporate executives moved their companies abroad to employ cheaper labor (56 cents per hour in El Salvador) and avoid paying taxes in the United States. Corporate America is a world made up of ruthless bosses, massive layoffs, favoritism and nepotism, health and safety violations, pension plan losses, union busting, tax evasions, unfair competition, and price gouging, as well as fast buck deals, financial speculation, and corporate wheeling and dealing that serve the interests of the corporate elite, but are generally wasteful and destructive to workers and the economy in general.

It is no wonder Americans cannot think straight about class. The mass media is neither objective, balanced, independent, nor neutral. Those who own and direct the mass media are themselves part of the upper class, and neither they nor the ruling class in general have to conspire to manipulate public opinion. Their interest is in preserving the status quo, and their view of society as fair and equitable comes naturally to them. But their ideology dominates our society and justifies what is in reality a perverse social order—one that perpetuates unprecedented elite privilege and power on the one hand and widespread deprivation on the other. A mass media that did not have its own class interests in preserving the status quo would acknowledge that inordinate wealth and power undermines democracy and that a "free market" economy can ravage a people and their communities.

Notes

1. American households with incomes of less than $10,000 give an average of 5.5 percent of their earnings to charity or to a religious organization, while those making more than $100,000 a year give only 2.9 percent. After changes in the 1986 tax code reduced the benefits of charitable giving, taxpayers earning $500,000 or more slashed their average donation by nearly one-third. Furthermore, many of these acts of benevolence do not help the needy. Rather than provide funding to social service agencies that aid the poor, the voluntary contributions of the wealthy go to places and institutions that entertain, inspire, cure, or educate wealthy Americans—art museums, opera houses, theaters, orchestras, ballet companies, private hospitals, and elite universities.

2. The number of private security guards in the United States now exceeds the number of public police officers. (Robert Reich, "Secession of the Successful." *New York Times Magazine,* February 1991.)

3. A total of $20 billion is spent on welfare when you include all state funding. But the average state funding also comes to only two cents per state dollar.

4. In 92 percent of the murders nationwide the assailant and the victim are of the same race (46 percent are white/white, 46 percent are black/black), 5.6 percent are black on white, and 2.4 percent are white on black. (FBI and Bureau of Justice Statistics, 1985–1986, quoted in Raymond S. Franklin. *Shadows of Race and Class,* University of Minnesota Press, Minneapolis, 1991, p. 108.)

THINKING ABOUT THE READING

What kinds of messages do people get about wealth and social position from the media? What do these messages suggest about who is deserving and who is not? If these messages are based on inaccurate stereotypes, where can people get more accurate information? Do you think that people in different social classes view themselves and their lives differently based on how they are portrayed in the news and on television? If these portrayals are a significant source of information about one's place in society, do you think these media images affect a person's sense of self-worth and opportunity?

The Compassion Gap in American Poverty Policy

Fred Block, Anna C. Korteweg, and Kerry Woodward, with Zach Schiller and Imrul Mazid

(2006)

Every 30 or 40 years, Americans seem to "discover" that millions of our citizens are living in horrible and degrading poverty. Jacob Riis shocked the nation in 1890 with a book entitled *How the Other Half Lives,* which helped to inspire a change in public opinion and the reforms of the Progressive Era. In the 1930s, the devastation of the Great Depression led FDR to place poverty at the top of the national agenda. In the early 1960s, Michael Harrington's *The Other America* made poverty visible and paved the way for Lyndon Johnson's brief War on Poverty. In 2005, an act of nature became the next muckraker—Hurricane Katrina, which shockingly revealed the human face of poverty among the displaced and helpless victims of the storm's devastation in New Orleans.

But what makes poverty so invisible between such episodes of discovery? The poor are always with us, but why do they repeatedly disappear from public view? Why do we stop seeing the pain that poverty causes?

Our society recognizes a moral obligation to provide a helping hand to those in need, but those in poverty have been getting only the back of the hand. They receive little or no public assistance. Instead, they are scolded and told that they have caused their own misfortunes. This is our "compassion gap"—a deep divide between our moral commitments and how we actually treat those in poverty.

The compassion gap does not just happen. It results from two key dynamics. First, powerful groups in American society insist that public help for the poor actually hurts them by making them weak and dependent. Every epoch in which poverty is rediscovered and generosity increases is followed by a backlash in which these arguments reemerge and lead to sharp reductions in public assistance. Second, the consequence of reduced help is that the assertions of welfare critics turn into self-fulfilling prophecies. They insist that immorality is the root cause of poverty. But when assistance becomes inadequate, the poor can no longer survive by obeying the rules; they are forced to break them. These infractions, in turn, become the necessary proof that "the poor" are truly intractable and that their desperate situations are rightly ignored.

The results are painfully clear in our official data. In 2004, 37 million people, including 13 million children, lived below the government's official poverty line of $15,219 for a family of three. The number of people in poverty has increased every year for the last four years, rising from 31.6 million in 2000. Moreover, our government's official poverty line is quite stingy by international standards. If we used the most common international measure, which counts people who live on less than half of a country's median income as poor, then almost 55 million people in the United States, or almost 20 percent of the population, would be counted as poor.

Most distressingly, the number of people living in catastrophic poverty—in households with incomes less than 50 percent of the official U.S. poverty line—has increased every year since 1999. There are now 15.6 million people living in this kind of desperate poverty. This is close to the highest number ever, and it is twice the number of extremely poor people that we had in the mid-1970s, before the cuts in poverty programs of the Reagan administration.

Children, single mothers with children, and people of color—particularly African Americans and Latinos—make up a disproportionate segment of the nation's poorest groups, with women in each group consistently more likely to be poor than men in that group. But poverty is not unusual or rare—as many as 68 percent of all Americans will spend a year or more living in poverty or near-poverty as adults. Nor is poverty always related to not working; there are still 9 million working-poor adults in the United States.

Moreover, poverty has become more devastating over the past generation. Thirty years ago, a family living at the poverty line—earning a living at low-wage work—could still see the American Dream as an achievable goal (see Figure 1). With a bit more hard work and some luck, they too could afford a single-family home, comprehensive health insurance, and a college education for their children. Today, for many of the poor, including many of the faces we saw at the New Orleans Superdome and Convention Center, that dream has become a distant and unattainable vision. Even a two-parent family working full-time at the minimum wage earns less than half of what is needed to realize the dream at today's prices. The old expectation that the poor would pull themselves up by their own bootstraps is increasingly unrealistic.

Figure 1 The Dream Line is an estimate of the cost for an urban or suburban family of four to enjoy a no-frills version of the American Dream that includes owning a single-family home, full health-insurance coverage, quality child care for a four-year-old, and enough annual savings to assure that both children can attend a public, four-year college or university. The Dream Line is not a wage figure because it includes the full cost of health insurance coverage that is often, but decreasingly, offered as a benefit by employers. The figures are national averages and are lower than what people would pay for these services in the largest and most expensive metropolitan areas on the East and West coasts. The housing figure reflects the cost of mortgage payments on the median-priced existing family home at current interest rates. The Dream Line rises so dramatically because the costs of the four H's—housing, health insurance, high-quality child care, and higher education—have risen so much more rapidly than other consumer prices [Table 1]. Dollar figures have not been adjusted for inflation. More details on the way the Dream Divide was calculated are available at http://www.longviewinstitute.org/research/block/amerdream/view.

When the Dream Line is compared to the federal poverty line or to the income that a two-parent family would earn if both parents were working full-time at the minimum wage, it is clear that the dream has become increasingly distant for millions.

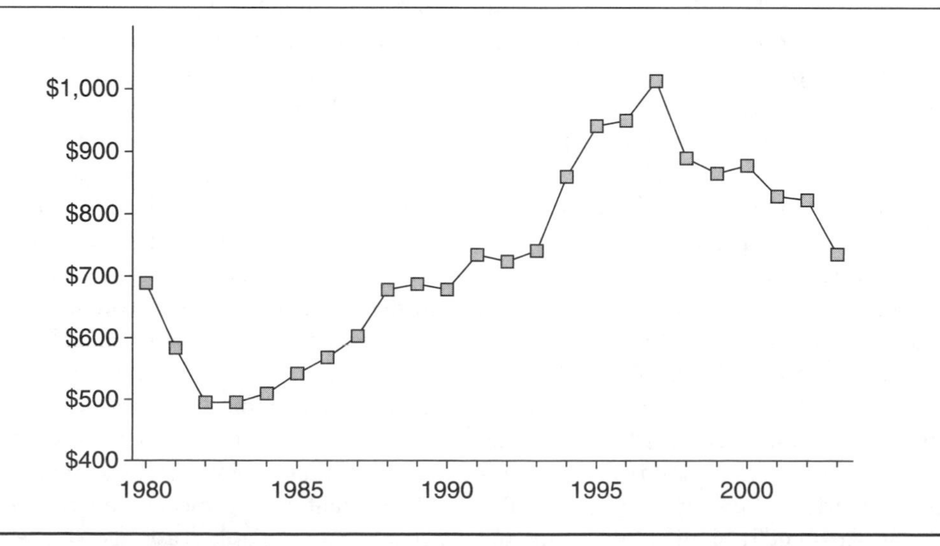

Table 1 Price rises for the four H's

	Housing	*High-quality child care*	*Higher education*	*Health insurance*
1973 (annual cost)	$1,989	$978	$736	$509
2003 (annual cost)	$10,245	$7,200	$5,000	$8,933
Percent increase	515%	736%	679%	1755%

Despite the growing poor population and the increasing difficulty of escaping poverty into economic security through paid work, the government has been doing less and less to help. Aid to Families with Dependent Children (AFDC) used to be our biggest program to help poor people, but federal legislation passed in 1996 ended AFDC and replaced it with Temporary Aid to Needy Families (TANF). TANF's focus on moving recipients from "welfare to work" has led to a major decline in the number of households receiving benefits and a huge drop in cash assistance to the poor. The average monthly TANF benefit was $393 in 2003, compared to $490 in 1997.

Not only are our programs miserly, they reach too few people among those who are eligible, further reducing the chances that those in poverty can achieve the American Dream. Only 60 percent of eligible households receive food stamps. Despite a commitment to provide health insurance to all children under 18, nearly 12 percent of those children remained without such insurance in 2004, and only 27 percent of all poor families received TANF in 2000. Finally, subsidized housing is provided to only 25 percent of those who need it, and current budget proposals would cut this program dramatically.

Against this backdrop of decreasing spending on most antipoverty measures, the Earned Income Tax Credit (EITC) has become our biggest antipoverty program for the working-age population. EITC aids the working poor by providing an income-tax refund to lift the poorest workers above the poverty line. But for families to benefit significantly from the EITC, someone in the household must be earning at least several thousand dollars per year. Each year, millions of households do not have such an earner because of unemployment, illness, lack of child care, or a mismatch between available skills and job demands. The consequence is a relentless increase in our rate of catastrophic poverty.

Figure 2 shows the combined spending for the two most important cash assistance programs—AFDC/TANF and the EITC. It demonstrates that despite increases in EITC outlays, our total spending on the poor peaked in 1997 and has dropped almost 20 percent since then. Figure 3 takes the further step of adjusting the annual spending for the impact of inflation and the shifting size of the poor population. Spending for each nonelderly poor person peaked at around $1,000 in 1997 and has dropped every year since, with a total decline of close to 30 percent. And if we added food stamps to this chart, the trend would be even stronger, since their real value has also fallen since 1997. There is no clearer evidence that our compassion gap has deepened poverty.

The compassion gap has been greatly increased by the revival in the 1980s and 1990s of the very old theory that the real source of poverty is bad behavior. Since African American and Hispanic women and men, as well as single mothers of all ethnicities and races, are disproportionately represented among the poor, this theory defines these people as morally deficient. Its proponents assume that anyone with enough grit and determination can escape poverty. They claim that giving people cash assistance worsens

Figure 2 Assistance to those in poverty from 1990 to 2004 in billions of dollars

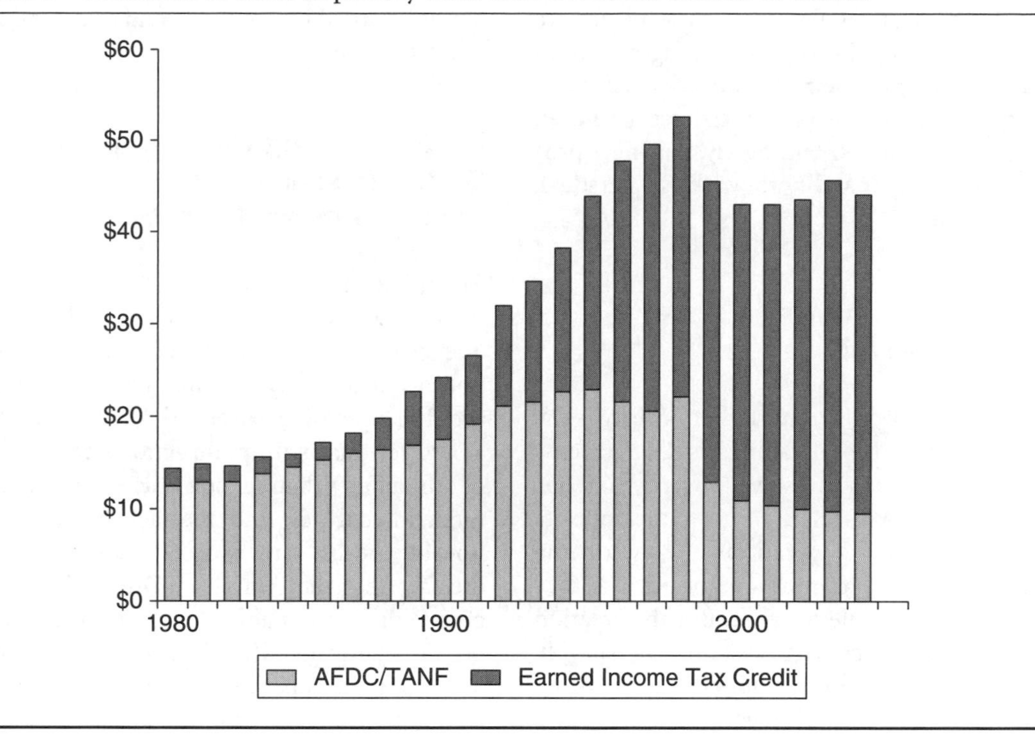

Figure 3 Spending on poor individuals per person

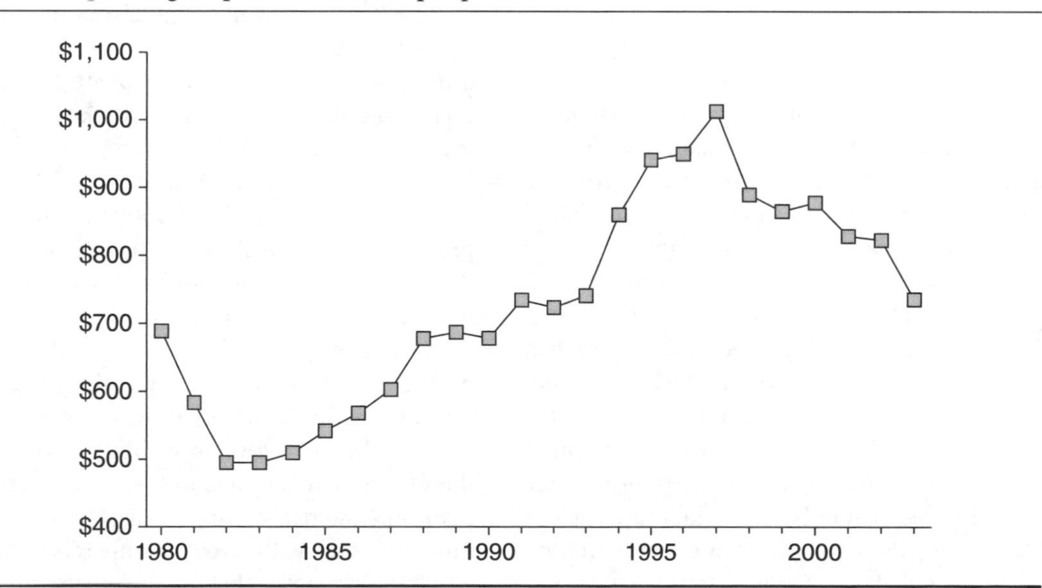

poverty by taking away their drive to improve their circumstances through work. Arguing that poor people bear children irresponsibly and that they lack the work ethic necessary for economic success, they have launched a sustained war on bad behavior that targets those groups most at risk of poverty.

One of the key events in this war was the passage in 1996 of the Personal Responsibility and Work Opportunities Reconciliation Act

(PRWORA), which replaced AFDC with TANF. TANF requires single mothers who receive welfare to find paid work, encourages them to marry, and limits their time on aid to a lifetime maximum of five years. Some states have even shorter time limits. Ultimately, this new program treats the inability to work as a personal, moral failing.

Can Governments Solve Poverty?

The flip side of the premise that poverty is the result of such moral failings is that government actions cannot solve poverty. Yet our own national experience points to the opposite conclusion. For generations, many of the elderly lived in extreme poverty because they were no longer able to work. But the creation of the Social Security system has sharply reduced poverty among seniors by recognizing that most people need government assistance as they age. Yet, rather than celebrating the compassion reflected in this program, the current administration is proposing destructive changes in Social Security that will make it less effective in preventing poverty among this group. And instead of recognizing that most young families also need assistance to survive and thrive, our major antipoverty program, the EITC, leaves out all those families who find themselves squeezed out of the labor market.

Looking abroad also shows that government policies can dramatically reduce poverty levels. The probability of living in poverty is more than twice as high for a child born in the United States than for children in Belgium, Germany, or the Netherlands. Children in single-mother households are four times more likely to be poor in the United States than in Norway [Table 2]. The fact that single-parent households are more common in the United States than in many of these countries where the poor receive greater assistance undermines the claim that more generous policies will encourage more single women to have children out of wedlock. These other countries all take a more comprehensive government approach to combating poverty, and they assume that it is caused by economic and structural factors rather than bad behavior.

Understanding the Compassion Gap: A Misguided Focus on Moral Poverty

The miserliness of our public assistance is justified by the claim that poverty is the consequence of personal moral failings. Most of our policies incorrectly assume that people can avoid or overcome poverty through hard work alone. Yet this assumption ignores the realities of our failing urban schools, increasing employment insecurities, and the lack of affordable housing, health care, and child care. It ignores the fact that the American Dream is rapidly becoming unattainable for an increasing number of Americans, whether employed or not.

The preoccupation with the moral failings of the poor disregards the structural problems underlying poverty. Instead, we see increasing numbers of policies that are obsessed with preventing "welfare fraud." This obsession creates barriers to help for those who need it. Welfare offices have always required recipients to "prove" their eligibility. Agency employees are in effect trained to begin with the presumption of guilt; every seemingly needy face they encounter is that of a cheater until the potential client can prove the contrary. With the passage of TANF, the rules have become so complex that even welfare caseworkers do not always understand them, let alone their clients. Some of those who need help choose to forego it rather than face this humiliating eligibility process.

But this system of suspicion also produces the very welfare cheaters that we fear. Adults in poor households are caught in a web of different programs, each with its own complex set of rules and requirements, that together provide less assistance than a family needs. Recipients have no choice but to break the rules—usually by not reporting all their income. A detailed study from ten years ago, conducted by Kathryn Edin and Laura Lein [1997], showed that most welfare mothers worked off the books or took money under the table from relatives because they could not make ends meet with only their welfare checks. Since then we have reduced benefits and added more rules, undoubtedly increasing such "cheating."

Those who lack compassion have made their own predictions come true. They begin by claiming that the poor lack moral character. They use stories of welfare cheaters to increase public concerns about people getting something for nothing. Consequently, our patchwork of poorly funded programs reaches only a fraction of the poor and gives them less than they need. Those who depend on these programs must cut corners and break rules to keep their families together. This "proves" the original proposition that the poor lack moral character, and the "discovery" is used to justify ever more stringent policies. The result is a vicious spiral of diminishing compassion and greater preoccupation with the moral failings of the poor.

The War on Bad Behavior

The moral focus on poverty shifts our gaze from the social forces that create material poverty to the perceived moral failings of the poor. This shift has led to a war on bad behavior, exemplified by PRWORA, that is not achieving its goals. This war focuses on social problems like teenage pregnancy, high dropout rates, and drug addiction. But research shows that it has been ineffective. Poverty has risen, and punitive measures have had little effect on the behaviors they were supposed to change.

The reduction of teen pregnancy through abstinence-only sex education was one of the main goals of the Personal Responsibility and Work Opportunity Act. Its drafters mistakenly believed that teen pregnancy is one of the root causes of poverty. In fact, if the teenagers who are having children were to wait until they were adults, their children would be just as likely to be born into poverty. But the drafters' other error was ignoring the fact that teen pregnancy rates had already been declining for years when the new law went into effect, primarily because teenagers were using more effective methods of birth-control. (These gains are now threatened by the dramatic expansion of "abstinence-only sex education," which provides no information on birth-control techniques.)

PRWORA also makes assistance to teen mothers contingent on "good behavior." Teen mothers must stay in school or be enrolled in a training program and live with their parents or under other adult supervision in order to receive aid. While it makes sense to help teens stay in school and learn skills, these coercive efforts are failing the children of teen parents. Teen mothers are just as likely today to drop out of school or live on their own as when the act was passed. The only change is that they are now much less likely to receive government assistance: ill-conceived reforms have ensured that children born to teen mothers experience deeper deprivation.

Neither have PRWORA's efforts to control the behavior of the poor had much impact on illicit drug use. Under TANF, states were required to deny benefits to anyone convicted of a drug crime. This was so obviously counterproductive that Congress amended the law in 1999 to allow states to opt out of this ban. Yet neither policy shift appears to have had much impact. According to Justice Department data, adult drug arrests have been increasing relentlessly, from 1 million per year in the early 1990s to 1.5 million in 2003.

But advocates of the war on bad behavior always have a convenient scapegoat for the failure of their punitive policies: they simply shift the blame to single mothers. TANF requires single mothers to work outside the home regardless of whether work gets them out of poverty. But long hours of work and inadequate child care mean that children are often left with inadequate supervision. When these children get into trouble, the mother gets the blame. Teen pregnancy, drug use, and delinquency are then attributed to the mother's lack of parenting skills. Poor single mothers cannot win; they are failures if they stay home with their kids—providing the full-time mothering that conservatives have long advocated for middle-class children. But they are also failures if they work and leave their children unsupervised. Viewing poverty as the result of bad behavior produces the conclusion that poor single mothers are bad by definition. Since a disproportionate number of these poor single mothers are African American or Hispanic, this rhetoric also hides the racial history that has excluded people of color from opportunities for generations and the systemic racism that persists today.

This war on bad behavior is a deeply mistaken approach to poverty. It ignores the lived reality of people who face crushing poverty every day. It ignores the fundamental wisdom that we should not judge people until we have walked a mile in their shoes. Most basically, it denies compassion to those who need it most.

What to Do? Revitalize the American Dream

Reversing the compassion gap will not happen overnight. We have to persuade our fellow citizens that the war on bad behavior violates our society's fundamental values. We have to show them how far reality has departed from the American Dream, which holds that a child born in poverty in a ghetto or a barrio has the same chance for success and happiness as a child born in suburban affluence. We have to focus national debate on what policy measures would revitalize the American Dream for all of our citizens.

The reason the American Dream is now beyond reach for so many families is that the price of four critical services has risen much more sharply than wages and the rate of inflation: health care, higher education, high-quality child care, and housing. These are not luxuries, but indispensable ingredients of the dream.

Over the last three decades, our society has relied largely on market solutions to organize delivery of these indispensable services, but these solutions have not increased their supply. Instead, we use the price mechanism to ration their distribution; poor and working-class people are at the end of the line, and they find themselves priced out of the market.

Table 2 Children living in poverty (counting all sources of income, including income from government programs)

Country	*Year*	*Percent of all children in poverty*	*Percent of all children in single-mother homes in poverty*	*Percent of all children living with single mothers*
US	2000	21.9	49.3	19.5
UK	1999	15.3	33.8	19.5
Canada	2000	14.9	40.7	13.1
Netherlands	1999	9.8	35.1	8.1
Germany	2000	9.0	37.8	12.5
Belgium	2000	6.7	24.5	10.6
Norway	2000	3.4	11.3	14.5

Source: Luxemburg Income Survey: www.lisproject.org/keyfigures/povertytable.htm.

Note: This table uses the international convention of measuring poverty as income less than 50 percent of the nation's median income.

We need new initiatives to expand the supply of these key services while assuring their quality. This requires accelerated movement toward universal health insurance and universal availability of quality child care and preschool programs. We need to move toward universal access to higher education for all students who meet the admissions criteria. (We also need to ensure that all our public schools are preparing students for the higher education and training that most will need in order to succeed in the labor market.) And we need to create new public-private partnerships to expand the supply of affordable housing for poor and working-class families. These efforts would restore the American Dream for millions of working-class and lower middle-class families, while also putting the dream within the reach of the poor.

But we also need new policies that target the poor more directly. This requires restoring the value of the minimum wage. Between 1968 and 2002, the purchasing power of the federal minimum wage fell by a third. We need to reverse this trend and assure that in the future the minimum wage continues to rise with inflation. Most fundamentally, we must do what most other developed nations do—provide a stable income floor for all poor families so that no children grow up in horrible and degrading poverty. We could establish such a floor by transforming our present Earned Income Tax Credit into a program that provided all poor families with sufficient income to cover food and shelter. Households would be eligible for a monthly payment even if they had no earnings. Since such payments would target the poorest individuals and families, this would be a cost-effective way to immediately rescue millions of people from catastrophic poverty. Moreover, since payments would be coordinated through the tax system, a household's income would definitely improve as its labor-force earnings rose.

The key to making these policy initiatives feasible is to remind our fellow citizens what true compassion requires. The war on bad behavior offers us an easy way out. It is easy to believe that those in poverty are responsible for their own problems and that ignoring their needs is the best thing for them. It absolves those of us who are better off from the responsibility of caring for others. However, if we want to live up to our national commitment to compassion, we need to recognize that we have a collective responsibility to ensure that in the wealthiest nation in the world there are not millions of people going hungry, millions without health insurance, and hundreds of thousands without homes. Sure, some of those in poverty have made bad choices, but who has not? It is deeply unfair that those who are not poor get second chances, while the poor do not. Rush Limbaugh pays no price for becoming addicted to painkillers, but millions of poor people go to jail and lose access to public housing and welfare benefits for the same offense.

True compassion requires that we build a society in which every person has a first chance, a second chance, and, if needed, a third and fourth chance, to achieve the American Dream. We are our brother's and our sister's keepers, and we need to use every instrument we have—faith groups, unions, community groups, and most of all government programs—to address the structural problems that reproduce poverty in our affluent society.

Dealing with the inadequacies of our current antipoverty programs is a first step in moving the debate in the right direction. Since the fall of 2002, Congress has been stalemated on reauthorizing the TANF legislation that was first passed in 1996. Action in the immediate future seems unlikely because many governors oppose the more stringent work requirements for TANF recipients proposed by the Bush administration and its conservative allies in the House, because those changes would require the states to pay for new work-experience programs.

A compassionate reauthorization of TANF requires four basic steps. First, we must increase assistance levels to rescue families from the deepest poverty and give them enough income to put them over the poverty line. Second, we must abandon the whole system of mandatory time limits on aid, so that families in poverty no longer find the doors to help closed in their faces. Eliminating time limits is particularly important in ensuring that programs serve the many poor women who are victims of domestic abuse. While TANF is supposed to protect such women, too often they are being forced back into the arms of their abusers. Third, we must recognize basic and postsecondary education and training as a "work activity," so that recipients can prepare for jobs that would get them out of poverty. Finally, we need to improve the child-care provisions in TANF. We must do more than provide child-care subsidies to only one out of seven children who are federally eligible. Moreover, we must ensure that TANF children get a head start and are not relegated to the lowest-quality child care.

By themselves, these reforms would not close the compassion gap, but they would mark an end to the futile and destructive war on bad behavior. They could represent an initial down payment on restoring the American Dream.

Postscript

For more than three years Congress was unable to agree on a reauthorization of the TANF legislation that was initially passed in 1996. In early 2006, however, the Republican leadership moved the legislation without debate or discussion by including TANF reauthorization in a large deficit-reduction bill that passed both houses by the narrowest of margins. In fact, the legislation might yet be overturned by the courts because the House and the Senate passed slightly different versions of the bill.

If implemented, the new legislation will widen the compassion gap even further because states are required either to place 50 percent of adult recipients in work-related activities or to reduce the number of families receiving benefits. Since many of those currently on the rolls face multiple barriers to employment, these artificial targets are likely to create considerable hardship. Moreover, the allocation for child care is not enough to maintain the current availability of child care, let alone to keep pace with the new participation requirements.

THINKING ABOUT THE READING

What is the "compassion gap"? Give some examples from recent news stories and your own experiences of this "gap." According to the authors, how does this social attitude contribute to the persistence of social inequality? How do current social policies perpetuate poverty? What is the "dream line"? What kinds of changes are needed both culturally and politically to address the increasing poverty rates in the United States?

Branded With Infamy

Inscriptions of Poverty and Class in America

Vivyan Adair

(2002)

> *"My kids and I been chopped up and spit out just like when I was a kid. My rotten teeth, my kids' twisted feet. My son's dull skin and blank stare. My oldest girl's stooped posture and the way she can't look no one in the eye no more. This all says we got nothing and we deserve what we got. On the street good families look at us and see right away what they'd be if they don't follow the rules. They're scared too, real scared."*
>
> —Welfare recipient and activist,
> Olympia, Washington, 1998

I begin with the words of a poor, White, single mother of three. Although officially she has only a tenth-grade education, she expertly reads and articulates a complex theory of power, bodily inscription, and socialization that arose directly from material conditions of her own life. She sees what many far more "educated" scholars and citizens fail to recognize: that the bodies of poor women and children are produced and positioned as texts that facilitate the mandates of a . . . profoundly brutal and mean-spirited political regime. . . .

Over the past decade or so, a host of inspired feminist welfare scholars and activists have addressed and examined the relationship between state power and the lives of poor women and children. As important and insightful as these exposés are, with few exceptions, they do not get at the closed circuit that fuses together systems of power, the material conditions of poverty, and the bodily experiences that allow for the perpetuation—and indeed the justification—of these systems. They fail to consider what the speaker of my opening passage recognized so astutely: that systems of power produce and patrol poverty through the reproduction of both social and bodily markers. . . .

. . . [In this article I employ the theory of Michel Foucault to describe how the body is] the product of historically specific power relations. Feminists have used this notion of social inscription to explain a range of bodily operations from cosmetic surgery (Brush 1998, Morgan 1991), prostitution (Bell 1994), and Anorexia Nervosa (Hopwood 1995, Bordo 1993) to motherhood (Chandler 1999, Smart 1992), race (Stoler 1995, Ford-Smith 1995), and cultural imperialism (Desmond 1991). As these analyses illustrate, Foucault allows us to consider and critique the body as it is invested with meaning and inserted into regimes of truth via the operations of power and knowledge. . . .

Foucault clarifies and expands on this process of bodily/social inscription in his early work. In "Nietzsche, Genealogy, History," he positions the physical body as virtual text, accounting for the fact that "the body is the inscribed surface of events that are traced by language and dissolved by ideas" (1977, 83). . . . For Foucault, the body and [power] are inseparable. In his logic, power constructs and holds bodies. . . .

In *Discipline and Punish* Foucault sets out to depict the genealogy of torture and discipline as it reflects a public display of power on the body of subjects in the 17th and 18th centuries. In graphic detail Foucault begins his book with the description of a criminal being tortured and then drawn and quartered in a public square. The crowds of good parents and their growing children watch and learn. The public spectacle works as a patrolling image, socializing and controlling bodies within the body politic. Eighteenth century torture "must mark the victim: it is intended, either by the scar it leaves on the body or by the spectacle that accompanies it, to brand the victim with infamy . . . it traces around or rather on the very body of the condemned man signs that can not be effaced" (1984, 179). For Foucault, public exhibitions of punishment served as a socializing process, writing culture's codes and values on the minds and bodies of its subjects. In the process punishment . . . rearranged bodies.

. . . Foucault's point in *Discipline and Punish* is . . . that public exhibition and inscription have been replaced in contemporary society by a much more effective process of socialization and self-inscription. According to Foucault, today discipline has replaced torture as the privileged punishment, but the body continues to be written on. Discipline produces "subjected and practiced bodies, docile bodies" (1984, 182). We become subjects . . . of ideology, disciplining and inscribing our own bodies/minds in the process of becoming stable and singular subjects. . . . The body continues to be the site and operation of ideology. . . .

Indeed, while we are all marked discursively by ideology in Foucault's paradigm, in the United States today poor women and children of all races are multiply marked with signs of both discipline and punishment that cannot be erased or effaced. They are systematically produced through both 20th century forces of socialization and discipline and 18th century exhibitions of public mutilation. In addition to coming into being as disciplined and docile bodies, poor single welfare mothers and their children are physically inscribed, punished, and displayed as dangerous and pathological "other." It is important to note when considering the contemporary inscription of poverty as moral pathology etched onto the bodies of profoundly poor women and children, that these are more than metaphoric and self-patrolling marks of discipline. Rather on myriad levels—sexual, social, material and physical—poor women and their children, like the "deviants" publicly punished in Foucault's scenes of torture, are marked, mutilated, and made to bear and transmit signs in a public spectacle that brands the victim with infamy. . . .

The (Not So) Hidden Injuries of Class

Recycled images of poor, welfare women permeate and shape our national consciousness.[1] Yet—as is so often the case—these images and narratives tell us more about the culture that spawned and embraced them than they do about the object of the culture's obsession. . . .

These productions orchestrate the story of poverty as one of moral and intellectual lack and of chaos, pathology, promiscuity, illogic, and sloth, juxtaposed always against the order, progress, and decency of "deserving" citizens. . . .

I am, and will probably always be, marked as a poor woman. I was raised by a poor, single, White mother who had to struggle to keep her four children fed, sheltered, and clothed by working at what seemed like an endless stream of minimum wage, exhausting, and demeaning jobs. As a child poverty was written onto and into my being at the level of private and public thought and body. At an early age my body bore witness to and emitted signs of the painful devaluation carved into my flesh; that same devaluation became integral to my being in the world. I came into being as disciplined body/mind while at the same time I was taught to read my abject body as the site of my own punishment and erasure. In this excess of meaning the space between private body and public sign was collapsed.

For many poor children this double exposure results in debilitating . . . shame and lack. As Carolyn Kay Steedman reminds us in *Landscape for a Good Woman,* the mental life of poor children flows from material deprivation. Steedman speaks of the "relentless laying down of guilt" she experienced as a poor child living in a world where identity was shaped through envy and unfulfilled desire and where her own body "told me stories of the terrible unfairness of things, of the subterranean culture of longing for that which one can never have" (1987, 8). For Steedman, public devaluation and punishment "demonstrated to us all the hierarchies of our illegality, the impropriety of our existence, our marginality within the social system" (1987, 9). Even as an adult she recalls that:

> . . . the baggage will never lighten for me or my sister. We were born, and had no choice in the matter; but we were social burdens, expensive, unworthy, never grateful enough. There was nothing we could do to pay back the debt of our existence. (1987, 19)

Indeed, poor children are often marked with bodily signs that cannot be forgotten or erased. Their bodies are physically inscribed as "other" and then read as pathological, dangerous, and undeserving. What I recall most vividly about being a child in a profoundly poor family was that we were constantly hurt and ill, and because we could not afford medical care, small illnesses and accidents spiraled into more dangerous illnesses and complications that became both a part of who we were and written proof that we were of no value in the world.

In spite of my mother's heroic efforts, at an early age my brothers and sister and I were stooped, bore scars that never healed properly, and limped with feet mangled by ill-fitting, used Salvation Army shoes. When my sister's forehead was split open by a door slammed in frustration, my mother "pasted" the angry wound together on her own, leaving a mark of our inability to afford medical attention, of our lack, on her very forehead. When I suffered from a concussion, my mother simply put borrowed ice on my head and tried to keep me awake for a night. And when throughout elementary school we were sent to the office for mandatory and very public yearly checks, the school nurse sucked air through her teeth as she donned surgical gloves to check only the hair of poor children for lice.

We were read as unworthy, laughable, and often dangerous. Our school mates laughed at our "ugly shoes," our crooked and ill-serviced teeth, and the way we "stank," as teachers excoriated us for inability to concentrate in school, our "refusal" to come to class prepared with proper school supplies, and our unethical behavior when we tried to take more than our allocated share of "free lunch."[2] Whenever backpacks or library books came up missing, we were publicly interrogated and sent home to "think about" our offences, often accompanied by notes that reminded my mother that as a poor single parent she should be working twice as hard to make up for the discipline that allegedly walked out the door with my father. When we sat glued to our seats, afraid to stand in front of the class in ragged and ill-fitting hand-me-downs, we were held up as examples of unprepared and uncooperative children. And when our grades reflected our otherness, they were used to justify even more elaborate punishment. . . .

Friends who were poor as children, and respondents to a survey I conducted in 1998,[3] tell similar stories of the branding they received at the hands of teachers, administrators, and peers. An African-American woman raised in Yesler Terrace, a public housing complex in Seattle, Washington, writes:

> Poor was all over our faces. My glasses were taped and too weak. My big brother had missing teeth. My mom was dull and ashy. It was like a story of how poor we were that anyone could see. My sister Evie's lip was bit by a dog and we just had dime store stuff to put on it. Her lip was a big scar. Then she never smiled and no one smiled at her cause she never smiled. Kids called her "Scarface." Teachers never smiled at her. The princip[al] put her in detention all the time because she was mean and bad (they said).

And, a White woman in the Utica, New York, area remembers:

> We lived in dilapidated and unsafe housing that had fleas no matter how clean my mom tried to be. We had bites all over us. Living in our car between evictions was even worse—then we didn't have a bathroom so I got kidney problems that I never had doctor's help for. When my teachers wouldn't let me got to the bathroom every hour or so I would wet my pants in class. You can imagine what the kids did to me about that. And the teachers would refuse to let me go to the bathroom because they said I was willful.

Material deprivation is publicly written on the bodies of poor children in the world. In the United States poor families experience violent crime, hunger, lack of medical and dental care, utility shut-offs, the effects of living in unsafe housing and/or of being homeless, chronic illness, and insufficient winter clothing (Lein and Edin 1996, 224–231). According to Jody Raphael of the Taylor Institute, poor women and their children are also at five times the risk of experiencing domestic violence (Raphael, 2000).

As children, our disheveled and broken bodies were produced and read as signs of our inferiority and undeservedness. As adults our mutilated bodies are read as signs of inner chaos, immaturity, and indecency as we are punished and then read as proof of need for further discipline and punishment. When my already bad teeth started to rot and I was out of my head with pain, my choices as an adult welfare recipient were to either let my teeth fall out or have them pulled out. In either case the culture would then read me as a "toothless illiterate," as a fearful joke. In order to pay my rent and to put shoes on my daughter's feet I sold blood at two or three different clinics on a monthly basis until I became so anemic that they refused to buy it from me. A neighbor of mine went back to the man who continued to beat her and her children after being denied welfare benefits, when she realized that she could not adequately feed, clothe and house her family on her own minimum wage income. My good friend sold her ovum to a fertility clinic in a painful and potentially damaging process. Other friends exposed themselves to all manner of danger and disease by selling their bodies for sex in order to feed and clothe their babies.

Exhaustion also marks the bodies of poor women in indelible script. Rest becomes a privilege we simply cannot afford. After working full shifts each day, poor mothers trying to support themselves at minimum wage jobs continue to work to a point of exhaustion that is inscribed on their faces, their bodies, their posture, and their diminishing sense of self and value in the world. My former neighbor recently recalled:

> I had to take connecting buses to bring and pick up my daughters at childcare after working on my feet all day. As soon as we arrived at home, we would head out again by bus to do laundry. Pick up groceries. Try to get to the food bank. Beg the electric company to not turn off our lights and heat again. Find free winter clothing. Sell my blood. I would be home at nine or ten o'clock at night. I was loaded down with one baby asleep and one crying. Carrying lots of heavy bags and ready to drop on my feet. I had bags under my eyes and no shampoo to wash my hair so I used soap. Anyway I had to stay up to wash diapers in the sink. Otherwise they wouldn't be dry when I left the house in the dark with my girls. In the morning I start all over again.

This bruised and lifeless body, hauling sniffling babies and bags of dirty laundry on the bus, was then read as a sign that she was a bad mother and a threat that needed to be disciplined and made to work even harder for her own good. Those who need the respite less go away for weekends, take drives in the woods, take their kids to the beach. Poor women without education are pushed into minimum wage jobs and have no money, no car, no time, no energy, and little support, as their bodies are made to display marks of their material deprivation as a socializing and patrolling force.

Ultimately, we come to recognize that our bodies are not our own; that they are rather public property. State mandated blood tests, interrogation of the most private aspects of our lives, the public humiliation of having to beg officials for food and medicine, and the loss of all right to privacy, teach us that our bodies are only useful as lessons, warnings, and signs of degradation that everyone loves to hate. In "From Welfare to Academe: Welfare Reform as College-Educated Welfare Mothers Know It," Sandy Smith-Madsen describes the erosion of her privacy as a poor welfare mother:

> I was investigated. I was spied upon. A welfare investigator came into my home and after thoughtful deliberation, granted me permission to keep my belongings. . . . Like the witch hunts of old, if a neighbor reports you as a welfare queen, the guardians of the state's compelling interest come into your home and interrogate you. While they do not have the right to set your body ablaze on the public square, they can forever devastate heart and soul by snatching away children. Just like a police officer, they may use whatever they happen to see against you, including sexual orientation. Full-fledged citizens have the right to deny an officer entry into their home unless they possess a search warrant; welfare mothers fork over citizenship rights for the price of a welfare check. In Tennessee, constitutional rights go for a cash value of $185 per month for a family of three. (2000, 185)

Welfare reform policy is designed to publicly expose, humiliate, punish and display "deviant" welfare mothers. "Workfare" and "Learnfare"—two alleged successes of welfare reform—require that landlords, teachers, and employers be made explicitly aware of the second class status of these very public bodies. In Ohio, the Department of Human Services uses tax dollars to pay for advertisements on the side of Cleveland's RTA busses that show a "Welfare Queen" behind bars with a logo that proclaims "Crime does not pay. Welfare fraud is a crime" (Robinson 1999). In Michigan a pilot program mandating drug tests for all welfare recipients began on October 1, 1999. Recipients who refuse the test will lose their benefits immediately (Simon 1999). In Buffalo, New York, a County Executive proudly announced that his county will begin intensive investigation of all parents who refuse minimum wage jobs that are offered to them by the state. He warned: "We have many ways of investigating and exposing these errant parents who choose to exploit their children in this way" (Anderson 1999). And, welfare reform legislation enacted in 1996 as the Personal Responsibility and Work Opportunities Reconciliation Act (PRWORA), requires that poor mothers work full-time, earning minimum wage salaries with which they cannot support their children. Often denied medical, dental, and childcare benefits, and unable to provide their families with adequate food, heat, or clothing, through this legislation the state mandates child neglect and abuse. The crowds of good parents and their growing children watch and learn. . . .

Reading and Rewriting the Body . . .

The bodies of poor women and children, scarred and mutilated by state mandated material deprivation and public exhibition, work as spectacles, as patrolling images socializing and controlling bodies within the body politic. . . .

Spectacular cover stories of the "Welfare Queen" play and re-play in the national mind's eye, becoming a prescriptive lens through which the American public as a whole reads the individual dramas of the bodies of poor women and their place and value in the world. These dramas produce "normative" citizens as singular, stable, rational, ordered, and free. In this dichotomous, hierarchical frame the poor welfare mother is juxtaposed against a logic of "normative" subjectivity as the embodiment of disorder, disarray, and other-ness. Her broken and scarred body becomes proof of her inner pathology and chaos, suggesting the need for further punishment and discipline.

In contemporary narrative welfare women are imagined to be dangerous because they refuse to sacrifice their desires and fail to participate in legally sanctioned heterosexual relationships; theirs is read, as a result, as a selfish, "unnatural," and immature sexuality. In this script, the bodies of poor women are viewed as being dangerously beyond the control of men and are as a result construed as the bearers of perverse desire. In this androcentric equation fathers become the sole bearers of order and of law, defending poor women and children against their own unchecked sexuality and lawlessness.

For Republican Senator [now Attorney General] John Ashcroft writing in *The St. Louis Dispatch,* the inner city is the site of "rampant illegitimacy" and a "space devoid of discipline" where all values are askew. For Ashcroft, what is insidious is not material poverty, but an entitlement system that has allowed "out-of-control" poor women to rupture traditional patriarchal authority, valuation, and boundaries (1995, A:23). Impoverished communities then become a site of chaos because without fathers they allegedly lack any organizing or patrolling principle. George Gilder agrees with Ashcroft when he writes in the conservative *American Spectator* that:

> The key problem of the welfare culture is not unemployed women and poor children. It is the women's skewed and traumatic relationships with men. In a reversal of the pattern of civilized societies, the women have the income and the ties to government authority and support.... This balance of power virtually prohibits marriage, which is everywhere based on the provider role of men, counterbalancing the sexual and domestic superiority of women. (1995, B:6)

For Gilder, the imprimatur of welfare women's sordid bodies unacceptably shifts the focus of the narrative from a male presence to a feminized absence.

In positioning welfare mothers as sexually chaotic, irrational, and unstable, their figures are temporarily immobilized and made to yield meaning as a space that must be brought under control and transformed through public displays of punishment. Poor single mothers and children who have been abandoned, have fled physical, sexual, and/or psychological abuse, or have in general refused to capitulate to male control within the home are mythologized as dangerous, pathological, out of control, and selfishly unable—or unwilling—to sacrifice their "naturally" unnatural desires. They are understood and punished as a danger to a culture resting on a foundation of inviolate male authority and absolute privilege in both public and private spheres.

William Raspberry disposes of poor women as selfish and immature, when in "Ms. Smith Goes After Washington," he warrants that:

> ... unfortunately AFDC is paid to an unaccountable, accidental and unprepared parent who has chosen her head of household status as a personal form of satisfaction, while lacking the simple life skills and maturity to achieve love and job fulfillment from any other source. I submit that all of our other social ills—crime, drugs, violence, failing schools ... are a direct result of the degradation of parenthood by emotionally immature recipients. (1995, A:19)

Raspberry goes on to assert that like poor children, poor mothers must be made visible reminders to the rest of the culture of the "poor choices" they have made. He claims that rather than "coddling" her, we have a responsibility to "shame her" and to use her failure to teach other young women that it is "morally wrong for unmarried women to bear children," as we "cast single motherhood as a selfish and immature act" (1995, A:19).

Continuous, multiple, and often seamless public inscription, punishing policy, and lives of unbearable material lack leave poor women and their children scarred, exhausted, and confused. As a result their bodies are imagined as an embodiment of decay and cultural dis-ease that threatens the health and progress of our nation.... In a 1995 *USA Today* article entitled "America at Risk: Can We Survive Without Moral Values?" for example, the inner city is portrayed as a "*dark*" realm of "*decay* rooted in the *loss* of values, the *death* of work ethics, and the *deterioration* of families and communities." Allegedly here, "all morality has *rotted* due to a *breakdown* in gender discipline." This space of disorder and disease is marked with tropes of race and gender. It is also associated with the imagery of "communities of women *without* male leadership, cultural values and initiative [emphasis added]" (1995, C:3). In George Will's *Newsweek* editorial he proclaims that "*illogical* feminist and racial *anger* coupled with *misplaced* American emotion may be part or a cause of the *irresponsible* behavior *rampant* in poor neighborhoods." Will continues, proclaiming that here "mothers *lack* control over their children and have *selfishly* taught them to embrace a *pathological* ethos that values *self-need* and *self-expression* over self-control [emphasis added]" (1995, 23).

Poor women and children's bodies, publicly scarred and mutilated by material deprivation, are read as expressions of an essential lack of discipline and order. In response to this perception, journalist Ronald Brownstein of the *L.A. Times* proposed that the *Republican Contract with America* will "*restore* America to its path, *enforcing* social *order* and common *standards* of behavior, and replacing *stagnation* and *decay* with *movement* and *forward* thinking *energy* [emphasis added]" (1994, A:20). In these rhetorical fields poverty is . . . linked to lack of progress that would allegedly otherwise order, stabilize, and restore the culture. What emerges from these diatribes is the positioning of patriarchal, racist, capitalist, hierarchical, and heterosexist "order" and movement against the alleged stagnation and decay of the body of the "Welfare Queen."

Race is clearly written on the body of the poor single mother. The welfare mother, imagined as young, never married, and Black (contrary to statistical evidence[4]), is framed as dangerous and in need of punishment because she "naturally" emasculates her own men, refuses to service White men, and passes on—rather than appropriate codes of subservience and submission—a disruptive culture of resistance, survival, and "misplaced" pride to her children (Collins 1991). In stark contrast, widowed women with social security and divorced women with child support and alimony are imaged as White, legal, and propertied mothers whose value rests on their abilities to stay in their homes, care for their own children, and impart traditional cultural morals to their offspring, all for the betterment of the culture. In this narrative welfare mothers have only an "outlaw" culture to impart. Here the welfare mother is read as both the product and the producer of a culture of disease and disorder. These narratives imagine poor women as powerful contagion capable of, perhaps even lying in wait to infect their own children as raced, gendered, and classed agents of their "diseased" nature. In contemporary discourses of poverty racial tropes position poor women's bodies as dangerous sites of "naturalized chaos" and as potentially valuable economic commodities who refuse their proper role.

Gary McDougal in "The Missing Half of the Welfare Debate" furthers this image by referring to the "crab effect of poverty" through which mothers and friends of individuals striving to break free of economic dependency allegedly "pull them back down." McDougal affirms—again despite statistical evidence to the contrary—that the mothers of welfare recipients are most often themselves "generational welfare freeloaders lacking traditional values and family ties who can not, and will not, teach their children right from wrong." "These women" he asserts "would be better off doing any kind of labor regardless of how little it pays, just to get them out of the house, to break their cycles of degeneracy" (1996, A:16).

In this plenitude of images of evil mothers, the poor welfare mother threatens not just her own children, but all children. The Welfare Queen is made to signify moral aberration and economic drain; her figure becomes even more impacted once responsibility for the destruction of the "American Way of Life" is attributed to her. Ronald Brownstein reads her "spider web of dependency" as a "crisis of character development that leads to a morally bankrupt American ideology" (1994, A:6).

These representations position welfare mothers' bodies as sites of destruction and as catalysts for a culture of depravity and disobedience; in the process they produce a reading of the writing on the body of the poor woman that calls for further punishment and discipline. In New York City, "Workfare" programs force *lazy* poor women to take a job—"any job"—including working for the city wearing orange surplus prison uniforms picking up garbage on the highway and in parks for about $1.10 per hour (Dreier 1999). "Bridefare" programs in Wisconsin give added benefits to *licentious* welfare women who marry a man—"any man"—and publish a celebration of their "reform" in local newspapers (Dresang 1996). "Tidyfare" programs across the nation allow state workers to enter and inspect the homes of poor *slovenly* women so that they can monetarily sanction families whose homes are deemed to be appropriately tidied.[5] "Learnfare" programs in many states publicly expose and fine *undisciplined* mothers who for any reason have children who don't (or can't) attend school on a regular basis (Muir 1993). All of these welfare reform programs are designed to expose and publicly punish the *misfits* whose bodies are read as proof of their refusal or inability to capitulate to androcentric, capitalist, racist, and heterosexism values and mores.

The Power of Poor Women's Communal Resistance

Despite the rhetoric and policy that mark and mutilate our bodies, poor women survive. Hundreds of thousands of us are somehow good parents despite the systems that are designed to prohibit us from being so. We live on the unlivable and teach our children love, strength, and grace. We network, solve irresolvable dilemmas, and support each other and our families. If we somehow manage to find a decent pair of shoes, or save our foodstamps to buy our children a birthday cake, we are accused of being cheats or living too high. If our children suffer, it is read as proof of our inferiority and bad mothering; if they succeed we are suspect for being too pushy, for taking more than our share of free services, or for having too much free time to devote to them. Yet, as former welfare recipient Janet Diamond says in the introduction to *For Crying Out Loud:*

> In spite of public censure, welfare mothers graduate from school, get decent jobs, watch their children achieve, make good lives for themselves . . . welfare mothers continue to be my inspiration, not because they survive, but because they dare to dream. Because when you are a welfare recipient, laughter is an act of rebellion. (1986, 1)

. . . Because power is diffuse, heterogeneous, and contradictory, poor women struggle against the marks of their degradation. . . .

Poor women rebel by organizing for physical and emotional respite, and eventually for political power. My own resistance was born in the space between self-loathing and my love of and respect for poor women who were fighting together against oppression. In the throes of political activism (at first I was dragged blindly into such actions, ironically, in a protest that required, according to the organizer, just so many poor women's bodies) I became caught up in the contradiction between my body's meaning as despised public sign, and our shared sense of communal power, knowledge, authority, and beauty. Learning about labor movements, fighting for rent control, demanding fair treatment at the welfare office, sharing the costs, burdens, and joys of raising children, forming good cooperatives, working with other poor women to go to college, and organizing for political change, became addictive and life affirming acts of resistance.

Communal affiliation among poor women is discouraged, indeed in many cases prohibited, by those with power over our lives. Welfare offices, for example, are designed to prevent poor women from talking together; uncomfortable plastic chairs are secured to the ground in arrangements that make it difficult to communicate, silence is maintained in waiting rooms, case workers are rotated so that they do not become too "attached" to their clients, and, reinforced by "Welfare Fraud" signs covering industrially painted walls, we are daily reminded not to trust anyone with the details of our lives for fear of further exposure and punishment. And so, like most poor women, I had remained isolated, ashamed, and convinced that I was alone in, and responsible for, my suffering.

Through shared activism we became increasingly aware of our individual bodies as sites of contestation and of our collective body as a site of resistance and as a source power.

Noemy Vides in "Together We Are Getting Freedom," reminds us that "by talking and writing about learned shame together, [poor women] pursue their own liberation" (305). Vides adds that it is through this process that she learned to challenge the dominant explanations that decreed her value in the world,

> provoking an awareness that the labels—ignorant peasant, abandoned woman, broken-English speaker, welfare cheat—have nothing to do with who one really is, but serve to keep women subjugated and divided. [This communal process] gives women tools to understand the uses of power; it emboldens us to move beyond the imposed shame that silences, to speak out and join together in a common liberatory struggle. (305)

In struggling together we contest the marks of our bodily inscription, disrupt the use of our bodies as public sign, change the conditions of our lives, and survive. In the process we come to understand that the shaping of our bodies is not coterminous with our beings or abilities as a whole. Contestation and the deployment of new truths cannot erase the marks of our poverty, but the process does transform the ways in which we are able to interrogate and critique our bodies and the systems that have branded them with infamy. As a result these signs are rendered fragile, unstable, and ultimately malleable.

Notes

1. Throughout this paper I use the terms "welfare recipients," and "poor working women" interchangeably because as the recent *Urban Institute* study made clear, today these populations are, in fact, one and the same. (Loprest 1999)

2. As recently as 1995, in my daughter's public elementary school cafeteria, "free lunchers" (poor children who could not otherwise afford to eat lunch, including my daughter) were reminded with a large and colorful sign to "line up last."

3. The goal of my survey was to measure the impact of the 1996 welfare reform legislation on the lives of profoundly poor women and children in the United States. Early in 1998 I sent fifty questionnaires and narrative surveys to four groups of poor women on the West and the East coasts; thirty-nine were returned to me. I followed these surveys with forty-five minute interviews with twenty of the surveyed women.

4. In the two years directly preceding the passage of the PRWORA, as a part of sweeping welfare reform, in the United States the largest percentage of people on welfare were white (39%) and fewer than 10% were teen mothers. (1994. U.S. Department of Health and Human Services, "An Overview of Entitlement Programs")

5. *Tidyfare* programs additionally required that caseworkers inventory the belongings of AFDC recipients so that they could require them to "sell-down" their assets. In my own case, in 1994 a HUD inspector came into my home, counted my daughter's books, checked them against his list to see that as a nine year old she was only entitled to have twelve books, calculated what he perceived to be the value of the excess books, and then had my AFDC check reduced by that amount in the following month.

REFERENCES

Abramovitz, Mimi. 1989. *Regulating the lives of women: Social welfare policy from colonial times to the present.* Boston: South End Press.

———. 2000. *Under attack, fighting back.* New York: Monthly Review Press.

Albelda, Randy. 1997. *Glass ceilings and bottomless pits: Women's work, women's poverty.* Boston: South End Press.

"America at risk: Can we survive without moral values?" 1995. *USA Today.* October, Sec. C: 3.

Amott, Teresa. 1993. *Caught in the crises: Women and the U.S. economy today.* New York: Monthly Review Press.

Anderson, Dale. 1999. "County to investigate some welfare recipients." *The Buffalo News.* August 18, Sec. B: 5.

Ashcroft, John. 1995. "Illegitimacy rampant." *The St. Louis Dispatch.* July 2, Sec. A: 23.

Bell, Shannon. 1994. *Reading, writing and rewriting the prostitute body.* Bloomington and Indianapolis: Indiana University Press.

Bordo, Susan, 1993. *Unbearable Weight: Feminism, western culture and the body.* Berkeley: University of California Press.

Brownstein, Ronald. 1994. "GOP welfare proposals more conservative." *Los Angeles Times,* May 20, Sec. A: 20.

———. 1994. "Latest welfare reform plan reflects liberals' priorities." *Los Angeles Times.* May 20, Sec. A: 6.

Chandler, Mielle. 1999. "Queering maternity." *Journal of the Association for Research on Mothering.* Vol. 1, no. 2, (21–32).

Collins, Patricia Hill. 2000. *Black feminist thought: Knowledge, consciousness, and the politics of empowerment.* New York: Routledge.

Crompton, Rosemary. 1986. *Gender and stratification.* New York: Polity Press.

Desmond, Jane. 1991. "Dancing out the difference; cultural imperialism and Ruth St. Denis's Radna of 1906." *Signs.* Vol. 17, no. 1, Autumn, (28–49).

Diamond, Janet. 1986. *For crying out loud: Women and poverty in the United States.* Boston: Pilgrim Press.

Dreier, Peter. 1999. "Treat welfare recipients like workers." *Los Angeles Times.* August 29, Sec. M: 6.

Dresang, Joel. 1996. "Bridefare designer, reform beneficiary have role in governor's address." *Milwaukee Journal Sentinel.* August 14, Sec. 9.

Dujon, Diane and Ann Withorn. 1996. *For crying out loud: Women's poverty in the Unites States.* South End Press.

Edin, Kathryn and Laura Lein. 1997. *Making ends meet: How single mothers survive welfare and low wage work.* Russell Sage Foundation.

Ford-Smith, Honor. 1995. "Making white ladies: race, gender and the production of identity in late colonial Jamaica." *Resources for Feminist Research,* Vol. 23, no. 4, Winter, (55–67).

Foucault, Michel. 1984. Discipline and punish. In P. Rabinow (ed.) *The Foucault reader.* New York: Pantheon Books.

———. 1978. *The history of sexuality: An introduction.* Trans. R. Hurley. Harmondsworth: Penguin.

———. 1984. "Nietzsche, genealogy, history." In P. Rabinow (ed.) *The Foucault reader.* New York: Pantheon Books.

———. 1980. *Power/knowledge: Selected interviews and other writings 1972–1977.* C. Gordon (ed.) Brighton: Harvester.

Funiciello, Theresa. 1998. "The brutality of bureaucracy." *Race, class and gender: An anthology,* 3rd ed. Eds. Margaret L. Andersen and Patricia Hill Collins. Belmont: Wadsworth Publishing Company, (377–381).

Gilder, George. 1995. "Welfare fraud today." *American Spectator.* September 5, Sec. B: 6.

Gordon, Linda. 1995. *Pitied, but not entitled: Single mothers and the history of welfare.* New York: Belknap Press, 1995.

hooks, bell. "Thinking about race, class, gender and ethics" 1999. Presentation at Hamilton College, Clinton, New York.

Hopwood, Catherine. 1995. "My discourse/myself: therapy as possibility (for women who eat compulsively)." *Feminist Review.* No. 49, Spring, (66–82).

Langston, Donna. 1998. "Tired of playing monopoly?" *In Race, class and gender: An anthology,* 3rd ed. Eds. Margaret L. Andersen and Patricia Hill Collins. Belmont: Wadsworth Publishing Company, (126–136).

Lerman, Robert. 1995. "And for fathers?" *The Washington Post.* August 7, Sec. A: 19.

Loprest, Pamela. 1999. "Families who left welfare: Who are they and how are they doing?" *The Urban Institute,* Washington, D.C. August, No. B-1.

McDougal, Gary. 1996. "The missing half of the welfare debate." *The Wall Street Journal.* September 6, Sec. A: 16 (W).

McNay, Lois. 1992. *Foucault and feminism: Power, gender and the self.* Boston: Northeastern University Press.

Mink, Gwendoly. 1998. *Welfare's end.* Cornell University Press.

———. 1996. *The wages of motherhood: Inequality in the welfare state 1917–1942.* Cornell University Press.

Morgan, Kathryn. 1991. "Women and the knife: Cosmetic surgery and the colonization of women's bodies." *Hepatia.* V6, No 3. Fall, (25–53).

Muir, Kate. 1993. "Runaway fathers at welfare's final frontier. *The Times.* Times Newspapers Limited. July 19, Sec. A: 2.

"An overview of entitlement programs." 1994. U.S. Department of Health and Human Services. Washington, DC: U.S. Government Printing Office.

Piven, Frances Fox and Richard Cloward. 1993. *Regulating the poor: The functions of public welfare.* New York: Vintage Books.

Raspberry, William. 1995. "Ms. Smith goes after Washington." *The Washington Post.* February 1, Sec. A: 19.

———. 1996. "Uplifting the human spirit." *The Washington Post.* August 8, Sec. A: 31.
Robinson, Valerie. 1999. "State's ad attacks the poor." *The Plain Dealer,* November 2, Sec. B: 8.
Sennett, Richard and Jonathan Cobb. 1972. *The hidden injuries of class.* New York: Vintage Books.
Sidel, Ruth. 1998. *Keeping women and children last: America's war on the poor.* New York: Penguin Books.
Simon, Stephanie. 1999. "Drug tests for welfare applicants." *The Los Angeles Times.* December 18, Sec. A: 1. National Desk.
Smart, Carol. 1997. *Regulating womanhood: Essays on marriage, motherhood and sexuality.* New York: Routledge.
———. "Disruptive bodies and unruly sex: the regulation of reproduction and sexuality in the nineteenth century." New York: Routledge, (7–32).
Smith-Madsen, Sandy. 2000. "From welfare to academe: Welfare reform as college-educated welfare mothers know it." *And still we rise: women, poverty and the promise of education in America.* Forthcoming. Vivyan Adair and Sandra Dahlber (eds.). Philadelphia: Temple University Press, (160–186).
Steedman, Carolyn Kay. 1987. *Landscape for a good woman.* New Brunswick, N. J., Rutgers University Press.
Stoler, Ann Laura. 1995. *Race and the education of desire: Foucault's history of sexuality and the colonial order of things.* Durham: Duke University Press.
Sylvester, Kathleen. 1995. "Welfare debate." *The Washington Post.* September 3, Sec. E: 15.
Tanner, Michael. 1995. "Why welfare pays." *The Wall Street Journal.* September 28, Sec. A: 18 (W).
Vides, Noemy and Victoria Steinitz. 1996. "Together we are getting freedom." *For crying out loud.* Diane Dujon and Ann Withorn (eds.). Boston: South End Press, (295–306).
Will, George. 1995. "Welfare gate." *Newsweek.* February 5, Sec. 23.

THINKING ABOUT THE READING

When we think of people's bodies being labeled as deviant, we usually assume the bodies in question either deviate from cultural standards of shape and size or are marked by some noticeable physical handicap. However, Adair shows us that poor women's and children's bodies are tagged as undesirable in ways that are just as profound and just as hard to erase. What does she mean when she says that the illnesses and accidents of youth became part of a visible reminder of who poor people are in the eyes of others? How do the public degradations suffered by poor people (for instance, having a school nurse wear surgical gloves to check only the hair of poor children for lice) reinforce their subordinate status in society? Why do you think Adair continually evokes the images of "danger," "discipline," and "punishment" in describing the ways non-poor people perceive and respond to the physical appearance of poor people? Explain how focusing on the "deviance" of poor people deflects public attention away from the harmful acts committed by more affluent citizens.

The Architecture of Inequality

Race and Ethnicity

The history of race in the United States is an ambivalent one. Cultural beliefs about equality conflict with the experiences of most racial and ethnic minorities: oppression, violence, and exploitation. Opportunities for life, liberty, and the pursuit of happiness have always been distributed along racial and ethnic lines. U.S. society is built on the assumption that different immigrant groups will ultimately assimilate, changing their way of life to conform to that of the dominant culture. But the increasing diversity of the population has shaped people's ideas about what it means to be an American and has influenced our relationships with one another and with our social institutions.

Sociologists tell us that race is not a biological characteristic but rather a social construction that can change across time and from culture to culture. The socially constructed nature of race is illustrated in "Racial and Ethnic Formation" by Michael Omi and Howard Winant. However, the authors are quick to point out that just because race is socially created doesn't mean it is insignificant. Indeed, our definitions of race are related to inequality, discrimination, and cultural dominance and resistance. Race may not be a purely biological trait, but it is an important part of every social institution.

It has been said that white people in the United States have the luxury of "having no color." When people are described with no mention of race, the default assumption is that they are white. In other words, *white* is used far less often as a modifying adjective than *black, Asian,* or *Latino.* As a result, "whiteness" is rarely questioned or examined as a racial or ethnic category. In her article, "Optional Ethnicities," Mary C. Waters argues that unlike members of other groups, U.S. whites can choose whether or not to include their specific ancestry in descriptions of their own identities. For whites of European descent, claiming an ethnic identity is a voluntary "leisure-time activity" with few social implications. Indeed, the option of being able *not* to claim any ethnicity is available only to the majority group in a society.

Racial inequality is both a personal and structural phenomenon. On one hand, it is lodged in individual prejudice and discrimination. On the other hand, it resides in our language, collective beliefs, and important social institutions. This latter manifestation of racism is more difficult to detect than personal racism, and hence it is more difficult to change. Because such racism exists at a level beyond personal attitudes, it will not disappear simply by reducing people's prejudices.

In "Silent Racism: Passivity in the Well-Meaning White People," Barbara Trepagnier discusses the "bystander effect" that is marked by detachment and passivity in white people when they observe racism around them. She points out that despite a tangible level of race awareness, different types of racism (silent, everyday, color-blind) still contribute to the maintenance of racism at individual and institutional levels. The racial privilege that allows whites to quietly watch racist processes occur and gain benefits from their outcomes demonstrates how passivity works against the ultimate goal of racial equality.

Something to Consider as You Read

As you read these selections, consider the differences between individual prejudice and institutional racism. Is it possible for someone not to be racist and still participate in practices that perpetuate racism? Compare these readings with those in other sections. Consider the connections between access to economic resources, social class, and race. How might socioeconomic status influence attitudes and behaviors toward others who may share your ethnicity but not your class position? Think also about how you identify your own race or ethnicity. When you fill out a questionnaire that asks you to select a racial/ethnic category, do you think the category adequately reflects you? When you go somewhere, do you assume you will easily find others of your own race or ethnicity? When you watch television or a movie, how likely is it that the central characters will be people who share your racial background? Practice asking yourself similar questions as a way of enhancing your racial awareness.

Racial and Ethnic Formation

Michael Omi and Howard Winant

(1994)

In 1982–83, Susie Guillory Phipps unsuccessfully sued the Louisiana Bureau of Vital Records to change her racial classification from black to white. The descendant of an 18th-century white planter and a black slave, Phipps was designated "black" in her birth certificate in accordance with a 1970 state law which declared anyone with at least 1/32nd "Negro blood" to be black.

The Phipps case raised intriguing questions about the concept of race, its meaning in contemporary society, and its use (and abuse) in public policy. Assistant Attorney General Ron Davis defended the law by pointing out that some type of racial classification was necessary to comply with federal record-keeping requirements and to facilitate programs for the prevention of genetic diseases. Phipps's attorney, Brian Begue, argued that the assignment of racial categories on birth certificates was unconstitutional and that the 1/32nd designation was inaccurate. He called on a retired Tulane University professor who cited research indicating that most Louisiana whites have at least 1/20th "Negro" ancestry.

In the end, Phipps lost. The court upheld the state's right to classify and quantify racial identity.[1]

Phipps's problematic racial identity, and her effort to resolve it through state action, is in many ways a parable of America's unsolved racial dilemma. It illustrates the difficulties of defining race and assigning individuals or groups to racial categories. It shows how the racial legacies of the past—slavery and bigotry—continue to shape the present. It reveals both the deep involvement of the state in the organization and interpretation of race, and the inadequacy of state institutions to carry out these functions. It demonstrates how deeply Americans both as individuals and as a civilization are shaped, and indeed haunted, by race.

Having lived her whole life thinking that she was white, Phipps suddenly discovers that by legal definition she is not. In U.S. society, such an event is indeed catastrophic.[2] But if she is not white, of what race is she? The *state* claims that she is black, based on its rules of classification,[3] and another state agency, the court, upholds this judgment. But despite these classificatory standards which have imposed an either-or logic on racial identity, Phipps will not in fact "change color." Unlike what would have happened during slavery times if one's claim to whiteness was successfully challenged, we can assume that despite the outcome of her legal challenge, Phipps will remain in most of the social relationships she had occupied before the trial. Her socialization, her familial and friendship networks, her cultural orientation, will not change. She will simply have to wrestle with her newly acquired "hybridized" condition. She will have to confront the "Other" within.

The designation of racial categories and the determination of racial identity is no simple task. For centuries, this question has precipitated intense debates and conflicts, particularly in the U.S.—disputes over natural and legal rights, over the distribution of resources, and indeed, over who shall live and who shall die.

A crucial dimension of the Phipps case is that it illustrates the inadequacy of claims that race is a mere matter of variations in human physiognomy, that it is simply a matter of skin color. But if race cannot be understood in this manner, how *can* it be understood? We cannot fully hope to address this topic—no less than the meaning of race, its role in society, and the forces which shape it—in one chapter, nor indeed in one book. Our goal in this chapter, however, is far from modest: we wish to offer at least the outlines of a theory of race and racism.

What Is Race?

There is a continuous temptation to think of race as an *essence,* as something fixed, concrete, and objective. And there is also an opposite temptation: to imagine race as a mere *illusion,* a purely ideological construct which some ideal non-racist social order would eliminate. It is necessary to challenge both these positions, to disrupt and reframe the rigid and bipolar manner in which they are posed and debated, and to transcend the presumably irreconcilable relationship between them.

The effort must be made to understand race as an unstable and "decentered" complex of social meanings constantly being transformed by political struggle. With this in mind, let us propose a definition: *race is a concept which signifies and symbolizes social conflicts and interests by referring to different types of human bodies.* Although the concept of race invokes biologically based human characteristics (so-called "phenotypes"), selection of these particular human features for purposes of racial signification is always and necessarily a social and historical process. In contrast to the other major distinction of this type, that of gender, there is no biological basis for distinguishing among human groups along the lines of race.[4] Indeed, the categories employed to differentiate among human groups along racial lines reveal themselves, upon serious examination, to be at best imprecise, and at worst completely arbitrary.

If the concept of race is so nebulous, can we not dispense with it? Can we not "do without" race, at least in the "enlightened" present? This question has been posed often, and with greater frequency in recent years.[5] An affirmative answer would of course present obvious practical difficulties: it is rather difficult to jettison widely held beliefs, beliefs which moreover are central to everyone's identity and understanding of the social world. So the attempt to banish the concept as an archaism is at best counterintuitive. But a deeper difficulty, we believe, is inherent in the very formulation of this schema, in its way of posing race as a *problem,* a misconception left over from the past, and suitable now only for the dustbin of history.

A more effective starting point is the recognition that despite its uncertainties and contradictions, the concept of race continues to play a fundamental role in structuring and representing the social world. The task for theory is to explain this situation. It is to avoid both the utopian framework which sees race as an illusion we can somehow "get beyond," and also the essentialist formulation which sees race as something objective and fixed, a biological datum.[6] Thus we should think of race as an element of social structure rather than as an irregularity within it; we should see race as a dimension of human representation rather than as an illusion. These perspectives inform the theoretical approach we call racial formation.

Racial Formation

We define *racial formation* as the sociohistorical process by which racial categories are created, inhabited, transformed, and destroyed. Our attempt to elaborate a theory of racial formation will proceed in two steps. First, we argue that racial formation is a process of historically situated *projects* in which human bodies and social structures are represented and organized. Next we link racial formation to the evolution of hegemony, the way in which society is organized and ruled. Such an approach, we believe, can facilitate understanding of a whole range of contemporary controversies and dilemmas involving race, including the nature of racism, the relationship of race to other forms of differences, inequalities, and oppression such as sexism and nationalism, and the dilemmas of racial identity today.

From a racial formation perspective, race is a matter of both social structure and cultural representation. Too often, the attempt is made to understand race simply or primarily in terms of only one of these two analytical dimensions.[7] For example, efforts to explain racial inequality as a purely social structural phenomenon are unable to account for the origins, patterning, and transformation of racial difference.

Conversely, many examinations of racial difference—understood as a matter of cultural attributes á la ethnicity theory, or as a society-wide signification system á la some poststructuralist accounts—cannot comprehend such structural phenomena as racial stratification in the labor market or patterns of residential segregation.

An alternative approach is to think of racial formation processes as occurring through a linkage between structure and representation. Racial *projects* do the ideological "work" of making these links. A *racial project is simultaneously an interpretation, representation, or explanation of racial dynamics, and an effort to reorganize and redistribute resources along particular racial lines.* Racial projects connect what race *means* in a particular discursive practice and the ways in which both social structures and everyday experiences are racially *organized*, based upon that meaning. Let us consider this proposition, first in terms of large-scale or macro-level social processes, and then in terms of other dimensions of the racial formation process.

Racial Formation as a Macro-Level Social Process

To interpret the meaning of race is to frame it social structurally. Consider for example, this statement by Charles Murray on welfare reform:

> My proposal for dealing with the racial issue in social welfare is to repeal every bit of legislation and reverse every court decision that in any way requires, recommends, or awards differential treatment according to race, and thereby put us back onto the track that we left in 1965. We may argue about the appropriate limits of government intervention in trying to enforce the ideal, but at least it should be possible to identify the ideal: Race is not a morally admissible reason for treating one person differently from another. Period.[8]

Here there is a partial but significant analysis of the meaning of race: it is not a morally valid basis upon which to treat people "differently from one another." We may notice someone's race, but we cannot act upon that awareness. We must act in a "color-blind" fashion. This analysis of the meaning of race is immediately linked to a specific conception of the role of race in the social structure: it can play no part in government action, save in "the enforcement of the ideal." No state policy can legitimately require, recommend, or award different status according to race. This example can be classified as a particular type of racial project in the present-day U.S.—a "neoconservative" one.

Conversely, *to recognize the racial dimension in social structure is to interpret the meaning of race.* Consider the following statement by the late Supreme Court Justice Thurgood Marshall on minority "set-aside" programs:

> A profound difference separates governmental actions that themselves are racist, and governmental actions that seek to remedy the effects of prior racism or to prevent neutral government activity from perpetuating the effects of such racism.[9]

Here the focus is on the racial dimensions of *social structure*—in this case of state activity and policy. The argument is that state actions in the past and present have treated people in very different ways according to their race, and thus the government cannot retreat from its policy responsibilities in this area. It cannot suddenly declare itself "color-blind" without in fact perpetuating the same type of differential, racist treatment.[10] Thus, race continues to signify difference and structure inequality. Here, racialized social structure is immediately linked to an interpretation of the meaning of race. This example too can be classified as a particular type of racial project in the present-day U.S.—a "liberal" one.

To be sure, such political labels as "neoconservative" or "liberal" cannot fully capture the complexity of racial projects, for these are always multiply determined, politically contested, and deeply shaped by their historical context. Thus, encapsulated within the neoconservative example cited here are certain egalitarian commitments which derive from a previous historical context in which they played a very different role, and which are rearticulated in neoconservative racial discourse precisely to oppose a more open-ended, more capacious conception of the meaning of equality. Similarly, in the liberal example, Justice Marshall recognizes that the contemporary state, which was formerly the architect of segregation and the chief enforcer of racial difference, has a tendency to reproduce those patterns of inequality in a new guise. Thus he admonishes it (in dissent, significantly) to fulfill its responsibilities to uphold a robust conception of equality. These particular instances, then, demonstrate how racial projects are always concretely framed, and thus are always contested and unstable. The social structures they uphold or attack, and the representations of race they articulate, are never invented out of the air, but exist in a definite historical context, having descended from previous conflicts. This contestation appears to be permanent in respect to race.

These two examples of contemporary racial projects are drawn from mainstream political debate; they may be characterized as center-right and center-left expressions of contemporary racial politics.[11] We can, however, expand the discussion of racial formation processes far beyond these familiar examples. In fact, we can identify racial projects in at least three other analytical dimensions: first, the political spectrum can be broadened to include radical projects, on both the left and right, as well as along other political axes. Second, analysis of racial projects can take place not only at the macro-level of racial policy-making, state activity, and collective action, but also at the micro-level of everyday experience. Third, the concept of racial projects can be applied across historical time, to identify racial formation dynamics in the past. We shall now offer examples of each of these types of racial projects.

The Political Spectrum of Racial Formation

We have encountered examples of a neoconservative racial project, in which the significance of race is denied, leading to a "color-blind" racial politics and "hands off' policy orientation; and of a "liberal" racial project, in which the significance of race is affirmed, leading to an egalitarian and "activist" state policy. But these by no means exhaust the political possibilities. Other racial projects can be readily identified on the contemporary U.S. scene. For example, "far right" projects, which uphold biologistic and racist views of difference, explicitly argue for white supremacist policies. "New right" projects overtly claim to hold "color-blind" views, but covertly manipulate racial fears in order to achieve political gains.[12] On the left, "radical democratic" projects invoke notions of racial "difference" in combination with egalitarian politics and policy.

Further variations can also be noted. For example, "nationalist" projects, both conservative and radical, stress the incompatibility of racially defined group identity with the legacy of white supremacy, and therefore advocate a social structural solution of separation, either complete or partial.[13] . . . Nationalist currents represent a profound legacy of the centuries of racial absolutism that initially defined the meaning of race in the U.S. Nationalist concerns continue to influence racial debate in the form of Afrocentrism and other expressions of identity politics.

Taking the range of politically organized racial projects as a whole, we can "map" the current pattern of racial formation at the level of the public sphere, the "macro-level" in which public debate and mobilization takes place.[14] But important as this is, the terrain on which racial formation occurs is broader yet.

Racial Formation as Everyday Experience

At the micro-social level, racial projects also link signification and structure, not so much as efforts to shape policy or define large-scale meaning, but as the applications of "common sense." To see racial projects operating at the level of everyday life, we have only to examine the many ways in which, often unconsciously, we "notice" race.

One of the first things we notice about people when we meet them (along with their sex) is their race. We utilize race to provide clues about *who* a person is. This fact is made painfully obvious when we encounter someone whom we cannot conveniently racially categorize—someone who is, for example, racially "mixed" or of an ethnic/racial group we are not familiar with. Such an encounter becomes a source of discomfort and momentarily a crisis of racial meaning.

Our ability to interpret racial meanings depends on preconceived notions of a racialized social structure. Comments such as, "Funny, you don't look black," betray an underlying image of what black should be. We expect people to act out their apparent racial identities; indeed we become disoriented when they do not. The black banker harassed by police while walking in casual clothes through his own well-off neighborhood, the Latino or white kid rapping in perfect Afro patois, the unending *faux pas* committed by whites who assume that the non-whites they encounter are servants or tradespeople, the belief that non-white colleagues are less qualified persons hired to fulfill affirmative action guidelines, indeed the whole gamut of racial stereotypes—that "white men can't jump," that Asians can't dance, etc., etc.—all testify to the way a racialized social structure shapes racial experience and conditions meaning. Analysis of such stereotypes reveals the always present, already active link between our view of the social structure—its demography, its laws, its customs, its threats—and our conception of what race means.

Conversely, our ongoing interpretation of our experience in racial terms shapes our relations to the institutions and organizations through which we are imbedded in social structure. Thus we expect differences in skin color, or other racially coded characteristics, to explain social differences. Temperament, sexuality, intelligence, athletic ability, aesthetic preferences, and so on are presumed to be fixed and discernible from the palpable mark of race. Such diverse questions as our confidence and trust in others (for example, clerks or salespeople, media figures, neighbors), our sexual preferences and romantic images, our tastes in music, films, dance, or sports, and our very ways of talking, walking, eating, and dreaming become racially coded simply because we live in a society where racial awareness is so pervasive. Thus in ways too comprehensive even to monitor consciously, and despite periodic calls—neoconservative and otherwise—for us to ignore race and adopt "color-blind" racial attitudes, skin color "differences" continue to rationalize distinct treatment of racially identified individuals and groups.

To summarize the argument so far: the theory of racial formation suggests that society is suffused with racial projects, large and small, to which all are subjected. This racial "subjection" is quintessentially ideological. Everybody learns some combination, some version, of the rules of racial classification, and of her own racial identity, often without obvious teaching or conscious inculcation. Thus are we inserted in a comprehensively racialized social structure. Race becomes "common sense"—a way of comprehending, explaining, and acting in the world. A vast web of racial projects mediates between the discursive or representational means in which race is identified and signified on the one hand, and the institutional and organizational forms in which it is routinized and standardized on the other. These projects are the heart of the racial formation process.

Under such circumstances, it is not possible to represent race discursively without simultaneously locating it, explicitly or implicitly, in a social structural (and historical) context. Nor is it possible to organize, maintain, or transform social structures without simultaneously engaging, once more either explicitly or implicitly, in racial signification. Racial formation, therefore, is a kind of synthesis, an outcome, of the interaction of racial projects on a society-wide level. These projects are, of course, vastly different in scope and effect. They include large-scale public action, state activities, and interpretations of racial conditions in artistic, journalistic, or academic fora,[15] as well as the seemingly infinite number of racial judgments and practices we carry out at the level of individual experience.

Since racial formation is always historically situated, our understanding of the significance of race, and of the way race structures society, has changed enormously over time. The processes of racial formation we encounter today, the racial projects large and small which structure U.S. society in so many ways, are merely the present-day outcomes of a complex historical evolution. The contemporary racial order remains transient. By knowing something of how it evolved, we can perhaps better discern where it is heading. . . .

Notes

1. *San Francisco Chronicle,* 14 September 1982, 19 May 1983. Ironically, the 1970 Louisiana law was enacted to supersede an old Jim Crow statute which relied on the idea of "common report" in determining an infant's race. Following Phipps's unsuccessful attempt to change her classification and have the law declared unconstitutional, a legislative effort arose which culminated in the repeal of the law. See *San Francisco Chronicle,* 23 June 1983.

2. Compare the Phipps case to Andrew Hacker's well-known "parable" in which a white person is informed by a mysterious official that "the organization he represents has made a mistake" and that ". . . [a]ccording to their records . . . you were to have been born black: to another set of parents, far from where you were raised." How much compensation, Hacker's official asks, would "you" require to undo the damage of this unfortunate error? See Hacker, *Two Nations: Black and White, Separate, Hostile, Unequal* (New York: Charles Scribner's Sons, 1992) pp. 31–32.

3. On the evolution of Louisiana's racial classification system, see Virginia Dominguez, *White By Definition: Social Classification in Creole Louisiana* (New Brunswick: Rutgers University Press, 1986).

4. This is not to suggest that gender is a biological category while race is not. Gender, like race, is a social construct. However, the biological division of humans into sexes—two at least, and possibly intermediate ones as well—is not in dispute. This provides a basis for argument over gender divisions—how "natural," etc.—which does not exist with regard to race. To ground an argument for the "natural" existence of race, one must resort to philosophical anthropology.

5. "The truth is that there are no races, there is nothing in the world that can do all we ask race to do for us. . . . The evil that is done is done by the concept, and by easy—yet impossible—assumptions as to its application." (Kwame Anthony Appiah, *In My Father's House: Africa in the Philosophy of Culture* [New York: Oxford University Press, 1992].) Appiah's eloquent and learned book fails, in our view, to dispense with the race concept, despite its anguished attempt to do so; this indeed is the source of its author's anguish. We agree with him as to the non-objective character of race, but fail to see how this recognition justifies its abandonment. This argument is developed below.

6. We understand essentialism as *belief in real, true human, essences, existing outside or impervious to social and historical context.* We draw this definition, with some small modifications, from Diana Fuss, *Essentially Speaking: Feminism, Nature, & Difference* (New York: Routledge, 1989) p. xi.

7. Michael Omi and Howard Winant, "On the Theoretical Status of the Concept of Race," in Warren Crichlow and Cameron McCarthy, eds., *Race, Identity, and Representation in Education* (New York: Routledge, 1993).

8. Charles Murray, *Losing Ground: American Social Policy, 1950–1980* (New York: Basic Books, 1984) p. 223.

9. Justice Thurgood Marshall, dissenting in *City of Richmond v. J. A. Croson Co.,* 488 U.S. 469 (1989).

10. See, for example, Derrick Bell, "Remembrances of Racism Past: Getting Past the Civil Rights Decline," in Herbert Hill and James E. Jones, Jr., eds., *Race in America: The Struggle for Equality* (Madison: The University of Wisconsin Press, 1993) pp. 75–76; Gertrude Ezorsky, *Racism and Justice: The Case for Affirmative Action* (Ithaca: Cornell University Press, 1991) pp. 109–111; David Kairys, *With Liberty and Justice for Some: A Critique of the Conservative Supreme Court* (New York: The New Press, 1993) pp. 138–41.

11. Howard Winant has developed a tentative "map" of the system of racial hegemony in the U.S. circa 1990, which focuses on the spectrum of racial projects running from the political right to the political left. See Winant, "Where Culture Meets Structure: Race in the 1990s," in idem, *Racial Conditions: Politics, Theory, Comparisons* (Minneapolis: University of Minnesota Press, 1994).

12. A familiar example is use of racial "code words." Recall George Bush's manipulations of racial fear in the 1988 "Willie Horton" ads, or Jesse Helms's use of the coded term "quota" in his 1990 campaign against Harvey Gantt.

13. From this perspective, far right racial projects can also be interpreted as "nationalist." See Ronald Walters, "White Racial Nationalism in the United States," *Without Prejudice*, vol. no. 1 (Fall 1987).

14. To be sure, any effort to divide racial formation patterns according to social structural location—"macro" vs. "micro," for example—is necessarily an analytic device. In the concrete, there is no such dividing line. See Winant, "Where Culture Meets Structure."

15. We are not unaware, for example, that publishing this work is in itself a racial project.

THINKING ABOUT THE READING

What do Omi and Winant mean when they say that "race is always historically situated"? What do they mean when they say that everyone learns a system of rules and routines about race that become common sense? Consider some examples of these commonsense rules in contemporary society. How do people learn these rules? How do they unlearn them? Is the idea that race is "natural" one of the rules of the current "race project" in this society? If so, how does this particular rule contribute to social inequality?

Optional Ethnicities

For Whites Only?

Mary C. Waters

(1996)

What does it mean to talk about ethnicity as an option for an individual? To argue that an individual has some degree of choice in their ethnic identity flies in the face of the commonsense notion of ethnicity many of us believe in—that one's ethnic identity is a fixed characteristic, reflective of blood ties and given at birth. However, social scientists who study ethnicity have long concluded that while ethnicity is based on a *belief* in a common ancestry, ethnicity is primarily a *social* phenomenon, not a biological one (Alba 1985, 1990; Barth 1969; Weber [1921] 1968, p. 389). The belief that members of an ethnic group have that they share a common ancestry may not be a fact. There is a great deal of change in ethnic identities across generations through intermarriage, changing allegiances, and changing social categories. There is also a much larger amount of change in the identities of individuals over their lives than is commonly believed. While most people are aware of the phenomenon known as "passing"—people raised as one race who change at some point and claim a different race as their identity—there are similar life course changes in ethnicity that happen all the time and are not given the same degree of attention as "racial passing."

White Americans of European ancestry can be described as having a great deal of choice in terms of their ethnic identities. The two major types of options White Americans can exercise are (1) the option of whether to claim any specific ancestry, or to just be "White" or American, (Lieberson [1985] called these people "unhyphenated Whites") and (2) the choice of which of their European ancestries to choose to include in their description of their own identities. In both cases, the option of choosing how to present yourself on surveys and in everyday social interactions exists for Whites because of social changes and societal conditions that have created a great deal of social mobility, immigrant assimilation, and political and economic power for Whites in the United States. Specifically, the option of being able to not claim any ethnic identity exists for Whites of European background in the United States because they are the majority group—in terms of holding political and social power, as well as being a numerical majority. The option of choosing among different ethnicities in their family backgrounds exists because the degree of discrimination and social distance attached to specific European backgrounds has diminished over time. . . .

Symbolic Ethnicities for White Americans

What do these ethnic identities mean to people and why do they cling to them rather than just abandoning the tie and calling themselves American? My own field research with suburban Whites in California and Pennsylvania found that later-generation descendants of European origin maintain what are called "symbolic ethnicities." Symbolic ethnicity is a term coined by Herbert Gans (1979) to refer to ethnicity that is individualistic in nature and without real social cost for the individual. These symbolic identifications are essentially leisure-time activities, rooted in nuclear family traditions and reinforced by the voluntary enjoyable aspects of being ethnic (Waters 1990). Richard Alba (1990) also found later-generation Whites in Albany, New York, who chose to keep a tie with an ethnic identity because of the enjoyable and voluntary aspects to those identities, along with the feelings of specialness they entailed. An example of symbolic ethnicity is individuals who identify as Irish, for example, on occasions such as Saint Patrick's Day, on family holidays, or for vacations. They do not usually belong to Irish American organizations, live in Irish neighborhoods, work in Irish jobs, or marry other Irish people. The symbolic meaning of being Irish American can be constructed by individuals from mass media images, family traditions, or other intermittent social activities. In other words, for later-generation White ethnics, ethnicity is not something that influences their lives unless they want it to. In the world of work and school and neighborhood, individuals do not have to admit to being ethnic unless they choose to. And for an increasing number of European-origin individuals whose parents and grandparents have intermarried, the ethnicity they claim is largely a matter of personal choice as they sort through all of the possible combinations of groups in their genealogies. . . .

Race Relations and Symbolic Ethnicity

However much symbolic ethnicity is without cost for the individual, there is a cost associated with symbolic ethnicity for the society. That is because symbolic ethnicities of the type described here are confined to White Americans of European origin. Black Americans, Hispanic Americans, Asian Americans, and American Indians do not have the option of a symbolic ethnicity at present in the United States. For all of the ways in which ethnicity does not matter for White Americans, it does matter for non-Whites. Who your ancestors are does affect your choice of spouse, where you live, what job you have, who your friends are, and what your chances are for success in American society, if those ancestors happen not to be from Europe. The reality is that White ethnics have a lot more choice and room to maneuver than they themselves think they do. The situation is very different for members of racial minorities, whose lives are strongly influenced by their race or national origin regardless of how much they may choose not to identify themselves in terms of their ancestries.

When White Americans learn the stories of how their grandparents and great-grandparents triumphed in the United States over adversity, they are usually told in terms of their individual efforts and triumphs. The important role of labor unions and other organized political and economic actors in their social and economic successes are left out of the story in favor of a generational story of individual Americans rising up against communitarian, Old World intolerance, and New World resistance. As a result, the "individualized" voluntary, cultural view of ethnicity for Whites is what is remembered.

One important implication of these identities is that they tend to be very individualistic. There is a tendency to view valuing diversity in a pluralist environment as equating all groups. The symbolic ethnic tends to think that all groups are equal; everyone has a background that is their right to celebrate and pass on to their children. This leads to the conclusion that all identities are equal and all identities in some sense are interchangeable—"I'm Italian American, you're Polish American. I'm Irish American, you're African American." The important thing is to treat people as individuals and all equally. However, this assumption ignores the very big difference between an individualistic symbolic ethnic identity and a socially enforced and imposed racial identity.

My favorite example of how this type of thinking can lead to some severe misunderstandings between people of different backgrounds is from the *Dear Abby* advice column. A few years back a person wrote in who had asked an acquaintance of Asian background where his family was from. His acquaintance answered that this was a rude question and he would not reply. The bewildered White asked Abby why it was rude, since he thought it was a sign of respect to wonder where people were from, and he certainly would not mind anyone asking HIM about where his family was from. Abby asked her readers to write in to say whether it was rude to ask about a person's ethnic background. She reported that she got a large response, that most non-Whites thought it was a sign of disrespect, and Whites thought it was flattering:

> Dear Abby,
>
> I am 100 percent American and because I am of Asian ancestry I am often asked "What are you?" It's not the personal nature of this question that bothers me, it's the question itself. This query seems to question my very humanity. "What am I? Why I am a person like everyone else!"
>
> Signed, A REAL AMERICAN
>
> Dear Abby,
>
> Why do people resent being asked what they are? The Irish are so proud of being Irish, they tell you before you even ask. Tip O'Neill has never tried to hide his Irish ancestry.
>
> Signed, JIMMY.

(Reprinted by permission of Universal Press Syndicate)

In this exchange Jimmy cannot understand why Asians are not as happy to be asked about their ethnicity as he is, because he understands his ethnicity and theirs to be separate but equal. Everyone has to come from somewhere—his family from Ireland, another's family from Asia—each has a history and each should be proud of it. But the reason he cannot understand the perspective of the Asian American is that all ethnicities are not equal; all are not symbolic, costless, and voluntary. When White Americans equate their own symbolic ethnicities with the socially enforced identities of non-White Americans, they obscure the fact that the experiences of Whites and non-Whites have been qualitatively different in the United States and that the current identities of individuals partly reflect that unequal history.

In the next section I describe how relations between Black and White students on college campuses reflect some of these asymmetries in the understanding of what a racial or ethnic identity means. While I focus on Black and White students in the following discussion, you should be aware that the myriad other groups in the United States—Mexican Americans, American Indians, Japanese Americans—all have some degree of social and individual influences on their identities, which reflect the group's social and economic history and present circumstance.

Relations on College Campuses

Both Black and White students face the task of developing their race and ethnic identities. Sociologists and psychologists note that at the time people leave home and begin to live independently from their parents, often ages eighteen to twenty-two, they report a heightened sense of racial and ethnic identity as they sort through how much of their beliefs and behaviors are idiosyncratic to their families and how much are shared with other people. It is not until one comes in close contact with many people who are different from oneself that individuals realize the ways in which their backgrounds may influence their individual personality. This involves coming into contact with people who are different in terms of their ethnicity, class, religion, region, and race. For White students, the ethnicity they claim is more often than not a symbolic one—with all of the voluntary, enjoyable, and intermittent characteristics I have described above.

Black students at the university are also developing identities through interactions with others who are different from them. Their identity development is more complicated than that of Whites because of the added element of racial discrimination and racism, along with the "ethnic" developments of finding others who share their background. Thus Black students have the positive attraction of being around other Black students who share some cultural elements, as well as the need to band together with other students in a reactive and oppositional way in the face of racist incidents on campus.

Colleges and universities across the country have been increasing diversity among their student bodies in the last few decades. This has led in many cases to strained relations among students from different racial and ethnic backgrounds. The 1980s and 1990s produced a great number of racial incidents and high racial tensions on campuses. While there were a number of racial incidents that were due to bigotry, unlawful behavior, and violent or vicious attacks, much of what happens among students on campuses involves a low level of tension and awkwardness in social interaction.

Many Black students experience racism personally for the first time on campus. The upper-middle-class students from White suburbs were often isolated enough that their presence was not threatening to racists in their high schools. Also, their class background was known by their residence and this may have prevented attacks being directed at them. Often Black students at the university who begin talking with other students and recognizing racial slights will remember incidents that happened to them earlier that they might not have thought were related to race.

Black college students across the country experience a sizeable number of incidents that are clearly the result of racism. Many of the most blatant ones that occur between students are the result of drinking. Sometimes late at night, drunken groups of White students coming home from parties will yell slurs at single Black students on the street. The other types of incidents that happen include being singled out for special treatment by employees, such as being followed when shopping at the campus bookstore, or going to the art museum with your class and the guard stops you and asks for your I.D. Others involve impersonal encounters on the street—being called a nigger by a truck driver while crossing the street, or seeing old ladies clutch their pocketbooks and shake in terror as you pass them on the street. For the most part these incidents are not specific to the university environment, they are the types of incidents middle-class Blacks face every day throughout American society, and they have been documented by sociologists (Feagin 1991).

In such a climate, however, with students experiencing these types of incidents and talking with each other about them, Black students do experience a tension and a feeling of being singled out. It is unfair that this is part of their college experience and not that of White students. Dealing with incidents like this, or the ever-present threat of such incidents, is an ongoing developmental task for Black students that takes energy, attention, and strength of character. It should be clearly understood that this is an asymmetry in the "college experience" for Black and White students. It is one of the unfair aspects of life that results from living in a society with ongoing racial prejudice and discrimination. It is also very understandable that it makes some students angry at the unfairness of it all, even if there is no one to blame specifically. It is also very troubling because, while most Whites do not create these incidents, some do, and it is never clear until you know someone well whether they are the type of person who could do something like this. So one of the reactions of Black students to these incidents is to band together.

In some sense then, as Blauner (1992) has argued, you can see Black students coming together on campus as both an "ethnic" pull of wanting to be together to share common experiences and community, and a "racial" push of banding together defensively because of perceived rejection and tension from Whites. In this way the ethnic identities of Black students are in some sense similar to, say, Korean students wanting to be together to share experiences. And it is an ethnicity that is generally much stronger than, say, Italian Americans. But for Koreans who come together there is generally a definition of themselves as "different from" Whites. For Blacks reacting to exclusion there is a tendency for the coming together to involve both being "different from" but also "opposed to" Whites.

The anthropologist John Ogbu (1990) has documented the tendency of minorities in a variety of societies around the world, who have experienced severe blocked mobility for long periods of time, to develop such oppositional identities. An important component of having such an identity is to describe others of your group who do not join in the group solidarity as devaluing and denying their very core identity. This is why it is not common for successful Asians to be accused by others of "acting White" in the United States, but it is quite common for such a term to be used by Blacks and Latinos. The oppositional component of a Black identity also explains how Black people can question whether others are acting "Black enough." On campus, it explains some of the intense pressures felt by Black students who do not make their racial identity central and who choose to hang out primarily with non-Blacks. This pressure from the group, which is partly defining itself by not being White, is exacerbated by the fact that race is a physical marker in American society. No one immediately notices the Jewish students sitting together in the dining hall, or the one Jewish student sitting surrounded by non-Jews, or the Texan sitting with the Californians, but everyone notices the Black student who is or is not at the "Black table" in the cafeteria.

An example of the kinds of misunderstandings that can arise because of different understandings of the meanings and implications of symbolic versus oppositional identities concerns questions students ask one another in the dorms about personal appearances and customs. A very common type of interaction in the dorm concerns questions Whites ask Blacks about their hair. Because Whites tend to know little about Blacks, and Blacks know a lot about Whites, there is a general asymmetry in the level of curiosity people have about one another. Whites, as the numerical majority, have had little contact with Black culture; Blacks, especially those who are in college, have had to develop bicultural skills—knowledge about the social worlds of both Whites and Blacks. Miscommunication and hurt feelings about White students' questions about Black students' hair illustrate this point. One of the things that happens freshman year is that White students are around Black students as they fix their hair. White students are generally quite curious about Black students' hair—they have basic questions such as how often Blacks wash their hair, how they get it straightened or curled, what products they use on their hair, how they comb it, etc. Whites often wonder to themselves whether they should ask these questions. One thought experiment Whites perform is to ask themselves whether a particular question would upset them. Adopting the "do unto others" rule, they ask themselves, "If a Black person was curious about my hair would I get upset?" The answer usually is "No, I would be happy to tell them." Another example is an Italian American student wondering to herself, "Would I be upset if someone asked me about calamari?" The answer is no, so she asks her Black roommate about collard greens, and the roommate explodes with an angry response such as, "Do you think all Black people eat watermelon too?" Note that if this Italian American knew her friend was Trinidadian American and asked about peas and rice the situation would be more similar and would not necessarily ignite underlying tensions.

Like the debate in *Dear Abby*, these innocent questions are likely to lead to resentment. The issue of stereotypes about Black Americans and the assumption that all Blacks are alike and have the same stereotypical cultural traits has more power to hurt or offend a Black person than vice versa. The innocent questions about Black hair also bring up a number of asymmetries between the Black and White experience. Because Blacks tend to have more knowledge about Whites than vice versa, there is not an even exchange going on; the Black freshman is likely to have fewer basic questions about his White roommate than his White roommate has about him. Because of the differences historically in the group experiences of Blacks and Whites there are some connotations to Black hair that don't exist about White hair. (For instance, is straightening your hair a form of assimilation, do some people distinguish between women having "good hair" and "bad hair" in terms of beauty and how is that related to looking "White"?) Finally, even a Black freshman who cheerfully disregards or is unaware that there are these asymmetries will soon slam into another asymmetry if she willingly answers every innocent question asked of her. In a situation where Blacks make up only 10 percent of the student body, if every non-Black needs to be educated about hair, she will have to explain it to nine other students. As one Black student explained to me, after you've been asked a couple of times about something so personal you begin to feel like you are an attraction in a zoo, that you are at the university for the education of the White students.

Institutional Responses

Our society asks a lot of young people. We ask young people to do something that no one else does as successfully on such a wide scale—that is to live together with people from very different backgrounds, to respect one another, to appreciate one another, and to enjoy and learn from one another. The successes that occur every day in this endeavor are many, and they are too often overlooked. However, the problems and tensions are also real, and they will not vanish on their own. We tend to see pluralism working in the United States in much the same way some people expect capitalism to work. If you put together people with various interests and abilities and resources, the "invisible hand" of capitalism is supposed to make all the parts work together in an economy for the common good.

There is much to be said for such a model—the invisible hand of the market can solve complicated problems of production and distribution better than any "visible hand" of a state plan. However, we have learned that unequal power relations among the actors in the capitalist marketplace, as well as "externalities" that the market cannot account for, such as long-term pollution, or collusion between corporations, or the exploitation of child labor, means that state regulation is often needed. Pluralism and the relations between groups are very similar. There is a lot to be said for the idea that bringing people who belong to different ethnic or racial groups together in institutions with no interference will have good consequences. Students from different backgrounds will make friends if they share a dorm room or corridor, and there is no need for the institution to do any more than provide the locale. But like capitalism, the invisible hand of pluralism does not do well when power relations and externalities are ignored. When you bring together individuals from groups that are differentially valued in the wider society and provide no guidance, there will be problems. In these cases the "invisible hand" of pluralist relations does not work, and tensions and disagreements can arise without any particular individual or group of individuals being "to blame." On college campuses in the 1990s some of the tensions between students are of this sort. They arise from honest misunderstandings, lack of a common background, and very different experiences of what race and ethnicity mean to the individual.

The implications of symbolic ethnicities for thinking about race relations are subtle but consequential. If your understanding of your own ethnicity and its relationship to society and politics is one of individual choice, it becomes harder to understand the need for programs like affirmative action, which recognize the ongoing need for group struggle and group recognition, in order to bring about social change. It also is hard for a White college student to understand the need that minority students feel to band together against discrimination. It also is easy, on the individual level, to expect everyone else to be able to turn their ethnicity on and off at will, the way you are able to, without understanding that ongoing discrimination and societal attention to minority status makes that impossible for individuals from minority groups to do. The paradox of symbolic ethnicity is that it depends upon the ultimate goal of a pluralist society, and at the same time makes it more difficult to achieve that ultimate goal. It is dependent upon the concept that all ethnicities mean the same thing, that enjoying the traditions of one's heritage is an option available to a group or an individual, but that such a heritage should not have any social costs associated with it.

As the Asian Americans who wrote to *Dear Abby* make clear, there are many societal issues and involuntary ascriptions associated with non-White identities. The developments necessary for this to change are not individual but societal in nature. Social mobility and declining racial and ethnic sensitivity are closely associated. The legacy and the present reality of discrimination on the basis of race or ethnicity must be overcome before the ideal of a pluralist society, where all heritages are treated equally and are equally available for individuals to choose or discard at will, is realized.

REFERENCES

Alba, Richard D. 1985. *Italian Americans: Into the Twilight of Ethnicity.* Englewood Cliffs, NJ: Prentice Hall.

———. 1990. *Ethnic Identity: The Transformation of White America.* New Haven: Yale University Press.

Barth, Frederick. 1969. *Ethnic Groups and Boundaries.* Boston: Little, Brown.

Blauner, Robert. 1992. "Talking Past Each Other: Black and White Languages of Race." *American Prospect* (Summer):55–64.

Feagin, Joe R. 1991. "The Continuing Significance of Race: Anti-Black Discrimination in Public Places." *American Sociological Review* 56: 101–17.

Gans, Herbert. 1979. "Symbolic Ethnicity: The Future of Ethnic Groups and Cultures in America." *Ethnic and Racial Studies* 2:1–20.

Lieberson, Stanley. 1985. *Making It Count: The Improvement of Social Research and Theory.* Berkeley: University of California Press.

Ogbu, John. 1990. "Minority Status and Literacy in Comparative Perspective." *Daedalus* 119: 141–69.

Waters, Mary C. 1990. *Ethnic Options: Choosing Identities in America.* Berkeley: University of California Press.

Weber, Max. [1921]/1968. *Economy and Society: An Outline of Interpretive Sociology.* Eds. Guenther Roth and Claus Wittich, trans. Ephraim Fischoff. New York: Bedminister Press.

THINKING ABOUT THE READING

What is "symbolic ethnicity," according to Waters? Why is this form of ethnic expression optional for some and not others? Based on Waters's thesis, would a campus club for Norwegian Americans be the same as one for African Americans? Consider the slogan "different but equal." Do you think this idea can be applied to racial and ethnic relations in contemporary society? Why are some ethnic and racial groups the subject of discrimination and oppression while others are a source of group membership and belonging? When might an ethnic identity be both? How would you describe the ethnic and racial climate of your college campus?

Silent Racism

Passivity in Well-Meaning White People

Barbara Trepagnier

(2010)

Silent racism is a cultural phenomenon, not a psychological one. This does not imply that all white people are affected in a similar way, but it does imply that all whites are infected. No one is immune to ideas that permeate the culture in which he or she is raised. *Silent racism* here refers to unspoken negative thoughts, emotions, and assumptions about black Americans that dwell in the minds of white Americans, including well-meaning whites that care about racial equality, some of which are called "new abolitionists." Limited to ideas in people's minds, silent racism is unspoken. It therefore does not refer to racist statements or actions referred to as everyday racism (Essed 1991); rather, *silent racism* refers to the negative thoughts and beliefs that fuel everyday racism and other racist action.

Everyday Racism

The concept of everyday racism (Essed 1991) refers to routine actions that go unquestioned by members of the dominant group, which in some way discriminate against members of a racial or ethnic category. The concept of silent racism is closely linked to this concept in that silent racism precedes everyday racism; silent racism constitutes the platform on which everyday racism is enacted. Silent racism predisposes white people to commit or collude with routine practices that are perceived by blacks as everyday racism. Silent racism is the cognitive aspect of everyday racism—in contrast to the behavioral aspect. Silent racism underpins the "broad white-racist worldview" (Feagin 2001:34) and inhabits the minds of all white Americans.

Color-Blind Racism

Eduardo Bonilla-Silva's (2003) concept of color-blind racism is perhaps the idea most closely related to silent racism. Bonilla-Silva posits that the viewpoint that race is no longer important is an attempt to maintain white privilege without appearing racist. He identifies four frames of color-blind racism: *abstract liberalism*—the idea that liberal notions such as equality, which was a cornerstone of the civil rights movement, are now used to oppose affirmation action on the grounds that all groups should be treated the same; *naturalization*—the idea that racial patterns such as informal segregation are natural, for example, that blacks prefer to live in black neighborhoods; *cultural racism*—the idea that norms within the black culture account for racial inequality; and the *minimization of racism*—the idea that racism ended with passage of civil rights legislation.

Passivity

Passivity regarding racism is not well documented in the race literature. The exception is Joe Feagin (2001), who briefly mentions "bystanders" as a category of white racists that "provide support for others' racism" (p. 140), and who in his work with Hemán Vera and Pinar Batur (2001) suggests that "passivity is a first step in learning to ally oneself with white victimizers against black victims" (p. 49). The latter study deals primarily with passivity in the face of antiblack violence. This follows much of the literature on bystanders, which is based largely on the "anonymous crowd" that colluded with the atrocities of the World War II Holocaust (Barnett 1999:109). Our interest concerns the passivity of well-meaning white people who collude not with violent acts but with subtle forms of racism.

Implementing the Study

I used small discussion groups, also called focus groups, as a format for data gathering. An advantage of this method over individual interviews is that using small, homogeneous groups is a more appropriate model when sensitive topics are discussed (Aaker and Day 1986; Churchill 1988). Because a frank discussion about racism relies upon a context of safety, focus groups were preferred for this study. I cannot know what data I might have collected in individual interviews, but I am confident that participants shared openly and honestly about their own racism in the small group format used here. I also sensed a feeling of group unity when, although joining the study entailed participating in only one discussion group, participants in several groups joked about when the group would meet again, an indication that they were open to such an idea.

After the participants for a particular group arrived, I explained that a discussion group differs from a support group in that interaction is encouraged rather than discouraged. Often, a participant was reminded of a childhood memory by another participant's comment. Furthermore, group members were urged to engage with other participants, asking for clarification and even disagreeing with others' ideas if the occasion arose. In this way, I hoped to ensure that the groups would be dynamic, which would enhance the data. In addition, interaction within a group—called synergism—produces especially meaningful responses. For example, participants occasionally responded to another's comments by examining their own commonsense explanations. This occurred in one group when a participant, prompted by what another member of her group said, asked herself, "God, do I have any prejudices like that?" This example of synergism illustrates not only that group members are likely to be reminded of events by other members but also that participants have time in a group discussion to think about what has been said because they are not constantly under scrutiny. Observing interaction among the participants also contributed to an important finding: the importance of race awareness became evident as I noticed that several participants interrupted racism when they perceived it in their respective groups. This evidence was instrumental in the finding that race awareness is more important than whether well-meaning white people are racist.

Because I am a white woman and would be facilitating all of the focus groups, I limited the study to white women in order to ensure homogeneity in terms of gender and race/ethnicity. In addition, I was interested in looking at racism from the point of view of those performing it, and women seemed to be a logical choice because of their relative ease in engaging in open self-reflection and in articulating their emotions (Belenky et al. 1986; Spacks 1981). In addition, I wanted to explore subtle forms of racism in people who would ordinarily not be considered racist. The study flier, headlined "Women Against Racism," was expected to attract participants who were progressive in terms of race politics.

Facilitation of the groups consisted of asking questions intended to draw out information about topics such as early messages concerning race matters, experiences with black Americans, thoughts about the participants' own racism, and comments about their commitment to lessening racism. The questions increased the likelihood that the content of the eight discussions would be somewhat consistent. However, each focus group had the flexibility to differ as participants introduced unique topics in their respective groups.

Detachment from Race Issues

The "not racist" category distances well-meaning white people from racism by implication: White people who see themselves as "not racist" are unlikely to see their connection to race or racism. Sharon expressed a sense of detachment from race issues several times during the discussion in her group. The first example was in response to the question, "What do you think needs to happen in order for racism to end?" Sharon said, "Racism has no connection to my life." Later, when asked if she had ever been told by someone else, or realized herself, that she had said or done something racist, Sharon again appears detached. She said, "I can't think of anything. I'm sure there must be, but I can't think of anything. It didn't hit me." Sharon would not regard her indifference as problematic in any way. Rather, as she stated, "Racism has no connection to my life." But Sharon's thinking is faulty: we are all intimately connected with issues of race (Frankenberg 1993).

Sharon's detachment from race issues makes her a passive bystander when confronted with others' racism. For example, when asked, "What do you do when you are around someone who has made a racist remark or tells a racist joke?" Sharon responded, "Nothing, usually." The indifference characterized by Sharon is akin to willful blindness, a term used in reference to the perpetrators of white-collar crime such as Ken Lay, the president of Enron. Lay claims no knowledge of criminal behavior that he and others greatly profited from. Similarly, detachment from race matters serves white people who benefit from the racial status quo. Sharon's detachment from race issues is more striking than any other participant's, although others demonstrated disconnections as well. For example, Karen, in Sharon's focus group, also said that she usually does nothing when confronted with others' racism.

Detachment from racism is not limited to people like Sharon, who came to this study accidentally and who knows very little about racism. Penny is more representative of well-meaning white people who are concerned, yet passive. Penny senses that she should interrupt racism, but she openly admits that often she does not. When Penny answered the question about what she does if someone tells a racist joke or makes a racist comment, her answer illustrates passivity. Penny said, "Ideally, I would say, 'I don't laugh at that.' Do I say it? [That] depends on how grounded I'm feeling that day or what my relationship, my role in the group, is. . . . Then you get into the whole thing about, 'Oh, I didn't say it.' And 'I'm complicit.' It can be quite a conundrum." Penny, unlike Sharon, has good intentions about interrupting racism and feels bad about not doing it. Penny mentions that her role in the group could affect her reaction as a bystander. Bystanders who identify with the perpetrator or inhabit a subordinate role in relation to the perpetrator are less likely to take an active role for fear of disapproval or alienation (Staub 2003).

Vanessa said in response to the question about being around someone telling a racist joke, "I probably just don't laugh," an interesting response because the word "probably" casts her answer as a hypothetical statement rather than a statement of fact. A hypothetical answer instead of a factual one about one's behavior is likely to indicate avoidance of the question, perhaps due to being unsure about how the inquiring party might react. Nevertheless, whether Vanessa "just doesn't laugh" or laughs politely, her answer appears to indicate a measure of detachment.

Racist comments and jokes that go uninterrupted implicate the listener as well as the actor. The only way to not comply with racism when it occurs is to interrupt it. It is not correct to think that racism only occurs in interactions between whites and blacks or other people of color. To the contrary, those interactions may demonstrate less racism than comments that occur between or among white people when no blacks are present. Interrupting racism is as important at these times as it is when blacks are present, primarily because not to do so is perceived by perpetrators as encouragement of their racism.

Unintended Consequences

The "not racist" category appears to produce two unintended consequences: apprehension about being perceived as racist, and confusion about what constitutes racism. Both of these consequences result in passive behavior in white people.

Apprehension About Being Seen as Racist. Everyday rules regarding race matters, known as "racial etiquette" (Omi and Winant 1986:62), are imbued with myriad meanings regarding race and racial difference that produce apprehension in white people. Several participants said that they felt apprehension about being perceived as racist. Elaine articulated her self-consciousness in dealing with black/white difference when she shared a story about meeting Dorothy, the friend of a friend, at a barbecue. Elaine said, "I opened the door and she's *black*. Oh! And I was just so mad at myself, and embarrassed for thinking that. I mean like, 'Oh, did that show?' Really worrying about it; just never getting past that."

Elaine's surprise that Dorothy was black was only exceeded by her embarrassment about being surprised. Based on her past experience, Elaine expected to see only white people at the barbecue. The racial etiquette that Elaine learned in her "all white upbringing seems to have left her unsure about how to navigate a social setting that included both whites and blacks. The phrase "Did that show?" indicates that Elaine was afraid Dorothy might have noticed her surprise and interpreted it as racist. Apprehension about being perceived as racist troubled Elaine quite a bit, as evidenced by the comment, "Really worrying about it; just never getting past that." Elaine elaborated her discomfort by explaining how she makes sense of her reluctance to initiate friendships with black women. She said, "I do tend to socialize with people that are like me. . . . It's comfortable, it's easy, the knowns outweigh the unknowns. I think working against racism includes that fear of offending someone or fear of saying/doing the wrong thing and not being conscious of this. . . . I'm gonna make a mistake and I don't want to have to worry about that."

Elaine's comments do not imply that she thinks it is right to avoid situations in which she might make a misstep, as in her response to Dorothy. Nevertheless, she acknowledges that she often takes the easier path in developing friendships rather than the path that is more likely to provoke her anxiety about race difference. Her apprehension is important because of its own consequences: a tendency to avoid interactions with people of color. Ironically, having close ties with blacks and other people of color is important in developing race awareness—something that would lessen Elaine's apprehension.

Elaine added, "Racism has such a stigma attached to it that yes, we fear it. We don't want to be associated with [it]—we are not supposed to be making any mistakes." The "not racist" category produces fear of losing one's status as not racist and, in the process, lessens the tendency to question ideas about racism.

Karen made a related point in her group when she said, "I sometimes feel a barrier in approaching black women, in that I feel that they don't want to deal with me, and so I feel like I'm being respectful by keeping my distance, or something. I feel more comfortable letting them make the first move instead of me going over and starting conversations." Karen's reluctance to initiate friendships, or even conversations, with black women so they won't have to "deal with her" may relate to another incident when Karen's black friend, Belle, rebuffed Karen's attempt to order her ice cream for her. Belle had not said why she was upset about the incident, and Karen did not ask. Consequently, Karen assumed it was simply because she was white, not realizing that it was because she had expressed a paternalistic assumption.

Apprehension about being perceived as racist keeps well-meaning white people from finding out more about racism. Anita made this point when she said, "[The] fear of saying anything that's going to label you racist . . . you're not really dealing with. Well, is it or isn't it [racist], and why do I feel like that?" Lucy makes a similar point when she says, "Something that gets in my way [of dealing with my own racism] is feeling that I've got to be cool, or good, or maybe it's feeling like I try too hard or I care too much. I think it gets in my way because it prevents me from . . . acknowledging that I am human." I think what Lucy means by "acknowledging that [she is] human" refers to the inevitability that she will at times be unwittingly racist. Humans make mistakes, and sometimes those mistakes are because of misconceptions or ignorance regarding racism. The need to be seen by oneself and others as "not racist" hinders becoming more aware of race matters. Moreover, people with low race awareness are not likely to be active bystanders who interrupt others' racism; rather, people with low race awareness are likely to be passive bystanders, encouraging racism.

Loretta also indicated apprehension about being perceived as racist. She said, "People silencing themselves out of the fear of not saying the right thing [means] not being able to talk, and therefore not being able to change. Making actual change may mean making a mistake, saying the wrong thing, and having somebody call you on it and having to own that." Loretta's statement shows insight into the paradox of being unable to discuss racism for fear of being perceived as racist. Loretta's comment also shows insight into the danger of seeing racism as deviant. The original definition of *political correctness,* now known as PC, was "internal self-criticism" among liberals (Berube 1994:94). For example, liberals hoped to raise awareness about biases in language—such as the use of sexist language—because biases in language reinforce biases in society (Hofstadter 1985). Conservatives co-opted the term, mocking liberals by casting political correctness as an attempt to limit the freedom of speech. Today political correctness is widely perceived as destructive, rather than as it was originally intended: an attempt not to be offensive (Feldstein 1997).

Passivity resulting from the apprehension about being perceived as racist is evident in the preceding stories of the well-meaning white women. The fear of being seen as racist paralyzes some well-meaning white people, causing them to avoid meeting and interacting with blacks. This is significant—and ironic—because forming close relationships with blacks and other people of color is the most important step they can take to lessen their apprehension.

Confusion About What Is Racist. Confusion about racism is epitomized by uncertainty and embarrassment and is sometimes related to being apprehensive about being seen as racist. People who see themselves as not being racist often presume that they should know what is racist and what is not, even when they are not sure. Confusion about what is racist is closely related to passivity in that it suppresses action. In the following comments, participants share experiences demonstrating confusion.

Anne spoke of her confusion about whether referring to people as "black" is in itself racist. Anne reported a conversation she had with her mother in reference to a baseball announcer during a New York Yankees game. When her mom asked who announced the game, Anne said that it was Bill White. "My mom asked me, 'Who's Bill White?' I didn't want to say he was black—I thought it would be racist." Anne attempted to avoid using color as a marker for distinguishing among the sports announcers, believing that mentioning his race would have been racist. After describing many details about Bill White—color of hair, size, and so on—Anne could not think of any other way to distinguish him and finally told her mother that he was "the black announcer." This raises an important point of discussion: Was Anne's telling her mother that Bill White was "the black announcer" racist? Was it the same as Ruth, who earlier said she had a bright "black" student in her class?

I classified Ruth's comment as racist, as did the friend that interrupted it, pointing out that Ruth's reason for mentioning that the student was black was related to the fact that he was bright. However, that is not the case in

Anne's situation; saying that Bill White was the black announcer was not related to any negative stereotype but was instrumental in identifying him to her mother.

Some would argue that using "black" as an identifying characteristic is always racist because it reinforces the notion that blacks are racialized and whites are not. This view is called *otherizing* and is thought to marginalize blacks and other minorities. However, sometimes identifying someone as black is pertinent to the context of a situation. To say he was "the black announcer" was not racist in Anne's situation because it was instrumental in that context and in no way reproduced a stereotype about blacks. I agree that white people virtually never use white in the same way. Nevertheless, avoiding the word *black* simply because its use is not equivalent to the use of the word *white* seems like faulty logic to me.

While discussing this issue with a black colleague, I was given this response: "Sometimes a person will apologize for saying the word *black* even when it is appropriate to include for clarity. Very often I have had whites apologize for even uttering the word. It's as if, for them, the word *black* is gaining status with *nigger* as a racially sensitive word."

Avoiding the use of *black* because it might be racist results from confusion about what is racist. Using *black* as an identifying characteristic is racist when its use is associated with a racist stereotype or if it is tacked on solely because a person is not white. However, rigidly avoiding *black* unnecessarily when its use would serve a purpose is tantamount to pretending that race does not exist or was not noticed, a prime example of racial etiquette (Omi and Winant 1986).

Anne's reluctance to "utter the word" *black* indicates some hesitation about saying the word at all. Anne may have received a message as a child similar to the one Lisa received from her parents. Lisa said that she was told explicitly not to notice race differences. Lisa said that in addition to telling her "colors don't matter," her parents added, "[but] don't ever say the word *black*, don't say the word *Mexican*, and don't ever refer to a person's color. It's offensive to say those words." Lisa said that when she was ten, she and her family moved into a housing project where she would be in close proximity to black children. Since Lisa would undoubtedly play with black children—her new neighbors—Lisa's parents were perhaps trying to prepare her for that experience. By cautioning Lisa to ignore difference—a difference they also denied was there—Lisa's parents wanted to both protect her from any repercussions they thought might occur from pointing out difference *and* teach her about equality. However, parents' double messages about race and racism can cause confusion in their children in terms of what is and what is not racist. In Anne's case, confusion contributed to her apprehension about being racist, which had a paralyzing effect. Avoiding any mention of race or the word *black* rather than acknowledging one's confusion keeps people from understanding what is and what is not racist.

In a related incident, Penny, who grew up in the 1960s, spoke of asking her mother about a house that looked "different" from the ones in their neighborhood—she said that it was pink and had iron grillwork across the front. Penny stated, "My mother said, 'Oh, that's where Egyptians live.' [My mother] didn't think that I'd ever meet Egyptians, and so it was okay for me to think that Egyptians were different." The logic that Penny attributes to her mother's comment—that it was okay to think that Egyptians were different because it was unlikely that Penny would meet any—indicates the lengths to which Penny's mother went in avoiding a discussion about race difference with her children. Although Penny did not recall receiving an explicit message to "not notice" race, it appears that her mother saw acknowledgment of difference *itself* as problematic and perhaps racist, a confusion of what is racist and what is not.

Although the reluctance to mention race can be referred to as being color-blind, the instances here do not meet the definition of *color-blind racism*, a racial ideology that "explains contemporary [racial] inequality as the outcome of nonracial dynamics" (Bonilla-Silva 2003:2). Color blindness derives from a racist ideology that is at times racist but that is not necessarily always racist. The individuals described in this section are only color-blind in that they did not want to draw attention to race difference for fear it would be racist to do so. However, I would characterize the reluctance to mention race as confusion about what is racist rather than as racism per se.

Confusion was also evident in Heather's description of an incident that occurred in her high school circle of friends. However, Heather's confusion is not coupled with apprehension. Heather said,

> I just remembered a very good friend of mine in high school who was half black—his dad was black and his mom was white—and he was blond, with blue eyes. There was an incident [in high school] that was really sticky. One of our friends didn't even know that David's father was black, and she made a very bad mistake by telling a joke about a black man and a Jewish man in an airplane—an awful, awful joke that just did not go over [well]. . . . I think part of it was that [David] was such a blond guy. And his father had a Ph.D. in some hard science and has taught at [a major university]; he was on the faculty and then went to work at a laboratory. [David's] mom is a nurse.

Heather characterized the "sticky" incident as "a mistake" and that the friend telling the racist joke "didn't even know that David's father was black." What seems to be problematic for Heather is that David had inadvertently heard the racist joke, not the fact that the joke was racist. This interpretation is substantiated by Heather's comment that the "mistake" resulted from David being "such a blond guy" whose father has a Ph.D. and whose mother is a nurse. Heather's confusion about what is racist concerning the joke incident is likely to result from her not thinking the incident through, and, as a result, excusing her friend's racism by seeing it as harmless rather than as racist.

Confusion was also evident in a statement Alyssa made in her focus group:

> I think that everyone should be noted for their differences and celebrate their differences, instead of just ignoring, and looking through them and saying, "You know, I don't see color." Because you do [see color]. Everyone sees it. You may not think negatively of it, but when you think of the fact that you notice that a person is black, you think it's something bad. But I don't see that as something being bad—you can celebrate a difference.

The confusion in Alyssa's notion of celebrating difference becomes evident when she states inconsistent views centered on the pronoun you: "you may not think negatively of it" and "you think it's something bad." Alyssa seems to notice the apparent contradiction between these two thoughts when she quickly distances herself from the second statement by adding, "*I* don't see it as something bad." The confusion in Alyssa's thinking (that noticing race difference is "good" and that being black is seen as "something bad") presumably remains intact in her thinking, perhaps below her awareness. Holding contradictory beliefs without scrutinizing them may explain how many white people harbor racist thoughts about blacks and other people of color without being aware of it.

Loretta talked about the celebration of difference, but without the confusion exhibited by Alyssa. She said, "We can have a kind of 'feel good' cultural diversity yet not be antiracist. We [can] all talk the same talk, isn't this great, and cultural diversity is great. I [can] go to a food fair and taste [different food] and that's great, on one level. But if the reality is that economically only certain people are getting jobs . . . and people of color are getting paid less than white people . . . then there is still going to be racism." Loretta does not embrace the celebration of difference uncritically, as Alyssa does. Her critical assessment of the concept exposes the danger of celebrating the different cultural traditions of black and white Americans without acknowledging the history of racial oppression in the United States and the current racial inequality that continues today.

Confusion in well-meaning white people does not produce passivity as directly as detachment from race issues does. Neither does confusion produce passivity in the same sense that apprehension about being racist does, through the avoidance of contact. However, confusion is linked to passivity indirectly in that white people who are confused about racism are not likely to take a stand against it; one must be able to conclusively define an act as racist in order to feel justified in contesting it. Only white people who are clear about the historical legacy of racism in the United States, who understand how institutional racism operates, and who sense their own complicity with a system that benefits them to the detriment of people of color are likely to be active in interrupting racism when they encounter it. In this way, confusion along with detachment and apprehension is the antithesis of antiracism. For this reason, I consider it racist and place it just inside the midpoint toward the less racist end of the racism continuum.

The aforementioned data support the claim that the "not racist" category itself produces several latent effects that bring about passivity in well-meaning white people. Just as silent racism produces institutional racism, passivity produces collusion with racism. Said differently, everyday racism could not stand without the participation and cooperation of well-meaning white people.

Conclusion

Passivity is common in well-meaning white people. It is marked by detachment that produces a bystander effect in white people who find themselves in the face of others' racism. In addition, passivity results from apprehension about being seen as racist and from confusion about what is racist—both unintended but direct effects of the "not racist" category. This chapter's central thesis is that *passivity works against racial equality.* Well-meaning white people who are passive bystanders quietly watch America grow more divided over race issues. Yet, these are not innocent bystanders. They profit from the racial divide; they reap the same advantages received by those performing racist acts that they silently witness. The well-meaning whites who are the least aware of this fact feel little or no discomfort about the situation—they do not recognize the benefits that institutional racism affords them. The well-meaning whites who are detached have a measure of race awareness and feel bad about the situation as well as about their own passivity. Other passive bystanders are apprehensive—afraid to make a move for fear that they may be seen as racist. And still others are confused or misinformed, even though most do not recognize their confusion.

REFERENCES

Aaker, David, and George Day. 1986. *Marketing Research.* New York: John Wiley and Sons.

Barnett, Victoria. 1999. *Bystanders: Conscience and Complicity during the Holocaust.* Westport, CT: Praeger.

Belenky, Mary, Blythe Clinchy, Nancy Goldberger, and Jill Tarule. 1986. *Women's Ways of Knowing: The Development of Self, Voice, and Mind.* New York: Basic.

Berube, Michael. 1994. *Public Access: Literary Theory and American Cultural Politics.* New York: Verso.

Bonilla-Silva, Eduardo. 2003. *Racism Without Racists: Color-Blind Racism and the Persistence of Racial Inequality in the United States.* Lanham, MD: Rowman and Littlefield.

Churchill, Gilbert. 1988. *Basic Marketing Research.* Chicago: Dryden Press.

Essed, Philomena. 1991. *Understanding Everyday Racism: An Interdisciplinary Theory.* Newbury Park, CA: Sage.

Feagin, Joe. 2001. *Racist America: Roots, Current Realities, and Future Reparations,* New York: Routledge.

Feagin, Joe, Hernán Vera, and Pinar Batur. 2001. *White Racism: The Basics,* 2nd ed. New York: Routledge.

Feldstein, Richard. 1997. *Political Correctness: A Response from the Cultural Left.* Minneapolis: University of Minnesota Press.

Frankenberg, Ruth. 1993. *White Women, Race Matters: The Social Construction of Whiteness.* Minneapolis: University of Minnesota Press.

Omi, Michael, and Howard Winant. 1986. *Racial Formation in the United States*. New York: Routledge.

Spacks, Patricia. 1981. "The Difference It Makes." In *A Feminist Perspective in the Academy*, edited by E. Langland and W. Gove. Chicago: University of Chicago Press.

Staub, Ervin. 2003. *The Psychology of Good and Evil: Why Children, Adults, and Groups Help and Harm Others*. New York: Cambridge University Press.

THINKING ABOUT THE READING

A popular notion is recent years has been that the United States is now a "post-racial society," free of racial preferences, discrimination, and prejudice. The election of President Barack Obama is often used as the symbolic event that ushered in this post-racial society. What does Trepagnier's research suggest about this post-racial idea? What is silent racism, according to Trepagnier? Why is it vital to understand silent racism in order to understand the outcomes and consequences of racist actions? What did Trepagnier mean when she said "passivity works against racial equality?" In terms of understanding racism in everyday life, what was *the* most important finding from Trepagnier's focus groups? While many discussions of racism focus on important evidence of unequal outcomes between racial groups, why is it important to understand some of the cognitive aspects of racism discussed in this article? Where, in your everyday examples in your various spheres of interaction (work, education, family, social events), have you observed some of the themes in Trepagnier's findings?

The Architecture of Inequality

12

Sex and Gender

In addition to racial and class inequality, gender inequality—and the struggle against it—has been a fundamental part of the historical development of our national identity. Gender ideology has influenced the lives and dreams of individual people, shaped popular culture, and created or maintained social institutions. Gender is a major criterion for the distribution of important economic, political, and educational resources in most societies. Gender inequality is perpetuated by a dominant cultural ideology that devalues women on the basis of presumed biological differences between men and women. This ideology overlooks the equally important role of social forces in determining male and female behavior.

Bart Landry explores the intersections of race and gender in "Black Women and a New Definition of Womanhood." Landry examines the difficulties black women have faced throughout history in being seen by others as virtuous and moral. This article provides a fascinating picture of women's struggle for equality from the perspective of black women, a group that is often ignored and marginalized in discussions of the women's movement. Although much of the article focuses on black women's activism in the 19th century, it provides important insight into the intersection of race and gender today. Landry raises an important contrast between the way in which 19th-century middle-class white women and middle-class black women framed the relationship between family and public life.

Gender inequality exists at the institutional level as well, in the law, in the family (in terms of such things as the domestic division of labor), and in economics. Not only are social institutions sexist in that women are systematically segregated, exploited, and excluded, but they are also gendered. Institutions themselves are structured along gender lines so that traits associated with success are usually stereotypically male characteristics: tough-mindedness, rationality, assertiveness, competitiveness, and so forth.

Women have made significant advances politically, economically, educationally, and socially over the past decades. The traditional obstacles to advancement continue to fall. Women have entered the labor force in unprecedented numbers. Despite their growing presence in the labor force and their entry into historically male occupations, however, rarely do women work alongside men or perform the same tasks and functions.

Jobs within an occupation still tend to be divided into "men's work" and "women's work." Such gender segregation has serious consequences for women in the form of blocked advancement and lower salaries. But looking at gender segregation on the job as something that happens only to women gives us an incomplete picture of the situation. It is just as important to examine what keeps men out of "female" jobs as it is to examine what keeps women out of "male" jobs. The proportion of women in male jobs has increased over the past several decades, but the proportion of men in female jobs has remained virtually unchanged. In "Still a Man's World," Christine L. Williams looks at the experiences of male nurses, social workers, elementary school teachers, and librarians. She finds that although these men do feel somewhat stigmatized by their nontraditional career choices, they still enjoy significant gender advantages.

Bodily transformation practices like plastic surgery, tattooing, and weight lifting have become widespread practices across society, but new technologies are taking this transformation process to another level. In "New Biomedical Technologies, New Scripts, New Genders," Eve Shapiro uses case studies to explore the connection between new biomedical technologies and gendered bodies and identities. She illustrates how the somatechnic frontier is giving individuals the opportunity to construct new bodies to fit *and* contest existing social scripts for women and men.

Something to Consider as You Read

While reading these selections, think about the significance of gender as a social category. A child's gender is the single most important thing people want to know when it is born. "What is it?" is a commonly understood shorthand for "Is it a boy or a girl?" From the time children are born, they learn that certain behaviors, feelings, and expectations are associated with the gender category to which they have been assigned. Think about some of the behaviors associated with specific gender categories. Make a list of stereotypical gender expectations. Upon reflection, do these seem reasonable to you? What are some recollections you have about doing something that was considered inappropriate for your gender? Think about ways in which these stereotypical expectations affect people's perceptions, especially in settings such as school or jobs.

Black Women and a New Definition of Womanhood

Bart Landry

(2000)

A popular novel of 1852 chirped that the white heroine, Eoline, "with her fair hair, and celestial blue eyes bending over the harp . . . really seemed 'little lower than the angels,' and an aureola of purity and piety appeared to beam around her brow."[1] By contrast, in another popular antebellum novel, *Maum Guinea and Her Plantation Children* (1861), black women are excluded from the category of true womanhood without debate: "The idea of modesty and virtue in a Louisiana colored-girl might well be ridiculed; as a general thing, she has neither."[2] Decades later, in 1902, a commentator for the popular magazine *The Independent* noted, "I sometimes hear of a virtuous Negro woman, but the idea is absolutely inconceivable to me. . . . I cannot imagine such a creature as a virtuous Negro woman."[3] Another writer, reflecting early-twentieth-century white male stereotypes of black and white women, remarked that, like white women, "Black women had the brains of a child, [and] the passions of a woman" but, unlike white women, were "steeped in centuries of ignorance and savagery, and wrapped about with immoral vices."[4]

Faced with the prevailing views of white society that placed them outside the boundaries of true womanhood, black women had no choice but to defend their virtue. Middle-class black women led this defense, communicating their response in words and in the actions of their daily lives. In doing so they went well beyond defending their own virtue to espouse a broader conception of womanhood that anticipated modern views by more than half a century. Their vision of womanhood combined the public and the private spheres and eventually took for granted a role for women as paid workers outside the home. More than merely an abstract vision, it was a philosophy of womanhood embodied in the lives of countless middle-class black women in both the late nineteenth and the early twentieth centuries.

Virtue Defended

Although black women were seen as devoid of all four of the cardinal virtues of true womanhood—piety, purity, submissiveness, and domesticity—white attention centered on purity. As Hazel Carby suggests, this stemmed in part from the role assigned to black women in the plantation economy. She argues that "two very different but interdependent codes of sexuality operated in the antebellum South, producing opposite definitions of motherhood and womanhood for white and black women which coalesce in the figures of the slave and the mistress."[5] In this scheme, white mistresses gave birth to heirs, slave women to property. A slave woman who attempted to preserve her virtue or sexual autonomy was a threat to the plantation economy. In the words of Harriet Jacobs's slave narrative, *Incidents in the Life of a Slave Girl* (1861), it was "deemed a crime in her [the slave woman] to wish to be virtuous."[6]

Linda Brent, the pseudonym Jacobs used to portray her own life, was an ex-slave struggling to survive economically and protect herself and her daughter from sexual exploitation. In telling her story, she recounts the difficulty all black women faced in practicing the virtues of true womanhood. The contrasting contexts of black and white women's lives called for different, even opposite, responses. While submissiveness and passivity brought protection to the white mistress, these characteristics merely exposed black women to sexual and economic exploitation. Black women, therefore, had to develop strength rather than glory in fragility, and had to be active and assertive rather than passive and submissive. . . .

Three decades later, in the 1890s, black women found reasons to defend their moral integrity with new urgency against attacks from all sides. Views such as those in *The Independent* noted earlier were given respectability by a report of the Slater Fund, a foundation that supported welfare projects for blacks in this period. The foundation asserted without argument, "The negro women of the South are subject to temptations . . . which come to them from the days of their race enslavement. . . . To meet such temptations the negro woman can only offer the resistance of a low moral standard, an inheritance from the system of slavery, made still lower from a lifelong residence in a one-room cabin."[7]

At the 1893 World Columbian Exposition in Chicago, where black women were effectively barred from the exhibits on the achievements of American women, the few black women allowed to address a women's convention there felt compelled to publicly challenge these views. One speaker, Fannie Barrier Williams, shocked her audience by her forthrightness. "I regret the necessity of speaking of the moral question of our women," but "the morality of our home life has been commented on so disparagingly and meanly that we are placed in the unfortunate position of being defenders of our name."[8] She went on to emphasize that black women continued to be the victims of sexual harassment by white men and chided her white female audience for failing to protect their black sisters. In the same vein, black activist and educator Anna Julia Cooper told the audience that it was not a question of "temptations" as much as it was "the painful, patient, and silent toil of mothers to gain title to the bodies of their daughters."[9] Williams was later to write on the same theme. "It is a significant and shameful fact that I am constantly in receipt of letters from the still unprotected women in the South, begging me to find employment for their daughters . . . to save them from going into the homes of the South as servants as there is nothing to save them from dishonor and degradation."[10] Another black male writer was moved to reveal in *The Independent:* "I know of more than one colored woman who was openly importuned by White women to become the mistress of their husbands, on the ground that they, the white wives, were afraid that, if their husbands did not associate with colored women they would certainly do so with outside white women. . . . And the white wives, for reasons which ought to be perfectly obvious, preferred to have all their husbands do wrong with colored women in order to keep their husbands *straight!*"[11] The attacks on black women's virtue came to a head with a letter written by James Jacks, president of the Missouri Press Association, in which he alleged, "The Negroes in this country were wholly devoid of morality, the women were prostitutes and all were natural thieves and liars."[12] These remarks, coming from such a prominent individual, drew an immediate reaction from black women throughout the country. The most visible was Josephine St. Pierre Ruffin's invitation to black club women to a national convention in Boston in 1895; one hundred women from ten states came to Boston in response. In a memorable address to representatives of some twenty clubs, Ruffin directly attacked the scurrilous accusations:

> Now for the sake of the thousands of self-sacrificing young women teaching and preaching in lonely southern backwoods, for the noble army of mothers who gave birth to these girls, mothers whose intelligence is only limited by their opportunity to get at books, for the cultured women who have carried off the honors at school here and often abroad, for the sake of our own dignity, the dignity of our race and the future good name of our children, it is "meet, right and our bounden duty" to stand forth and declare ourselves and our principles, to teach an ignorant and suspicious world that our aims and interests are identical with those of all good, aspiring women. Too long have we been silent under unjust and unholy charges. . . . It is to break this silence, not by noisy protestations of what we are not, but by a dignified showing of what we are and hope to become, that we are impelled to take this step, to make of this gathering an object lesson to the world.[13]

At the end of three days of meetings, the National Federation of Afro-American Women was founded, uniting thirty-six black women's clubs in twelve states.[14] The following year, the National Federation merged with the National League of Colored Women to form the National Association of Colored Women (NACW).

Racial Uplift: In Defense of the Black Community

While the catalyst for these national organizations was in part the felt need of black women to defend themselves against moral attacks by whites, they soon went beyond this narrow goal. Twenty years after its founding, the NACW had grown to fifty thousand members in twenty-eight federations and more than one thousand clubs.[15] The founding of these organizations represented a steady movement by middle-class black women to assume more active roles in the community. Historian Deborah Gray White argues that black club women "insisted that only black women could save the black race," a position that inspired them to pursue an almost feverish pace of activities.[16]

These clubs, however, were not the first attempts by black women to participate actively in their communities. Since the late 1700s black women had been active in mutual-aid societies in the North, and in the 1830s northern black women organized anti-slavery societies. In 1880 Mary Ann Shadd Cary and six other women founded the Colored Women's Progressive Franchise Association in Washington, D.C. Among its stated goals were equal rights for women, including the vote, and the even broader feminist objective of taking "an aggressive stand against the assumption that men only begin and conduct industrial and other things."[17] Giving expression to this goal were a growing number of black women professionals, including the first female physicians to practice in the South.[18] By the turn of the twentieth century, the National Business League, founded by Booker T. Washington, could report that there were "160 Black female physicians, seven dentists, ten lawyers, 164 ministers, assorted journalists, writers, artists, 1,185 musicians and teachers of music, and 13,525 school instructors."[19]

Black women's activism was spurred by the urgency of the struggle for equality, which had led to a greater acceptance of black female involvement in the abolitionist movement. At a time when patriarchal notions of women's domestic role dominated, historian Paula Giddings asserts, "There is no question that there was greater acceptance among black men of women in activist roles than there was in the broader society."[20] This is not to say that all black men accepted women as equals or the activist roles that many were taking. But when faced with resistance, black women often *demanded* acceptance of their involvement. In 1849, for example, at a black convention in Ohio, "Black women, led by Jane P. Merritt, threatened to boycott the meetings if they were not given a more substantial voice in the proceedings."[21]

In the postbellum period black women continued their struggle for an equal voice in activities for racial uplift in both secular and religious organizations. . . . These women's organizations then played a significant role not only in missionary activities, but also in general racial uplift activities in both rural and urban areas.[22] . . .

Black Women and the Suffrage Movement

In their struggle for their own rights, black women moved into the political fray and eagerly joined the movement for passage of a constitutional amendment giving women the right to vote. Unlike white women suffragists, who focused exclusively on the benefits of the vote for their sex, black women saw the franchise as a means of improving the condition of the black community generally. For them, race and gender issues were inseparable. As historian Rosalyn Terborg-Penn emphasizes, black feminists believed that by "increasing the black electorate" they "would not only uplift the women of the race, but help the children and the men as well."[23]

Prominent black women leaders as well as national and regional organizations threw their support behind the suffrage movement. At least twenty black suffrage organizations were founded, and black women participated in rallies and demonstrations and gave public speeches.[24] Ironically, they often found themselves battling white women suffragists as well as men. Southern white women opposed including black women under a federal suffrage as a matter of principle. Northern white women suffragists, eager to retain the support of southern white women, leaned toward accepting a wording of the amendment that would have allowed the southern states to determine their own position on giving black women the vote, a move that would have certainly led to their exclusion.[25]

After the Nineteenth Amendment was ratified in 1920 in its original form, black women braved formidable obstacles in registering to vote. All across the South white registrars used "subterfuge and trickery" to hinder them from registering, including a "grandmother clause" in North Carolina, literacy tests in Virginia, and a $300 poll tax in Columbia, South Carolina. In Columbia, black women "waited up to twelve hours to register" while white women were registered first.[26] In their struggle to register, black women appealed to the NAACP [National Association for the Advancement of Colored People], signed affidavits against registrars who disqualified them, and finally asked for assistance from national white women suffrage leaders. They were especially disappointed in this last attempt. After fighting side by side with white women suffragists for passage of the Nineteenth Amendment, they were rebuffed by the National Woman's Party leadership with the argument that theirs was a race rather than a women's rights issue.[27] Thus, white women continued to separate issues of race and sex that black women saw as inseparable.

Challenging the Primacy of Domesticity

A conflicting conception of the relationship between gender and race issues was not the only major difference in the approaches of black and white women to their roles in the family and society. For most white women, their domestic roles as wives and mothers remained primary. In the late nineteenth century, as they began increasingly to argue for acceptance of their involvement on behalf of child-labor reform and growing urban problems, white women often defended these activities as extensions of their housekeeping role. Historian Barbara Harris comments, "The [white women] pioneers in women's education, who probably did more than anyone else in this period to effect change in the female sphere, advocated education for women and their entrance into the teaching profession on the basis of the values proclaimed by the cult of true womanhood. In a similar way, females defended their careers as authors and their involvement in charitable, religious, temperance, and moral reform societies."[28] Paula Giddings notes that in this way white women were able "to become more active outside the home while still preserving the probity of 'true womanhood.'"[29] From the birth of white feminism at the Seneca Falls Convention in 1848, white feminists had a difficult time advancing their goals. Their numbers were few and their members often divided over the propriety of challenging the cult of domesticity. . . .

In the late nineteenth century the cult of domesticity remained primary even for white women graduates of progressive women's colleges such as Vassar, Smith, and Wellesley. For them, no less than for those with only a high-school education, "A Woman's Kingdom" was "a well-ordered home."[30] In a student essay, one Vassar student answered her rhetorical question, "Has the educated woman a duty towards the kitchen?" by emphasizing that the kitchen was "exactly where the college woman belonged" for "the orderly, disciplined, independent graduate is the woman best prepared to manage the home, in which lies the salvation of the world."[31] This essay reflects the dilemma faced by these young white women graduates. They found little support in white society to combine marriage and career. . . .

Society sanctioned only three courses for the middle-class white woman in the Progressive period: "marriage, charity work or teaching."[32] Marriage and motherhood stood as the highest calling. If there were no economic need for them to work, single women were encouraged to do volunteer charity work. For those who needed an independent income, teaching was the only acceptable occupation.

Historian John Rousmaniere suggests that the white college-educated women involved in the early settlement house movement saw themselves as fulfilling the "service norm" so prominent among middle-class women of the day. At the same time, he argues, it was their sense of uniqueness as college-educated women and their felt isolation upon returning home that led them to this form of service. The settlement houses, located as they were in white immigrant, working-class slums, catered to these women's sense of noblesse oblige; they derived a sense of accomplishment from providing an example of genteel middle-class virtues to the poor. Yet the settlement houses also played into a sense of adventure, leading one resident to write, "We feel that we know life for the first time."[33] For all their felt uniqueness, however, with some notable exceptions these women's lives usually offered no fundamental challenge to the basic assumptions of true womanhood. Residency in settlement houses was for the most part of short duration, and most volunteers eventually embraced their true roles of wife and mother without significant outside involvement. The exceptions were women like Jane Addams, Florence Kelley, Julia Lathrop, and Grace Abbott, who became major figures in the public sphere. Although their lives disputed the doctrine of white women's confinement to the private sphere, the challenge was limited in that most of them did not themselves combine the two spheres of marriage and a public life. Although Florence Kelley was a divorced mother, she nevertheless upheld "the American tradition that men support their families, their wives throughout life," and bemoaned the "retrograde movement" against man as the breadwinner.[34]

Most college-educated black middle-class women also felt a unique sense of mission. They accepted Lucy Laney's 1899 challenge to lift up their race and saw themselves walking in the footsteps of black women activists and feminists of previous generations. But their efforts were not simply "charity work"; their focus was on "racial uplift" on behalf of themselves as well as of the economically less fortunate members of their race.[35] The black women's club movement, in contrast to the white women's, tended to concern themselves from the beginning with the "social and legal problems that confronted both black women and men."[36] While there was certainly some elitism in the NACW's motto, "Lifting as We Climb," these activists were always conscious that they shared a common experience of exploitation and discrimination with the masses and could not completely retreat to the safe haven of their middle-class homes.[37] On the way to meetings they shared the black experience of riding in segregated cars or of being ejected if they tried to do otherwise, as Ida B. Wells did in 1884.[38] Unlike white women for whom, as black feminist Frances Ellen Watkins Harper had emphasized in 1869, "the priorities in the struggle for human rights were sex, not race,"[39] black women could not separate these twin sources of their oppression. They understood that, together with their working-class sisters, they were assumed by whites to have "low, animalistic urges." Their exclusion from the category of true womanhood was no less complete than for their less educated black sisters.

It is not surprising, therefore, that the most independent and radical of black female activists led the way in challenging the icons of true womanhood, including on occasion motherhood and marriage. Not only did they chafe under their exclusion from true womanhood, they viewed its tenets as strictures to their efforts on behalf of racial uplift and their own freedom and integrity as women. In 1894 *The Woman's Era* (a black women's magazine) set forth the heretical opinion that "not all women are intended for mothers. Some of us have not the temperament for family life. . . .

Clubs will make women think seriously of their future lives, and not make girls think their only alternative is to marry."[40] Anna Julia Cooper, one of the most dynamic women of the period, who had been married and widowed, added that a woman was not "compelled to look to sexual love as the one sensation capable of giving tone and relish, movement and vim to the life she leads. Her horizon is extended."[41] Elsewhere Cooper advised black women that if they married they should seek egalitarian relationships. "The question is not now with the woman 'How shall I so cramp, stunt, and simplify and nullify myself as to make me eligible to the honor of being swallowed up into some little man?' but the problem . . . rests with the man as to how he can so develop . . . to reach the ideal of a generation of women who demand the noblest, grandest and best achievements of which he is capable."[42]

. . . Black activists were far more likely to combine marriage and activism than white activists. . . . Historian Linda Gordon found this to be the case in her study of sixty-nine black and seventy-six white activists in national welfare reform between 1890 and 1945. Only 34 percent of the white activists had ever been married, compared to 85 percent of the black activists. Most of these women (83 percent of blacks and 86 percent of whites) were college educated.[43] She also found that "The white women [reformers], with few exceptions, tended to view married women's economic dependence on men as desirable, and their employment as a misfortune. . . ."[44] On the other hand, although there were exceptions, Gordon writes, ". . . most black women activists projected a favorable view of working women and women's professional aspirations."[45] Nor could it be claimed that these black activists worked out of necessity, since the majority were married to prominent men "who could support them."[46]

Witness Ida B. Wells-Barnett (married to the publisher of Chicago's leading black newspaper) in 1896, her six-month-old son in tow, stumping from city to city making political speeches on behalf of the Illinois Women's State Central Committee. And Mary Church Terrell dismissing the opinion of those who suggested that studying higher mathematics would make her unappealing as a marriage partner with a curt, "I'd take a chance and run the risk."[47] She did eventually marry and raised a daughter and an adopted child. Her husband, Robert Terrell, a Harvard graduate, was a school principal, a lawyer, and eventually a municipal court judge in Washington, D.C. A biographer later wrote of Mary Terrell's life, "But absorbing as motherhood was, it never became a full-time occupation."[48] While this could also be said of Stanton, perhaps what most distinguished black from white feminists and activists was the larger number of the former who unequivocally challenged domesticity and the greater receptivity they found for their views in the black community. As a result, while the cult of domesticity remained dominant in the white community at the turn of the twentieth century, it did not hold sway within the black community.

Rejection of the Public/Private Dichotomy

Black women of the nineteenth and early twentieth centuries saw their efforts on behalf of the black community as necessary for their own survival, rather than as noblesse oblige. "Self preservation," wrote Mary Church Terrell in 1902, "demands that [black women] go among the lowly, illiterate and even the vicious, to whom they are bound by ties of race and sex . . . to reclaim them."[49] These women rejected the confinement to the private sphere mandated by the cult of domesticity. They felt women could enter the public sphere without detriment to the home. As historian Elsa Barkley Brown has emphasized, black women believed that "Only a strong and unified community made up of both women and men could wield the power necessary to allow black people to shape their own lives. Therefore, only when women were able to exercise their full strength would the community be at its full strength. . . ."[50]

In her study of black communities in Illinois during the late Victorian era (1880–1910), historian Shirley Carlson contrasts the black and white communities' expectations of the "ideal woman" at that time:

> The black community's appreciation for and development of the feminine intellect contrasted sharply with the views of the larger society. In the latter, intelligence was regarded as a masculine quality that would "defeminize" women. The ideal white woman, being married, confined herself almost exclusively to the private domain of the household. She was demure, perhaps even self-effacing. She often deferred to her husband's presumably superior judgment, rather than formulating her own views and vocally expressing them, as black women often did. A woman in the larger society might skillfully manipulate her husband for her own purposes, but she was not supposed to confront or challenge him directly. Black women were often direct, and frequently won community approval for this quality, especially when such a characteristic was directed toward achieving racial uplift. Further, even after her marriage, a black woman might remain in the public domain, possibly in paid employment. The ideal black woman's domain, then, was both the private and public spheres. She was wife and mother, but she could also assume other roles such as schoolteacher, social activist, or businesswoman, among others. And she was intelligent.[51]

. . . Although many black males, like most white males, opposed the expansion of black women's roles, many other black males supported women's activism and even criticized their brethren for their opposition. Echoing Maggie Walker's sentiments, T. Thomas Fortune wrote, "The race could not succeed nor build strong citizens, until we have a race of women competent to do more than hear a brood of negative men."[52] Support for women's suffrage was especially strong among black males. . . . Black men saw women's suffrage as advancing the political empowerment of the race. For black women, suffrage promised to be a potent weapon in their fight for their rights, for education and jobs.[53]

A Threefold Commitment

An expanded role for black women did not end at the ballot box or in activities promoting racial uplift. Black middle-class women demanded a place for themselves in the paid labor force. Theirs was a threefold commitment to family, career, and social movements. According to historian Rosalyn Terborg-Penn, "most black feminists and leaders had been wives and mothers who worked yet found time not only to struggle for the good of their sex, but for their race." Such a threefold commitment "was not common among white women."[54]

In her study of eighty African American women throughout the country who worked in "the feminized professions" (such as teaching) between the 1880s and the 1950s, historian Stephanie Shaw comments on the way they were socialized to lives dedicated to home, work, and community. When these women were children, she indicates, "the model of womanhood held before [them] was one of achievement in *both* public and private spheres. Parents cast domesticity as a complement rather than a contradiction to success in public arenas."[55] . . .

An analysis of the lives of 108 of the first generation of black clubwomen bears this out. "The career-oriented clubwomen," comments Paula Giddings, "seemed to have no ambivalence concerning their right to work, whether necessity dictated it or not."[56] According to Giddings, three-quarters of these 108 early clubwomen were married, and almost three-quarters worked outside the home, while one-quarter had children.

A number of these clubwomen and other black women activists not only had careers but also spoke forcefully about the importance of work, demonstrating surprisingly progressive attitudes with a very modern ring. "The old doctrine that a man marries a woman to support her," quipped Walker, "is pretty nearly thread-bare to-day."[57] "Every dollar a woman makes," she declared in a 1912 speech to the Federation of Colored Women's Clubs, "some man gets the direct benefit of same. Every woman was by Divine Providence created for some man; not for some man to marry, take home and support, but for the purpose of using her powers, ability, health and strength, to forward the financial . . . success of the partnership into which she may go, if she will. . . ."[58] Being married with three sons and an adopted daughter did not in any way dampen her commitment to gender equality and an expanded role for wives.

Such views were not new. In a pamphlet entitled *The Awakening of the Afro-American Woman*, written in 1897 to celebrate the earlier founding of the National Association of Colored Women, Victoria Earle Matthews referred to black women as "co-breadwinners in their families."[59] Almost twenty years earlier, in 1878, feminist writer and activist Frances Ellen Harper sounded a similar theme of equality when she insisted, "The women as a class are quite equal to the men in energy and executive ability." She went on to recount instances of black women managing small and large farms in the postbellum period.[60]

It is clear that in the process of racial uplift work, black middle-class women also included membership in the labor force as part of their identity. They were well ahead of their time in realizing that their membership in the paid labor force was critical to achieving true equality with men. For this reason, the National Association of Wage Earners insisted that all black women should be able to support themselves.[61] . . .

As W. E. B. DuBois commented as early as 1924, "Negro women more than the women of any other group in America are the protagonists in the fight for an economically independent womanhood in modern countries. . . . The matter of economic independence is, of course, the central fact in the struggle of women for equality."[62]

Defining Black Womanhood

In the late 1930s when Mary McLeod Bethune, the acknowledged leader of black women at the time and an adviser to President Franklin Roosevelt on matters affecting the black community, referred to herself as the representative of "Negro womanhood" and asserted that black women had "room in their lives to be wives and mothers as well as to have careers," she was not announcing a new idea.[63] As Terborg-Penn emphasizes:

> . . . most black feminists and leaders had been wives and mothers who worked yet found time not only to struggle for the good of their sex, but for their race. Until the 1970s, however, this threefold commitment—to family and to career and to one or more social movements—was not common among white women. The key to the uniqueness among black feminists of this period appears to be their link with the past. The generation of the woman suffrage era had learned from their late nineteenth-century foremothers in the black women's club movement, just as the generation of the post World War I era had learned and accepted the experiences of the preceding generation. Theirs was a sense of continuity, a sense of group consciousness that transcended class.[64]

This "sense of continuity" with past generations of black women was clearly articulated in 1917 by Mary Talbert, president of the NACW. Launching an NACW campaign to save the home of the late Frederick Douglass, she said, "We realize today is the psychological moment for us women to show our true worth and prove the Negro women of today measure up to those sainted women of our race, who passed through the fire of slavery and its galling remembrances."[65] Talbert certainly lived up to her words, going on to direct the NAACP's antilynching campaign and becoming the first woman to receive the NAACP's Spingarn Medal for her achievements.

What then is the expanded definition of true womanhood found in these black middle-class women's words and embodied in their lives? First, they tended to define womanhood in an inclusive rather than exclusive sense. Within white society, true womanhood was defined so narrowly that it excluded all but a small minority of white upper- and upper-middle-class women with husbands who were able to support them economically. Immigrant women and poor women—of any color—did not fit this definition. Nor did black women as a whole, regardless of class, because they were all seen as lacking an essential characteristic of true womanhood—virtue. For black women, however, true womanhood transcended class and race boundaries. Anna Julia Cooper called for "reverence for woman as woman regardless of rank, wealth, or culture."[66] Unlike white women, black women refused to isolate gender issues from other forms of oppression such as race and nationality, including the struggles of colonized nations of Africa and other parts of the world. Women's issues, they suggested, were tied to issues of oppression, whatever form that oppression might assume. . . .

The traditional white ideology of true womanhood separated the active world of men from the passive world of women. As we have seen, women's activities were confined to the home, where their greatest achievement was maintaining their own virtue and decorum and rearing future generations of male leaders. Although elite black women did not reject their domestic roles as such, many expanded permissible public activities beyond charity work to encompass employment and participation in social progress. They founded such organizations as the Atlanta Congress of Colored Women, which historian Erlene Stetson claims was the first grassroots women's movement organized "for social and political good."[67]

The tendency of black women to define womanhood inclusively and to see their roles extending beyond the boundaries of the home led them naturally to include other characteristics in their vision. One of these was intellectual equality. While the "true" woman was portrayed as submissive ("conscious of inferiority, and therefore grateful for support"),[68] according to literary scholar Hazel Carby, black women such as Anna Julia Cooper argued for a "partnership with husbands on a plane of intellectual equality."[69] Such equality could not exist without the pursuit of education, particularly higher education, and participation in the labor force. Cooper, like many other black women, saw men's opposition to higher education for women as an attempt to make them conform to a narrow view of women as "sexual objects for exchange in the marriage market."[70] Education for women at all levels became a preoccupation for many black feminists and activists. Not a few—like Anna Cooper, Mary L. Europe, and Estelle Pinckney Webster—devoted their entire lives to promoting it, especially among young girls. Womanhood, as conceived by black women, was compatible with—indeed, required—intellectual equality. In this they were supported by the black community. While expansion of educational opportunities for women was a preoccupation of white feminists in the nineteenth century, as I noted above, a college education tended to create a dilemma in the lives of white women who found little community support for combining marriage and career. In contrast, as Shirley Carlson emphasizes, "The black community did not regard intelligence and femininity as conflicting values, as the larger society did. That society often expressed the fear that intelligent women would develop masculine characteristics—a thickening waist, a diminution of breasts and hips, and finally, even the growth of facial hair. Blacks seemed to have had no such trepidations, or at least they were willing to have their women take these risks."[71]

In addition to women's rights to an education, Cooper, Walker, Alexander, Terrell, the leaders of the National Association of Wage Earners, and countless other black feminists and activists insisted on their right to work outside the home. They dared to continue very active lives after marriage. Middle-class black women's insistence on the right to pursue careers paralleled their view that a true woman could move in both the private and the public spheres and that marriage did not require submissiveness or subordination. In fact, as Shirley Carlson has observed in her study of black women in Illinois in the late Victorian period, many activist black women "continued to be identified by their maiden names—usually as their middle names or as part of their hyphenated surnames—indicating that their own identities were not subsumed in their husbands."[72]

While the views of black women on womanhood were all unusual for their time, their insistence on the right of all women—including wives and mothers—to work outside the home was the most revolutionary. In their view the need for paid work was not merely a response to economic circumstances, but the fulfillment of women's right to self-actualization. Middle-class black women like Ida B. Wells-Barnett, Margaret Washington, and Mary Church Terrell, married to men who were well able to support them, continued to pursue careers throughout their lives, and some did so even as they reared children. These women were far ahead of their time, foreshadowing societal changes that would not occur within the white community for several generations. . . .

Rather than accepting white society's views of paid work outside the home as deviant, therefore, black women fashioned a competing ideology of womanhood—one that supported the needs of an oppressed black community and their own desire for gender equality. Middle-class black women, especially, often supported by the black community, developed a consciousness of themselves as persons who were competent and capable of being influential. They believed in higher education as a means of sharpening their talents, and in a sexist world that looked on men as superior, they dared to see themselves as equals both in and out of marriage.

This new ideology of womanhood came to have a profound impact on the conception of black families and gender roles. Black women's insistence on their role as co-breadwinners clearly foreshadows today's dual-career and dual-worker families. Since our conception of the family is inseparably tied to our views of women's and men's roles, the broader definition of womanhood advocated by black women was also an argument against the traditional family. The cult of domesticity was anchored in a patriarchal notion of women as subordinate to men in both the family and the larger society. The broader definition of womanhood championed by black middle-class women struck a blow for an expansion of women's rights in society and a more egalitarian position in the home, making for a far more progressive system among blacks at this time than among whites.

Notes

1. Quoted in Hazel V. Carby, *Reconstructing Womanhood: The Emergence of the Afro-American Woman Novelist* (New York: Oxford University Press, 1987), p. 26.
2. Ibid.
3. Quoted in Paula Giddings, *When and Where I Enter: The Impact of Black Women and Race and Sex in America* (New York: Bantam Books, 1985), p. 82.
4. Ibid., p. 82.
5. Carby, *Reconstructing Womanhood,* p. 20.
6. Harriet Jacobs, *Incidents in the Life of a Slave Girl,* L. Baria Child, ed. (1861; paperback reprint, New York: Harcourt Brace Jovanovich, 1973), p. 29.
7. Quoted in Giddings, *When and Where I Enter,* p. 82.
8. Ibid., p. 86.
9. Ibid., p. 87.
10. Ibid., pp. 86–87.
11. Ibid., p. 87.
12. Quoted in Sharon Harley, "Black Women in a Southern City: Washington, D.C., 1890–1920," pp. 59–78 in Joanne V. Hawks and Sheila L. Skemp, eds., *Sex, Race, and the Role of Women in the South* (Jackson, Miss.: University Press of Mississippi, 1983), p. 72.
13. Eleanor Flexner, *Century of Struggle: The Woman's Rights Movement in the United States* (Cambridge: Harvard University Press, 1959), p. 194.
14. Giddings, *When and Where I Enter,* p. 93.
15. Ibid., p. 95. For a discussion of elitism in the "uplift" movement and organizations, see Kevin K. Gains, *Uplifting the Race: Black Leadership, Politics, and Culture in the Twentieth Century* (Chapel Hill, N.C.: University of North Carolina Press, 1996). Black reformers, enlightened as they were, could not entirely escape being influenced by Social Darwinist currents of the times.

16. Deborah Gray White, *Too Heavy a Load: Black Women in Defense of Themselves, 1894–1994* (New York: W. W. Norton & Company, 1999), p. 36.

17. Quoted in Giddings, *When and Where I Enter*, p. 75.

18. Ibid.

19. Ibid.

20. Ibid., p. 59.

21. Ibid.

22. Evelyn Brooks Higginbotham, *Righteous Discontent: The Women's Movement in the Black Baptist Church, 1880–1920* (Cambridge: Harvard University Press, 1993).

23. Rosalyn Terborg-Penn, "Discontented Black Feminists: Prelude and Postscript to the Passage of the Nineteenth Amendment," pp. 261–278 in Lois Scharf and Joan M. Jensen, eds., *Decades of Discontent: The Woman's Movement, 1920–1940* (Westport, Conn.: Greenwood Press, 1983), p. 264.

24. Ibid., p. 261.

25. Ibid., p. 264.

26. Ibid., p. 266.

27. Ibid., pp. 266–267.

28. Barbara J. Harris, *Beyond Her Sphere: Women and the Professions in American History* (Westport, Conn.: Greenwood Press, 1978), pp. 85–86.

29. Giddings, *When and Where I Enter*, p. 81.

30. John P. Rousmaniere, "Cultural Hybrid in the Slums: The College Woman and the Settlement House, 1889–1984," *American Quarterly* 22 (Spring 1970): p. 56.

31. Ibid., p. 55.

32. Rousmaniere, "Cultural Hybrid in the Slums," p. 56.

33. Ibid., p. 61.

34. Quoted in Linda Gordon, "Black and White Visions of Welfare: Women's Welfare Activism, 1890–1945," *Journal of American History* 78 (September 1991):583.

35. Giddings, *When and Where I Enter*, p. 97.

36. Estelle Freedman, "Separatism as Strategy: Female Institution Building and American Feminism, 1870–1930," pp. 445–462 in Nancy F. Cott, ed., *Women Together: Organizational Life* (New Providence, RI: K. G. Saur, 1994), p. 450; Nancy Forderhase, "'Limited Only by Earth and Sky': The Louisville Woman's Club and Progressive Reform, 1900–1910," pp. 365–381 in Cott, ed. *Women Together: Organizational Life* (New Providence, RI: K. G. Saur, 1994); . . . Mary Dell Brady, "Kansas Federation of Colored Women's Clubs, 1900–1930," pp. 382–408 in Nancy F. Cott, *Women Together.*

37. Higginbotham, *Righteous Discontent*, pp. 206–207.

38. Giddings, *When and Where I Enter*, p. 22.

39. Terborg-Penn, "Discontented Black Feminists," p. 267.

40. Giddings, *When and Where I Enter*, p. 108.

41. Ibid., pp. 108–109.

42. Ibid., p. 113.

43. Linda Gordon, "Black and Whites Visions of Welfare," p. 583.

44. Ibid., p. 582.

45. Ibid., p. 585.

46. Ibid., pp. 568–69.

47. Ibid., p. 109.

48. Quoted in Giddings, ibid., p. 110.

49. Ibid., p. 97.

50. Elsa Barkley Brown, "Womanist Consciousness: Maggie Lena Walker and the Independent Order of Saint Luke," *Signs: Journal of Women in Culture and Society* 14, no. 3 (1989):188.

51. Shirley J. Carlson, "Black Ideals of Womanhood in the Late Victorian Era," *Journal of Negro History* 77, no. 2 (Spring 1992):62. Carlson notes that these black women of the late Victorian era also observed the proprieties of Victorian womanhood in their deportment and appearance but combined them with the expectations of the black community for intelligence, education, and active involvement in racial uplift.

52. Quoted in Giddings, *When and Where I Enter*, p. 117.

53. See Rosalyn Terborg-Penn, *African American Women in the Struggle for the Vote, 1850–1920* (Bloomington, Ind.: Indiana University Press, 1998).

54. Rosalyn Terborg-Penn, "Discontented Black Feminists," p. 274.

55. Stephanie J. Shaw, *What a Woman Ought to Be and to Do: Black Professional Women Workers During the Jim Crow Era* (Chicago: University of Chicago Press, 1996), p. 29. Shaw details the efforts of family and community to socialize these women for both personal achievement and community service. The sacrifices some families made included sending them to private schools and sometimes relocating the entire family near a desired school.

56. Giddings, *When and Where I Enter*, p. 108.

57. Brown, "Womanist Consciousness," p. 622.

58. Ibid., p. 623.

59. Carby, *Reconstructing Womanhood*, p. 117.

60. Quoted in Giddings, *When and Where I Enter*, p. 72.

61. Brown, "Womanist Consciousness," p. 182.

62. Quoted in Giddings, *Where and When I Enter,* p. 197.

63. Quoted in Terborg-Penn, "Discontented Black Feminists," p. 274.

64. Ibid., p. 274.

65. Quoted in Giddings, *Where and When I Enter,* p. 138.

66. Quoted in Carby, *Reconstructing Womanhood,* p. 98.

67. Erlene Stetson, "Black Feminism in Indiana, 1893–1933," *Phylon* 44 (December 1983):294.

68. Quoted in Barbara Welter, "The Cult of True Womanhood: 1820–1860," p. 318.

69. Carby, *Reconstructing Womanhood,* p. 100.

70. Ibid., p. 99.

71. Carlson, "Black Ideals of Womanhood in the Late Victorian Era," p. 69. This view is supported by historian Evelyn Brooks Higginbotham's analysis of schools for blacks established by northern Baptists in the postbellum period, schools that encouraged the attendance of both girls and boys. Although, as Higginbotham observes, northern Baptists founded these schools in part to spread white middle-class values among blacks, blacks nevertheless came to see higher education as an instrument of their own liberation (*Righteous Discontent,* p. 20).

72. Ibid., p. 67.

THINKING ABOUT THE READING

How were the needs and goals of black women during the 19th-century movement for gender equality different from those of white women? How did their lives differ with regard to the importance of marriage, motherhood, and employment? What does Landry mean when he says that for these women, "race and gender are inseparable"? What was the significance of the "clubs" for these black women? How does this article change what you previously thought about the contemporary women's movement?

Still a Man's World

Men Who Do "Women's Work"

Christine L. Williams

(1995)

Gendered Jobs and Gendered Workers

A 1959 article in *Library Journal* entitled "The Male Librarian—An Anomaly?" begins this way:

> My friends keep trying to get me out of the library. . . . Library work is fine, they agree, but they smile and shake their heads benevolently and charitably, as if it were unnecessary to add that it is one of the dullest, most poorly paid, unrewarding, off-beat activities any man could be consigned to. If you have a heart condition, if you're physically handicapped in other ways, well, such a job is a blessing. And for women there's no question library work is fine; there are some wonderful women in libraries and we all ought to be thankful to them. But let's face it, no healthy man of normal intelligence should go into it.[1]

Male librarians still face this treatment today, as do other men who work in predominantly female occupations. In 1990, my local newspaper featured a story entitled "Men Still Avoiding Women's Work" that described my research on men in nursing, librarianship, teaching, and social work. Soon afterwards, a humor columnist for the same paper wrote a spoof on the story that he titled, "Most Men Avoid Women's Work Because It Is Usually So Boring."[2] The columnist poked fun at hairdressing, librarianship, nursing, and babysitting—in his view, all "lousy" jobs requiring low intelligence and a high tolerance for boredom. Evidently people still wonder why any "healthy man of normal intelligence" would willingly work in a "woman's occupation."

In fact, not very many men do work in these fields, although their numbers are growing. In 1990, over 500,000 men were employed in these four occupations, constituting approximately 6 percent of all registered nurses, 15 percent of all elementary school teachers, 17 percent of all librarians, and 32 percent of all social workers. These percentages have fluctuated in recent years: As Table 1 indicates, librarianship and social work have undergone slight declines in the proportions of men since 1975; teaching has remained somewhat stable; while nursing has experienced noticeable gains. The number of men in nursing actually doubled between 1980 and 1990; however, their overall proportional representation remains very low.

Very little is known about these men who "cross over" into these nontraditional occupations. While numerous books have been written about women entering male-dominated occupations, few have asked why men are underrepresented in traditionally female jobs.[3] The underlying assumption in most research on gender and work is that, given a free choice, both men and women would work in predominantly male occupations, as they are generally better paying and more prestigious than predominantly female occupations. The few men who willingly "cross over" must be, as the 1959 article suggests, "anomalies."

Popular culture reinforces the belief that these men are "anomalies." Men are rarely portrayed working in these occupations, and when they are, they are represented in extremely stereotypical ways. For example, in the 1990 movie *Kindergarten Cop*, muscle-man Arnold Schwarzenegger played a detective forced to

Table 1 Men in the "Women's Professions": Number (in thousands) and Distribution of Men Employed in the Occupations, Selected Years

Profession	*1975*	*1980*	*1990*
Registered Nurses			
Number of men	28	46	92
% men	3.0	3.5	5.5
Elementary Teachers[a]			
Number of men	194	225	223
% men	14.6	16.3	14.8
Librarians			
Number of men	34	27	32
% men	18.9	14.8	16.7
Social Workers			
Number of men	116	134	179
% men	39.2	35.0	21.8

Sources: U.S. Department of Labor, Bureau of Labor Statistics, Employment and Earnings 38, no. 1 (January 1991), table 22 (employed civilians by detailed occupation), p. 185; vol. 28, no. 1 (January 1981), table 23 (employed persons by detailed occupation), p. 180; vol. 22, no. 7 (January 1976), table 2 (employed persons by detailed occupation), p. 11.

[a]Excludes kindergarten teachers.

work undercover as a kindergarten teacher; the otherwise competent Schwarzenegger was completely overwhelmed by the five-year-old children in his class. . . .

[I] challenge these stereotypes about men who do "women's work" through case studies of men in four predominantly female occupations: nursing, elementary school teaching, librarianship, and social work. I show that men maintain their masculinity in these occupations, despite the popular stereotypes. Moreover, male power and privilege is preserved and reproduced in these occupations through a complex interplay between gendered expectations embedded in organizations, and the gendered interests workers bring with them to their jobs. Each of these occupations is "still a man's world" even though mostly women work in them.

I selected these four professions as case studies of men who do "women's work" for a variety of reasons. First, because they are so strongly associated with women and femininity in our popular culture, these professions highlight and perhaps even exaggerate the barriers and advantages men face when entering predominantly female environments. Second, they each require extended periods of educational training and apprenticeship, requiring individuals in these occupations to be at least somewhat committed to their work (unlike those employed in, say, clerical or domestic work). Therefore I thought they would be reflective about their decisions to join these "nontraditional" occupations, making them "acute observers" and, hence, ideal informants about the sort of social and psychological processes I am interested in describing.[4] Third, these occupations vary a great deal in the proportion of men working in them. Although my aim was not to engage in between-group comparisons, I believed that the proportions of men in a work setting would strongly influence the degree to which they felt accepted and satisfied with their jobs.[5]

I traveled across the United States conducting in-depth interviews with seventy-six men and twenty-three women who work in nursing, teaching, librarianship, and social work. Like the people employed in these professions generally, those in my sample were predominantly white (90 percent). Their ages ranged from twenty to sixty-six, and the average age was thirty-eight. I interviewed women as well as men to gauge their feelings and reactions to men's entry into "their" professions. Respondents were intentionally selected to represent a wide range of specialties and levels of education and experience. I interviewed students in professional schools, "front line" practitioners, administrators, and retirees, asking them about their motivations to enter

Table 2 Median Weekly Earnings of Full-Time Professional Workers, by Sex, and Ratio of Female: Male Earnings, 1990

Occupation	*Both*	*Men*	*Women*	*Ratio*
Registered Nurses	608	616	608	.99
Elementary Teachers	519	575	513	.89
Librarians	489	—*	479	—
Social Workers	445	483	427	.88
Engineers	814	822	736	.90
Physicians	892	978	802	.82
College Teachers	747	808	620	.77
Lawyers	1,045	1,178	875	.74

Source: U.S. Department of Labor, Bureau of Labor Statistics, Employment and Earnings 38, no. 1 (January 1991), table 56, p. 223.

*The Labor Department does not report income averages for base sample sizes consisting of fewer than 50,000 individuals.

these professions, their on-the-job experiences, and their opinions about men's status and prospects in these fields. . . .

Riding the Glass Escalator

Men earn more money than women in every occupation—even in predominantly female jobs (with the possible exceptions of fashion modeling and prostitution).[6] Table 2 shows that men outearn women in teaching, librarianship, and social work; their salaries in nursing are virtually identical. The ratios between women's and men's earnings in these occupations are higher than those found in the "male" professions, where women earn 74 to 90 percent of men's salaries. That there is a wage gap at all in predominantly female professions, however, attests to asymmetries in the workplace experiences of male and female tokens. These salary figures indicate that the men who do "women's work" fare as well as, and often better than, the women who work in these fields. . . .

Hiring Decisions

Contrary to the experience of many women in the male-dominated professions, many of the men and women I spoke to indicated that there is a *preference* for hiring men in these four occupations. A Texas librarian at a junior high school said that his school district "would hire a male over a female":

> [CW: Why do you think that is?]
>
> Because there are so few, and the . . . ones that they do have, the library directors seem to really . . . think they're doing great jobs. I don't know, maybe they just feel they're being progressive or something, [but] I have had a real sense that they really appreciate having a male, particularly at the junior high. . . . As I said, when seven of us lost our jobs from the high schools and were redistributed, there were only four positions at junior high, and I got one of them. Three of the librarians, some who had been here longer than I had with the school district, were put down in elementary school as librarians. And I definitely think that being male made a difference in my being moved to the junior high rather than an elementary school.

Many of the men perceived their token status as males in predominantly female occupations as an *advantage* in hiring and promotions. When I asked an Arizona teacher whether his specialty (elementary special education) was an unusual area for men compared to other areas within education, he said,

> Much more so. I am extremely marketable in special education. That's not why I got into the field. But I am extremely marketable because I am a man.

. . . Sometimes the preference for men in these occupations is institutionalized. One man landed his first job in teaching before he earned the appropriate credential "because I was a wrestler and they wanted a wrestling coach." A female math teacher similarly told of her inability to find a full-time teaching position because the schools she applied to reserved the math jobs for people (presumably men) who could double as coaches. . . .

. . . Some men described being "tracked" into practice areas within their professions which were considered more legitimate for men. For example, one Texas man described how he was pushed into administration and planning in social work, even though "I'm not interested in writing policy; I'm much more interested in research and clinical stuff." A nurse who is interested in pursuing graduate study in family and child health in Boston said he was dissuaded from entering the program specialty in favor of a concentration in "adult nursing." And a kindergarten teacher described his difficulty finding a job in his specialty after graduation: "I was recruited immediately to start getting into a track to become an administrator. And it was men who recruited me. It was men that ran the system at that time, especially in Los Angeles."

This tracking may bar men from the most female-identified specialties within thesepro-fessions. But men are effectively being "kicked upstairs" in the process. Those specialties considered more legitimate practice areas for men also tend to be the most prestigious, and better-paying specialties as well. For example, men in nursing are overrepresented in critical care and psychiatric specialties, which tend to be higher paying than the others.[7] The highest paying and most prestigious library types are the academic libraries (where men are 35 percent of librarians) and the special libraries which are typically associated with businesses or other private organizations (where men constitute 20 percent of librarians).[8]

A distinguished kindergarten teacher, who had been voted citywide "Teacher of the Year," described the informal pressures he faced to advance in his field. He told me that even though people were pleased to see him in the classroom, "there's been some encouragement to think about administration, and there's been some encouragement to think about teaching at the university level or something like that, or supervisory-type position."

The effect of this "tracking" is the opposite of that experienced by women in male-dominated occupations. Researchers have reported that many women encounter "glass ceilings" in their efforts to scale organizational and professional hierarchies. That is, they reach invisible barriers to promotion in their careers, caused mainly by the sexist attitudes of men in the highest positions.[9] In contrast to this "glass ceiling," many of the men I interviewed seem to encounter a "glass escalator." Often, despite their intentions, they face invisible pressures to move up in their professions. Like being on a moving escalator, they have to work to stay in place. . . .

Supervisors and Colleagues: The Working Environment

. . . Respondents in this study were asked about their relationships with supervisors and female colleagues to ascertain whether men also experienced "poisoned" work environments when entering nontraditional occupations.

A major difference in the experience of men and women in nontraditional occupations is that men are far more likely to be supervised by a member of their own sex. In each of the four professions I studied, men are overrepresented in administrative and managerial capacities, or, as in the case of nursing, the organizational hierarchy is governed by men. For example, 15 percent of all elementary school teachers are men, but men make up over 80 percent of all elementary school principals and 96 percent of all public school superintendents and assistant superintendents.[10] Likewise, over 40 percent of all male social workers hold administrative or managerial positions, compared to 30 percent of all female social workers.[11] And 50 percent of male librarians hold administrative positions, compared to 30 percent of female librarians, and the majority of deans and directors of major university and public libraries are men.[12] Thus, unlike women who enter "male fields," the men in these professions often work under the direct supervision of other men.

Many of the men interviewed reported that they had good rapport with their male supervisors. It was not uncommon in education, for example, for the male principal to informally socialize with the male staff, as a Texas special education teacher describes:

> Occasionally I've had a principal who would regard me as "the other man on the campus" and "it's us against them," you know? I mean, nothing really that extreme, except that some male principals feel like there's nobody there to talk to except the other man. So I've been in that position.

These personal ties can have important consequences for men's careers. For example, one California nurse, whose performance was judged marginal by his nursing superiors, was transferred to the emergency room staff (a prestigious promotion) due to his personal friendship with the physician in charge. And a Massachusetts teacher acknowledged that his principal's personal interest in him landed him his current job:

> [CW: You had mentioned that your principal had sort of spotted you at your previous job and had wanted to bring you here [to this school]. Do you think that has anything to do with the fact that you're a man, aside from your skills as a teacher?]
>
> Yes, I would say in that particular case, that was part of it. . . . We have certain things in common, certain interests that really lined up.
>
> [CW: Vis-à-vis teaching?]
>
> Well, more extraneous things—running specifically, and music. And we just seemed to get along real well right off the bat. It is just kind of a guy thing; we just liked each other. . . .

Interviewees did not report many instances of male supervisors discriminating against them, or refusing to accept them because they were male. Indeed, these men were much more likely to report that their male bosses discriminated against the *females* in their professions. . . .

Of course, not all the men who work in these occupations are supervised by men. Many of the men interviewed who had female bosses also reported high levels of acceptance—although the level of intimacy they achieved with women did not seem as great as with other men. But in some cases, men reported feeling shut-out from decision making when the higher administration was constituted entirely by women. I asked this Arizona librarian whether men in the library profession were discriminated against in hiring because of their sex:

> Professionally speaking, people go to considerable lengths to keep that kind of thing out of their [hiring] deliberations. Personally, is another matter. It's pretty common around here to talk about the "old girl network." This is one of the few libraries that I've had any intimate knowledge of which is actually controlled by women. . . . Most of the department heads and upper level administrators are women. And there's an "old girl network" that works just like the "old boy network," except that the important conferences take place in the women's room rather than on the golf course. But the political mechanism is the same, the exclusion of the other sex from decision making is the same. The reasons are the same. It's somewhat discouraging. . . .

Although I did not interview many supervisors, I did include twenty-three women in my sample to ascertain their perspectives about the presence of men in their professions. All of the women I interviewed claimed to be supportive of their male colleagues, but some conveyed ambivalence. For example, a social work professor said she would like to see more men enter the social work profession, particularly in the clinical specialty (where they are underrepresented). She said she would favor affirmative action hiring guidelines for men in the profession, and yet, she resented the fact that her department hired "another white male" during a recent search. I confronted her about this apparent ambivalence:

> [CW: I find it very interesting that, on the one hand, you sort of perceive this preference and perhaps even sexism with regard to how men are evaluated and how they achieve higher positions within the profession, yet, on the other hand, you would be encouraging of more men to enter the field. Is that contradictory to you, or . . . ?]
>
> Yeah, it's contradictory. . . .

Men's reception by their female colleagues is thus somewhat mixed. It appears that women are generally eager to see men enter "their" occupations, and the women I interviewed claimed they were supportive of their male peers. Indeed, several men agreed with this social worker that their female colleagues had facilitated their careers in various ways (including college mentorship). At the same time, however, women often resent the apparent ease with which men seem to advance within these professions, sensing that men at the higher levels receive preferential treatment, and thus close off advancement opportunities for women.

But this ambivalence does not seem to translate into the "poisoned" work environment described by many women who work in male-dominated occupations. Among the male interviewees, there were no accounts of sexual harassment (indeed, one man claimed this was a disappointment to him!). However, women do treat their male colleagues differently on occasion. It is not uncommon in nursing, for example, for men to be called upon to help catheterize male patients, or to lift especially heavy patients. Some librarians also said that women asked them to lift and move heavy boxes of books because they were men. . . .

Another stereotype confronting men, in nursing and social work in particular, is the expectation that they are better able than women to handle aggressive individuals and diffuse violent situations. An Arizona social worker who was the first male caseworker in a rural district, described this preference for men:

> They welcomed a man, particularly in child welfare. Sometimes you have to go into some tough parts of towns and cities, and they felt it was nice to have a man around to accompany them or be present when they were dealing with a difficult client. Or just doing things that males can do. I always felt very welcomed.

But this special treatment bothered some respondents: Getting assigned all the violent patients or discipline problems can make for difficult and unpleasant working conditions. Nurses, for example, described how they were called upon to subdue violent patients. A traveling psychiatric nurse I interviewed in Texas told how his female colleagues gave him "plenty of opportunities" to use his wrestling skills. . . .

But many men claimed that this differential treatment did not distress them. In fact, several said they liked being appreciated for the special traits and abilities (such as strength) they could contribute to their professions.

Furthermore, women's special treatment of men sometimes enhanced—rather than detracted from—the men's work environments. One Texas librarian said he felt "more comfortable working with women than men" because "I think it has something to do with control. Maybe it's that women will let me take control more than men will." Several men reported that their female colleagues often cast them into leadership roles. . . .

The interviews suggest that the working environment encountered by "nontraditional" male workers is quite unlike that faced by women who work in traditionally male fields. Because it is not uncommon for men in predominantly female professions to be supervised by other men, they tend to have closer rapport and more intimate social relationships with people in management. These ties can facilitate men's careers by smoothing the way for future promotions. Relationships with female supervisors were also described for the most part in positive terms, although in some cases, men perceived an "old girls'" network in place that excluded them from decision making. But in sharp contrast to the reports of women in nontraditional occupations, men in these fields did not complain of feeling discriminated against because they were men. If anything, they felt that being male was an asset that enhanced their career prospects.

Those men interviewed for this study also described congenial workplaces, and a very high level of acceptance from their female colleagues. The sentiment was echoed by women I spoke to who said that they were pleased to see more men enter "their" professions. Some women, however, did express resentment over the "fast-tracking" that their male colleagues seem to experience. But this ambivalence did not translate into a hostile work environment for men: Women generally included men in their informal social events and, in some ways, even facilitated men's careers. By casting men into leadership roles, presuming they were more knowledgeable and qualified, or relying on them to perform certain critical tasks, women unwittingly contributed to the "glass escalator effect" facing men who do "women's work."

Relationships with Clients

Workers in these service-oriented occupations come into frequent contact with the public during the course of their work day. Nurses treat patients; social workers usually have client case loads; librarians serve patrons; and teachers are in constant contact with children, and often with parents as well. Many of those interviewed claimed that the clients they served had different expectations of men and women in these occupations, and often treated them differently.

People react with surprise and often disbelief when they encounter a man in nursing, elementary school teaching, and, to a lesser extent, librarianship. (Usually people have no clear expectations about the sex of social workers.) The stereotypes men face are often negative. For example, according to this Massachusetts nurse, it is frequently assumed that male nurses are gay:

> Fortunately, I carry one thing with me that protects me from [the stereotype that male nurses are gay], and the one thing I carry with me is a wedding ring, and it makes a big difference. The perfect example was conversations before I was married. . . . [People would ask], "Oh, do you have a girlfriend?" Or you'd hear patients asking questions along that idea, and they were simply implying, "Why is this guy in nursing? Is it because he's gay and he's a pervert?" And I'm not associating the two by any means, but this is the thought process.

. . . It is not uncommon for both gay and straight men in these occupations to encounter people who believe that they are "gay 'til proven otherwise," as one nurse put it. In fact, there are many gay men employed in these occupations. But gender stereotypes are at least as responsible for this general belief as any "empirical" assessment of men's sexual lifestyles. To the degree that men in these professions are perceived as not "measuring up" to the supposedly more challenging occupational roles and standards demanded of "real" men, they are immediately suspected of being effeminate—"like women"—and thus, homosexual.

An equally prevalent sexual stereotype about men in these occupations is that they are potentially dangerous and abusive. Several men described special rules they followed to guard against the widespread presumption of sexual abuse. For example, nurses were sometimes required to have a female "chaperone" present when performing certain procedures or working with specific populations. This psychiatric nurse described a former workplace:

> I worked on a floor for the criminally insane. Pretty threatening work. So you have to have a certain number of females on the floor just to balance out. Because there were female patients on the floor too. And you didn't want to be accused of rape or any sex crimes.

Teachers and librarians described the steps they took to protect themselves from suspicions of sexual impropriety. A kindergarten teacher said:

> I know that I'm careful about how I respond to students. I'm careful in a number of ways—in my physical interaction with students. It's mainly to reassure parents. . . . For example, a little girl was very affectionate, very anxious to give me a hug. She'll just throw herself at me. I need to tell her very carefully: "Sonia, you need to tell me when you want to hug me." That way I can come down, crouch down. Because you don't want a child giving you a hug on your hip. You just don't want to do that. So I'm very careful about body position.

. . . Although negative stereotypes about men who do "women's work" can push men out of specific jobs, their effects can actually benefit men. Instead of being a source of negative discrimination, these prejudices can add to the "glass escalator effect" by pressuring men to move *out* of the most feminine-identified areas and *up* to those regarded as more legitimate for men.

The public's reactions to men working in these occupations, however, are by no means always negative. Several men and women reported that people often assume that men in these occupations are more competent than women, or that they bring special skills and expertise to their professional practice. For example, a female academic librarian told me that patrons usually address their questions to the male reference librarian when there is a choice between asking a male or a female. A male clinical social worker in private practice claimed that both men and women generally preferred male psychotherapists. And several male nurses told me that people often assume that they are physicians and direct their medical inquiries to them instead of to the female nurses.[13]

The presumption that men are more competent than women is another difference in the experience of token men and women. Women who work in nontraditional occupations are often suspected of being incompetent, unable to survive the pressures of "men's work." As a consequence, these women often report feeling compelled to prove themselves and, as the saying goes, "work twice as hard as men to be considered half as good." To the degree that men are assumed to be competent and in control, they may have to be twice as incompetent to be considered half as bad. One man claimed that "if you're a mediocre male teacher, you're considered a better teacher than if you're a female and a mediocre teacher. I think there's that prejudice there." . . .

There are different standards and assumptions about men's competence that follow them into nontraditional occupations. In contrast, women in both traditional and nontraditional occupations must contend with the presumption that they are neither competent nor qualified. . . .

The reasons that clients give for preferring or rejecting men reflect the complexity of our society's stereotypes about masculinity and femininity. Masculinity is often associated with competence and mastery, in contrast to femininity, which is often associated with instrumental incompetence. Because of these stereotypes, men are perceived as being stricter disciplinarians and stronger than women, and thus better able to handle violent or potentially violent situations. . . .

Conclusion

Both men and women who work in nontraditional occupations encounter discrimination, but the forms and the consequences of this discrimination are very different for the two groups. Unlike "nontraditional" women workers, most of the discrimination and prejudice facing men in the "female" professions comes from clients. For the most part, the men and women I interviewed believed that men are given fair—if not preferential—treatment in hiring and promotion decisions, are accepted by their supervisors and colleagues, and are well-integrated into the workplace subculture. Indeed, there seem to be subtle mechanisms in place that enhance men's positions in these professions—a phenomenon I refer to as a "glass escalator effect."

Men encounter their most "mixed" reception in their dealings with clients, who often react negatively to male nurses, teachers, and to a lesser extent, librarians. Many people assume that the men are sexually suspect if they are employed in these "feminine" occupations either because they do or they do not conform to stereotypical masculine characteristics.

Dealing with the stress of these negative stereotypes can be overwhelming, and it probably pushes some men out of these occupations.[14] The challenge facing the men who stay in these fields is to accentuate their positive contribution to what our society defines as essentially "women's work." . . .

Notes

1. Allan Angoff, "The Male Librarian—An Anomaly?" *Library Journal,* February 15, 1959, p. 553.

2. *Austin-American Statesman,* January 16, 1990; response by John Kelso, January 18, 1990.

3. Some of the most important studies of women in male-dominated occupations are: Rosabeth Moss Kanter, *Men and Women of the Corporation* (New York: Basic Books, 1977); Susan Martin, *Breaking and Entering: Policewomen on Patrol* (Berkeley: University of California Press, 1980); Cynthia Fuchs Epstein, *Women in Law* (New York: Basic Books, 1981); Kay Deaux and Joseph Ullman, *Women of Steel* (New York: Praeger, 1983); Judith Hicks Stiehm, *Arms and the Enlisted Woman* (Philadelphia: Temple University Press, 1989); Jerry Jacobs, *Revolving Doors: Sex Segregation and Women's Careers* (Stanford: Stanford University Press, 1989); Barbara Reskin and Patricia Roos, *Job Queues, Gender Queues: Explaining Women's Inroads into Male Occupations* (Philadelphia: Temple University Press, 1990).

Among the few books that do examine men's status in predominantly female occupations are Carol Tropp Schreiber, *Changing Places: Men and Women in Transitional Occupations* (Cambridge: MIT Press, 1979); Christine L. Williams, *Gender Differences at Work: Women and Men in Nontraditional Occupations* (Berkeley: University of California Press, 1989); and Christine L. Williams, ed., *Doing "Women's Work": Men in Nontraditional Occupations* (Newbury Park, CA: Sage Publications, 1993).

4. In an influential essay on methodological principles, Herbert Blumer counseled sociologists to "sedulously seek participants in the sphere of life who are acute observers and who are well informed. One such person is worth a hundred others who are merely unobservant participants." See "The Methodological Position of Symbolic Interactionism," in *Symbolic Interactionism: Perspective and Method* (Berkeley: University of California Press, 1969), p. 41.

5. The overall proportions in the population do not necessarily represent the experiences of individuals in my sample. Some nurses, for example, worked in groups that were composed almost entirely of men, while some social workers had the experience of being the only man in their group. The overall statistics provide a general guide, but relying on them exclusively can distort the actual experiences of individuals in the workplace. The statistics available for research on occupational sex segregation are not specific enough to measure internal divisions among workers. Research that uses firm-level data finds a far greater degree of segregation than research that uses national data. See William T. Bielby and James N. Baron, "A Woman's Place Is with Other Women: Sex Segregation within Organizations," in *Sex Segregation in the Workplace: Trends, Explanations, Remedies,* ed. Barbara Reskin (Washington, D.C.: National Academy Press, 1984), pp. 27–55.

6. Catharine MacKinnon, *Feminism Unmodified* (Cambridge: Harvard University Press, 1987), pp. 24–25.

7. Howard S. Rowland, *The Nurse's Almanac,* 2d ed. (Rockville, MD: Aspen Systems Corp., 1984), p. 153; John W. Wright, *The American Almanac of*

Jobs and Salaries, 2d ed. (New York: Avon, 1984), p. 639.

8. King Research, Inc., *Library Human Resources: A Study of Supply and Demand* (Chicago: American Library Association, 1983), p. 41.

9. See, for example, Sue J. M. Freeman, *Managing Lives: Corporate Women and Social Change* (Amherst: University of Massachusetts Press, 1990).

10. Patricia A. Schmuck, "Women School Employees in the United States," in *Women Educators: Employees of Schools in Western Countries* (Albany: State University of New York Press, 1987), p. 85; James W. Grimm and Robert N. Stern, "Sex Roles and Internal Labor Market Structures: The Female Semi-Professions," *Social Problems* 21(1974):690–705.

11. David A. Hardcastle and Arthur J. Katz, *Employment and Unemployment in Social Work: A Study of NASW Members* (Washington, D.C.: NASW, 1979), p. 41; Reginald O. York, H. Carl Henley and Dorothy N. Gamble, "Sexual Discrimination in Social Work: Is It Salary or Advancement?" *Social Work* 32 (1987):336–340; Grimm and Stern, "Sex Roles and Internal Labor Market Structures."

12. Leigh Estabrook, "Women's Work in the Library/Information Sector," in *My Troubles Are Going to Have Trouble with Me*, ed. Karen Brodkin Sacks and Dorothy Remy (New Brunswick, NJ: Rutgers University Press, 1984), p. 165.

13. Liliane Floge and D. M. Merrill found a similar phenomenon in their study of male nurses. See "Tokenism Reconsidered: Male Nurses and Female Physicians in a Hospital Setting," *Social Forces* 64 (1986):931–932.

14. Jim Allan makes this argument in "Male Elementary Teachers: Experiences and Perspectives," in *Doing "Women's Work": Men in Nontraditional Occupations*, ed. Christine L. Williams (Newbury Park, CA: Sage Publications, 1993), pp. 113–127.

THINKING ABOUT THE READING

Compare the discrimination men experience in traditionally female occupations to that experienced by women in traditionally male occupations. What is the "glass escalator effect"? In what ways can the glass escalator actually be harmful to men? What do you suppose might happen to the structure of the American labor force if men did in fact begin to enter predominantly female occupations in the same proportion as women entering predominantly male occupations?

New Biomedical Technologies, New Scripts, New Genders

Eve Shapiro

(2010)

In 1939, soon after graduating from St Anne's College of the University of Oxford, England, Lawrence Michael Dillon became the first transsexual man to undergo physical transition from female to male. Dillon had lived as a masculine woman during college and experienced discrimination for years because of his gender presentation. While Dillon came to a masculine identity during his college years, he had long looked and acted masculine and expressed desires to be a man. Even though Dillon knew himself to be a man, albeit one hidden within a female body, he had no social support and no social scripts with which he could make sense of his situation. He was without any language to talk about gender non-conformity or transgenderism; indeed the word 'transsexual' had yet to be coined, and there was certainly no discussion of any difference between sex and gender. Dillon was unable to find anyone who would either support his gender identity or enable the physical changes he required to live as a man, and he struggled to make sense of these desires.

Michael Dillon's life story stands in stark contrast to the contemporary experiences of many young female-to-male transgender people. A feature article in the *New York Times Magazine* on March 16, 2008, for example, included a profile of Rey, an 18-year-old White female-to-male transgender college student (Quart 2008). In this *New York Times Magazine* article Rey reflects on growing up a masculine child, being mistaken for a boy, and coming out as transgender to himself at 14 and to his family at the age of 17.

Rey's story is similar in part to Michael Dillon's; both grew up masculine and developed a gender identity as a boy/man, at a young age. But while Michael Dillon negotiated an identity without a language for gender non-conformity, Rey not only gained language and learned social scripts to describe who he was from other transgender people, he was able to do so at a relatively young age. Specifically, he heard a transgender man speak at a Gay Straight Alliance meeting at his high school and immediately went home and ran a Google Internet search for the word 'transgender.' In another illustration of the online identity work, Rey elaborated that, "The Internet is the best thing for trans people . . . Living in the suburbs, online groups were an access point [for me]" (Quart 2008:34). Unlike Michael Dillon, who had no access to information about transgenderism, Rey was able to find and use a wide array of information and support resources to validate, define, and negotiate his own masculine identity and female body.

The ability to communicate with other transgender individuals, learn about treatments for transsexualism, and engage with others as a boy online, all helped Rey redefine his identity. Advances in communication helped Rey understand and define his gender identity, while developments in medical technology, allowed Rey to reshape his body to reflect his gendered identity. Whether and how new social scripts emerge in response to technological innovations in fields dealing with human anatomy can be illuminated by examining how individuals use biomedical means to know and construct their bodies.

New Body Technologies

Notwithstanding issues of transgenderism, the ability for and acceptability of body modification has also changed, as demonstrated by the rise and social acceptance of bodily transformation practices such as plastic surgery, use of pharmaceuticals, weightlifting, tattooing, shaving, and hair dyeing. More so now than ever before, it is common practice for individuals to produce and refine their gendered bodies in ways that both reinforce and contest normative social scripts for women's and men's bodies.

While body work is transformative and often purposeful on the part of individuals, these changes, whether by chance, social structure, or agency of the individual, are always already shaped by social norms and historical context. When an individual chooses to get a tattoo, for example, they may do so for any of a variety of reasons including an effort to adorn their body, mark a significant event, or signal participation in a community or identity category. But this agentic choice is informed by and given meaning through gendered societal beliefs about tattoos and their significance, body and beauty scripts, and the dominant societal paradigms. Approaching the body from the perspective of being something both shaped by and actively shaping identity as well as society allows a better understanding of how new technologies are dynamically engaged with gendered bodies.

In this [article] I examine a number of case studies to make sense of how gendered bodies and identities both inform and respond to these new biomedical technologies. Using a sociological approach to map the intricate, multiple connections between embodied identities, technologies, and social gender paradigms, I examine how the ability to construct new bodies is changing who people think they are and can be.

Somatechnics: Technologies and the Body

Biomedical technology has become the medium through which we know and intervene into our bodies, and genetic testing, body scans, surgery, and medication are just a few examples. A term that emerged in the 1970s to reflect the increasingly technological approaches to biology and medicine, 'biotechnology' brings engineering and technological theories in disciplines including agriculture, medicine, genetics, and physiology to bear on natural systems. In other words, biotechnology refers to any intentional manipulation of organic processes/organisms.

Of particular relevance to this discussion is biomedical technology, or technologies that are directed at maintaining and/or transforming the human body. This includes genetic testing and manipulation, pharmacology, surgery including microsurgery, imaging, cloning, synthetic drugs, hormones and vaccines, prosthetics, and implants, to name a few. Moving beyond discovery for its own sake as a motivation for scientific research, the profitability of biotechnology has led to the development of numerous attendant industries centered on body work. Rates of plastic surgery, 'lifestyle drug' use, genetic engineering, and 'medi-spa' treatments have increased dramatically over the last 20 years. The hugely profitable biotechnological industry in the United States generated 58.8 billion U.S. dollars in health care revenue in 2006 alone. In Canada biotechnology firms generated 4.2 billion Canadian dollars of revenue in 2005. This exponential growth in the techniques of biomedical intervention and their acceptance also signals a source of social change for institutions and individuals.

Recently, some scholars have used the term *somatechnics* to describe human-body-focused technologies and to distinguish them from agricultural- and/or animal-focused biotechnologies.

Somatechnics

Technologies of the body. More specifically, an understanding that the body and technology are always and already interrelated and mutually constitutive. Technologies shape how we know, understand, and shape the body, and the body is always a product of historically and culturally specific transformative practices.

Nikki Sullivan, one of the pioneers of this concept, has focused on body modification like tattooing to make sense of how body technologies are both shaped by and an intentional engagement with social scripts for gendered bodies. Others have done work in a similar vein, including Susan Stryker, who has explored the social and technological history of transsexualism, and Samantha Murray, whose work focuses on fatness and the emergence of bariatric surgery. Research by each of these scholars demonstrates that modern embodied identities are always already in dynamic relationship to technologies, and more specifically that technologies are used to construct, maintain, and transform gendered bodies and identities.

Somatechnics and Social Norms

One very timely example of the dynamic relationship between somatechnics and social norms is bariatric (i.e., weight loss) surgery. With weight loss surgery, individuals—mostly women—are using biomedical technologies to reshape their bodies in dramatically increasing numbers. Many individuals benefit from this surgery, which can reduce health problems, raise self-esteem, and facilitate alignment between body and identity.

Even though versions of the surgery have been used for more than 50 years, widespread access to this biomedical technology was limited. During the 11-year period from 1995–2006 bariatric surgery rates skyrocketed by 800 percent; estimates suggest that in 2008 more than 200,000 surgeries were performed in the United States alone. What makes bariatric surgery such an interesting case study is that the highly contentious debates about the surgery engage directly with contemporary body and gender paradigms and social scripts. These debates take place within both medical circles and larger society as they manifest in and through the bodies and identities of individuals. Dominant body paradigms posit the idea that fat bodies are inherently unhealthy, undesirable, and a sign of internal character failings, which legitimates biomedical intervention.

Societal body norms have a direct effect on what individuals do to reshape their bodies. Both men and women are pressured to change their bodies, and are stigmatized if they do not conform to these demands. Although the scientific research and development of weight loss technologies may be seen as unbiased, the emphasis on that objective revolves around contemporary body and gender paradigms. These societal forces are joined with contemporary advances in medical technology to form the foundation of a phenomenally profitable weight-loss industry—just think of how many diet and body shaping products you can name—and that industry puts even more pressure on individuals to conform to physical ideals.

The debate playing out around bariatric surgery is over the nature of the body. Dominant paradigms view thinness as natural and achievable through discipline and in turn discount the need for surgery. Slowly challenging this (aided by bariatric surgery, weight loss drugs like Alli, and the hunt for a 'fat gene') is a paradigm that views fatness as disease and therefore a malady worthy of medical intervention and treatment. Finally, the recent emergence of fat-positive activism has challenged social paradigms regarding body size. Groups like the National Association to Advance Fat Acceptance (NAAFA) take issue with dominant medical and social paradigms that link health to thinness and fat to disease. These organizations point toward historical and cultural variation in body size scripts, and stress that research reveals a wide range of differences in health and body size; not all large bodies are unhealthy and not all thin bodies are healthy. For many individuals whose bodies do not conform to normative body scripts, the presence of a counter-hegemonic paradigm reinforces their own positive social body and identity scripts. In other words, the debate about 'normal' bodies is in fact a debate over the dominant body paradigm, and it is taking place in part through debates about biomedical 'treatments' for obesity. This new technology of bariatric surgery is reshaping individual bodies and identities while the application of the technology is simultaneously responding to and reshaping societal body paradigms (paradigms such as what constitutes a healthy body) and social scripts for 'normal' embodied identities.

A number of social scientists have studied this emerging somatechnical phenomenon and found that weight loss surgery is altering individual gendered identities and bodies. For example, as Patricia Drew documents in her research, part of how bariatric advocates have tried to legitimize this new biomedical intervention has been to first create scripts for the ideal patient that draw on and reinforce hegemonic body paradigms and then require the adoption of these scripts in order to access bariatric surgery. This gate-keeping requires that individuals adopt (or, at the very least, pretend to adopt) particular physiological, behavioral, and attitude scripts (much like transsexual scripts to access 'sex-reassignment surgery'). This ideal patient script is shaped by the controversial history of the technology, a history in which early versions led to high rates of complication and death. It is, in turn, reshaping dominant body paradigms and scripts, offering a fine example of how technological development can interact with social scripts and bodies. Paradigm shifts in the perception of fatness as a disease can both be shaped by the increase in bariatric surgery, and, simultaneously, further legitimize the biomedical intervention. Similarly, the dominant belief that the internal self is reflected in the body compels individuals to change their body to match their internal identity, and simultaneously reinforces body scripts that devalue fat bodies (often decreasing people's estimation of the worth of their own inner selves). These dynamics are also gendered; fat male bodies are viewed as feminized while fat female bodies are de-feminized, particularly in terms of sexuality. Body scripts shape the identities and bodies of individuals by demanding particular gendered scripts and body practices from individuals, as in turn the bodies and identities present in a context reinforce or challenge existing social scripts.

In her study, Patricia Drew found that the very public medical debates about weight loss surgery shaped the ways one could be an acceptable patient by constructing ideal patient scripts, while these same scripts shaped individuals in significant ways. Drew's interviews with patients revealed that most individuals incorporated into their own story the key narrative elements of the dominant script, elements such as viewing themselves as empowered through the use of weight loss surgery, and as responsible for their body. At the same time, those whose narrative did not contain the key elements of the acceptable script still used it strategically to access surgery. What Drew concludes is that these ideal patient scripts, or discourses-in-practice, learned in part through mandatory support group meetings, which afforded discursive practice, helped individuals negotiate between larger social body paradigms and individual identity. Most patients adopted the ideal patient scripts in part or full, and in the process, hegemonic body paradigms. Concomitantly, the social scripts rooted in those ideologies shaped the bodies and identities of participants.

In addition to this clear example of how technologies are in dynamic relationship to social scripts, ideologies, bodies and identities, bariatric surgery is a compelling case study for another reason: it is deeply gendered. According to the U.S. Centers for Disease Control, while women make up 59 percent of the obese population they account for 85 percent of weight loss surgery patients. If surgery was simply the product of obesity, then men and women would be accessing surgery at rates equal to the ratio of obesity in the general population, that is, statistically only 59 percent of patients should be women. These numbers suggest that people use this new technology of weight loss surgery based on gendered ideologies and gendered social scripts for ideal patients. Thus the technology is gendered and it produces gendered outcomes and societal changes.

Patricia Drew concludes that weight loss surgery is deeply gendered because of four intersecting gendered paradigms and scripts. First, as many scholars have documented, North American societies' gendered body and beauty paradigms place higher demands on women, and place more stringent sanctions on them for deviating from normative beauty standards. Women are expected to go to greater lengths and exercise more discipline upon themselves and their bodies than men are (McKinley 1999). Second, as Nelly Oudshoorn argued with regard to birth control, part of why women are held more accountable is that when the male body is held as a normal baseline, the female body is resultantly seen as more in need of intercession, and as a more legitimate target for biomedical intervention. Women are more likely to seek any medical care, which is, in itself, a product of gendered body and health paradigms, and this holds true for weight loss surgery. Third, weight loss surgery requires participation in support groups, groups that our society views as largely the domain of women. Finally, these gendered dynamics shape the social scripts disseminated by medical and media sources about weight loss surgery. In her analysis of hundreds of brochures, advertisements, and websites about weight loss surgery, Drew found that publicity materials pictured women much more often than men. For example, in 21 issues of *Obesity Help*, with a total of 80 advertisements, only nine of the ads featured men as patients. Drew concludes that not only do dominant ideologies shape the ideal patient scripts, but they also shape whether and how individuals use the new technologies. This, in turn, inspires change in both men's and women's bodies and identities, and reinforces the ideologies and scripts that produced these bodies and identities in the first place. These dynamically intertwined relationships are just one example of how individuals both reinforce and contest paradigms and scripts for femininity and masculinity as they use somatic technologies.

Technology and Body Work

It is now possible to alter the look of one's body through myriad technologies, just a few of which are plastic surgery, steroids, growth hormones, hair dye, permanent makeup, hair transplants, subcultural body modification practices like tattoos and scarification, laser hair removal, machine-enhanced exercise regimens, and spa treatments.

For individuals who want and have the means to engage in this transformative work, the ability to embody new identities, in order to either manifest what was previously consigned to one's existing inner selfhood or produce a body that matches a sought after inner identity, is increasingly possible. This holds true for both normative and non-normative bodily changes; individuals can become more masculine men (for example through steroid use or testosterone shots), more feminine women (through breast augmentation and laser hair removal, as examples), as well as transverse gender norms to become more feminine men, masculine women, or more androgynous male, female, or transgender individuals. All of this work is *body work*. 'Body work' refers to both the intentional nature of interventions into the body and to the technological and personal labor involved in those transformations.

Gendered Selves, Gendered Bodies

Anne Balsamo published a groundbreaking book in 1996, *Technologies of the Gendered Body*, which explored how body technologies in the late twentieth century were shaped by, and in turn reproduced, dominant gender paradigms and inequalities. Examining primarily media and cultural products, she analyzed technological interventions into the body, and concluded that these technologies are "ideologically shaped by the operation of gender interests, and consequently . . . serve to reinforce traditional gendered patterns of power and authority" (Balsamo 1996:10).

In other words, what Balsamo is saying is that body technologies are developed and used in tandem with hegemonic gender paradigms to reproduce gender inequality and maintain the status quo. Later work has taken both a more empirical approach to studying gendered technologies by relying more on examination of individuals' lived experiences rather than on textual analysis, and a more liberatory view of technological intervention. However, Balsamo's scholarship captures the central connection between gender ideologies and scripts and somatechnics that I have been examining. In her analysis of body technologies and gender, she asserts that technologies shape and are shaped by dominant gender paradigms and that these together reshape gendered bodies and identities.

How Are Biomedical Technologies Shaping Gendered and Raced Bodies?

When scholars speak about biomedical technologies and gender, they are referring to a wide range of bodily interventions that are a subset of the range of biomedical technologies we discussed earlier. Gendered technologies include hormone manipulation (estrogen and testosterone for both men and women, birth control pills, hormone blockers, synthetic thyroid medications, steroids, etc.), non-surgical body modification (tattoos, hair dye, weight lifting, dieting, piercing, dress, etc.), and surgical body modification (plastic surgery, weight-loss surgery, sex-reassignment surgery, breast augmentation, etc.). These technologies can be used, as I explore below, in both liberating and regressive ways. In all of the cases that follow, many individuals benefit from biomedical technologies like plastic and bariatric surgery. My intent here is not to argue these technologies are good or bad, but to bring complexity to their analysis.

While both men and women are using gendered technologies to shape their bodies in a variety of ways, these changes are neither evenly distributed among men and women, nor gender neutral in their consequences. By way of illustrating this uneven distribution, consider the example of gender distribution among plastic surgery recipients. According to 2008 data from the American Society of Plastic Surgeons, almost 11 million cosmetic procedures in the United States were performed on women, compared to 1.1 million procedures on men. This amounts to women comprising a staggering 91 percent of all plastic surgery cases. While the rates of invasive cosmetic procedures like liposuction have held relatively stable over the last few years, the rise in minimally invasive procedures such as Botox injection has been astronomical. This increase marks not only a remarkable increase in the overall number of cosmetic procedures, but also a significant statistical increase of women as recipients in proportion to men. In 2000, women comprised 86 percent of all procedures, but between 2000 and 2008 there was a 72 percent increase in procedures for women whereas there was only a 9 percent increase in rates for men.

Examining this demographic data alongside ethnographic accounts of plastic surgery use, it is evident that plastic surgery is being used to construct explicitly gendered bodies and identities. These are products of social scripts, gender paradigms, and available technologies, and are often hyper-normative. For example, the most common surgical cosmetic procedures for women are breast augmentation and liposuction, both of which are invasive methods to produce hyper-normative femininity: thinness, and large breasted-ness. This gendered aspect is not lost on patients; in her interviews with women patients, Debra Gimlin found that plastic surgery was a deeply gendered endeavor deployed by women to "make do" within a sexist and beauty-obsessed culture. In the personal narratives Gimlin collected, she found that the body work women engaged in was a conscious part of negotiating a gendered identity within the constraints of gender, class, and race norms.

The ability to produce socially valued bodies, bodies that possess the ideal skin color, facial features, and so forth, rests not only in the production of normative gender, but also requires race- and class-based privileges. Indeed, women of color in North America face unattainable expectations because social scripts include very racialized ideal beauty norms. As societies, North America prizes White features, and this list of prized features is limited to characteristics natural only in some White phenotypes. Similarly, body size is intertwined with social class; a well toned body is often a mark of wealth since cheap food is more fattening and promotes poor health, and the time and means to exercise is often a class-based privilege. When women use plastic surgery they are constructing a racialized, gendered, and classed body and they often do so in line with a narrow ideal characterized by features such as blond flowing hair, a thin nose, almond-shaped eyes, large breasts, a small waist, and broad hips. And as Balsamo pointed out, just as gender inequality affects somatechnics, racism affects the technologies that are developed and used.

Women of color are increasingly turning to cosmetic surgery; in 2008 White men and women made up 73 percent of patients, which was a significant decrease from 2000 when 86 percent of patients were White. In fact, while cosmetic procedures decreased 2 percent for White people in 2008, they increased 11 percent for men and women of color. Looking at trends over the last eight years, in the United States between 2000 and 2008 there was a 161 percent increase in cosmetic procedures among African-Americans, 227 percent among Hispanics, and 281 percent for Asian Americans compared to an increase of 63 percent among White individuals. Moreover, the most common cosmetic surgery procedures for people of color are nose reshaping, eyelid surgery, and breast augmentation, which are all procedures that alter racialized facial and body features to better match White norms (American Society of Plastic Surgeons 2009).

The racial disparities in the statistics among cosmetic procedures suggest a trend by women of color toward using these technologies to mediate radicalized gender beauty norms. In this process, these women reaffirm the hegemony of White body and beauty paradigms. Eugenia Kaw's 1991 study of plastic surgery and race in San Francisco is a strong example of these processes. Kaw interviewed Asian American women, asking questions about why they used plastic surgery and what it meant to them. In these interviews women described plastic surgery as a way to better meet societal beauty scripts. In her interviews it was also clear that, like Gimlin found in her study of mostly White women, these Asian American women were conscious about what they were doing and how it mattered. For example, 'Jane' commented,

> Especially if you go into business, whatever, you kind of have to have a Western facial type and you have to have like their features and stature—you know, be tall and stuff. In a way you can see it is an investment in your future. (Kaw 1993:78)

While the women Kaw spoke with were all vocal about their pride at being Asian, they also understood, as 'Jane' summarized, that White features were viewed more positively in society. The plastic surgeons that Kaw interviewed expressed very similar views, while also revealing how racialized gender scripts not only shape individuals, but whether and how technologies may be used. For instance, Kaw notes that doctors couched their racialized cosmetic procedures as efforts to help women achieve a look that is 'naturally' more beautiful, implying that White features are objectively more attractive. For example, one doctor stated that, "90 percent of people look better with double eyelids. It makes the eye look more spiritually alive" (Kaw 1993:81). Through these and other compelling examples Kaw builds a substantial analysis of how plastic surgery is being used to produce particular raced and gendered bodies concurrently.

Based on these interviews, Kaw suggests that social and ideological changes have coincided with the increased acceptance of plastic surgery in recent years to encourage surgical body work among Asian women and that this body work, in turn, constrains available scripts for femininity by erasing racialized differences among women's bodies. What Kaw concluded was that plastic surgery is, "a means by which the women can attempt to permanently acquire not only a feminine look considered more attractive by society, but also a certain set of racial features considered more prestigious." In other words, experiences of body work were gendered and racialized in such a way that while plastic surgery was simultaneously liberating on the individual level, it was detrimental on the societal level as social scripts for normatively gendered bodies became even more ethnocentric.

Hegemonic Race and Gender Norms Are Reproduced Through Body Work

Across the board women's bodies are more subject to body work than men's are. Many cultural critics have argued that new media technologies are creating unrealistic ideals for bodies and that these unattainable body scripts affect women disproportionately.

Although photographic images are still commonly viewed as factual evidence, recent technological advancements in print and film now allow imperceptible alterations to these images. Because of the ability to alter media images to create features like smaller pores, bigger eyes, thinner legs, larger breasts, and more defined muscles, published and broadcast representations of idealized beauty are themselves fictions. Recent resistance to this manipulation on the part of some actresses has made public how even thin and normatively beautiful actresses are subject to body-editing. For example, Keira Knightley, whose breasts were digitally enhanced in publicity for the 2004 movie *King Arthur,* refused similar manipulation for the 2008 movie *The Duchess,* and the ensuing tension between the actress and the movie studio was played out in the media. Kate Winslet publicly critiqued the manipulated images of her legs in *GQ* magazine in 2003, an edit she was not consulted about. These and similar examples point to how no bodies—not even famous ones prized for their sex appeal—meet the ideal without somatechnic manipulation.

Although beauty scripts place a greater burden on women to meet bodily expectations, men are also subject to gendered scripts that suggest the need to technologically enhance their masculinity. Recent revelations about the seemingly omnipresent use of steroids by male athletes are signs of scripts that declare that men's bodies are inadequate in their unenhanced state. The use of steroids in U.S. Major League Baseball has become so expected that revelations of use do little to damage the careers of players like Alex "A-Rod" Rodriguez and Barry Bonds. The investigatory "Mitchell Report," submitted to the Commissioner of Major League Baseball, quotes National League Most Valuable Player Ken Caminiti as stating in 1992 that in his estimate, "at least half" of Major League players were using anabolic steroids (Mitchell 2007:60–61). This widespread use of steroids and the subsequent bodily changes in baseball players have shifted body scripts for athletes so much that unenhanced bodies stand little chance of competing.

Similarly, the increasing attention paid to men's bodies and the rising rates of eating disorders among boys suggest that boys and men are increasingly subject to gendered body pressures. Television shows like "Queer Eye for the Straight Guy" and men's magazines such as *GQ* all capitalize on the rise of the 'metrosexual,' a masculinity rooted in high levels of body work. This body work encompasses not only pursuits of traditional male attributes by means such as working out and sculpting efforts, but also includes practices formerly confined to the pursuit of feminine ideals, such as shaving, waxing, dyeing, plucking, and renewed attention to clothing.

Recent scholarship by Jennifer Wesely offers a rich example of how individuals are intentionally using biomedical technologies to construct hegemonically gendered and raced bodies. Wesely interviewed 20 women in the southwest of the United States to examine how women working in a strip club used body technologies to construct profitable bodies, and to negotiate multiple identities: for example to demarcate their true self as separate from their stripper self. What she found was that the women engaged in a wide variety of often dangerous and painful technologies like drug use, plastic surgery, waxing, and diuretics in order to produce the idealized femininity they felt was expected of them. Moreover, this gendered body work became a central focus of their lives. Wesely found that, "As dancers, these women relied on their bodies in ways that necessitated their constant critique, attention, and maintenance, leading to more body technologies." The pervasive use of these body technologies erased differences in bodies through implants, hair dye, tanning, and dieting, and reinforced hegemonic beauty scripts such that the ideal to which the women held themselves accountable was one which is now biomedically constructed. Samantha Kwan and Mary Nell Trautner summarize this process as it functions in society at large and conclude that, "Women's effortless authentic beauty is thus far from it. Beauty work is in large part this process of transforming the natural body to fit the cultural ideal, altogether while concealing the process and making it seem natural." In the case of Wesely's study, the intentionally constructed nature of gendered bodies was rendered invisible and assumed to be natural because body work was ubiquitous at the strip club, and produced bodies that aligned with idealized femininities.

One particularly insightful part of Wesely's research is her investigation of how these bodily changes function in conversation with the multiple layers of identity that the dancers (and everyone else) construct and employ through body technologies. Wesely found that the dancers' bodies and identities were in dynamic relationship to one another. What is key here is the complexity by which this happens. First of all, these women are not dupes; they are intentionally crafting their bodies because it makes dancing more profitable. By the same token, however, these choices, which make sense within the world of strip clubs, set these women apart from mainstream society. The choices the strippers make about body work are shaped and constrained by their context. Further, their choices have meaning and import beyond the personal level; the more the women shape their bodies to match an unrealistic feminine ideal, the more masked the constructed nature of femininity becomes, and the more normative, or, rather, hyper-normative the feminine body and identity scripts supported at the clubs become. The technologically enhanced bodies that the women who work at the strip club construct, shaped in line with the particular norms within that narrow context, are more feminine, more sexual, and more gendered than our broader society's normative scripts demand.

Through her ethnographic research, Wesely is able to document how the women experienced identity changes as the product of these technological interventions. The more technologies the women used to produce ideal bodies, the more wedded they became to their 'stripper' identities. Even though the women often wanted to separate their 'true identity' from their 'dancer identity,' body technologies such as breast enhancement, genital piercing, and hair dyeing would not allow them to leave the dancer-life behind. As one dancer commented, "In real life, when we're dressing in clothes . . . if you've got huge tits you look awful during the day. They look good only in a G-string in a strip club." In other words, some body technologies used by the women met beauty scripts only in the strip club, but the women had to 'wear' them all the time, which limited their ability to cast off a 'stripper identity' at the end of the day. Simultaneously, Wesely found that the women engaged in other technological interventions in an effort to cordon off their 'true' identities from their 'stripper' identities (for example through different clothing, by shaving, and through drug use).

Along with altering their bodies, then, the women tried to walk the line between producing a marketable body and maintaining a body that was a meaningful reflection of their internal sense of self. The women made choices about their bodies, but did so within a context that limited their options and as a result often were unable to embody their 'inner selves.' As Wesely concludes:

> Although body technologies have the potential to destabilize or challenge constructions of gendered bodies and related identity, this is even more difficult in a context that capitalizes on very limited constructions of the fantasy feminine body. Indeed, the women in the study felt tremendous pressure to conform to body constructions that revolve around extreme thinness, large breasts, and other features that conform to a "Barbie doll" image. (Wesely 2003:655)

The consequences of these choices, as Wesely suggests, are significant. A number of scholars have documented how women who embody hegemonic femininity earn more money for stripping, and the women Wesely talked with acknowledged that normative gender scripts alongside financial, peer, and managerial pressure, directly informed the changes they made in their bodies.

On the personal level, this body work affects the identities of the women. They engage in body work that is encouraged within the context of their occupation, and which is aimed at producing femininities in line with the dominant gender paradigms of the strip club. In due course, this body work, in tandem with each individual's personal biography, shapes their identity. On an institutional level, the outcome of the biomedical construction of hyper-normative femininities by the women was an erasure of difference. By producing a very narrow set of femininities in line with hegemonic paradigms and gendered body scripts, the women naturalized a feminine body that was virtually unattainable without the use of body technologies, and in this process they erased the very real differences that had existed between each of their bodies. Predictably, the somatechnical changes the women manifested were not only gendered, but also raced; the women of color at the clubs Wesely studied spoke about how they had to look *more* sexy, and produce a *more* ideal femininity than White women to be seen as acceptable by both management and customers. These findings are in line with what Eugenia Kaw found in her study of Asian American women. A consequence of this body work, then, was the reproduction of racist beauty norms, and the re-entrenchment of phenotypically White bodies as the only ideal body type.

The Complexities of Body Work

It is important to remember, however, that while each technology may have the possibility of reifying gender scripts, it can also open up potential for new gendered bodies. Females can lift weights, play sports, and cut their hair; males can don makeup, wear high heels, and dance ballet. Multiple mundane technologies can be, and are, deployed to create new masculinities and femininities. Technologies can and do have multiple, contradictory personal and social implications. For instance, hair removal and surgical technologies are used by members of the transgender community in order to manipulate public perception of their bodies so that this perception matches their gender identities. Plastic surgery is neither good nor bad; it is a technology engaged by individuals in complex ways within particular social contexts.

Alongside these circumstantial changes, new technologies are allowing people to intervene into the shape, function, and appearance of their bodies in transformative ways. The ability to manifest, in an embodied fashion, chosen identities and/or appearance norms is significant, and these technologies are working hand-in-hand with existing body and gender paradigms and scripts to refashion people's lives. Returning to the stories of Michael Dillon and Rey demonstrates how these dynamics bear on the lives of individuals; these two men came of age in two very different historical moments, and the gender paradigms, scripts, and technologies of their day and the social contexts within which they were situated crafted radically different paths for each of them.

Michael Dillon came of age in the early 1930s, in England. At the same moment that Michael Dillon was struggling to make sense of his own gender non-conformity Radcliffe Hall was embroiled in an obscenity trial that catapulted language and knowledge of lesbianism and gender non-conformity into the public sphere. In Radcliffe Hall's *Well of Loneliness*, the gender and sexuality of the main character, Steven, are conflated such that Steven was understood as lesbian because of his gender non-conformity. This became one of the only places Dillon saw himself reflected and it was through this public debate that he learned about gender non-conformity. But, just as Radcliffe Hall's *Well of Loneliness* was about gender non-conformity that was culturally understood as homosexuality, Michael Dillon was told to make sense of his own gender non-conformity as homosexuality by the few people in whom he confided.

Michael Dillon spent years trying to situate himself within society and ultimately sought medical intervention so he could manifest socially his internal gender identity. His quest for help, however, was thwarted, in part because there was no gender paradigm within which transgenderism could fit. When Dillon's search for medical help failed, he became a doctor in his own right in order to support his own and others' bodily changes. He began taking testosterone in 1939 and by 1944 had legally changed his gender after both hormonal and surgical 'sex-reassignment' efforts. Just eight years after Lib Elbe's publicized sex-reassignment surgery (she is credited with being the first male-to-female person to medically change her sex) and 11 years before Christine Jorgensen's public coming out after her surgery, Dillon became the first female-to-male (FTM) person on record to change his sex. He was finally able to bring his gender identity as a man into more alignment with his public role and body. Over the next 20 years, Dillon wrote about what would eventually be termed transsexuality (see, for example, his book *Self: A Study in Endocrinology and Ethics*), and struggled to make a life for himself. Dillon intentionally cultivated a hetero-normative life, and in fact took on a misogynist persona as part of constructing his masculinity. After being publicly outed as transsexual in 1958, Dillon retreated to a life of monasticism in Tibet, and died in 1962 at age 47.

Rey's story is not yet fully written—he is, after all, only 18—but already there is much more to tell about his path toward social masculinity than there was for Dillon. At age 18, after coming out to his family and starting college, Rey pursued hormone therapy and began to live his life in his chosen gender. Within a few months he began taking testosterone to produce masculine secondary sex characteristics like facial and body hair and a deeper voice. He also had 'top surgery' which included a double mastectomy alongside the construction of a male-appearing chest. Compared to Michael Dillon's long wait and multiple surgeries (surgeries which were often failures—Dillon endured more than 13), Rey was able to engage in body-altering procedures with relative ease. Rey is part of a growing population of young transgender and transsexual individuals who have both the ability and social support to reshape their bodies and identities.

A number of things are significant about Rey's experience and the magazine article that profiled it. First, Rey's ability to manifest his chosen gender, in a bodily fashion, is remarkable. Compared to Michael Dillon's multi-year struggle to physically change his sex, Rey's ability to do so as soon as he turned 18 (the point at which he no longer needed parental consent) marks a significant shift in accessibility, education, and legitimacy. Second, the respect and acumen with which Alissa Quart constructed her story on Rey and other young transgender individuals is heartening. In the span of 50 years, social scripts have expanded significantly such that they reflect a familiarity with the language and complexity of gender non-conformity; for example the *New York Times Magazine* used terms like transgender, transmale, and genderqueer that were unfamiliar or non-existent during Dillon's lifetime.

These changes suggest that the possible ways of being sexed and gendered in the world have expanded. While I am not claiming that transgenderism has been incorporated as normative into North American cultures, I am suggesting that progress has been made. New technologies have been developed that range from the simple expansion of language to cutting-edge surgeries that allow and facilitate precise bodily changes. Dominant gender paradigms have shifted to include transgenderism as a possibility hand-in-hand with these technologies. Alternative gender scripts have proliferated making it possible for individuals—including young people like Rey—to access information about transgenderism more readily and to construct more diverse gender identities and sexed bodies than ever before. Indeed, the life-stories of Michael Dillon and Rey reveal significant change in gender scripts over the past 50 years. And, Dillon's and Rey's experiences reveal how these changes in gender paradigms, scripts, technologies, and embodied selves matter in the everyday lives of individuals.

Like the stories of Dillon and Rey, all of the case studies I have discussed in this [reading] have demonstrated significant relationships between social scripts, individual bodies and identities, and social paradigms. As embodied gender continues to change, I suspect that it will fuel ongoing transformation of social scripts and paradigms. I would expect, for example, a shift in gender norms alongside more diversity of bodies. But, technology is neither Utopian nor regressive. Technologies are being used to transform bodies in both non-normative ways and in ways that reinforce expectations about gendered bodies. Further, new bodies and identities can both support and inhibit social change, provoke normative identity re-entrenchment and spark an expansion in social scripts, regardless of the desire or intention of individuals. Personal meaning making around one's body or identity does not exist in a vacuum.

We are clearly living in a moment where gender paradigms, scripts, bodies, and identities are all being simultaneously refined and renegotiated. New technologies are being deployed to re-entrench hegemonic masculinities and femininities and erase race and gender differences in bodies. Hormonal birth control places the burdens of sexual decisions on women and genital surgeries such as 'hymenorrhaphy' (hymen reconstruction) reinforce the importance of virginity in women. Conversely, these same biomedical technologies such as testosterone and estrogen regimens and genital construction methods are allowing individuals to shape their bodies in new ways that create more diverse pairings of sex and gender, and these new embodied genders are significant.

We must recognize the social gender paradigms and scripts tied up with biomedical innovation and attune ourselves to whether, and how, these new technologies are disciplining, regulating, and transforming the gendered body in new ways. Are we on the brink of a new gender order? Somatechnic frontiers are certainly reshaping the body in previously unknown ways, and this process challenges gender norms and scripts to make space accordingly. The documented expansion of gender possibilities—for both transgender and cisgender individuals—certainly suggests that gender ideologies and scripts are being reworked. But just as information technologies are not moving North American societies unidirectionally toward expanded identity possibilities, biomedical technologies are used in some ways that encourage expansion of gender possibilities while in others they help to resist this process. If, however, we take as true the dynamic and reciprocal relationships between technology, ideology, scripts, bodies and identities, then gender is now and will continue to transform itself alongside technological innovation.

REFERENCES

American Society of Plastic Surgeons. 2009. "Cosmetic Procedures Up in All Ethnic Groups Except Caucasians in 2008." Arlington Heights, IL: Society of Plastic Surgeons. Retrieved May 26, 2009 (http://www.plasticsurgury.org/Media/Press_Realease/Cosmetic_Procedures_Up_in_All_Ethnic_Groups_Excpet_Caucasians_in_2008.html).

Balsamo, Anne. 1996. *Technologies of the Gendered Body: Reading Cyborg Women*. Durham, NC: Duke University Press.

Kaw, Eugenia. 1993. "Medicalization of Racial Features: Asian American Women and Cosmetic Surgery." *Medical Anthropology Quarterly* 7(1):74–89.

McKinley, Nita Mary. 1999. "Women and Objectified Body Consciousness: Mothers' and Daughters' Body Experience in Culture, Developmental, and Familial Context." *Developmental Psychology* 35:760–769.

Mitchell, George J. 2007." Report to the Commissioner of Baseball of an Independent Investigation into the Illegal Use of Steroids and Other Performance Enhancing Substances by Players in Major League Baseball." New York, NY: Office of the Commissioner of Baseball. Retrieved March 8, 2009 (http://files.mlb.com/mitchrpt.pdf).

Quart, Alissa. 2008. "When Girls Will Be Boys." *New York Times Magazine*, March 16, pp. 32–37.

Wesely, Jennifer. 2003. "Exotic Dancing and the Negotiation of Identity: The Multiple Use of Body Technologies." *Journal of Contemporary Ethnography* 32(6):643–669.

THINKING ABOUT THE READING

According to Shapiro, how have biomedical technologies affected gender paradigms, scripts, bodies, and identities? How has Shapiro's discussion of her case studies and other research impacted your understanding of gender? Shapiro notes that forms of body transformation have become very socially acceptable in society. Besides reinforcing and contesting gendered bodies, how have race and class privileges operated in body transformation procedures to produce "natural beauty?" What roles have the media and business played in the production of socially valued bodies and scripts?

Global Dynamics and Population Demographic Trends

13

In the past several chapters, we have examined the various interrelated sources of social stratification. Race, class, and gender continue to determine access to cultural, economic, and political opportunities. Another source of inequality that we don't think much about, but one that has enormous local, national, and global significance, is the changing size and shape of the human population and how people are distributed around the planet. Globally, population imbalances between richer and poorer societies underlie most if not all of the other important forces for change that are taking place today. Poor, developing countries are expanding rapidly, while the populations in wealthy, developed countries have either stabilized or, in some cases, declined. When the population of a country grows rapidly, the age structure is increasingly dominated by young people. In slow-growth countries with low birthrates and high life expectancy, the population is much older. Countries with different age structures face different challenges regarding the allocation of important resources.

One form of segregation that people may be less aware of is age segregation—the culture and institutional separation of people of different ages. Social demographers point out that cultural survival is dependent not only on older people sharing traditions and knowledge with younger people, but also on reverse knowledge sharing whereby young people help older people keep up with cultural changes. Current social processes of work/education separation, high rates of mobility, etc., have resulted in a pattern of extreme age segregation in developed countries. Young people rarely interact in a sustained way with older people unless they are related through family ties. Peter Uhlenberg and Jenny de Jong Gierveld ask how integrated we are across age differences. Using a study based on a Dutch survey, they explore this question by examining personal networks. How many people of varying ages are in your personal network? Although this study is based in the Netherlands, it has strong relevance to most Western nations.

Some other large-scale demographic phenomena affect people regardless of their age. Take, for instance, immigration. As social and demographic conditions in poor, developing countries grow worse, pressures to migrate increase. Countries on the receiving end of this migration often experience high levels of cultural, political, and economic fear. Immigration—both legal and illegal—has become one of the most contentious political issues in the United States today. While politicians debate proposed immigration restrictions, people from all corners of the globe continue to come to this country looking for a better life. An informed understanding of this phenomenon requires an awareness of the reasons for migration and the connection between the choices individuals make to immigrate and larger economic conditions that reflect global markets.

As Arlie Russell Hochschild points out in "Love and Gold," immigration can create serious problems in the families people leave behind. Many destitute mothers in places such as the Philippines, Mexico, and Sri Lanka leave their children for long periods of time to work abroad because they cannot make ends meet at home. Ironically, the jobs these women typically take when they leave their families—nannies, maids, service workers—involve caring for and nurturing other people's families. So while migrant women provide much-needed income for their own families and valuable "care work" for their employers, they leave an emotional vacuum in their home countries. Hochschild asks us to consider the toll this phenomenon is taking on the children of these absent mothers. Not surprisingly, most of the women feel a profound sense of guilt and remorse that is largely invisible to the families they work for.

While Hochschild points out the disruption in families that immigration can bring, Felicity Schaeffer-Grabiel's research on the cyberbride industry explores how some Mexican women use matchmaking services to create family in their search for U.S. husbands. Larger structural changes in Mexico have been associated with changes to traditional gender relations and to intimate relationships there and have led some middle-class Mexican women to seek out marriage partners abroad. Ironically, Schaeffer-Grabiel finds that the marriage goals of these Mexican women are in stark contrast with the goals of the U.S. men who are searching for brides through the cyber-bride industry. The convergence of technology with economic and social opportunities demonstrates how the architecture of married life can be understood in conflicting ways in the global marriage market.

Something to Consider as You Read

Global or demographic perspectives are big-picture perspectives. As you read these selections, practice thinking about the ways that demographic and global processes may shape individual experiences and choices. For example, consider your personal networks: Do they show signs of age segregation? How has immigration affected your everyday life? Do you know the story of how your family arrived in this country? How many generations have they been here? Is there a substantial immigrant population in your hometown? How has their presence been received by others? How do your personal experiences with immigrants compare to the largely negative images that are often presented in the media? Beyond immigration, think about the ways in which big economic and political changes affect the choices individuals make. Now, add wealth and technology to the equation and consider which countries are going to be in the best position to adjust to these global changes. Who is going to be most affected, possibly even exploited, in this global adjustment?

Age-Segregation in Later Life

An Examination of Personal Networks

Peter Uhlenberg and Jenny de Jong Gierveld

(2004)

Introduction

Margaret Mead (1970) argued that in societies where change is slow and imperceptible, knowledge and culture are passed on from older generations to younger ones. In these traditional settings, she suggested, it is essential for older people to teach newcomers how to function in the society. In contrast, in modern societies where social and technological change is pervasive, it also is necessary for younger people to teach the old. If older people do not interact with and learn from younger people, they risk becoming increasingly excluded from contemporary social developments as they age through later life. Older people may not need or want to know everything that younger ones know, but acquiring some new knowledge is essential to avoid becoming marginalised in later life. The most common example of what the young can currently teach the old is how to use e-mail and the Internet, but many other areas of new knowledge created by cultural change could be described. In either traditional or modern societies, therefore, age-integration is needed if all generations are to be productive participants in the society. Of course there are additional reasons why it would be mutually beneficial for older and younger people to interact with each other. Older people may have resources that could promote the well-being of younger people (and *vice versa*). The absence of interaction, or age-segregation, promotes ageism and insensitivity to the challenges faced by others who differ in age. In general, it seems likely that age-integration promotes a more civil society. In this paper we take the perspective of older people and explore the level of their integration with, or segregation from, younger adults.

One way to examine the level of age-segregation of older people from younger ones in contemporary society is to examine the age-composition of personal social networks. How diverse are the ages of those with whom individuals interact most frequently and most significantly? Age-integration at the level of personal networks is relevant because network members play an important role in integrating individuals (of any age) into the larger society. Through network members, information and ideas are shared, new ways of thinking and living are discussed, and advice is exchanged. Network members exchange social, emotional, material and informational support that promotes well-being. Through networks individuals are recruited into social movements and organisations, which provide further opportunities for developing personal bonds (Marsden 1988; McPherson, Smith-Lovin and Cook 2001). Thus it is likely that older people whose personal networks lack younger members may be excluded from full participation in the society in which they live.

Forces Promoting Age Homophily in Networks

The social forces that have produced the institutionalisation and age-related stages of the life course over the past two centuries are also likely to have led to widespread age-segregation in social networks (Kohli 1988). Consider, for example, the structured social contexts from which network members might be drawn. A structured pattern of age-segregation begins early in life, for educational institutions use single years of age to group most children throughout childhood, while nurseries and day-care anticipate the age-homogeneity of the school environment from soon after birth. Sports and music for children are often tied to school, and result in age-segregated activities after school and on weekends. Churches imitate schools by establishing Sunday schools, where children are taught in age-homogeneous groups. Laws forbid children to participate in work settings. Specialised doctors see children; specialised therapists counsel and work with children; and special courts deal with children. The separation into homogeneous age groupings is further promoted by television, movies and other forms of entertainment that target children of particular ages. Quite similar institutional forces now largely segregate adolescents and young adults to age-homogeneous networks and activities (Lofland 1968). In these ways a culture that emphasises age-homogeneous groups is established early in life, so that one expects to find a deficit of older people in the personal networks of children and young adults, and *vice versa*. In somewhat similar ways, the age-segregated social institutions encountered by older people encourage age-homogeneity in personal networks through later life.

Work organisations tend to exclude people past age 60 or 65 years from a significant life activity, excluding them from one mechanism that promotes integration and some cross-age interactions with younger adults. Old people continue to be excluded from mainline educational settings (Hamil-Luker and Uhlenberg 2002). When efforts are made to involve older people in educational activities, they often operate from an age-segregationist principle, with separate programmes for old people. Many older people report that participating in church or other religious activity is their most significant social activity outside the family. But in church people often are grouped on the basis of age for activities, so older churchgoers interact with other old people, and their social networks remain age-homogeneous. Participating in a senior centre or other age-restricted organisation may increase social activity and help expand social networks, but also reinforce age-segregated interactions. Similarly, nursing homes, retirement homes and retirement communities promote extreme age-segregation towards the end of life. In many ways, therefore, older people encounter a society that restricts opportunities for developing age-integrated personal social networks.

Although age-composition has seldom been the focus of studies of personal social networks, several report interesting findings on age homophily (and homogeneity) in networks. A recent review of the literature on homophily in social networks concludes that age consistently creates strong divisions in personal networks (McPherson et al. 2001). In his studies of Detroit men and Northern California residents, Fischer (1977, 1982) reported striking age-homogeneity in non-kin friendship networks. Indeed, 72 percent of the close friends of the Detroit men were within eight years of their own ages. Similarly, Feld (1984), analysing the Northern California data, found that approximately half of all non-family associates with whom respondents were sociable or discussed problems were within five years of their age. In her analysis of friendship structure, Verbrugge (1977) reported that half of the friends identified by Detroit men occupied the same 10-year age category as the respondent, as did over 40 percent of the friends of respondents in a German survey. And, as noted above, the GSS [General Social Survey] study of discussion-partner networks found most non-kin partners to be similar in age (Burt 1991; Marsden 1988). In general, studies have found age-homogeneity in non-kin networks across respondents of all ages, although it is stronger among younger than older people.

As already suggested, however, much less age-homogeneity is observed in kin networks (Burt 1991). This is not surprising, because older people often identify the relationships with their adult children, who tend to be 20 to 40 years younger than themselves, as very important. The 1988 *National Survey of Families and Households* showed that two-thirds of older women in the United States who had children visited a child at least once a week, and over 80 percent had weekly contact with a child (Uhlenberg and Cooney 1990). Not only do intergenerational ties involve a high level of communication, but also these relationships are generally reported to be emotionally close and significant for instrumental support (for a review see Lye 1996). Furthermore, other kin (parents, aunts and uncles, siblings, cousins, grandchildren, and nieces and nephews) of diverse ages are frequently cited as significant network members. Thus one would expect the age-heterogeneity of personal networks to vary by the number of kin who are included in the network. The primary factor affecting the number of kin in a network is kinship composition. Other family-related events may affect how often older people include kin in their personal networks. In particular, partner status and partner history are relevant, e.g., adult children tend to intensify social interactions with a recently widowed parent who had been in a first marriage (Lopata 1996; Wolf, Freedman and Soldo 1997), and an earlier parental divorce reduces the likelihood that adult children interact frequently with their fathers in later life (Doherty, Kouneski and Erickson 1998; Dykstra 1998; Furstenberg, Hoffman and Shrestha 1995; Jong Gierveld and Dykstra 2002; Lye et al. 1995).

One would expect, of course, that the probability of a network including younger non-kin would increase with the total number of non-kin in the network. More interesting, it is likely that older people have more opportunities to recruit network members of diverse ages when they are active in social contexts that include younger adults. Therefore we anticipate that employed people are more likely than the retired to identify younger non-kin as network members. Similarly, attending church regularly or engaging in volunteer activities might promote greater age-integration, if these occur in age-heterogeneous contexts. The age-composition of the neighbourhood could also be a factor influencing the likelihood of interacting with younger adults. In addition to these structured settings for recruiting non-kin network members, current and past family context may also be relevant. Marital and partner status might be related to the size and intensity of non-kin network relationships. Older adults who are embedded in a large kinship circle, including a partner, children, children-in-law, grandchildren and siblings, need to invest a lot of time in maintaining these social and supportive relationships. In general, therefore, they have less time and energy than others to invest in a varied set of non-kin contacts (Dykstra 1995). Some widowed older adults who live without a partner may intensify contacts with their children, but others may revive latent bonds with others. The latter are to an extent building a new social network of people outside their own household that includes non-kin relationships. Indeed, success has been reported for a special training programme to support widowed older adults to begin new relationships (Stevens 2001). It is not yet known how age-heterogeneous the new relationships formed by widowed persons are.

Adults who divorce and remain without a partner may also compensate for the reduction in the size of the social networks. Personal contacts with new friends, with people "in the same boat," may be established in order to rebuild a social network. Those who never formed a partner union and the childless are however in a different position and do not experience the same transition. They often rely on siblings, friends, neighbours and other kin and acquaintances (such as colleagues and co-members of sport and hobby clubs) to maintain social participation and integration (Dykstra 1995). The never-married especially have been found to have a varied network of long-standing non-kin relationships (Wagner, Schütze and Lang 1999).

This interpretation of the literature on networks, kinship and ageing leads to several hypotheses. First, we expect that young adults are under-represented in the personal networks of older people. Second, that the presence of young adults in the personal networks of older people becomes increasingly rare at the more advanced ages. Third, it is expected that a disproportionate number of the younger network members of older people will be kin rather than non-kin. Fourth, the number of living children should be positively associated with having younger kin network members, but not with having younger non-kin network members. Fifth, the likelihood of having younger non-kin network members is higher for those who are employed, attend church, do volunteer work or live in age-integrated neighbourhoods. Sixth, the likelihood of having younger non-kin network members is higher for currently widowed and divorced older adults, who may have renewed and broadened their personal networks, than for those who are currently married, who tend to maintain their past couple-oriented social contacts. Seventh, the larger the number of friends, neighbours and other non-kin in an older person's network, the more likely that there will be young non-kin in the network.

As this study is exploratory, we also include in the analysis two variables of interest but without hypotheses of their effect, namely sex and the educational level of the respondent. One might expect older women from these Dutch cohorts to have less non-family social interaction than men, and hence to have less age diversity in their non-kin networks, but it is also possible that women possess superior social skills that allow them to bridge age barriers more easily than men. Higher levels of educational attainment are associated with higher levels of geographical mobility, so may reduce the breadth of network members that develop over time in a small community. But more education could also be associated with less ageism and greater acceptance of cross-age relationships.

Discussion

Despite the potentially significant implications, previous research has not examined the extent to which people in later life regularly interact with young adults. Using data from The Netherlands, this study has provided evidence on the extent to which older people have age-integrated or age-segregated personal social networks. Further, it has explored the factors associated with diversity in the age-composition of the networks of older people. Several interesting and provocative findings have emerged, and it is hoped that they will stimulate further research.

First, there clearly is a deficit of young adults in the networks of older people. People aged 55–64 years have significantly fewer young adult network members than would be expected if age were not a factor in selection, and the deficit grows even larger for people over the age of 65 years. For example, those aged 75–89 years had only about one-fifth of the number of network members aged less than 35 years that would be expected with complete age-integration. In fact, 68 percent of the population older than 75 years did not identify any network member younger than 35 years of age.

Second, an overwhelming proportion of the younger network members identified by older people were kin. About 90 per cent of the network members aged less than 45 years old who were reported by people past age 65 years were kin, and a large majority of older people reported no non-kin less than 45 years of age in their networks. Most neighbours, friends and other non-kin associates of older people were old themselves. Thus the most crucial determinant of having younger network members is the size of the kin group, and especially the number of living children. Family building in the young adult phase of the life course turns out to be the major determinant of age-integrated or age-segregated personal networks in late life.

Third, although no segment of the older population appeared to be well integrated with younger adults outside of family relationships, several factors did increase the likelihood that an older person had some significant cross-age interactions. These included participation in organisations that had members of different ages (e.g., work and volunteer settings), and living in a neighbourhood with a high proportion of non-old adults. A plausible explanation for the significance of these factors is that a necessary condition for forming cross-age associations is the opportunity for meeting people of different ages. The failure of church activity to foster more age-heterogeneous relationships may be because church attendance in The Netherlands is much higher among older than younger age groups. In other words, churches may not be strongly age-integrated settings. It also may be that simply occupying common space is insufficient to promote the development of cross-age relationships. Relationships develop when structures promote mutual interaction around a meaningful activity, so while sitting side-by-side in a church service may have no effect, working together on a common project may be highly effective. Further, cultural norms are almost certainly important. When age differences are emphasised and age-stereotypes are prevalent, a significant barrier exists for forming friendships and close associations between young and old people.

Fourth, specific life course events, in particular divorce followed by living alone, increased the likelihood that an older person had some significant cross-age interactions with non-kin. Several studies have shown that shortly after divorce there tends to be a reduction in the number of personal relationships (DeGarmo and Kitson 1996). As time passes after a divorce, however, new relationships are formed. In this process of forming replacement relationships, there is an opportunity for younger non-kin to join the network.

Looking ahead, we anticipate two changes that could significantly increase the age-segregation of the personal networks of older people in The Netherlands. First is the ageing of the population, which will decrease the relative supply of younger adults as potential network members and increase the relative supply of older ones. Around the time of the NESTOR [Netherlands Program for Research on Aging] survey, about 34 percent of the population aged over 20 years was in the age group 20–35 years, while 17 per cent was aged 65 or more years. By 2050, these two percentages will be reversed—21 per cent of the adult population will be aged 20–35 years, and 33 per cent will be 65 or more years. The second and related change in future cohorts will be a significant decline in the average number of adult children. Because children are the major source of young adult network members, a decline in the number of children could have a large effect. Those aged 65 or more years in 1992 lived out their reproductive years when the Total Fertility Rate exceeded 3.0, but the cohorts entering old age in the near future will have completed family sizes of only about half that level. Further, the increasing prevalence of divorce in future cohorts entering old age may lead to a weakening of the tie between parent and adult child for an increasing proportion of older people (Cooney and Uhlenberg 1990; Dykstra 1998; Jong Gierveld and Peeters 2003). The increase in the number of younger non-kin that is associated with divorce is far smaller than the loss of children from the network. Thus, unless other changes occur, older people in the future are likely to have even less interaction with young adults than they currently do—and as shown above, current levels of interaction are extremely low.

This prospect provokes the question of what changes might divert a trend towards even greater age-segregation of older people. If, as argued in this paper, non-kin network members tend to be recruited from structured social contexts such as workplaces, volunteer settings, educational organisations and neighbourhoods, more attention might be given to increasing the involvement of older people in social structures that include people of various ages. This line of thinking leads directly to the issue of institutional age-segregation, as occurs when chronological age is used as a criterion for participation. Matilda Riley called attention to the structural lags in major social institutions which denied opportunities to healthy and skilled people reaching old age to engage productively in society (Riley, Kahn and Foner 1994). The institutions which are most clearly structured by age are schools and places of work, but the rules and practices of many others create age-group separation. Age is embedded in the formulation and implementation of many social welfare policies and programmes, e.g., nutrition, housing, protective services and recreation. Concerns related to the old often fall under different government programmes and offices than do matters related to children and youth (Hagestad 2002). Even academic disciplines (such as gerontology) tend to sustain separation by age. There is, however, some evidence that the use of chronological age to structure the life course may have peaked.

A recent tendency to break down structural age barriers has been noted in both work and education (Riley and Riley 2000). Retirement in the United States has recently become more flexible, allowing an increasing number of older people to participate in the labour force. The long trend towards earlier age at retirement stopped in the mid 1980s in the United States, and since then labour force participation rates among those aged 55 or more years have been gradually increasing (Clark and Quinn 2002). The long-discussed idea of lifelong learning may now be happening, as an increasing number of people in mid and later life learn alongside younger people (Davey 2002). There are interesting examples in the United States of breaking down the age barriers around schools and creating community learning-centres open to all ages (US Department of Education 2000). In academic programmes, traditional gerontological approaches are being challenged by a life course perspective that views ageing as a lifelong process. If, as suggested by these examples, institutional age-segregation is declining, opportunities for cross-age interaction should increase.

Related to institutional age-segregation is cultural age-segregation, as reflected in age stereotypes and ageist language. In addition to removing the barriers to cross-age interaction, a reduction in ageism and cultural age-stereotyping could facilitate age-integration. The prevalence of age-stereotypes in society hinders the formation of close non-kin relationships between older and younger people (Bytheway 1995; Hummert et al. 1994; Nelson 2002). There is of course some circularity in this association, because age-segregation is a root cause of age-stereotypes. Nevertheless, educational programmes and media efforts to combat ageist stereotypes and language might play a role in increasing understanding and empathy between disparate age groups. Similar efforts to reduce racism and sexism are generally considered to have produced positive results.

Attention is being given not only to ways of reducing structural and cultural barriers between older and younger people, but also to inter-generational programmes that purposely bring diverse ages together. In The Netherlands, a co-ordinated effort to bring older people into age-integrated settings is occurring through an inter-generational neighbourhood development programme at *The Netherlands Institute for Care and Welfare* (Penninx 1999). A notable initiative from this inter-generational programme has involved the Dutch Guilds that exist in about 90 municipalities. People who are aged 50 or more years and who are willing to share their knowledge and skills can form a guild that anyone can contact for assistance free of charge. A request for help, e.g., with car repair, tutoring in school, business advice or care for a disabled child, is referred to an appropriate guild member who then responds directly to the individual needing assistance. Through this matching process, older volunteers and younger people are brought together in a context that is likely to promote positive inter-generational interaction. Other inter-generational programmes described by Penninx include: children visiting older people living in age-segregated institutional settings, older people helping children in local schools, adolescent chore-teams helping older neighbourhood residents with various household chores, and older people meeting with immigrant youth to promote their successful integration into Dutch society. Similar inter-generational programmes are developing in other countries. Careful evaluations of the various types of deliberate efforts to bridge age gaps would provide useful information on what structures actually facilitate age-integration.

REFERENCES

Burt, R. S. 1991. Measuring age as a structural concept. *Social Networks, 13,* 1–34.

Bytheway, B. 1995. *Ageism.* Open University Press, Buckingham.

Clark, R. L. and Quinn, J. F. 2002. Patterns of work and retirement for a new century. *Generations, 22,* 17–24.

Cooney, T. M. and Uhlenberg, P. 1990. The role of divorce in men's relations with their adult children after mid-life. *Journal of Marriage and the Family, 52,* 677–88.

Davey, J. A. 2002. Active aging and education in mid and later life. *Ageing & Society, 22,* 95–113.

DeGarmo, D. S. and Kitson, G. C. 1996. Identity relevance and disruption as predictors of psychological distress for widowed and divorced women. *Journal of Marriage and the Family, 58,* 983–97.

Doherty, W. J., Kouneski, E. F. and Erickson, M. F. 1998. Responsible fathering: an overview and conceptual framework. *Journal of Marriage and the Family, 60,* 277–92.

Dykstra, P. A. 1995. Network composition. In C. P. M. Knipscheer, J. de Jong Gierveld, T. G. van Tilburg and P. A. Dykstra (eds), *Living Arrangements and Social Networks of Older Adults.* VU University Press, Amsterdam, 97–114.

Dykstra, P. A. 1998. The effects of divorce on intergenerational exchanges in families. *The Netherlands Journal of Social Sciences, 33,* 77–93.

Feld, S. L. 1984. The structured use of personal associates. *Social Forces, 62,* 640–52.

Fischer, C. S. 1977. *Networks and Places: Social Relations in the Urban Setting.* Free Press, New York.

Fischer, C. S. 1982. *To Dwell Among Friends: Personal Networks in Town and City.* University of Chicago Press, Chicago.

Furstenberg, F. F. Jr., Hoffman, S. D. and Shrestha, L. 1995. The effect of divorce on intergenerational transfers: new evidence. *Demography, 32,* 319–33.

Hagestad, G. O. 2002. Personal communication.

Hamil-Luker, J. and Uhlenberg, P. 2002. Later life education in the 1990s: increasing involvement and continuing disparity. *Journal of Gerontology: Social Sciences, 57B,* S324–31.

Hummert, M. L., Garsta, T. A., Shaner, J. L. and Strahm, S. 1994. Stereotypes of the elderly held by young, middle-aged, and elderly adults. *Journal of Gerontology: Psychological Sciences, 49,* P240–9.

Jong Gierveld, J. de and Dykstra, P. A. 2002. The long-term rewards of parenting: older adults' marital history and the likelihood of receiving support from adult children. *Ageing International, 27,* 49–69.

Jong Gierveld, J. de and Peeters, A. 2003. The interweaving of repartnered older adults' lives with their children and siblings. *Ageing & Society, 22,* 1–19.

Kohli, M. L. 1988. Social organization and subjective construction of the life course. In A. B. Sorensen, F. E. Weiner and L. R. Sherrod (eds), *Human Development and the Life Cycle.* Erlbaum, Hillsdale, New Jersey, 271–92.

Lofland, J. 1968. The youth ghetto. *Journal of Higher Education, 39,* 121–43.

Lopata, H. Z. 1996. *Current Widowhood: Myths and Realities.* Sage, Thousand Oaks, California.

Lye, D. N. 1996. Adult child-parent relationships. *Annual Review of Sociology, 22,* 79–102.

Lye, D. N., Klepinger, D. H., Hyle, P. D. and Nelson, A. 1995. Childhood living arrangements and adult children's relations with their parents. *Demography, 32,* 261–80.

Marsden, P. V. 1988. Homogeneity in confiding relationships. *Social Networks, 10,* 57–76.

McPherson, M., Smith-Lovin, L. and Cook, J. M. 2001. Birds of a feather: homophily in social networks. *Annual Review of Sociology, 27,* 415–44.

Mead, M. 1970. *Culture and Commitment: A Study of the Generation Gap.* Natural History Press, Garden City, New York.

Nelson, T. D. (ed.) 2002. *Ageism, Stereotyping and Prejudice against Older Persons.* MIT Press, Cambridge, Massachusetts.

Penninx, K. 1999. *DeBuurt voor Alle Leeftijden [The Neighbourhood of All Ages].* NIZW Uitgeverij, Utrecht, The Netherlands.

Riley, M. W. and Riley, J. W. Jr. 2000. Age-integration: conceptual and historical background. *The Gerontologist, 40,* 266–70.

Riley, M. W., Kahn, R. L. and Foner, A. 1994. *Age and Structural Lag: Society's Failure to Provide Meaningful Opportunities in Work, Family, and Leisure.* Wiley, New York.

Stevens, N. 2001. Combating loneliness: a friendship enrichment programme for older women. *Ageing & Society, 21,* 183–202.

Uhlenberg, P. and Cooney, T. M. 1990. Family size and mother-child relations in later life. *The Gerontologist, 30,* 618–25.

US Department of Education 2000. *Schools as Centers of Community: A Citizen's Guide for Planning and Design.* US Department of Education, Washington, DC.

Verbrugge, L. M. 1977. The structure of adult friendship choices. *Social Forces, 56,* 576–97.

Wagner, M., Schütze, Y. and Lang, F. R. 1999. Social relationships in old age. In P. B. Baltes and K. U. Mayer (eds), *The Berlin Aging Study: Aging from 70 to 100.* Cambridge University Press, Cambridge, 282–301.

Wolf, D. A., Freedman, V. and Soldo, B. J. 1997. The division of family labor: care for elderly parents. *The Journals of Gerontology, 52B,* special issue, 102–9.

THINKING ABOUT THE READING

What is age segregation? According to the authors, what are some of the reasons for age segregation? What are some of the everyday consequences of age segregation? Draw a diagram of your personal networks (e.g., the people you see daily, people you spend holidays with, people you work with). What is the age range of the people in your networks? How many older people do you know who are not your relatives? This reading uses information from a study of Dutch people. How would the findings compare to other cultures? In which social settings would you expect to find the *least* age segregation?

Love and Gold

Arlie Russell Hochschild

(2002)

Whether they know it or not, Clinton and Princela Bautista, two children growing up in a small town in the Philippines apart from their two migrant parents, are the recipients of an international pledge. It says that a child "should grow up in a family environment, in an atmosphere of happiness, love, and understanding," and "not be separated from his or her parents against their will. . . ." Part of Article 9 of the United Nations Declaration on the Rights of the Child (1959), these words stand now as a fairy-tale ideal, the promise of a shield between children and the costs of globalization.

At the moment this shield is not protecting the Bautista family from those human costs. In the basement bedroom of her employer's home in Washington, D.C., Rowena Bautista keeps four pictures on her dresser: two of her own children, back in Camiling, a Philippine farming village, and two of her children she has cared for as a nanny in the United States. The pictures of her own children, Clinton and Princela, are from five years ago. As she recently told *Wall Street Journal* reporter Robert Frank, the recent photos "remind me how much I've missed." She has missed the last two Christmases, and on her last visit home, her son Clinton, now eight, refused to touch his mother. "Why," he asked, "did you come back?"

The daughter of a teacher and an engineer, Rowena Bautista worked three years toward an engineering degree before she quit and went abroad for work and adventure. A few years later, during her travels, she fell in love with a Ghanaian construction worker, had two children with him, and returned to the Philippines with them. Unable to find a job in the Philippines, the father of her children went to Korea in search of work and, over time, he faded from his children's lives.

Rowena again traveled north, joining the growing ranks of Third World mothers who work abroad for long periods of time because they cannot make ends meet at home. She left her children with her mother, hired a nanny to help out at home, and flew to Washington, D.C., where she took a job as a nanny for the same pay that a small-town doctor would make in the Philippines. Of the 792,000 legal household workers in the United States, 40 percent were born abroad, like Rowena. Of Filipino migrants, 70 percent, like Rowena, are women.

Rowena calls Noa, the American child she tends, "my baby." One of Noa's first words was "Ena," short for Rowena. And Noa has started babbling in Tagalog, the language Rowena spoke in the Philippines. Rowena lifts Noa from her crib mornings at 7:00 A.M., takes her to the library, pushes her on the swing at the playground, and curls up with her for naps. As Rowena explained to Frank, "I give Noa what I can't give to my children." In turn, the American child gives Rowena what she doesn't get at home. As Rowena puts it, "She makes me feel like a mother."

Rowena's own children live in a four-bedroom house with her parents and twelve other family members—eight of them children, some of whom also have mothers who work abroad. The central figure in the children's lives—the person they call "Mama"—is Grandma, Rowena's mother. But Grandma works surprisingly long hours as a teacher—from 7:00 A.M. to 9:00 P.M. As Rowena tells her story to Frank, she says little about her father, the children's grandfather (men are discouraged from participating actively in child rearing in the Philippines). And Rowena's father is not much involved with his grandchildren. So, she has hired Anna de la Cruz, who arrives daily at 8:00 A.M. to cook, clean, and care for the children. Meanwhile, Anna de la Cruz leaves her teenage son in the care of her eighty-year-old mother-in-law.

Rowena's life reflects an important and growing global trend: the importation of care and love from poor countries to rich ones. For some time now, promising and highly trained professionals have been moving from ill-equipped hospitals, impoverished schools, antiquated banks, and other beleaguered workplaces of the Third World to better opportunities and higher pay in the First World. As rich nations become richer and poor nations become poorer, this one-way flow of talent and training continuously widens the gap between the two. But in addition to this brain drain, there is now a parallel but more hidden and wrenching trend, as women who normally care for the young, the old, and the sick in their own poor countries move to care for the young, the old, and the sick in rich countries, whether as maids and nannies or as day-care and nursing-home aides. It's a care drain.

The movement of care workers from south to north is not altogether new. What is unprecedented, however, is the scope and speed of women's migration to these jobs. Many factors contribute to the growing feminization of migration. One is the growing split between the global rich and poor. . . .

[For example] domestic workers [who] migrated from the Philippines to the United States and Italy [in the 1990s] had averaged $176 a month, often as teachers, nurses, and administrative and clerical workers. But by doing less skilled—though no less difficult—work as nannies, maids, and care-service workers, they can earn $200 a month in Singapore, $410 a month in Hong Kong, $700 a month in Italy, or $1,400 a month in Los Angeles. To take one example, as a fifth-grade dropout in Colombo, Sri Lanka, a woman could earn $30 a month plus room and board as a housemaid, or she could earn $30 a month as a salesgirl in a shop, without food or lodging. But as a nanny in Athens she could earn $500 a month, plus room and board.

The remittances these women send home provide food and shelter for their families and often a nest egg with which to start a small business. Of the $750 Rowena Bautista earns each month in the United States, she mails $400 home for her children's food, clothes, and schooling, and $50 to Anna de la Cruz, who shares some of that with her mother-in-law and her children. As Rowena's story demonstrates, one way to respond to the gap between rich and poor countries is to close it privately—by moving to a better paying job. . . .

The International Organization for Migration estimates that 120 million people moved from one country to another, legally or illegally, in 1994. Of this group, about 2 percent of the world's population, 15 to 23 million are refugees and asylum seekers. Of the rest, some move to join family members who have previously migrated. But most move to find work.

As a number of studies show, most migration takes place through personal contact with networks of migrants composed of relatives and friends and relatives and friends of relatives and friends. One migrant inducts another. Whole networks and neighborhoods leave to work abroad, bringing back stories, money, know-how, and contacts. Just as men form networks along which information about jobs are passed, so one domestic worker in New York, Dubai, or Paris passes on information to female relatives or friends about how to arrange papers, travel, find a job, and settle. Today, half of all the world's migrants are women. . . .

The trends outlined above—global polarization, increasing contact, and the establishment of transcontinental female networks—have caused more women to migrate. They have also changed women's motives for migrating. Fewer women move for "family reunification" and more move in search of work. And when they find work, it is often within the growing "care sector," which, according to the economist Nancy Folbre, currently encompasses 20 percent of all American jobs.

A good number of the women who migrate to fill these positions seem to be single mothers. After all, about a fifth of the world's households are headed by women: 24 percent in the industrial world, 19 percent in Africa, 18 percent in Latin America and the Caribbean, and 13 percent in Asia and the Pacific. . . .

Many if not most women migrants have children. The average age of women migrants into the United States is twenty-nine, and most come from countries, such as the Philippines and Sri Lanka, where female identity centers on motherhood, and where the birth rate is high. Often migrants, especially the undocumented ones, cannot bring their children with them. Most mothers try to leave their children in the care of grandmothers, aunts, and fathers, in roughly that order. An orphanage is a last resort. A number of nannies working in rich countries hire nannies to care for their own children back home either as solo caretakers or as aides to the female relatives left in charge back home. Carmen Ronquillo, for example, migrated from the Philippines to Rome to work as a maid for an architect and single mother of two. She left behind her husband, two teenagers—and a maid.

Whatever arrangements these mothers make for their children, however, most feel the separation acutely, expressing guilt and remorse to the researchers who interview them. Says one migrant mother who left her two-month-old baby in the care of a relative. "The first two years I felt like I was going crazy. You have to believe me when I say that it was like I was having intense psychological problems. I would catch myself gazing at nothing, thinking about my child." Recounted another migrant nanny through tears, "When I saw my children again, I thought, 'Oh children do grow up even without their mother.' I left my youngest when she was only five years old. She was already nine when I saw her again, but she still wanted me to carry her."

Many more migrant female workers than migrant male workers stay in their adopted countries—in fact, most do. In staying, these mothers remain separated from their children, a choice freighted, for many, with a terrible sadness. Some migrant nannies, isolated in their employers' homes and faced with what is often depressing work, find solace in lavishing their affluent charges with the love and care they wish they could provide their own children. In an interview with Rhacel Parreñas, Vicky Diaz, a college-educated school teacher who left behind five children in the Philippines, said, "the only thing you can do is to give all your love to the child [in your care]. In my absence from my children, the most I could do with my situation was to give all my love to that child." Without intending it, she has taken part in a global heart transplant.

As much as these mothers suffer, their children suffer more. And there are a lot of them. An estimated 30 percent of Filipino children—some eight million—live in households where at least one parent has gone overseas. These children have counterparts in Africa, India, Sri Lanka, Latin America, and the former Soviet Union. How are these children doing? Not very well, according to a survey Manila's Scalabrini Migration Center conducted with more than seven hundred children in 1996. Compared to their classmates, the children of migrant workers more frequently fell ill; they were more likely to express anger, confusion, and apathy; and they performed particularly poorly in school. Other studies of this population show a rise in delinquency and child suicide. When such children were asked whether they would also migrate when they grew up, leaving their own children in the care of others, they all said no.

Faced with these facts, one senses some sort of injustice at work, linking the emotional deprivation of these children with the surfeit of affection their First World counterparts enjoy. In her study of native-born women of color who do domestic work, Sau-Ling Wong argues that the time and energy these workers devote to the children of their employers is diverted from their own children. But time and energy are not all that's involved; so, too, is love. In this sense, we can speak about love as an unfairly distributed resource—extracted from one place and enjoyed somewhere else.

Is love really a "resource" to which a child has a right? Certainly the United Nations Declaration on the Rights of the Child asserts all children's right to an "atmosphere of happiness, love, and understanding." Yet in some ways, this claim is hard to make. The more we love and are loved, the more deeply we can love. Love is not fixed in the same way that most material resources are fixed. Put another way, if love is a resource, it's a *renewable* resource; it creates more of itself. And yet Rowena Bautista can't be in two places at once. Her day has only so many hours. It may also be true that the more love she gives to Noa, the less she gives to her own three children back in the Philippines. Noa in the First World gets more love, and Clinton and Princela in the Third World get less. In this sense, love does appear scarce and limited, like a mineral extracted from the earth.

Perhaps, then, feelings *are* distributable resources, but they behave somewhat differently from either scarce or renewable material resources. According to Freud, we don't "withdraw" and "invest" feeling but rather *displace* or redirect it. The process is an unconscious one, whereby we don't actually give up a feeling of, say, love or hate, so much as we find a new object for it—in the case of sexual feeling, a more appropriate object than the original one, whom Freud presumed to be our opposite-sex parent. While Freud applied the idea of displacement mainly to relationships within the nuclear family, it seems only a small stretch to apply it to relationships like Rowena's to Noa. As Rowena told Frank, the *Wall Street Journal* reporter, "I give Noa what I can't give my children."

Understandably, First World parents welcome and even invite nannies to redirect their love in this manner. The way some employers describe it, a nanny's love of her employer's child is a natural product of her more loving Third World culture, with its warm family ties, strong community life, and long tradition of patient maternal love of children. In hiring a nanny, many such employers implicitly hope to import a poor country's "native culture," thereby replenishing their own rich country's depleted culture of care. They import the benefits of Third World "family values." Says the director of a coop nursery in the San Francisco Bay Area, "This may be odd to say, but the teacher's aides we hire from Mexico and Guatemala know how to love a child better than the middle-class white parents. They are more relaxed, patient, and joyful. They enjoy the kids more. These professional parents are pressured for time and anxious to develop their kids' talents. I tell the parents that they can really learn how to love from the Latinas and the Filipinas."

When asked why Anglo mothers should relate to children so differently than do Filipina teacher's aides, the nursery director speculated, "The Filipinas are brought up in a more relaxed, loving environment. They aren't as rich as we are, but they aren't so pressured for time, so materialistic, so anxious. They have a more loving, family-oriented culture." One mother, an American lawyer, expressed a similar view:

> Carmen just enjoys my son. She doesn't worry whether . . . he's learning his letters, or whether he'll get into a good preschool. She just enjoys him. And actually, with anxious busy parents like us, that's really what Thomas needs. I love my son more than anyone in this world. But at this stage Carmen is better for him.

Filipina nannies I have interviewed in California paint a very different picture of the love they share with their First World charges. Theirs is not an import of happy peasant mothering but a love that partly develops on American shores, informed by an American ideology of mother-child bonding and fostered by intense loneliness and longing for their own children. If love is a precious resource, it is not one simply extracted from the Third World and implanted in the First; rather, it owes its very existence to a peculiar cultural alchemy that occurs in the land to which it is imported.

For María Gutierrez, who cares for the eight-month-old baby of two hardworking professionals (a lawyer and a doctor, born in the Philippines but now living in San Jose, California), loneliness and long work hours feed a love for her employers' child. "I love Ana more than my own two children. Yes, more! It's strange, I know. But I have time to be with her. I'm paid. I am lonely here. I work ten hours a day, with one day off. I don't know any neighbors on the block. And so this child gives me what I need."

Not only that, but she is able to provide her employer's child with a different sort of attention and nurturance than she could deliver to her own children. "I'm more patient," she explains, "more relaxed. I put the child first. My kids, I treated them the way my mother treated me."

I asked her how her mother had treated her and she replied:

> My mother grew up in a farming family. It was a hard life. My mother wasn't warm to me. She didn't touch me or say "I love you." She didn't think she should do that. Before I was born she had lost four babies—two in miscarriage and two died as babies. I think she was afraid to love me as a baby because she thought I might die too. Then she put me to work as a "little mother" caring for my four younger brothers and sisters. I didn't have time to play.

Fortunately, an older woman who lived next door took an affectionate interest in María, often feeding her and even taking her in overnight when she was sick. María felt closer to this woman's relatives than she did to her biological aunts and cousins. She had been, in some measure, informally adopted—a practice she describes as common in the Philippine countryside and even in some towns during the 1960s and 1970s.

In a sense, María experienced a premodern childhood, marked by high infant mortality, child labor, and an absence of sentimentality, set within a culture of strong family commitment and community support. Reminiscent of fifteenth-century France, as Philippe Ariès describes it in *Centuries of Childhood,* this was a childhood before the romanticization of the child and before the modern middle-class ideology of intensive mothering. Sentiment wasn't the point; commitment was.

María's commitment to her own children, aged twelve and thirteen when she left to work abroad, bears the mark of that upbringing. Through all of their anger and tears, María sends remittances and calls, come hell or high water. The commitment is there. The sentiment, she has to work at. When she calls home now, María says, "I tell my daughter 'I love you.' At first it sounded fake. But after a while it became natural. And now she says it back. It's strange, but I think I learned that it was okay to say that from being in the United States."

María's story points to a paradox. On the one hand, the First World extracts love from the Third World. But what is being extracted is partly produced or "assembled" here: the leisure, the money, the ideology of the child, the intense loneliness and yearning for one's own children. In María's case, a premodern childhood in the Philippines, a postmodern ideology of mothering and childhood in the United States, and the loneliness of migration blend to produce the love she gives to her employers' child. That love is also a product of the nanny's freedom from the time pressure and school anxiety parents feel in a culture that lacks a social safety net—one where both parent and child have to "make it" at work because no state policy, community, or marital tie is reliable enough to sustain them. In that sense, the love María gives as a nanny does not suffer from the disabling effects of the American version of late capitalism.

If all this is true—if, in fact, the nanny's love is something at least partially produced by the conditions under which it is given—is María's love of a First World child really being extracted from her own Third World children? Yes, because her daily presence has been removed, and with it the daily expression of her love. It is, of course, the nanny herself who is doing the extracting. Still, if her children suffer the loss of her affection, she suffers with them. This, indeed, is globalization's pound of flesh.

Curiously, the suffering of migrant women and their children is rarely visible to the First World beneficiaries of nanny love. Noa's mother focuses on her daughter's relationship with Rowena. Ana's mother focuses on her daughter's relationship with María. Rowena loves Noa, María loves Ana. That's all there is to it. The nanny's love is a thing in itself. It is unique, private—fetishized. Marx talked about the fetishization of things, not feelings. When we make a fetish of an object—an SUV, for example—we see that object as independent of its context. We disregard, he would argue, the men who harvested the rubber latex, the assembly-line workers who bolted on the tires, and so on. Just as we mentally isolate our idea of an object from the human scene within which it was made, so, too, we unwittingly separate the love between nanny and child from the global capitalist order of love to which it very much belongs.

The notion of extracting resources from the Third World in order to enrich the First World is hardly new. It harks back to imperialism in its most literal form: the nineteenth-century extraction of gold, ivory, and rubber from the Third World. . . . Today, as love and care become the "new gold," the female part of the story has grown in prominence. In both cases, through the death or displacement of their parents, Third World children pay the price.

Imperialism in its classic form involved the north's plunder of physical resources from the south. Its main protagonists were virtually all men: explorers, kings, missionaries, soldiers, and the local men who were forced at gunpoint to harvest wild rubber latex and the like. . . .

Today's north does not extract love from the south by force: there are no colonial officers in tan helmets, no invading armies, no ships bearing arms sailing off to the colonies. Instead, we see a benign scene of Third World women pushing baby carriages, elder care workers patiently walking, arms linked, with elderly clients on streets or sitting beside them in First World parks.

Today, coercion operates differently. While the sex trade and some domestic service is brutally enforced, in the main the new emotional imperialism does not issue from the barrel of a gun. Women choose to migrate for domestic work. But they choose it because economic pressures all but coerce them to. That yawning gap between rich and poor countries is itself a form of coercion, pushing Third World mothers to seek work in the First for lack of options closer to home. But given the prevailing free market ideology, migration is viewed as a "personal choice." Its consequences are seen as "personal problems." . . .

Some children of migrant mothers in the Philippines, Sri Lanka, Mexico, and elsewhere may be well cared for by loving kin in their communities. We need more data if we are to find out how such children are really doing. But if we discover that they aren't doing very well, how are we to respond? I can think of three possible approaches. First, we might say that all women everywhere should stay home and take care of their own families. The problem with Rowena is not migration but neglect of her traditional role. A second approach might be to deny that a problem exists: the care drain is an inevitable outcome of globalization, which is itself good for the world. A supply of labor has met a demand—what's the problem? If the first approach condemns global migration, the second celebrates it. Neither acknowledges its human costs.

According to a third approach—the one I take—loving, paid child care with reasonable hours is a very good thing. And globalization brings with it new opportunities, such as a nanny's access to good pay. But it also introduces painful new emotional realities for Third World children. We need to embrace the needs of Third World societies, including their children. We need to develop a global sense of ethics to match emerging global economic realities. If we go out to buy a pair of Nike shoes, we want to know how low the wage and how long the hours were for the Third World worker who made them. Likewise, if Rowena is taking care of a two-year-old six thousand miles from her home, we should want to know what is happening to her own children.

If we take this third approach, what should we or others in the Third World do? One obvious course would be to develop the Philippine and other Third World economies to such a degree that their citizens can earn as much money inside their countries as outside them. Then the Rowenas of the world could support their children in jobs they'd find at home. While such an obvious solution would seem ideal—if not easily achieved—Douglas Massey, a specialist in migration, points to some unexpected problems, at least in the short run. In Massey's view, it is not underdevelopment that sends migrants like Rowena off to the First World but development itself. The higher the percentage of women working in local manufacturing, he finds, the greater the chance that any one woman will leave on a first, undocumented trip abroad. Perhaps these women's horizons broaden. Perhaps they meet others who have gone abroad. Perhaps they come to want better jobs and more goods. Whatever the original motive, the more people in one's community migrate, the more likely one is to migrate too.

If development creates migration, and if we favor some form of development, we need to find more humane responses to the migration such development is likely to cause. For those women who migrate in order to flee abusive husbands, one part of the answer would be to create solutions to that problem closer to home—domestic-violence shelters in these women's home countries, for instance. Another might be to find ways to make it easier for migrating nannies to bring their children with them. Or as a last resort, employers could be required to finance a nanny's regular visits home.

A more basic solution, of course, is to raise the value of caring work itself, so that whoever does it gets more rewards for it. Care, in this case, would no longer be such a "pass-on" job. And now here's the rub: the value of the labor of raising a child—always low relative to the value of other kinds of labor—has, under the impact of globalization, sunk lower still. Children matter to their parents immeasurably, of course, but the labor of raising them does not earn much credit in the eyes of the world. When middle-class housewives raised children as an unpaid, full-time role, the work was dignified by its aura of middle-classness. That was the one upside to the otherwise confining cult of middle-class, nineteenth- and early-twentieth-century American womanhood. But when the unpaid work of raising a child became the paid work of child-care workers, its low market value revealed the abidingly low value of caring work generally—and further lowered it.

The low value placed on caring work results neither from an absence of a need for it nor from the simplicity or ease of doing it. Rather, the declining value of child care results from a cultural politics of inequality. It can be compared with the declining value of basic food crops relative to manufactured goods on the international market. Though clearly more necessary to life, crops such as wheat and rice fetch low and declining prices, while manufactured goods are more highly valued. Just as the market price of primary produce keeps the Third World low in the community of nations, so the low market value of care keeps the status of the women who do it—and, ultimately, all women—low.

One excellent way to raise the value of care is to involve fathers in it. If men shared the care of family members worldwide, care would spread laterally instead of being passed down a social class ladder. In Norway, for example, all employed men are eligible for a year's paternity leave at 90 percent pay. Some 80 percent of Norwegian men now take over a month of parental leave. In this way, Norway is a model to the world. For indeed it is men who have for the most part stepped aside from caring work, and it is with them that the "care drain" truly begins.

In all developed societies, women work at paid jobs. According to the International Labor Organization, half of the world's women between ages fifteen and sixty-four do paid work. Between 1960 and 1980, sixty-nine out of eighty-eight countries surveyed showed a growing proportion of women in paid work. Since 1950, the rate of increase has skyrocketed in the United States, while remaining high in Scandinavia and the United Kingdom and moderate in France and Germany. If we want developed societies with women doctors, political leaders, teachers, bus drivers, and computer programmers, we will need qualified people to give loving care to their children. And there is no reason why every society should not enjoy such loving paid child care. It may even be true that Rowena Bautista or María Guttierez are the people to provide it, so long as their own children either come with them or otherwise receive all the care they need. In the end, Article 9 of the United Nations Declaration on the Rights of the Child—which the United States has not yet signed—states an important goal. . . . It says we need to value care as our most precious resource, and to notice where it comes from and ends up. For, these days, the personal is global.

THINKING ABOUT THE READING

Why do women leave their own families to work in other countries? Why is there such great demand for nannies and other care workers in some countries? Discuss the concept of care work as a commodity available for sale on a global market. What other services are available on a global market that used to be considered something one got "for free" from family members? Before such services were hired out, who, traditionally, was expected to provide them? What has changed? Discuss some reasons why women make up so much of the global labor force today. If these trends in global labor continue, what do you think families will look like in the near future?

Cyberbrides and Global Imaginaries

Mexican Women's Turn from the National to the Foreign

Felicity Schaeffer-Grabiel

(2007)

As I approached the glitzy Presidente hotel where I would interview men and women at the transnational single's party—otherwise known as the "Romance Vacation Tour"—the bus veered into Plaza del Sol, one of the wealthiest, most well-manicured, and most tourist-populated areas of Guadalajara, Mexico. As women began to arrive, I realized they were not your typical "mail-order brides," popularly thought to marry men from the United States out of poverty and desperation. On the contrary, the majority of women were well educated and from a small but burgeoning professional Mexican middle class. They were confident, savvy, and cosmopolitan in their familiarity with U.S. culture through film, television, the Internet, encounters with tourists, stories from family living in the United States, as well as through their own travel abroad. The owner of The Latina Connection (TLC) Worldwide gave me permission to attend the tour for research, because my bicultural identity set me apart from the "feminist type" whom he assumed would write a scathing report on these interactions, whereas I spoke Spanish and was an offspring of a mixed Anglo-Mexican union.

The owner's distrust of feminist types had to do with critical activism by members of the National Organization for Women (NOW) and the Gabriela Network (Los Angeles) who have helped shut down mail-order-bride agencies that cater to the Philippines. While women's activism has helped to bring oftentimes abusive mail-order marriages into mainstream visibility, feminists and scholars alike have tended to situate all mail-order brides within the larger framework of the global trafficking of women, focusing on women's lack of agency or women's victimization in relation to global processes (Glodava and Onizuka 1994; Riddenhour-Levitt 1999; Tolentino 1997; Gibbons and Pretlow 1999). The trafficking of women has been defined as the underside of globalization that victimizes all Third World women's bodies as cheap labor for First World consumption—whether they are factory workers, domestics, sex workers, or "servile wives" in the Internet-bride industry. While it is important to make these gendered neocolonial and imperial legacies visible in terms of women's migration, scholarship and the media make certain assumptions about the ways in which globalization creates unequal gender, class, and racial norms across First and Third World countries: that women are the producers (and commodities) and men the consumers, that women travel as workers and men as pleasure seekers, that women are victims and men victimizers, and that U.S. culture dramatically alters local cultures, without taking into account the reverse phenomenon. In a growing trend, however, cosmopolitan middle-class women use global processes, such as the Internet and tourism circuits, to imagine and attain more stable and liberating lifestyles, equitable gender relations, and more opportunities than can be found in their local environments.

As new accounts of women who seek out international lifestyles slowly surface, these women emerge not as mail-order brides escaping poverty, but as middle-class women impacted by global fantasies of the "American way of life." Women come to realize their gender and racial differences through a barrage of daily encounters with "foreign," U.S. culture. Mexican women turn to foreign men and lifestyles as a way to escape "traditional" value systems in the family, a corrupt and unstable government, and confining definitions of gender and womanhood. As women articulate their hopes to leave what is "oppressive" about Mexican men (and Mexico) for a seemingly more open and liberating journey with foreign men (and the United States), they demonstrate how powerful such a shift in their imagery—from national to transnational citizenship—can be. The space of the foreign offers greater prospects for self-improvement and growth through a more intimate and equitable marriage partner, opportunities to travel, better education, and sometimes, careers.

Yet, in the process of seeking love and marriage women do not completely detach themselves from the nation-state or traditional roles; instead, they accentuate these exact notions of tradition in an attempt to attract male clients. Mexican women are aware of the national and cultural differences between themselves and U.S. women, which they utilize as the basis to accentuate and "sell" a version of traditional Mexican femininity that is desired by U.S. men. Furthermore, the family is still the most important institution women use to enter into U.S. culture, preventing more radical critiques of the legacy of neocolonialism perpetuated through global policies such as North Atlantic Free Trade Agreement that create an atmosphere of dependency and disadvantage for Mexico. Not only is this new phenomenon of love and migration propelled by both contemporary local and global processes, but it also represents a rupture in traditional gender expectations that has reverberated across the Americas and beyond. Cosmopolitan women from Russia, Japan, Brazil, Colombia, Cuba, and Vietnam also utilize creative means to maintain and improve themselves and the lives of their children through global circuits of products, tourists, and imaginaries (Fusco 1997; O'Dougherty 2002; Ong 1999; Kojima 2001; Thai 2002). Women's gender ideologies resonate with modern ideals of selfhood in the United States, as individuals seeking personal fulfillment (rather than adhering to social and familial commitments) through romantic encounters via the Internet or through matchmaking services. I interviewed women (and men) at the "Romance Vacation Tour," set up individual interviews with women whose e-mail addresses I bought from an Internet company, and translated and read e-mail correspondence shared with me by both men and women.

Many Mexican women no longer feel bound by the futility of traditional gender roles that position men as head of the family. During the twentieth century, various changes in Guadalajara, Mexico—the rapid influx of people from rural to urban areas, industrialization, the secular and global expansion of commerce and services, an increase in mass communications—impacted women directly. As Guadalajara transformed from a rural to an urban economy, women enjoyed better employment, education, and healthcare (Oliveira 1990). The peso crisis of the 1980s affected single, middle-class women in particular. Because this widespread economic crisis resulted in a loss of jobs, large sectors of men migrated to the United States, opening up more job opportunities for the women left behind. Female work was no longer temporary or a rarity, but was incorporated into women's lives as a rite of passage through which women could escape isolation in the home (Hondagneu-Sotelo 1994, 13). When their newly found independence was coupled with higher levels of education, women began to want more equitable gender roles; they waited longer to marry, divorce rates increased, and a greater use of contraception resulted in fewer children (Levine and Correa 1993).

I conducted interviews with thirty-two women at "Romance Vacation Tours" in Guadalajara and through e-mail correspondence. I worked with two marriage organizations with full website services. Mexican Matchmakers, a small agency owned by a North American, was located in one of the most affluent neighborhoods of Guadalajara. TLC Worldwide, based in Houston, Texas, offered tours to Mexico usually four times a year. These companies attracted hundreds of women through radio announcements, by placing ads in the back of *Cosmopolitan,* and by word of mouth. More than twenty-five Internet companies offered matchmaking services and marriage with women in Mexico. On signing up with an agency, women provided a photograph, e-mail or mail addresses, and a physical and personal description that the companies then sold to men for varying prices. In exchange, women were invited to the vacation tours for free, while men paid between $500 and $1,000. The women varied in age from eighteen to fifty five, most were well educated, and they worked in an array of professional jobs: doctors, accountants, teachers, business owners, secretaries, beauticians, and models. Some attended tours out of curiosity, to practice their English, to enjoy a free night out, while others were serious about finding true love and, eventually, a husband.

Many accounts of women's involvement in these industries assume the women are objectified by company websites and catalogs wherein Mexican women are advertised as superior commodities (compared to U.S. feminists) to be consumed by men in the global marketplace (Tolentino 1997; Glodava and Onizuka 1994). Such accounts are, in fact, accurate on the representational level, demonstrating companies' complicity in shaping men's expectations for docile, feminine, and sexualized Latinas. An analysis of the Web pages alone, however, cannot explain these relationships at the level of complexity that ethnographic methods provide. At the tours, for example, women show up with girlfriends and family members, and they display confidence and professional attire. The roles are also reversed at the tours: men's bodies are on display for women to consume, as they have to get up in front of a rowdy audience and describe themselves in idealized ways. Furthermore, TLC Worldwide often circulates small catalogs of men's photos and descriptions for women to peruse; the women are then prompted to initiate e-mail letters and courtship.

Mexican women might want to marry a man from the United States for many reasons, yet these desires often conflict with the types of men the agencies attract. Women in their late twenties and beyond hope to escape the stigma of being "older" and single in Mexican society, a society that generally assumes they are past their prime. While marriage symbolizes positive qualities such as happiness, achievement, opportunity, and advancement, the state of being single symbolizes the exact opposite: lack of achievement, solitude, stagnation, and failure (Salazar 2001, 147). U.S. men, however, are told on websites that they can expect to date and/or marry women who are twenty to thirty years younger than themselves—which accentuates the market for younger women. The majority of Mexican women who use matchmaking services are attractive, intelligent, and express feeling undervalued by local men. They idealize men from the United States as being appreciative of their commitment to the family, their femininity, and their intelligence, while they accuse local men of taking for granted their domestic labor and of being threatened by smart and beautiful women.

The majority of women I interviewed came from a small but privileged middle class. In Mexico middle-class status is based not merely on one's economic level. Other factors designating class status include higher levels of education, owning a car or home, having a job with a stable income, residence, having children in private schooling, technological access, and social and cultural expectations such as the desire for self-improvement. The acquisition of a tourist visa is also critical for traveling across the U.S.–Mexico border. Reviewing information collected at Mexican Matchmakers, I found that two-thirds of the women signed up with the agency had university or postgraduate levels of education. Almost half of the women had visas, while another one-fifth had had one in the past.[1] These women were not interested in merely migrating to the United States to work; they repeatedly described wanting to find a good, hardworking, and compatible partner with whom they could share their ideas and feelings. And most were not interested in migrating at a lower-class level, so they sought marriages that could protect and hopefully augment their way of life in Mexico and the United States.

Women and (Trans)Nationalism

As women described why they wanted to marry a man from the United States, it became difficult to distinguish their accounts of Mexican men from the body of the Mexican nation. They looked to men from the United States to embody Utopian marriages and lifestyles—egalitarian relationships with men who would share in household chores and offer a better way of life, more economic stability, and opportunities—qualities Mexico and Mexican men lacked. In interviews I conducted and in written accounts from agency books, women stated that they wanted a man who was loyal, understanding of and responsive to their needs, and hardworking. When I asked why they could not find a man like that in Guadalajara, they shook their heads and voiced their dislike for "macho men."

Anna was a thirty-four-year-old widowed mother who worked part time as an accountant.[2] Her children attended a private school, and she told me she juggled working and taking care of them on her own with the support of her family. According to Anna, men in Mexico are more *machista*—than, presumably, U.S. men—because they are threatened by the fact that women earn more than they do. She said, "Economically, they [Mexican women] are more stable than the men . . . they already have their own house, car and luxuries that many men cannot give them. And, even more curious, what angers men here in Mexico is that the woman—and for this reason they are more macho—that women are more successful than them. But, the good thing about people from other countries is that they admire this kind of woman."

While Anna's conception that Mexican men's machismo stems from threats to their power in the home and workplace, she considers foreign men from First World countries to be the kind of men who respect strong and successful women. Anna keenly asserted that men in Mexico need to subordinate women in order to feel like a man. Machismo, according to Anna, is a defensive state against women's elevated social and economic positions. Yet the majority of men from the United States come to Mexico to find a traditional minded woman in the hope of reasserting their masculinity and power in the home and workplace, which complicates this image of the foreign "feminist man."

When I asked women why they thought men from the United States differed from men in Mexico, they responded that Mexican men were coddled in the home by their mothers and expected the same from their wives. On the other hand, Anna explained,

> I have noticed that men from over there [United States] are well-disposed to share in the chores I've seen something that almost never occurs here . . . over there they have told me, "I will cook for you," not like what they say here, "What do you mean I'm going to cook for you?" [She laughs] Over there men are more independent from a younger age, I think that they learn to value all of these aspects, you know . . . and this gives them a little more maturity. It's liberating that they themselves feel this way and that they have fewer prejudices than men here.

For Anna, U.S. men's willingness to participate in "women's work" is liberating, as it opens relationships up to negotiation, flexibility, and communication. While processes such as urbanization and increases in education and employment for women contribute to changing gender roles, I observed that women are changing faster than men. For example, Guadalajara is a city in motion, as men cross into the United States to find work. Among those who remain, said Josefina, a fifty two-year-old divorced doctor with two grown children, "many want sex without commitment or sex in exchange for going out to eat or for going with him to the cinema." Josefina does not see a fair exchange between men whose earnings, lifestyles, and cosmopolitan outlook do not match her own. Like Anna, she characterizes machismo as a juvenile or childlike state when compared to the paternal father-husband to the north.

Women's complaints about machismo reveal contradictory critiques of men that are both problematic and significant. First, women reveal racial and class biases in associating all Mexican men with negative macho qualities, characterizations that also serve to critique an irresponsible, abusive, and overly patriarchal government and nation-state. Furthermore, women creatively verbalize dissatisfaction with their subordinate gender position in the patriarchal family, culture, and society in general. There is little support of women's new professional careers, nor are there more flexible and equitable gender roles in the family. Mexican feminists have recently brought attention to a backlash in television shows, newspapers, and even a popular talk-radio show in Monterrey. The host of this show, Oscar Muzquiz, solicits men to call in with stories of neglectful wives—in search of the "Female Slob of the Year"—or wives who are channeling their energy into careers rather than their families. Muzquiz attributes this shift to the "Americanization" of family values: "Mexican women are increasingly confusing 'liberty with licentiousness' and that Mexican women are turning us into '*mandelones*' [slang for browbeaten wimps or feminized men]." These popular discourses and images of women out of control are meant to morally pressure or discipline women's bodies back into the home and into traditional gender roles.

Recent feminist scholarship on gender and nationalism focuses on the ways in which women are marginally positioned within national agendas as well as how their reproductive roles are both biological and ideological. Women have historically garnered value for their reproductive role in populating the nation as well as for serving as teachers so youth will learn how to be good citizens. Thus, women's roles as wives and mothers in the heterosexual family have been mythically narrated to guard women's placement within the private spheres of the home. Women who stepped out of these roles were marked as outcasts, prostitutes, whores, or *mujeres mala* (bad women).

While national projects have historically targeted women's bodies as the focus of disciplinary control (McClintock 1995), few accounts take seriously how women themselves disrupt the moral body of the nation through negative characterizations of men. Conversant with global scripts of family behaviors and structures, women pollute the boundaries of their own nation by characterizing it as an overly macho male body. Women naturalize their defection from their own nation and highlight their affinity to another. They reverse the gender hierarchy by polluting the body of Mexico as a "spectacle of men out of control." Women see themselves as having to defect from Mexico, a nation they equate with immature, restless, noncommittal, and backward men.

Las Malinchistas

In characterizing Mexico as a machista or macho nation, women respond to negative reactions from mainstream Mexican society toward their involvement with foreign men. For this reason, many women keep secret from friends and family their interactions with the tours, e-mail exchanges, and dates. Lacking other outlets, many of the women I approached eagerly talked to me as a cultural outsider, yet my biracial identity reinforced their belief that I could relate to them as a woman who understood Mexican culture. Alicia, a single thirty-three year-old with green eyes and light skin, owned her own photography studio and had traveled to the United States through a previous career with American Airlines. She asked that we meet in one of the new Guadalajara hot spots, El Centro Magno, a hip, cosmopolitan, and expensive mall with a Hard Rock Café, a Chili's (with higher prices than in the States), Italian restaurants and cafés, clothing stores with trendy styles from around the world, and a multiplex cinema that primarily featured films from the United States. Alicia mentioned that most of her friends call her a *Malinchista*, "'Cause I only date foreign men—Europeans, Canadians, and Americans. . . . I just don't like the men here—short, fat and dark-skinned . . . no-o—I like them tall, slender, and well-dressed."

The term *Malinchista* has deep historical roots in Mexico. The union between the Spanish conquistador Hernán Cortés and his indigenous concubine, Malintzin, or La Malinche, has been mythologized as leading to the birth of the first mestizo, or mixed-race Mexicano. La Malinche has been narrated through Mexican and Chicano literature as the one who "sold out" her people to the colonizer, the enemy. This historical narrative of the origins of Mexico and Mexico's mixed racial heritage continues to infiltrate popular memory through colloquial language. A Malinchista is popularly known as a traitor, or *la chingada*, literally the one who has been "fucked over," sexually or figuratively, by the penetration of foreign imperialism and policies. Thus, the consumption of foreignness or foreign products is particularly intertwined with gender, race, and class, placing nationalism and Mexico's turn to modernity in constant conflict. The lighter one's skin color, the more one is associated with the upper class, the conqueror, modernity, wealth, and culture. While middle-class women equate freedoms and opportunities with foreign culture, those who benefit less—especially poor men and the indigenous—internalize the phallic intrusion of imperialist and global capital as an emasculating and neocolonial process. For the elite of Mexico, the United States and "things foreign" connote culture, professionalization, and status. This idea of boosting the economy of Mexico through foreign culture is further complicated by contemporary popular Mexican filmmakers, musicians, and artists who speak for the voiceless and condemn elite culture for selling out the country to foreign companies and buying into foreign cultures of taste. Contestation over the national image varies depending on one's gender, race, sexuality, class, and vision for the future.

Alicia characterized men from Mexico as short, balding, dark-skinned, and overweight (and thus lazy); they were therefore lower class, uneducated, and more likely indigenous. Conversely, she associated foreign men—tall, slender, well dressed, and light-skinned—with education, culture, and suit-wearing professionalism. Alicia internalized this dichotomy between First and Third World countries, between the United States and Mexico, as modern versus traditional, and she aligned herself with a more cosmopolitan class that extended national borders.

By describing men in Mexico as macho, women turned the moralizing discourse away from their own bodies, from the accusation that they are the Malinchistas. They instead degraded the national body with images of poor, uneducated, and emotionally abusive men. It is not that the women I interviewed had not suffered from a macho culture—their stories of neglect and abuse attest otherwise—yet they conflated their individual experiences with abusive, insensitive, immature, and adulterous men with popular images of the Mexican nation. Women from Colombia, Asia, Russia, and Japan similarly justified their searches for foreign men by degrading local men, which reveals how far the "personal" gender revolution has spread (Glodava and Onizuka 1994; Del Rosario 1994). In an e-mail letter, a woman from Colombia wrote to her U.S. suitor, "But, thanks to GOD there are good people who work hard, not like the bad people of Colombia." Men who work hard are moral and upright citizens, unlike those men in Latin America, who women envision as drug dealers, unemployed, or lazy. While women turn this discourse onto men as unfit fathers, husbands, providers, and role models, they do not discuss the lack of economic opportunities for men in Mexico, which limits their ability to be as economically stable, well traveled, and experienced as men from the United States.

It is inaccurate to say that all Mexican men are macho, but this stereotype has been widely perpetuated by popular culture and scholarship. As Matthew C. Gutmann demonstrates (1996), what is considered manly or macho must be understood as changing alongside history and across region, gender, and class. While Gutmann finds that many of his male interviewees highlight positive qualities of machismo—such as men's sense of caring and duty toward their children and families—he attributes some of the negative descriptions of abuse by women as practices exacerbated by the peso crisis in Mexico, which contributed to men's loss of status, worth, and identity. Women's opposing constructions of men from Mexico and the United States demonstrate the power of their increasing interpellation as consumers, where commodities—including men—become fetish objects or signs that promise a new self and alternative lifestyles. As Pierrette Hondagneu-Sotelo and Michael Messner (1994) argue, the image of the sensitive Anglo man in U.S. media is internalized as a softer and more open expression of masculinity constructed against the more aggressive display of masculinity expressed by racialized men such as Mexican immigrants and African American males.

How ironic that what women want and the types of men these services attract are almost always at odds. Many U.S. men are looking for the traditional wife and family relationship they believe existed during the 1950s, before the breakdown of the nuclear family due to the social movements of the 1960s and mainstream feminism. Men also idealize love as outside rational time and space: they must look outside the bounds of the nation, outside capitalism, to find true love. Likewise, women in Mexico must leave the Mexican nation, what one woman characterizes as the "cradle of machismo" (Biemann 2000). They equate marriage and relationships in the United States with Utopian ideals of capitalism, democracy, and freedom in the First World.

Love, Work, and the "New Self"

Transnational marriages offer women dual citizenship and the flexibility to combine Mexican traditions of the importance of the family and a strong work ethic and to enact their citizenship as consumers of the global marketplace. Women's commitment to their difference from norms within Mexican culture and society serves to mark their symbolic move away from the Mexican nation-state toward being a citizen and consumer in the transnational family. Néstor Garcia Canclini (2001) argues that through consumption, most Latin Americans experience sentiments of belonging and citizenship by forging similar taste cultures across national, rather than regional, borders. As more and more women join the professional workforce in Mexico and realize that through hard work they can buy what they need, they are less dependent on men to embody this role.

Laura is a hardworking single woman in her early thirties who works five to six days a week for a company that imports and exports goods to and from Mexico and the United States. She lives with a relative and still has a hard time making the payments on a small new car she recently bought. In an e-mail she wrote to a wealthy man from Texas, whom she is dating, she describes her view on relationships: "I'm not looking for a man to take care of me, I am looking for a man that is ready to share his life with me, that knows how to work and who desires to grow alongside his partner. For me, it would not be pleasant to live with a man that sits around and hopes for good luck so that things go well. . . . I like to work and I would like to work together with my partner so that between the two of us we could make something together for our future."

The kind of marriage Laura describes sounds more like a partnership wherein two people contribute equally to build an empire and to grow together. She does not make a distinction between love and the economy, the public and private, the individual and the collective. In bed and in the workplace, a couple should contribute equally and work hard toward uplifting themselves and the relationship. This understanding of love as work echoes the discourse promoted by U.S. magazines and psychological research on love. Eva Illouz's study on the parallels between love and capitalism looks to popular culture to trace these interconnections: "Women's magazines suggest that instead of being 'stricken' or 'smitten' by love, a woman is responsible for her romantic successes and failures, that she must work hard to secure a comfortable emotional future for herself, and that she should guarantee that a relationship will provide an equitable exchange" (1997, 195).

Magazines such as *Cosmopolitan* are very popular with middle- to upper-class women in Mexico. The Mexican publication of *Cosmopolitan* mixes articles written in the United States with articles that are locally produced. Through this mixture of discourses, elite readers are asked to "vicariously" participate in emancipation even though editors know that women are expected to abide by more traditional norms (Illouz 1997, 30). The middle-class women I interviewed are not satisfied with vicarious participation in new ideas of womanhood and marriage. Instead, they see themselves as active participants in new flexible identities between the traditional and the modern, and between Mexico and the United States.

Mexican women look to the United States to be freed from cultural norms and hope to become architects of their own lives. This is a liberating prospect and has the potential for subverting the gender hierarchy in Mexico. As women garner confidence and independence through professional careers and exposure to stories of love and marriages from abroad, they begin to imagine new possibilities for themselves. Yet women also do not accept everything about American culture or the capitalist framework. Aware that women in the United States are more liberal, that families are nuclear rather than extended, and that many U.S. women are more materialistic, most Mexican women state the importance of holding onto spiritual and family traditions. Many of them, especially those with children, know that they will have to "sacrifice" their professions and families in order to find happiness with a foreigner.

Internet Encounters

Fantasies, stereotypes, and Utopian desires commingle on the screen through the act of Internet letter writing. The Mexican woman writes herself into a script in which she finds a loving, supportive, and gentle husband in a faraway land. Popular stories and images of the United States as a land of opportunity—where men respect feminism and love strong, yet family-oriented, women—make their way into this script. The act of writing to a faceless man from the privacy of one's home or workplace adds an element of mystery. Away from strict families, the gossip of friends, and Catholic teachings of respectable codes of behavior, the Mexican woman finds herself alone and able to explore her new role with an audience that she hopes will interact with her with fresh eyes. With the spread of the Internet in Mexico, women can participate in the creation of new gender identities not only as consumers of images but also as actors forging new personas. Part of the lure of the Internet is that women can express themselves outside of local norms and customs and explore new aspects of themselves as their audience extends across national, cultural, and racial boundaries. The Internet is a springboard for acting out changing times, sexual desires, and new identities. Sherry Turkle describes the computer screen as the place where "we project ourselves into our own dramas, dramas in which we are producer, director and star. . . . Computer screens are the new location for our fantasies, both erotic and intellectual" (1995, 26). Women turn to the Internet to express their hopes, dreams, and intimate desires, and in the process, access information about other people and their lives.

The use of the Internet and matchmaking agencies rather than social networks to find relationships also marks a new way of thinking about love, courtship, and marriage. According to Mexican family traditions, a woman is expected to wait patiently and passively for a man to make the first move. Once a man publicly claims his desire for a woman, she is marked as his territory, and she may not see anyone else. The courtship period may last a couple of years, or longer. During this time, the woman, called a *novia,* must not appear in places where she might be a sexual target for another man's desire. On the other hand, because masculinity depends on expressions of independence and fraternity with other males, men are afforded the liberty to frequent bars, clubs, and other social spaces. Furthermore, the man (*novio*) can have numerous sexual adventures with a variety of available women (Carrier 1995).

For women, the Internet proves to be an ideal place for less-restrictive forms of courtship. While women's bodies are guarded and watched closely, the Internet affords them the opportunity to communicate or date multiple people and to develop sexual intimacy in a society that heavily moralizes women's sexual activities outside of marriage.

I interviewed Bianca in her beautiful home, which is tucked away in a heavily guarded, gated community in one of the nicer areas of Guadalajara. She was an attractive and fit fifty-two-year-old who had divorced her husband after discovering that he had cheated on her with one of their neighbors. As she showed me around her home, she mentioned that the many large-screen televisions were gifts from a Mexican doctor she had dated. In order to maintain her lifestyle and put her son through one of the more prestigious universities in the area, she had taken on various jobs, from working in the United States as a nanny to inviting foreign students to reside with her while they attended school. She explained how restricted she felt: "Right now I am very confined, I almost never go out. I go out once in a while into the street and they follow me, people speak to me, but I don't like to get to know people off the street because I think, I *think* that they think that I am easy, and I'm not easy, I'm not an easy kind of woman." When asked whether it was also difficult for women to meet people at bars, she responded, "Well, look . . . another time I went out with some friends, only one time, we went out at night. It's not difficult, they had come up to me, but in reality they are people that are drinking, that think that if a woman goes to a bar . . . the men think that if one goes to a bar alone, she is looking for a sexual encounter."

Opportunities for women to meet a partner are limited to introductions by family and friends, and thus it is extremely difficult for older women who do not have strong social or family networks to meet someone. Blanca had tried various e-mail dating services and had even hosted matchmaking events at her home. When I last spoke to her, she had given up on foreign men and was dating another doctor from Mexico.

Anna also described a sense of isolation and the difficulty of finding a partner: "The truth is I've parted from my friendships and all social contacts that I could have had. But time has gone by and apart from feeling alone—in spite of having my kids and family—I felt the need to have someone else who I could express my feelings to and my thoughts about what is going on in my daily life. I realized that I couldn't have a life as a hermit. Men that I have known, I had only known through work relations. And the truth is that due to my job, my work as a mother and as the head of the household, I don't have much time to have a social life."

Even though Anna lives at home and has the support and care of her family, she does not have the time or energy to build a social world that would allow her to meet and date people. In fact, almost all of the women I interviewed had weak social networks because of obligations to their families, children, and jobs. Not only is a woman's presence in public space questioned, but she may have little time outside work and family to develop close relationships with others.

Women also enjoy having more control over the selection process through Internet dating. Teresa and I met at a traditional outdoor café in downtown Guadalajara. She was a single and confident forty-two-year-old, taking a break from a stressful life as a journalist to nurture and develop herself and her personal life. She said, "At the bar, most people select each other by their looks rather than on intelligence. The atmosphere of the bar does not allow for more in-depth conversations where you really get to know a person. . . . Yet on the Internet I can specify the man I want. I ask them personal and political questions and if they are not interested in responding in this way, I know that they don't want a woman who is intelligent."

Rather than adhering to the concept of love at first sight, many women want to get to know the inner life of a prospective partner before delving into an emotional relationship. Teresa told me she would be playful and witty to see how men responded to her playful intelligence. She could read between the lines in Internet conversations and quickly judge whether someone was open-minded and whether they respected a woman's confidence and intelligence. Interestingly, through various e-mail relationships, Teresa came to find Europeans more cultured, liberal, and open-minded than U.S. men and opted to use various online dating agencies rather than attend the vacation-tour parties. Similar to the motives that inspire online dating in the United States, Mexican use of Internet dating draws from modern ideas of intimacy and selfhood based on talk, rather than from passion and the desire for the advancement of the self through contact with others.

Having a larger cultural context in which women can assess themselves contributes to the rising number of women who feel that they do not have to settle for traditional patterns of marriage. Through conversations with men on the tours and through Internet e-mails and chat rooms, women gather ammunition for constructing norms around love, relationships, and marriage not as natural but as culturally determined. Regarding her experiences with TLC Tours and e-mail conversations, Anna said, "I think my country is renowned for having people and customs very deep-rooted . . . and from here that machismo still remains to this day very strongly rooted in the values of men . . . but, at the same time, I like to know other people who already consider this as a lack of maturity and that it gives guidelines so the woman has her place in society and in her life with men."

According to Anna, her discussions with men who did not abide by the same cultural norms strengthened her convictions that men benefit from machismo while women do not. Anna also said that she had received good advice from people with whom she had been communicating. Because many women condone and perpetuate machista behavior, Anna was often unable to find others with whom she could share her inner thoughts and feelings.

Yet, women are not entirely free to create themselves in these cyber-exchanges. Men often write to multiple women, an expensive process given that it involves not just e-mail access and translating fees but also the cost of sending flowers and gifts, and even of visiting a select few. And because many men send women between $500 and $1,000 to take English classes, women feel they must give the man what he wants, to be the ideally docile and appreciative woman who is available when he needs her. Monica wrote an e-mail to a man who had sent her $500 for English lessons: "Regarding my English classes, I'm very proud because they named me as the honor student. . . . I still don't know much but as I told you before, I'm doing my best to learn fast . . . and also I don't want to disappoint you." Women can be constricted by the consumer's wants and needs and codes of reciprocity.

Conclusion

Such searches for U.S. husbands demonstrate how global imaginaries affect women's intimate lives. Global processes, by bringing people from unevenly developed areas into greater contact with one another, provide middle-class women from Third World countries new scenarios with which to reimagine gender roles and thus to extend what is possible in their local culture. Recent studies locate the "female underside of globalization" as the process whereby millions of women from "poor countries in the south migrate to do the 'women's work' of the north—work that affluent women (and men) are no longer able or willing to do" (Ehrenreich and Hochschild 2002, 3). Rather than extract raw resources from Third World countries, wealthy nations hope to import workers who provide better care, love, and sex. In a similar slant, U.S. men look to Mexican women as more capable wives and mothers—that is, more dedicated, feminine, and willing to serve their husbands—and to take on the role they say feminist, career-driven women no longer want. And Mexican women likewise look to U.S. men as better husbands and fathers than Mexican men and culture. Their perceptions of foreign men coincide with the image of the globetrotter—the sensitive, loyal businessman who is economically savvy, successful, and hardworking—an image that is not always realized. Contrasting expectations produce uneven results, especially because U.S. men hope to replace traditional family and gender arrangements, while some women hope to transcend them. As women increasingly create communities of belonging through consumption in a global marketplace, they see these marriages as an opportunity to solidify a transborder middle-class identity. While Mexican women may turn to global circuits (such as tourism and Internet communication) and Western culture to express modern notions of the individual and of relationships, as well as liberal capitalist notions of consumer power, women incorporate these ideas unevenly and alongside traditional notions of family unity and codes of femininity. Thus, these intimate exchanges produced by the Internet and matchmaking services complicate an easy binary between the United States and Mexico, between the traditional and the modern, and between the global and the national.

Cyberbride industries target increasingly diverse populations of women as Internet matchmaking services become a more accepted, accessible, and widespread means (mostly for the middle class in Mexico) of finding a partner that fulfills one's individual needs and desires. I hope to have disrupted an easy equation of the cyberbride industry, a global broker of love and marriage, as an institution that exploits poor, desperate, and unsuspecting women. Along with this is a hope for a more nuanced understanding of the ways emerging sectors of educated and misplaced women from the "third world" turn to foreign men to step outside the limits of what is possible at home. Women are savvy excavators of opportunities that offer more stable, open, and exciting relationships, marriages, and futures.

Notes

1. The middle to upper classes in Mexico are more likely to hold visas because they can prove their return to Mexico through stable jobs, bank accounts, and the ownership of cars and/or property. In order to move to the United States, women must obtain a fiancée visa. Matchmaking agencies provide detailed information on their websites or at the actual agency and sometimes even sell "immigration kits" with all of the relevant paperwork and information.

2. I have changed the names of all of the women interviewed. Unless specified, quotes are from personal interviews that were transcribed. All translations from Spanish into English are my own.

REFERENCES

Biemann, Ursula, director. 2000. *Writing Desire.* Film. New York: Women Make Movies.

Carrier, Joseph M. 1995. *De los otros: Intimacy and Homosexuality among Mexican Men.* New York: Columbia University Press.

Del Rosario, Virginia. 1994. "Lifting the Smoke Screen: Dynamics of Mail-Order Bride Migration from the Philippines." Ph.D. diss., Institute of Social Studies, The Hague, Netherlands.

Ehrenreich, Barbara, and Arlie R. Hochschild, eds. 2002. *Global Woman: Nannies, Maids, and Sex Workers in the New Global Economy.* New York: Metropolitan.

Fusco, Coco. 1997. "Adventures in the Skin Trade." *Utne Reader,* July–August, 67–69, 107–109.

García Canclini, Néstor. 2001. *Consumers and Citizens: Globalization and Multicultural Conflicts.* Minneapolis: University of Minnesota Press.

Gibbons, Leeza, and Jose Pretlow, prods. 1999. "The Trafficking of Philipino Mail-Order Brides." *Leeza Gibbons Show.* New York: NBC.

Glodava, M., and R. Onizuka. 1994. *Mail-order Brides: Women for Sale.* Fort Collins, Colo.: Alaken

Gutmann, Matthew C. 1996. *The Meaning of Macho: Being a Man in Mexico City.* Berkeley: University of Californian Press.

Hondagneu-Sotelo, Pierrette. 1994. *Gender Transitions: Mexican Experiences of Immigration.* Berkeley: University of California Press.

Hondagneu-Sotelo, Pierrette, and Michael A. Messner. 1994. "Gender Displays and Men's Power: The 'New' Man and the Mexican Immigrant Man." In *Theorizing Masculinities,* edited by Harry Brod and Michael Kaufman, 200–218. Thousand Oaks, Calif.: Sage Publications.

Illouz, Eva. 1997. *Consuming the Romantic Utopia: Love and the Cultural Contradictions of Capitalism.* Berkeley: University of California Press.

Kojima, Yu. 2001. "In the Business of Cultural Reproduction: Theoretical Implications of the Mail-order Bride Phenomenon." *Women's Studies International Forum* 24:199–210.

Levine, Sarah, and Clara Sunderland Correa. 1993. *Dolar y Algira: Women and Social Change in Urban Mexico.* Madison: University of Wisconsin Press.

McClintock, Anne. 1995. *Imperial Leather: Race, Gender, and Sexuality in the Colonial Conquest.* New York: Routledge.

O'Dougherty, Maureen. 2002. *Consumption Intensified: The Politics of Middle-Class Daily Life in Brazil.* Durham, N.C.: Duke University Press.

Oliveria, Orlandina de. 1990. "Empleo feminine en México en tiempos de recesión económica: Tendencia recientes." In *Mujeres y Crisis: Respuestas Ante la Recession,* edited by L.G. Ortega et al. Caracas, Venezuela: Editorial Nueva Sociedad.

Ong, Aihwa. 1999. *Flexible Citizenship: The Cultural Logics of Transnationality.* Durham, N.C.: Duke University Press.

Riddenhour-Levitt, Jennifer. 1999. "Constructing Gender, Race, and Ethnicity in a Globalized Context: The 'Mail-order Bride' Trade." *Critica: A Journal of Critical Essays,* Spring, 51–56.

Salazar, Tania Rodríguez. 2001. *Las razones del matrimonio: Representaciones, relatos de vida y sociedad.* Guadalajara: University of Guadalajara Press.

Thai, Hung Cam. 2002. "Clashing Dreams: Highly Educated Overseas Brides and Low-wage U.S. Husbands." In *Global Woman: Nannies, Maids, and Sex Workers in the New Global Economy,* edited by Barara Ehrenreich and Arlie R. Hochschild, 230–53. New York: Metropolitan.

Tolentino, Roland B. 1997. "Bodies, Letters, Catalogs: Filipinas in Transnational Space." *Social Text* 48, no. 14:3.

Turkle, Sherry. 1995. *Life on the Screen: Identity in the Age of the Internet.* New York: Touchstone.

THINKING ABOUT THE READING

Why did the Mexican women in Schaeffer-Grabiel's research use these matchmaking services to find U.S. husbands? What qualities were important to these women in potential husbands? How did this compare to the rationale of U.S. men who were seeking Mexican wives? In what way did Schaeffer-Grabiel's findings contrast with the larger body of research on the transnational cyberbride industry? What did these Mexican women think of Mexican men as potential husbands? What larger structural processes in Mexico have impacted intimate heterosexual relationships and marriages there?

The Architects of Change

14

Reconstructing Society

Throughout this book, you've seen examples of how society is socially constructed and how these social constructions, in turn, affect the lives of individuals. It's hard not to feel a little helpless when discussing the control that culture, massive bureaucratic organizations, social institutions, systems of social stratification, and population trends have over our individual lives. However, social change is as much a part of society as social stability. Whether at the personal, cultural, or institutional level, change is the preeminent feature of modern societies. Social change occurs in many ways and on many levels (e.g., through population shifts and immigration, as illustrated in the previous chapter). Sociologists are also interested in specific, goal-based social movements. Who participates in social movements? What motivates this participation? How successful are they? Social movements range from neighborhood organizers seeking better funding for schools to large-scale religious groups seeking to influence law and politics regarding issues such as abortion, same-sex marriage, and immigration. Social movements come in all shapes and sizes. The readings in this final chapter provide three examples of very different forms of social movements.

Different groups are affected differently by significant historical events. In the aftermath of 9/11, Muslim Americans found themselves the subject of extreme cultural vilification and harassment both from the public and from government agencies. Sociologist Pierrette Hondagneu-Sotelo provides a detailed account of the ways in which Los Angeles-based Muslim Americans have organized and are working collectively to protect their image as decent Americans who deserve the same civil rights as all Americans. Muslim American activists are working not only to correct the extreme images of terrorism portrayed in the media but also the fear and lack of education regarding civil rights in their own communities.

Indeed, local community groups like Hondagneu-Sotelo discussed above can serve as strong advocates for social change and have a profound impact on the needs of individuals and groups. A small community, working-class group sheds new light on the usefulness of a direct action model in comparison to the modes of resistance utilized by large, established worker unions. In "The Seattle Solidarity Network: A New Approach to Working Class Social Movements," Walter Winslow shows how SeaSol has taken a very specific philosophy and applied it towards the plight of workers and tenants who have experienced discrimination in their workplace and homes.

In the final selection, sociologist William I. Robinson suggests that the current immigrant labor protests reflect more than temporary opposition to immigration policies. According to Robinson, these protests are indicative of a growing awareness regarding global capitalism and the exploitation of immigrant labor. Robinson traces the necessity of immigrant labor in the new global markets and asks us to consider the possibility that a global social movement is forming based on the issue of immigrant labor rights.

Something to Consider as You Read

As you read these selections, consider the connection between people's ideas, beliefs, and goals and the motivation to become involved in social change. Participation in a social movement takes time and resources. What do you care enough about to contribute your time and money? In thinking about the near future, which groups do you think are "worked up" enough about something to give a lot of time and energy in trying to create social change? If these groups prevail, what do you think the future will look like?

Muslim American Immigrants After 9/11

The Struggle for Civil Rights

Pierrette Hondagneu-Sotelo

(2008)

There is a new movement to make people in Muslim, Arab, and South Asian immigrant communities become politically engaged and informed American citizens, but unlike the civil rights movement of the 1950s and 1960s, religion is delicately interwoven into these current efforts. I was introduced to part of this movement on a bright Saturday morning in December 2002 when two thousand people convened at the gargantuan Long Beach Convention Center for the annual convention of the Muslim Public Affairs Council (MPAC). The large convention halls, the registration desks, the speakers dressed in expensive suits and business attire, and the prominent MPAC banners—in red, white, and blue and featuring stars and stripes—prompted my student and me to think we had stumbled into a Democratic or Republican Party convention. All that was missing were balloons, booze, and major television media.

In the wake of 9/11, many non-Muslim South Asians and Christian Arab American immigrants became both victims and activists, as did some Latinos and Asian Americans of various religions. White, U.S.-born Christians and Jews were generally not targets, but many of them, particularly those in the clergy, also worked tirelessly in interfaith dialogues and formed new alliances with these groups. In this chapter I focus particularly on the Muslim response, but also on the collective South Asian and Arab American immigrant response to 9/11. I provide a snapshot of what these community organizations and leaders did in Los Angeles and Orange Counties to restore civil liberties and how, in this process, they renegotiated religious and racial identity with media and government realities. The leaders and organizations discussed in this chapter constitute part of a new movement for immigrant civil rights and for Muslim American identities in the United States.

Muslim citizens and immigrants are a growing and increasingly visible part of the population in all Western, industrial, and postindustrial societies. Syrian immigrants from what is today Jordan and Lebanon, most of them Christian, came to the United States, mostly to the Midwest, during the late nineteenth century as labor migrants and peddlers. Racist exclusionary laws in the 1920s curtailed midcentury immigration from Asia and the Middle East. But the 1965 immigration act, with its preference system for highly educated, skilled migrants and its lifting of the racist exclusionary immigration laws, reopened the doors. Consequently, in the 1970s highly educated, urban-origin Muslim immigrants began coming to the United States from nations as diverse as Pakistan, India, Iran, Indonesia, and Jordan. They were seeking economic and academic opportunities in the United States and fleeing political violence. Many came as students and started student organizations, such as the Muslim Student Association and the Islamic Society of North America, to keep their religion alive in their families and communities. Shared religious identity allowed them to forge connections even though they came from diverse nations. Since the census does not collect data on religious affiliation, the precise number of Muslims in the United States is disputed, but a population of 6 million Muslims is the figure most often cited. About one-third are African American, with the remainder split among Arabs and South Asians.

A plethora of Muslim organizations, most of them built in the 1980s and 1990s, emerged in the forefront of the response to post-9/11 backlash against Muslim, Arab, and South Asian American immigrant communities. I studied a handful of these organizations in Los Angeles and Orange Counties, which is where approximately six hundred thousand Muslim Americans reside. Most of these organizations are directed by first-generation immigrant men who were educated in U.S. universities, many of them in the sciences or business. These men are the antithesis, in substance and physical appearance, of the dominant media representations of bearded, bomb-throwing, foreign, Muslim masculinity. They are clean-shaven and telegenic, they wear exquisite business suits, and they appear equally adept at speaking at press conferences, on panels with the FBI or officials from the Department of Homeland Security, or with Christian interfaith groups. They are not formally trained Islamic religious scholars or imams but savvy, eloquent spokesmen for Islam in America, and they were already making optimistic headway into mainstream American politics before 9/11.

These Muslim American leaders seek to work within the system. They threw their support to the Bush-Cheney ticket in the 2000 presidential election, driven in part by Bush's campaign promise of less support for Israel and by Bush's pledge to repeal the 1995 Antiterrorism and Effective Death Penalty Act—which allows the government to use secret evidence against non-U.S. citizens. While these groups gained momentum in the 1990s, it is the post-9/11 assaults on their communities that propelled them headfirst into the struggle for civil rights. These Muslim organizations, built on the model of modern, professional organizations—with executive directors, public relations specialists, administrative support, newsletters, websites, boards of directors, and small but skilled staffs—were well positioned to take political action, and they were joined by other groups, as I detail further below.

As I see it, this collective effort constitutes a traditional struggle for civil rights and civil liberties. The goal of these activists is both discursive, to carve out an identity as American Muslims (or as American Arabs or South Asian Americans, as the case may be), and instrumental, to end racial and religious discrimination, detentions, profiling, and harassment based on religion, race, and nativity. In this regard, the struggle waged by Muslim, Arab, and South Asian immigrants in the United States runs parallel to the civil rights movement waged by African Americans in the 1950s and 1960s. It is also parallel to the experience of Japanese Americans during World War II internment. Here, it is instructive to pause for a moment and contrast the religious contours of these movements.

In the Steps of Black Americans and Japanese Americans?

There appears to be much in common between the experience of Japanese Americans during World War II and Muslim Americans in the current era. In both instances, the United States government responded to violent attacks from outside the nation by seeking to define and retaliate against an enemy within the nation. Both instances rely on racial discrimination against citizens and immigrants of Japanese or Muslim, Arab, or South Asian origins. And not surprisingly, in the post-9/11 period, there has been an outpouring of support from Japanese American organizations to Muslim, Arab, and South Asian American communities affected by the post-9/11 backlash and new affiliations between these groups.

What happened to Japanese Americans is well known. Soon after Japan attacked Pearl Harbor on December 7, 1941, President Franklin Delano Roosevelt signed Executive Order 9066. Everyone of Japanese ancestry on the West Coast was subjected to curfew and, eventually, forced removal from their homes, schools, and workplaces. Entire families were freighted into internment in camps in remotely located rural places in Utah, Idaho, and

Montana and deserts in California. Approximately 110,000 people of Japanese descent, 70,000 of them American citizens, spent the duration of World War II living in cheaply constructed wooden barracks in places like Manzanar or Topaz, with armed sentry guards posted along the barbed wire enclosure fences.

The contemporary Muslim struggle for civil rights shares much with the goals of the civil rights movement of the 1950s and 1960s. These Muslim American leaders want to put an end to unfair treatment and discrimination against their communities, and they want the right to claim a Muslim American identity, just as blacks sought to become fully enfranchised American citizens. Religion, however, gets used differently by these groups. In the contemporary instance, religion is the central basis for discrimination and is a primary means of mobilization, but religion does not serve as a rationale for making claims for the restoration of civil liberties. While civil rights leaders in the 1950s and 1960s regularly quoted the Bible to give religious relevance to social injustices, the contemporary Muslim, Arab, and South Asian civil rights leaders do not appeal to Islamic sacred scriptures to claim their rights. Instead, they evoke the American Constitution as a textual source of justice. The Quran is not the warrant for making claims about inclusion in the American polity, nor is it a means of motivating people to social action. Rather, as we will see, the Quran is used variously, as a text that helps unify Muslim organizations and Muslim collective identity, and when it is engaged in public discourse, to show that American values and political traditions are compatible with Islam's major tenets.

Religious freedom is a foundation American narrative. The central struggle for these groups, however, is not the right to practice Islam but to lay claim to rights and civil liberties as Americans and as immigrants who are racialized, "alienized," and oppressed because of their religious identification. While these groups claim to share experiences of minority subordination with other U.S. racial-ethnic minorities, especially with Japanese Americans who were also held suspect during World War II, their struggle is to disestablish Christianity as a precursor to American national identity. To be clear: They are not against Christianity. In this regard, their goals go right to the heart of the origins of the United States.

For the first- and second-generation immigrants active in this project, religion and ethnicity act as an organizing net. But in an era when being foreign, Islamic, and Middle Eastern is conflated with "terrorist," they cannot deploy religion in overt, highly visual public ways. Organizations such as the Council on American-Islamic Relations (CAIR) and MPAC seek to represent themselves as both Muslims and Americans, while the South Asian Network (SAN), the American-Arab Anti-discrimination Committee (ADC), and the Palestinian American Women's Association (PAWA) are also working to represent themselves as Americans. In this struggle for recognition and self-definition, they use established, institutional modes of political engagement. These include town hall meetings, conventions, press releases, and formal meetings and collaborations and informal meetings with federal, state, and local government and law enforcement authorities. They organize as members of racialized immigrant groups, and in the post-9/11 era, they establish coalitions and working relations with other groups, such as Japanese Americans and Christian clergy as well as government representatives from the FBI, the Department of Homeland Security, and the local sheriff's office. They seek to influence public opinion and the state, but they refrain from bringing highly visible expressions of Islam to the political arena.

"We Should Be Able to Define Ourselves"

Constructing and promoting an identity is at the heart of the struggle for all the ethnic-religious organization leaders I interviewed. For the leaders of CAIR, MPAC, PAWA, ADC, and SAN, civil rights is, at core, a discursive struggle. This is their primary challenge. At stake in the post-9/11 era is who will control the image of Muslim Americans, Arab Americans, and South Asian Americans. What will be included in the content of this identity? And will this identity be used to promote inclusion or justify exclusion? These groups want inclusion, and they are actively seeking a place in the American polity and society that reflects their position both as Americans and as immigrants who will no longer be racialized and persecuted because of religion and phenotype. They want to contest the images of them that circulate through the media.

These groups seek inclusion as Americans, but not an inclusion that compromises being Muslim. Their remedy focuses on educating the larger society, to show that they are American *and* Muslim. As Salam Al-Marayati, the Los Angeles director of MPAC, told me, "We're stressing the American Muslim identity. We're trying to be more vocal and set America straight."

Know Your Neighbor, Know Your Rights—and Show Yourself to the FBI

Immediately after 9/11, organizations like CAIR and MPAC were thrown into high gear, initially responding reactively, defending and protecting members of their communities, and then proactively, educating and informing members of the Muslim immigrant communities and other Americans as well. The aperture of collective self-definition opened up as it never had before, and the leaders saw this as a new opportunity and obligation. The Islamic-identified organizations were most deeply affected by these imperatives. They set about the task of educating Americans about Muslims and of informing their own ethnic communities about civil rights. We can think of these, respectively, as "Know Your Neighbors" and "Know Your Rights" campaigns.

The executive director of the Muslim Public Affairs Council told me that in the Los Angeles area, MPAC had sponsored or participated in over four hundred public forums and outreach events between September 11, 2001, and February 2002, the time of the interview. MPAC sought not only to inform and protect community members, but also to educate the government and the larger public.

How did organizations with small, already stretched paid staffs accomplish this? They dipped into their general membership to develop a new pool of leaders. As Samer Hathout, a lawyer, MPAC board member, and daughter of a key leader reflected, "We feel so behind, so overwhelmed. There's so much to do now. . . . Everyone wants to know about Islam, so there is this overwhelming demand for speakers and appearances." MPAC developed new spokespeople during this period, and she noted, "People that didn't necessarily want to do public speaking are finding that it's not as scary as they thought it was. So it's really brought out some more leaders for us."

At the Council on American-Islamic Relations office in Orange County, the response was initially reactive and service oriented—taking reports of hate crimes, employment discrimination, and school and workplace harassment—but it also did proactive work aimed at information and outreach. Speaking of the immediate post-9/11 months, the executive director of CAIR, Hussam Ayloush, said, "We've been doing the same thing for the last almost eight years nationwide, and the last six years in Southern California. But what's happened is the degree or the amount of what we were doing has changed—the intensity. In the past maybe we used to give one presentation at a church maybe every two months, at a school every two months; we would deal maybe with twenty cases of discrimination. Within a few months after September 11th, we had to deal with over—if I'm not mistaken—close to two hundred cases in our area of hate incidents." Like other organizations, CAIR was not equipped for this barrage of activity. "As a small office," explained Ayloush, "we weren't prepared to deal with a flood of phone calls." They brought in more volunteers and hired new staff, but this required devoting more resources to training.

More time and resources were subsequently devoted to civil liberties issues. CAIR, for example, has continued to issue "action alerts" through the Internet, alerting Listserv recipients to instances of prejudice, discrimination, and violence against Muslims. For affluent, literate, educated, professional-class immigrants, the Internet is a useful resource. One observer has called this "action alert activism."[1]

Instead of a policy of noncooperation with government authorities, the groups decided to participate in town hall meetings, which brought together Muslim, South Asian, and Arab American community leaders and members with FBI, INS, and Department of Justice functionaries. I attended the second in the Southern California series of town hall meetings in January 2002, four months after 9/11. It was officially sponsored by the U.S. Department of Justice Community Relations Service, through the efforts of Ron Wakabayashi, a Japanese American with a long history of civil rights activism. The meeting was held on a Saturday afternoon in a ballroom of a Holiday Inn in La Mirada, a city just off Interstate 5, near the industrial area where southern Los Angeles and northern Orange County meet. There were over one hundred people, most of them Arab American or Muslim, and a handful of Sikhs. In the lobby, where various groups set up tables to distribute leaflets, I saw newspapers in Arabic, but I was most struck by how prosperous the people looked. In fact, I remember wishing that I had dressed up a bit more. The men—and it was mostly men—wore suits and ties, and the women wore professional attire. Many women wore headscarves that matched their outfits. Once in the ballroom, the audience mostly listened attentively—but sometimes heckled—as speakers from the FBI, INS, and the Department of Justice addressed questions of concern. Joining the three white middle-aged men representing the government were four community representatives, including one woman, and the moderator, Tareef Nashashibi, who introduced himself as president of the Arab American Committee of the Republican Party of Orange County. He began on an upbeat note, celebrating and thanking the FBI for incarcerating Irv Rubin of the Jewish Defense League, who bombed offices in 1985, killing Alex Odeh, and he emphasized the rights of citizens and the importance of working together with government. "We are aware," he stated at the outset, "of the FBI looking closely at us, and we want to look back at them." He identified the use of secret evidence as a major threat, and he said, "These issues are important to us, the recent immigrant group. We are all citizens of this country, and we need to be treated alike." At the meeting, the government representatives addressed questions from the audience about the use of secret evidence, racial profiling, detentions, and visas. The INS representative claimed that the term "racial profiling" had been abused by the media in "unsettling ways," and he tried to allay fears by saying that less than one hundred people in the INS western region had been detained due to post-9/11 investigations. The audience response varied, from polite questions and nodding heads to outright heckling.

The diversity of views expressed by the community organization speakers and the audience was also evident among the leaders I interviewed. Some of them saw the town hall meetings as important for building relations with local government bureaucrats and for educating government officials about their communities. They saw these meetings as "building bridges," as ways to keep their own communities abreast of developments, but also as educational efforts, so that government officials "will know that Arabs and Muslims are not what they see on TV." As one leader said to me in an interview: "We wanted to make sure that people do not have this fear of the FBI, so we arranged several town hall meetings with the FBI, the INS. . . . They had very pleasant people working with them. . . . It helped us both, both communities. I think it helped them realize that as they attended those meetings they saw that the Muslim community was not just a bunch of bearded men shouting, 'Death to America!'"

Other community leaders found little to celebrate in these new collaborations. One critic had this to say: "We had three town hall meetings with the FBI, the INS, and the Justice Department, and I felt like it was group therapy. We talked, . . . they listened and they did not do anything. There's still a lot of people being detained, still a lot of people going to jail. The idea about democracy that we are innocent until proven guilty no longer stands. There is no due process for the Arabs or the Muslims."

These internal conflicts speak to the diversity of Muslim American, Arab American, and South Asian American immigrant communities. Just as there is no monolithic voice in the Muslim world, there is no monolithic voice among these various communities in the United States. One interviewee candidly noted that the diversity of the Muslim community makes the advocacy work a challenge. Some members favor traditional party affiliations and congressional causes, while others advocate grassroots connections with labor and civil rights organizations; others bitterly disagree about the relative merits and dangers of participating with the FBI, INS, and Department of Justice. Fighting domestic surveillance of Muslim American immigrant communities and yet working with the federal government is the tightrope these groups walk. The groups want to work with the government, but they want to stop government surveillance based on racial-religious profiling and unspecified standards. While there are disagreements on approaches, they all agreed that a big part of the problem is the United States' ignorance of their communities.

Of the organizations I examined, none were as explicitly focused on the project of addressing imagery in media and among opinion makers as CAIR and MPAC. Within MPAC, no one was out on the frontlines more than Sarah Eltantawi. Freshly out of graduate school, female, and still in her twenties, she had only been on the job for a few months before 9/11. Suddenly, she found herself on Fox News and CNN. By February 2002 she had debated Daniel Pipes on the *Greta Van Susteren Show* and had been on the *O'Reilly Factor* three times. "The first time was with John Gibson," she said. "That was absolutely horrific. . . . I was supposed to go on and talk about American Muslims' response to 9/11, and as soon as I got on there, he immediately started screaming at me about 'Why do you people have a problem with the United States after all we've done for the Palestinians? After all we've done for the peace process, you ungrateful, blah, blah, blah.' I mean, he really just screamed at me, wouldn't let me get a word in edgewise." From her experience, she concluded, "O'Reilly's people just want Muslims up there, like sitting ducks." That television appearance was followed by more where she was often pitted against so-called terrorism experts like journalist Steve Emerson and Daniel Pipes, editor of the *Middle East Quarterly.* Both of these men frequently write and speak about the dangers of radical Islam and promote the view that Arab and Muslim American communities harbor terrorist sleeper cells. As Eltantawi recalled of the news shows, "The question of who Muslims are and who Arabs are is never approached objectively, but more like, The Quran says this and this about infidels. What do you have to say?' We're always on the defensive, always having to answer questions that are posed with a certain kind of bias in mind."

Moderate, peaceful Muslim Americans do not fit the narrative or what is profitable to print, and this determines, in part, how these communities come to be viewed by society at large. During the first few months after 9/11, a counternarrative appeared in the media news, as we saw the debut of a series of "human interest" stories on Muslim American families. These constituted the mass media's approach to the "Muslim moment." On the one hand, these stories presented humanizing quotidian portraits of Muslim American families. The features focused on Muslims as average American families, were typically shot in the domestic sphere of kitchens and dining rooms, and showed glimpses of all-American mortgages, children with homework, and family members gathered around a dining table for an evening meal. On the other hand, these portraits may have played into the new American paranoia of sleeper cells. Regardless of how these "American family" narratives were ultimately read by viewers, they did present a significant departure from the media-as-usual representations of Muslim and Arab Americans. Ra'id Faraj, the public relations director of CAIR, was among the most charitable in his assessment of the media. "The mainstream media," he said, "has been okay. After all, they've definitely worked with us on all kinds of stories. We've assisted them on stories, we gave them numbers and statistics, and we helped them to find individuals in the communities [to feature]." Nader Abuljebain of the ADC was less sanguine with his succinct assessment: "So I'm glad at least they know we exist, and we don't all have tails, and we are not all terrorists or millionaires or belly dancers."

In 2002 CAIR debuted a series of billboards along Southern California freeways showing photos of smiling, multiracial Muslims—the photo was reminiscent of a Benetton ad—with the text "Even a Smile Is Charity," to suggest that Muslims might be kind and compassionate rather than dangerous. CAIR also invested in getting books and videos with accurate portrayals of Islam into public and school libraries. MPAC members were encouraged to write letters to the editor, and they did, and some of these were published in the *Los Angeles Times*. In the post-9/11 period, MPAC actively encouraged members to become media spokespeople. "We are learning to do the sound-bite thing," explained Samer Hathout.

A long-term route to remedy media distortion involved getting more Muslim Americans and Arab Americans into media jobs. This is part of the larger Muslim American project of cultivating leaders in the second generation. As one leader said, Muslims should be "encouraging more Muslims in those fields, fields of media, journalism, communication, educating members of the media, sensitizing them." First-generation Muslim immigrants and their children tend to concentrate in science and engineering jobs, so as one interviewee said, "encouraging them to be journalists or [in] any area of the liberal arts . . . to consider politics as a career" is part of the solution.

In at least one instance, Hollywood stars were mobilized to fight prevailing negative Hollywood images. In the immediate aftermath of 9/11, the Los Angeles County Commission on Human Relations sought to deter hate crimes with the help of Hollywood celebrities. Robin Toma said that after the movie star Patricia Arquette, herself the daughter of a Muslim American father, came forward to volunteer "to do something about what she saw was the anti-Muslim, anti-Arab backlash in this country," celebrities were recruited to do radio public service announcements against hate. Arquette visited public schools to talk with youth and apparently used her personal networks to recruit celebrities for the radio spots. Hollywood was also rewarded by MPAC for fair and non-stereotypical portrayals of Muslims. To encourage fair representations in film, MPAC had already introduced a media awards program. Past winners now include Denzel Washington, Morgan Freeman, Spike Lee, George Clooney, Kevin Costner, and Yusuf Islam (formerly known as Cat Stevens).

Leaders of these organizations worked hard to allay fears and anxieties and to inform their communities of their civil rights. It was a tough sell. Fear prompted people to stay away from mosques and Islamic centers and to withdraw their financial contributions to Islamic charities. Randall Hamud, a third-generation Arab American civil rights attorney and ADC board member who was defending detainees, reported at an ACLU-sponsored public forum his frustration with raising bail money—no one now wanted to take the risk of association with detainees, even though they were not proven to be guilty of anything. Hamud also reported accompanying clients from San Diego who were asked to come forward for questioning. The FBI asked them, "Why were you trying to change your license plate?" Neighbors had reported seeing Hamud's client changing his license plate at night, but the client had merely been tightening a license plate that was coming loose.[2] In this context of surveillance and accusations, community members were less likely to volunteer information and were reluctant to report hate crimes out of fear and stigma.

As Hamid Khan, executive director of SAN, put it, "Right now we feel besieged, because of detentions, because of dealing with the FBI, we are having extreme difficulty in documenting needs because people are unwilling to share stories, but they tell us, 'We just had a raid.'" Michel Shehadeh of ADC concurred: "The community is not coming out to join organizations and to fight back. This is a scared community, and the challenge is to empower the community." Most agreed that the fear was greatest among the foreign-born.

In the aftermath of 9/11, the organizations discussed here devoted a good deal of their outreach and educational efforts to their own communities, particularly first-generation immigrants. Large public forums at Islamic centers, churches, town hall meetings, and hotels attracted thousands of people. Informational materials were distributed at these meetings and in ethnic newspapers. Yet the community leaders reported that immigrants in the Muslim American, Arab American, and South Asian American communities presented particular challenges: ignorance of their rights and entitlements in the United States; the legacy of having grown up under despotic rulers and being unaccustomed to freedom; and intensified fear and anxiety due to government repression following 9/11.

If information is power, knowledge of basic civil rights and entitlements, the leaders reasoned, may help deter abuses. Toward this end, the organizations distributed thousands of "know your rights" brochures and cards. The pocket-sized, fold-up cards such as the ones distributed by CAIR, for example, included titles such as "Know Your Rights as an Airline Passenger," "If the FBI Contacts You," "Your Rights as an Employee," "Your Rights as a Student," and "Reacting to Anti-Muslim Hate Crimes." These were brilliantly prepared, informative, pithy documents and included simple, sequential steps to take in a variety of problematic circumstances. Similar brochures and cards were distributed by the other organizations, and they were translated into multiple languages, including Arabic, Farsi, and Hindi. Some of the organizations set up websites. MPAC, for example, featured one with information ranging from First Amendment rights, Miranda rights, and the rights of due process for noncitizens to the difference between hate crimes and hate incidents, including online forms for downloading hate crime reports. The document on hate crimes instructed the aggrieved to do the following:

- Report the crime to your local police station immediately. Ask that the incident be treated as a hate crime. Follow up with investigators. Inform CAIR even if you believe it is a "small incident."
- Document the incident. Write down exactly what was said and/or done by the offender. Save evidence. Take photographs.

"Know Your Rights as an Airline Passenger" advised people of their rights and told them what to do in instances of racial profiling:

- As an airline passenger, you are entitled to courteous, respectful and non-stigmatizing treatment by airline and security personnel.
- You have the right to complain about treatment that you believe is discriminatory. If you believe you have been treated in a discriminatory manner, immediately:
    - Ask to speak to a supervisor.
    - Ask if you have been singled out because of your name, looks, dress, race, ethnicity, faith or national origin.
    - Ask for the names and ID numbers of all persons involved in the incident.
    - Ask witnesses to give you their names and contact information.
    - Write down a statement of facts immediately after the incident. . . .
    - Contact CAIR to file a report.

As the reader will observe, this information does not make reference to God, the Quran, or religion. Not only do the materials urge nonviolent responses, they are all based on protections offered by the U.S. Constitution and U.S. laws. The materials instructed the aggrieved to take proactive steps, to remain calm and seek witnesses, to gather evidence and documentation that might be used in court, and to contact legal advocates and start a paper trail of documentation. These efforts, however, sometimes fell on frightened ears.

In spite of these obstacles, the crisis galvanized an upsurge of public engagement among Muslim American immigrants and the advancement of a particular collective identity.

Conclusion: The Moderate Mainstream

Muslim American immigrant organizations responded to the post-9/11 backlash leveled at their communities through public engagement, civic participation, and outreach to their own communities and beyond. In all these efforts, they put forth an image of community members as moderate mainstream, middle-of-the-road, middle-class Muslims. Based on what leaders of these organizations told me and what I observed of their organizations' activities, I came to see four dimensions to this collective Muslim American—and sometimes, more expansively, Middle Eastern, Arab, and South Asian—identity project: (1) showing involvement with national domestic issues; (2) promoting moderate political views; (3) avoiding overt forms of religious piety in the public sphere; and (4) regularly offering public declarations of American patriotism and denouncements of Islamic fundamentalist violence and terrorism.

Notes

1. www.mpac.org, accessed January 26, 2007.

2. The forum where Randall Hamud reported this information, "Racial Profiling after 9/11," was held at a Jewish venue, the University Synagogue, in Brentwood, California, on March 12, 2002.

THINKING ABOUT THE READING

What are some common myths and stereotypes about Muslim Americans, especially since 9/11? What are some of the similarities between contemporary Muslim Americans and Japanese Americans living in the U.S. after the bombing of Pearl Harbor? What are some of the ways in which Muslim Americans are organizing to dispel these stereotypes? What are some of their main concerns and strategies? What role does religion play in their organizing activities? Why are they so focused on education about civil liberties?

The Seattle Solidarity Network

A New Approach to Working Class Social Movements

*Walter Winslow**

(2011)

The Future of Working Class Social Movements

The winter of 2011 proved timely for examining social movements in the United States. As an active member of The Seattle Solidarity Network (SeaSol), a small Seattle-based network of working class individuals that fight for tenant and worker rights in their community, I saw firsthand the two very different approaches that large seasoned unions and small community working class support groups utilized to collectively fight for rights in the workforce. In Renton, Washington, SeaSol was in the midst a bitter campaign against a small Italian restaurant in an effort to recover a waitress' unpaid wages. At the same time, in Madison, Wisconsin, a Republican governor was trying to push through legislation intended to quash public sector unions in the State while tens of thousands of people protested at the Capitol building. The stark contrast between the all but unknown conflict SeaSol was battling in Renton and the much-publicized protests in Wisconsin at the same time can be seen as parallel harbingers of two very different futures for working-class social movements in the United States: extinction or rejuvenation. The juxtaposition of these two specific events provides a useful starting point for understanding how SeaSol's approach differs from that of the mainstream political Left in the United States.

The Future From Madison, Wisconsin

On February 11th, 2011, Governor Scott Walker of Wisconsin introduced a new "budget repair bill" designed to strip Wisconsin's 283,351 public employees (WTA 2011) of their collective bargaining rights and greatly weaken public sector unions in the state. The proposed bill was designed to eliminate the automatic deduction of union dues from union employees' pay and mandatory union membership, limit labor contracts to one year, remove the right to collective bargaining entirely in some industries while strictly limiting it in others, and require public unions to run a campaign to be successfully re-certified in a National Labor Relations Board election every year (State Legislature of Wisconsin 2011).

The proposed bill was a litmus test for the strength of public sector unions across the nation as much of the country watched and waited to see what would happen. Despite their strength in numbers, Wisconsin's public sector unions and their supporters struggled to prevent the bill from becoming law. Thousands of protesters occupied the Capitol Rotunda in Madison, demanding that the bill be scrapped. Wisconsin's Democratic state lawmakers physically fled to Illinois in order to prevent the state senate from having the necessary quorum to vote on the bill, and a variety of private and

*Editors' note: This article is based on original research conducted by the author for his honors thesis in sociology. This version is extracted from the original thesis and has been edited by Jennifer Hamann for this reader.

public unions organized solidarity rallies at every state capital in the country. The cross-country reactions to the bill captured major media attention that pushed unions into the spotlight, questioning their effectiveness not only in Wisconsin, but across the United States. Though the demonstrations and media frenzy made for good television, it gave the protestors no real leverage. On March 11, 2011, Governor Walker signed the bill into law (Bauer 2011, Davey 2011, Haas and Stanely 2011, Ramirez 2011).

As a result of the new law, public sector unions across the country may soon be faced with similar measures—possibly marking another step towards the complete extinction of unions as a serious social force in the United States. In the wake of this disaster for organized labor, Wisconsin's protesters and their sympathizers across the nation have been left wondering: is there anything else they could have done?

There is at least one compelling answer to this question: the unions and their supporters could have used sustained direct action tactics to put pressure on Governor Walker. In the ensuing protests, teachers from across the State called in "sick" to attend demonstrations at the Capitol building in a brash wildcat strike. In Madison alone, forty percent of the districts' teachers phoned in sick, causing the entire school district to cancel classes (DeFour 2011). However, the unions' own leaders moved quickly to stop the strikes. The Madison Teacher's union, Madison Teacher's Inc. (MTI), The Milwaukee Teacher's Association, and the State's largest teacher's union, the Wisconsin Education Council Association (WEAC) consistently issued statements urging union workers to continue to report for work and reassuring them that they had the situation well in hand (Bell 2011, WISC-TV 2011). Union leaders made it clear that determining appropriate strategies for resisting the bill was their purview.

Instead of direct action, union leaders funneled popular anger into more passive modes of resistance; they urged supporters to sign petitions outside the capitol and brought in speakers from across the country to condemn Republicans and extol the Democrats. The Reverend Jesse Jackson led one group of protestors in a chant: "When we vote, we win! When we vote, we win!" (Wisconsin Reporter 2011). Jackson's message was simple: Governor Walker's bill should be dealt with using only the proper channels of legal and electoral processes.

Public sector unions utilized the same tired strategies the Left has been relying on for decades: public demonstrations, legal battles, and continued support of the Democratic Party. Any divergent strategies that were utilized were proactively suppressed and union members were urged to remain at work. These traditional approaches failed to stop the budget repair bill in Wisconsin and they have failed to slow the general decline of unions at large across the country. As unions have become less participatory and increasingly executive, they have not been able to mobilize the popular support necessary to stop corporate offensives at the bargaining table or in Congress. A union's greatest strength is the ability to unite workers to take action on the job that directly disrupts business. Today however, unions fail to utilize this power. In its place, America's unions continue to languish in a willful state of institutional bondage. As the most historically significant form of working-class social movement, the obvious impotency of unions as demonstrated in Wisconsin has grim implications for the future.

The Seattle Solidarity Network: An Alternative Approach

The Seattle Solidarity Network (SeaSol) is a small but growing grassroots mutual support organization for workers and tenants. The five young men who founded SeaSol in December of 2007, all members of the Industrial Workers of the World (IWW), wanted to find a way to contribute to rebuilding a revolutionary working-class social movement by winning tangible victories with a small number of supporters. Ultimately, they were interested in the potential of unions to serve as a mechanism to eventually overturn capitalist social relations entirely.

As their idea of forming some sort of mutual support network began to take shape in late 2007, they decided to include tenants' issues in their project, an issue that is often left out of a union's scope of support. Their class politics prompted them to view tenants' and workers' issues as inextricably linked and they hoped that by engaging with both, they would be able to ensure a higher and broader level of activity for their new organization.

Despite the revolutionary ambitions of many of its members, SeaSol does not base its day-to-day activities on any grandiose vision of the future and does not have any official political agenda or affiliation. Instead, SeaSol exists to achieve immediate material gains for low-wage workers and tenants in the here and now. Since its formation three and a half years ago, SeaSol has successfully resolved approximately twenty-five specific housing and job-related issues while growing to nearly one hundred members. In the absence of effective legal remedies and strong workers' or tenants' unions, SeaSol members work to protect one another from employer or landlord abuses by carrying out escalating campaigns of public protest.

This research examines both why this unique organization is experiencing growing success and how its members are politicized as a direct result of their participation. SeaSol's approach is especially notable because it defies prevailing ideologies surrounding social change by operating outside the paradigm of contemporary progressive organizations. SeaSol is unique in that it:

- Is entirely supported by volunteers
- Adheres to no explicit political ideology
- Does not rely on lawyers or other professionals
- Does not involve itself in electoral politics
- Is not a legally recognized non-profit organization
- Is funded exclusively by small individual donations

This article reveals how working-class people do not necessarily need to depend on politicians or non-profit organizations to improve their lives. In Seattle, they are coming together as equals to directly improve their lives using only their own collective power and imagination. SeaSol's present activities are limited, but the wider implications of the organization's strategy for social change are boundless.

The Future From Renton, Washington

The winter of 2011 also tested worker's rights in the suburb of Renton, Washington just outside of Seattle. SeaSol was leading their last picket outside of a Bella Napoli Italian restaurant because its owner, Ciro Donofrio, had fired a waitress named Ramona and was refusing to pay her for her last month of work. Ramona describes what happened to her in an article on SeaSol's website:

> For the entire month of September I worked for Ciro Donofrio at his Italian Restaurant in Renton, Bella Napoli. During this time, Ciro was verbally abusive towards his employees and even customers. He would throw temper tantrums in front of tables and claim we were out of things on the menu simply because he did not feel like making them. . . . I still had to pay rent so I continued to work for Ciro. Things got hairy when I had $110 of my bank "disappeared" one night when only he and I were working. Also, I needed my check and Ciro claimed that he only paid his employees at the end of every month. I thought this was strange, especially after I had seen him give a check to the cook, but I dismissed it. What was he going to do, not pay me? (The Seattle Solidarity Network, 2011)

Refusing to pay her was exactly what Donofrio ended up doing after firing Ramona. Ramona decided to file a claim against Donofrio with the Department of Labor and Industries (L&I), but quickly became frustrated with the slow and impersonal nature of the process. As a result, when Ramona's friend told her about a poster she had seen promoting SeaSol, Ramona decided to contact the organization.

Ramona brought her case to the next weekly SeaSol meeting where members voted to initiate a direct action campaign against Donofrio. Shortly after, forty SeaSol supporters marched into Bella Napoli restaurant with Ramona and delivered a letter to Donofrio telling him that he had 14 days to pay her the wages he owed before they would take further action. When he failed to pay Ramona's wages after two weeks, SeaSol began an escalating campaign that involved flyer distribution, posters and two and half months of picketing Donofrio and his restaurant. Despite the rainy cold winter weather, roughly thirty individuals consistently showed up for the evening pickets at Bella Napoli, a location that was also a good twenty minute drive from Seattle. We paraded up and down the sidewalk in front of the restaurant carrying signs and chanting, "Work for Ciro, get paid zero!" while he eyed us angrily from inside his empty restaurant. After about half an hour, as it became obvious that no one was going to cross the picket line that night, Donofrio decided to close his restaurant for the night. A week later, Donofrio was forced to close the doors to his business.

The fight was a milestone for SeaSol in many ways. For the first time, the network had the strength and numbers to force an employer to choose between paying what he owed or closing the doors of his business permanently. The amount of money at stake was small and the number of people involved was nothing compared to the tens of thousands who were protesting in Wisconsin at about the same time. However, despite the insignificance of the campaign on a grand scale, SeaSol's victory in Renton clearly demonstrated the organization's growing power.

Ramona did eventually receive a check through the Department of Labor and Industries for approximately half of the amount she was actually owed, but when she attempted to cash it the check bounced. Ramona remained unclear as to exactly why this was, but she said that after speaking with L&I it seemed to have something to do with the fact that L&I had not actually secured payment from Donofrio before issuing the check. SeaSol was also unable to secure payment from Donofrio, but when asked if she would remain involved in SeaSol anyway Ramona said:

> Definitely. It's just the justice, it's just seeing a group of people stand beside you and support you and tell you it's ok, I've been through this, it'll get better and we'll stand up to them and they won't win.

The sort of success SeaSol experienced in Ramona's campaign, small though it may be, provides on example of the possibility that powerful working class social movements can still be rebuilt in the United States. Moreover, the campaign proved that organized members of the working-class are capable of identifying and defeating their own enemies without legal or professional assistance.

An Egalitarian Organization

SeaSol is comprised of three groups of volunteers denoting three different levels of activity and commitment to the organization: organizers, members, and supporters. Anyone may become an organizer if they are willing, at a minimum, to commit to attend SeaSol's weekly planning meetings and to call ten to twenty SeaSol members every time the organization utilizes its phone-tree. People who volunteer at this level are more likely to be ideologically motivated. SeaSol's membership is defined as everyone who has agreed to receive notification about every SeaSol action. Individuals who may want to work with SeaSol in order to get help dealing with a specific job or housing issue are required to become involved at this level. Supporters are those who are interested in the organization but want to be notified less frequently, typically only about SeaSol's largest or most important actions. The organization is presently comprised of approximately sixteen organizers, ninety members, has over two hundred people on its phone-tree, and over five hundred supporters on its largest email list.

SeaSol has no formalized leadership structure and the organization's weekly meetings are open to the public. Typically twenty to thirty people attend these weekly meetings. Many members who are not organizers regularly attend, but it is less common for supporters to attend. The group makes all of its decisions during these weekly meetings by taking a simple majority vote after a period of discussion; anyone present at a meeting is permitted to vote. The primary purpose of these meetings is to plan SeaSol's activities for the coming week and to delegate logistical responsibilities.

The organization has successfully taken on a variety of workers' and tenants' issues including wage theft, landlord neglect, deposit theft, unfair fees, and predatory lawsuits. SeaSol carries out public campaigns designed to force the employer or landlord to meet a specific demand using escalating amounts of social and economic pressure. In the past three and a half years SeaSol has undertaken a wide array of tactics as part of these campaigns. A far-from-comprehensive list of direct action tactics includes storming into offices en masse to deliver written demands, putting up posters telling would-be renters or customers not to do business with a given company, picketing storefronts and other businesses, picketing public events connected with the employer or landlord, and putting up posters around the employer's or landlord's neighborhood or workplace.

SeaSol's tactics are intended not just to achieve the desired results, but also to do so in a way that empowers those who participate in SeaSol campaigns. This is one reason the organization does not seek legal help with its campaigns. The most active participants in SeaSol believe that legal processes are slow, biased, ineffectual, costly, and passive for the participant. SeaSol has only sought legal representation one time in order to defend itself in court against a lawsuit brought against them by a major Seattle real estate developer that they were pressuring. Additionally, as a rule of thumb, the organization only takes on cases that they feel are going to be empowering for the participants. For the same reason, SeaSol only takes on fights with people who are willing and able to take a leading role in their own campaign and seem genuinely interested in joining and helping others.

Their overall success rate is notably high, winning twenty-one of twenty-five fights in the past three years. It is no accident that SeaSol finds success in such a high percentage of their campaigns, rather, it is the direct result of one of the organization's most basic principles: "winnability."

Winnability

When asked why SeaSol had decided to take on Ramona's wage-theft, one organizer noted the following:

> The fact that it was very winnable, we had a lot of leverage on the business, we had the power to put this company out of business—so we ought to be able to win this fight!

Another SeaSol organizer expanded on the same idea in greater detail:

> Winnability is one of our basic principles. It is this concept that is really important and kind of straightforward and seems kind of silly to talk so explicitly about, but really I think it is kind of ignored by other activist groups generally and that is: can you win what you are trying to get? Can you get your demand? Could you do it? Is it possible? And while you can never know that concretely, you never know for sure, but you can use rational thinking about what that person [the employer or landlord] values and how they've been acting in the past.

This is probably the single most important principle SeaSol emphasizes in its internal trainings and public presentations. Every SeaSol fight is based around a specific concrete demand. If the organization does not feel like this concrete demand can be met, it will not take on the case. As simple and "silly to talk so explicitly about" as this idea may seem, the truth is that winnability is something many activists rarely consider. Many activist organizations rally around a specific issue, such as globalization, but never take the time to honestly ask themselves, for example: what would it really take to transform or dismantle the IMF [International Monetary Fund] and World Bank? What sort of popular movement would be needed to force the U.S. to restructure how it conducts world trade? Does the Left in this country really have the power to achieve this outcome?

SeaSol organizers seriously consider relevant power dynamics before they decide to begin a new campaign. The organization is open about its unwillingness to take on fights they do not believe they can win, regularly voting not to take on certain campaigns because they do not feel that they are winnable. One organizer explained how just recently the group had voted to take on a small-time landlord who had stolen several tenants' deposits but then changed their mind when further research indicated that it was probably unwinnable:

> [S]he had stolen their deposit and we really wanted to take on the fight and we thought she had this moving company we might be able to target, but even when we took it on we weren't sure. Then after doing more research and finding out she actually isn't even in the State three weeks out of the month and she has no other economic targets and no vacancies and has no reputation in the neighborhood—it made it seem like a very unwinnable campaign so we decided not to take it on after all.

These kind of pragmatic ideas about what SeaSol can and cannot accomplish form a major part of the organization's culture. Multiple informants reported that this simple pragmatism was part of what made SeaSol so distinct when compared to other activist groups. One organizer who also works as a paid union organizer said that compared to her paid work:

> Working with SeaSol has just kind of kept me sane. . . . I don't really think that a lot of activism is really leading to anything whereas with SeaSol I feel it can be very empowering for people.

Several other informants reported that SeaSol's pragmatic approach was actually part of a conscious strategy to build a larger and more powerful movement to accomplish greater goals. Many informants reported that they had become demoralized by the repeated failures they had experienced working with other organizations in the past that had tried to implement sweeping social changes that they did not have the power to make—like ending the U.S. wars in the Middle East or stopping governmental budget cuts to health and human services.

While SeaSol organizers believe that it is useful to understand social problems on a systemic level, they also understand that it is foolhardy at this point in time to think the Left can attack those systems directly with any success. To be able to do that successfully SeaSol organizers reason that they need to work on dramatically increasing their numbers through practical activity rather than through propaganda. All four SeaSol organizers interviewed said that this was why they thought winning campaigns was so crucial, every victory proves that SeaSol's approach really works. All six SeaSol contacts, including two members who did not initially get involved with SeaSol for any sort of ideological reasons, reported that they wanted SeaSol to continue to grow in order to successfully take on larger and more significant social problems such as Chase bank, "capitalism," "the State," and even "industrial aqua-culture." In one SeaSol organizer's words, "the basic motif of SeaSol that I know is we do what we can today so we can do what we want to tomorrow."

To quote another SeaSol organizer:

> The question, the difficult part, is how do you get from nobody to hundreds of thousands of people? How do you get that force so that it can operate well? So that it can operate sustainably and in a progressively better way? The answer to that for me, is what we're doing.

There is no doubt that SeaSol is effectively winning the small fights. However, the more important question is whether these victories will actually spawn the larger movement everyone involved in SeaSol hopes for?

Strength and Numbers

Between the spring of 2010 and the spring of 2011, SeaSol's organizing committee grew from eight members to sixteen. Attendance at weekly meetings also increased from the low teens to the mid-twenties during the same period. One of the most interesting ways SeaSol continues to grow is based around the organization's concept of mutual aid. A strong willingness to join SeaSol and a verbal agreement to continue to support the organization in the future is required before SeaSol will agree to take on a campaign for an outsider. One organizer's description of her first impression of a woman dealing with a landlord issue after their first meeting illustrates this point:

> She said, "I don't want them [the property management company] to do this to someone else," which is something that is really important for me to hear from someone. There is some enlightened self-interest involved, or a lot actually, but the fact that she's thinking about other people and recognizes that she is connected to other people, that others are like me, is a really good sign. She just really wanted to fight back, so it wasn't just, oh, I feel sorry for this woman, it was like, oh, I really feel for her but I also have a lot of respect for her. She's ready to fight back against this huge company. She doesn't have any experience that I ever got out of her doing this, so I had a lot of admiration for that.

SeaSol is making a conscious effort to distinguish itself from social service agencies. They don't want to provide direct action casework for someone who is not interested in passing on the support to the next worker or tenant in need. SeaSol wants to retain their permanent involvement. The woman made it clear that she did not want to have a passive role in her campaign. The organizer did not only feel sorry for the woman—she respected her—and left the meeting feeling excited to work with her side by side. In this case, SeaSol ended up launching a successful six-week campaign with the woman in the spring of 2011 to force her landlord to drop several hundred dollars in unjust fees and return her stolen security deposit. The woman has remained a SeaSol member and continues to attend actions when she can.

SeaSol is successfully retaining the participation of people who initially become involved in order to resolve their own problems. However, it is self-evident that a mass movement will never be built one person at a time and this is not the primary way SeaSol has grown in the past few years. SeaSol's success or failure at building a larger working-class social movement depends on its ability to get more than one person involved at a time. One organizer described his thoughts about SeaSol's growth this way:

> When I say "gathering people" there are sort of two things that go on with that in any given fight. There are the people that come into it because they are at the center of a fight and then there are the people that come into it because there is a fight going on and they want to help out. For that latter group, I have seen more people come on from that group in labor fights—because labor fights involve big actions that you want to have as many people as possible at and really landlord fights don't.

Another organizer echoed the idea that labor fights are ideal for strengthening SeaSol. When I asked him if he thought Ramona's campaign helped the organization grow, he commented:

> It brought in Ramona and some of Ramona's friends, but mainly it was a great fight because it gave us a lot of picketing opportunities. It gave us opportunities for fun and exciting actions that lots of people can participate in and that had an immediate and powerful impact—and people could see the power in that it actually destroyed the business. It gave people an opportunity to come out and picket that was real, not just symbolic.

Getting the maximum number of people involved is something SeaSol considers when deciding what campaigns to take on and what tactics to use. The more SeaSol has taken on compelling campaigns that have required multiple mass actions, the more the organization has grown. One organizer described how while in one campaign against a landlord SeaSol had largely relied on smaller groups informally heading out to put up "Do Not Rent Here" posters around properties owned by the landlord, the organization adapted its strategy in a later fight to involve more people:

> Everything in Nelson [the later campaign] was just a better job of what we did in George's fight [the previous campaign]. By having different groups go poster around different neighborhoods as one big action instead of just informally mobilizing for it. . . .

In this case SeaSol intentionally adapted its strategy to involve more people not because it was necessarily more effective at getting the posters put up, but rather because it was a way of allowing more people to help take action against the landlord. SeaSol's continued growth is very much dependent on how the organization can effectively mobilize larger groups of people in a meaningful way.

One of the most important elements of recruitment involves basic training. SeaSol provides first hand training on how to effectively fight back against employers and landlords. One organizer described the importance of this kind of growth like this:

> Well in the short term obviously, we have these very small issue-based economic fights, and you know it's helping people tackle [the problem], engaging people in struggle in their own life and then helping them actually win. In the long term I see it as helping people develop themselves as organizers, develop organizing skills, both for themselves and then just for everybody that is involved because it is such a collaborative and cooperative effort.

Every organizer hopes that SeaSol can provide practical training for people that will stay with them for the rest of their lives. This means training not only the individuals at the center of a specific campaign in how they can successfully face down their employer or landlord, but also providing useful experience for everyone involved in the campaign. A different organizer described what this might look like in more specific terms:

> My hope is that we can build it [SeaSol] into a stronger and stronger force and it can lead to having a large number of people who are competent and confident at organizing and doing direct action, we can hopefully branch out from the types of fights we are doing and organize groups of workers in workplaces and tenants in apartment buildings.

This sort of transition is vital to SeaSol's future growth. Everyone who is heavily involved in SeaSol recognizes that the organization's present model is not going to be able to build a truly mass movement. Instead, they want to use SeaSol's current activities as a springboard to expose people to direct action and inspire people to want to organize on their job at work or as tenants in their buildings. The organization hopes to continue evolving not only so that it can become more effective at what it is already doing, but more importantly so it can increasingly transition to taking actions that involve larger and larger numbers of people—perhaps into new kinds of workers' and tenants' unions. As one of SeaSol's founders put it, he hoped SeaSol could, "serve as the foundation for a broader working class movement."

The Process of Politicization

Nearly all of the most active participants in SeaSol identify as revolutionaries. In its only written statement regarding its long-term political ambitions SeaSol states in a pamphlet describing the organization that it hopes to someday create "a world without bosses or landlords." As one SeaSol organizer described it, she believed that SeaSol is ultimately trying to:

> build up enough people who are serious about taking control of their lives and who don't think bosses and landlords are necessary. To build up a militant, conscious, organized Left to take over, immediately, the sources of capitalism and the State that interfere the most directly in our own lives and to take control of our own lives

SeaSol is not only concerned with delivering immediate material victories to people. The organization is also passionately concerned with achieving these goals in a way that transforms people's opinions about society and empowers them to feel that the working-class could one day actually overturn the power relations that so utterly define their lives. First and foremost, SeaSol attempts to prefigure how such a society might work by focusing on how the organization is internally structured. This is why SeaSol is all volunteer and directly democratic. There is little difference between the ideological and practical reasoning for this. One of SeaSol's founding organizers explained the importance of SeaSol's decision-making process this way:

> To avoid authoritarianism is practical. It's sort of an ideological way of putting it, but it is a shorthand way of saying something that's practical that's much harder to describe in words. If we got in a situation where some individual or clique who isn't accountable to anyone else was able to force their will on the majority, force other people to do things that they didn't want to do rather than being free and democratic, then I don't think it would be possible to pursue the type of organizing we are trying to do. I think it would change the organizing model because our whole model is based on encouraging people to take action on their own because they want to.

SeaSol's organizers want to build a cooperative and egalitarian working-class movement to do away with those who they believe exercise illegitimate authority over other people's lives—namely bosses and landlords. Not only SeaSol's decision-making process, but also its entire strategic approach is intended to empower and politicize the people who become involved in the organization. This is one reason why SeaSol relies on direct action instead of legal or political action. They want to prove that when people are well organized, they can solve their own problems directly and effectively. The overwhelming majority of SeaSol's organizers have backgrounds in various types of activism and became involved in SeaSol for precisely these reasons. But to what extent does SeaSol actually transform how those people who become involved for entirely practical reasons feel about contemporary society? Ramona from Renton said:

> I felt like to L&I I am just like another case number and it's very impersonal and with SeaSol I just met a lot of people that I just really related to, that made me feel welcome, that made me feel like my voice was important, and really supported me.

Another informant named George, joined SeaSol to fight unjust charges brought against him by his former landlord regarding a bedbug infestation. He had similar reasons for deciding to join SeaSol:

> I know a little bit about the legal system and I know that attorneys are expensive and the legal process is [also] unfortunately. The landlord has a lot of money and a mansion you know and I can't afford to put myself in court against this man. It ain't gonna happen, I'm not gonna win. I had no resources to fight someone like that.

Both Ramona and George felt that their legal options were entirely inadequate. This frustration with their "official" options is undoubtedly what initially made SeaSol's approach so attractive to them, but their subsequent participation in a SeaSol campaign had a major impact on their personal beliefs about their own position in society. This does not mean that George or Ramona would now describe themselves as anarchists, as many SeaSol organizers and members do, and this is certainly not what SeaSol is trying to accomplish. Instead, SeaSol believes that taking action is a much more radicalizing experience than talking about politics in the abstract. This shared willingness to take direct action does not require everyone to share all of the same political beliefs. SeaSol's priority at this time is simply to build a shared culture of resistance for working-class power. One SeaSol organizer and anarchist who had confronted her own boss in the past as part of a union drive described the power of that kind of confrontational experience this way:

> I mean once you've marched on your own boss for instance, and I imagine it is the same for anybody who goes and confronts their landlord, it doesn't sound like a big deal handing this letter and saying, look, I demand what's right and I'm going to claim my right as another individual who should have equal power to you. It's definitely transformative. It is scary as hell and it's a huge moment of growth for people and it stays with you. It really does stay with you forever.

It was clear that Ramona and George were first brought into the organization by desperate circumstances and a willingness to try a different approach, but both of them said that they plan to remain permanently involved. Ramona described her opinion of SeaSol in these words:

> It's a really amazing organization that's really changed my whole perspective on things . . . it is like a family, I love it and I will always remain involved in SeaSol. I feel like I belong and I feel like it helps everyone feel like they belong, it's like a home.

It is significant that even though SeaSol was unable to recover her stolen wages, this did not taint her opinion of the organization. What was important to her was how her participation in the group made her feel. It made her feel that she belonged and that she does not have to face the injustices in her life alone. Ramona did not join SeaSol because she had read the theories of Bakunin or Marx and was inspired by their ideas, Ramona joined SeaSol because Ciro Donofrio stole her wages and she thought SeaSol could help. She remains involved with SeaSol not because she has adopted an anarchist or a Marxist position, but because she wants to be part of SeaSol where she feels a sense of empowerment and belonging. George recalled similar sentiments in his experience with SeaSol:

> It really saved my ass because the landlord would have sent it to collections and it made me believe in other human beings in the world. . . . I was very happy to get help from SeaSol and you know I feel like I can help and that's the nice thing about Seattle Solidarity. They helped me and I'm trying to help, what I can, back, because I like what they're doing number one and plus I feel like I owe Seattle Solidarity for the help.

One organizer noted:

> It [SeaSol] gives me a sense of something I've always been wanting . . . it's like we are making better lives for ourselves as immediately as possible and for people after us. To me that is meaning in itself and it's also a group of people who is also ready to be solid for you . . . to me building that up and making it more powerful is the most important thing I can think of to do to change what I think is wrong with the world.

SeaSol gives people at all levels of participation hope and a sense that they just might be able to change working-class issues that are close to home. Whether SeaSol's long-term strategy will work or not, there is little doubt that SeaSol has greatly influenced how its participants view the world. SeaSol is not a mass revolutionary working-class social movement at the moment, but it is undoubtedly reaching new people and exposing them to experiences that makes them believe in the power of their own united effort.

Conclusion

SeaSol is an extraordinary example of radical political praxis. The success of SeaSol's practical approach should give pause to those who believe revolutionary ideologies are irrelevant in the contemporary United States. SeaSol's revolutionary members are not only finding common ground with typical members of the working-class, they are taking common action, action that is delivering real material gains to SeaSol's members while forging new social relationships based on shared struggles. SeaSol has had enough success in the past three years to begin attracting the attention of other activists both nationally and internationally. Solidarity networks have begun to emerge across the country as a direct result of people learning about SeaSol's activities. Presently, solidarity networks have been formed in the cities of Atlanta, Boston, Iowa City, New York, Oakland, San Diego, San Francisco, and Santa Cruz. Internationally, people have been inspired by SeaSol's work to start their own solidarity networks in Canada, England, Scotland, Australia, and New Zealand. It is much too soon to say whether this model will grow into an actual popular movement in the United States or remain simply the obscure activity of a few scattered groups of like-minded individuals. Nevertheless, the continuing development of this new kind of working-class social movement demands further research and deserves to be followed closely in the coming years.

REFERENCES

Bauer, Scott, and Todd Richmond. 2011. "Thousands protest Wisconsin's anti-union bill." MSNBC Online, February 17. Retrieved April 9, 2011 (http://www.msnbc.msn.com/id/41624142/ns/politics-more_politics/)

Bauer, Scott. 2011. "Scott Walker Signs Wisconsin Union Bill Into Law." The Huffington Post, March 11. Retrieved June 29, 2011 (http://www.huffingtonpost.com/2011/03/11/scott-walker-signs-wiscon_n_834508.html)

Bell, Mary. 2011. "Statement from WEAC President Mary Bell." Washington Education Association Council, March 9, 2011. Retrieved June 29, 2011 (http://www.weac.Org/news_and_publications/11 -03- 09/Statement_from_WEAC_President_Mary_Bell_on_Senate_vote.aspx)

Davey, Monica. 2011. "Republican Tactics End Wisconsin Stalemate." *The New York Times*, March 9. Retrieved June 30, 2011 (http://www.nytimes.com/2011/03/10/us/10wisconsin.html?_r=1)

DeFour, Matthew. 2011. "Madison schools closed Wednesday due to district wide teacher sickout." Wisconsin State Journal, February 16. Retrieved June 29, 2011 (http://host.madison.com/wsj/news/local/education/local_schools/article e3cfe584–3953–11eO-9284–001cc4c03286.html)

Haas, Kevin, and Greg Stanely. 2011. "Wisconsin Democrats Flee to Clock Tower." Rockford Register Star, February 17. Retrieved April 9, 2011 (http.7/www.rrstar.com/carousel/x435 22562/Wisconsin-Democrats-flee-to-Rockford-to-block-anti-union-bill).

Mayers, Jeff. 2011. "Divided Supreme Court Upholds Wisconsin Law." Reuters, June 14. Retrieved June 29, 2011 (http://www.reuters.com/article/2011/06/15/us-wisconsin-unions-idUSTRE75D60520110615)

MSNBC. 2011. "Wisconsin Protesters Vacate Capitol After Judge Orders Them Out" Msnbc.com. Retrieved June 29, 2011 (http://www.msnbc.msn.com/id/41884135/)

Ramirez, Antonio. 2011. "Wisconsin Solidarity Rallies Today in all 50 States." Change.org. Retrieved April 9, 2011 (http://news.change.org/stories/wisconsin-solidarity-rallies-today-in-all-50-states).

Seattle Solidarity Network. 2010. "Problems with Your Boss or Landlord?" Outreach poster.

Seattle Solidarity Network. 2011. "Three month fight puts thieving restaurant out of business." Retrieved May 30th, 2011 (www.seasol.net)

State of Wisconsin. 2011–2012 State Legislature. 2011. "Bill." Madison, Wisconsin: January 2011 Special Session. Retrieved April 9, 2011 (http://bloximages.chicag02.vip.townnews.com/host.madison.com/content/tncms/assets/editorial/f/8c/e99/f8ce991a-3612–11e0–97f9–001cc4c03286 revisions/4d558a5d60362.pdf.pdf)

WISC-TV, 2011. "Senate Votes to Strip Collective Bargaining Rights." Channe1300.com (http://www.channe13000.com/politics/27138601/detail.html)

Wisconsin Reporter. 2011. "Jesse Jackson Outside Capitol Part One." Retrieved June 29, 2011 (http://www.wisconsinreporter.com/jesse-jackson-outside-capitol-pt-2)

Wisconsin Taxpayers Alliance. 2011. "The Wisconsin Taxpayer." The Wisconsin Taxpayer's Alliance. Retrieved April 9, 2011 (http://www.wistax.org/taxpayer/10Wiw0283.pdf)

THINKING ABOUT THE READING

Describe the principles and strategy of the working class social movement known as The Seattle Solidarity Network (SeaSol). How does Winslow contrast SeaSol with larger working class social movements like unions? What are their similarities? What are their differences? According to Winslow, why is SeaSol experiencing success and how are its members politicized from their participation? What was the motivation for the founding of SeaSol? How has SeaSol influenced its participants in the architecture of their own lives?

"Aquí estamos y no nos vamos!"

Global Capital and Immigrant Rights

William I. Robinson

(2006)

A spectre is haunting global capitalism—the spectre of a transnational immigrant workers' uprising. An immigrant rights movement is spreading around the world, spearheaded by Latino immigrants in the US, who have launched an all-out fight-back against the repression, exploitation and racism they routinely face with a series of unparalleled strikes and demonstrations. The immediate message of immigrants and their allies in the United States is clear, with marchers shouting: "*aquí estamos y no nos vamos!*" (we're here and we're not leaving!). However, beyond immediate demands, the emerging movement challenges the very structural changes bound up with capitalist globalisation that have generated an upsurge in global labour migration, thrown up a new global working class, and placed that working class in increasingly direct confrontation with transnational capital.

The US mobilisations began when over half a million immigrants and their supporters took to the streets in Chicago on 10 March 2006. It was the largest single protest in that city's history. Following the Chicago action, rolling strikes and protests spread to other cities, large and small, organised through expanding networks of churches, immigrant clubs and rights groups, community associations, Spanish-language and progressive media, trade unions and social justice organisations. Millions came out on 25 March for a "national day of action." Between one and two million people demonstrated in Los Angeles—the single biggest public protest in the city's history—and millions more followed suit in Chicago, New York, Atlanta, Washington DC, Phoenix, Dallas, Houston, Tucson, Denver and dozens of other cities. Again, on 10 April, millions heeded the call for another day of protest. In addition, hundreds of thousands of high school students in Los Angeles and around the country staged walk-outs in support of their families and communities, braving police repression and legal sanctions.

Then on the first of May, International Workers' Day, trade unionists and social justice activists joined immigrants in "The Great American Boycott 2006/A Day Without an Immigrant." Millions—perhaps tens of millions—in over 200 cities from across the country skipped work and school, commercial activity and daily routines in order to participate in a national boycott, general strike, rallies and symbolic actions. The May 1 action was a resounding success. Hundreds of local communities in the south, midwest, north-west and elsewhere, far away from the "gateway cities" where Latino populations are concentrated, experienced mass public mobilisations that placed them on the political map. Agribusiness in the California and Florida heartlands—nearly 100 percent dependent on immigrant labour—came to a standstill, leaving supermarket produce shelves empty for the next several days. In the landscaping industry, nine out of ten workers boycotted work, according to the American Nursery and Landscape Association. The construction industry suffered major disruptions. Latino truckers who move 70 percent of the goods in Los Angeles ports did not work. Caregiver referral agencies in major cities saw a sharp increase in calls from parents who needed last-minute nannies or baby-sitters. In order to avoid a total shutdown of the casino mecca in Las Vegas—highly dependent on immigrant labour—casino owners were forced to set up tables in employee lunch-rooms and hold meetings to allow their workers to circulate petitions in favour of immigrant demands. International commerce between Mexico and the United States ground to a temporary halt as protesters closed Tijuana, Juarez-El Paso and several other crossings along the 2,000-mile border.

These protests have no precedent in the history of the US. The immediate trigger was the passage in mid-March by the House of Representatives of HR4437, a bill introduced by Republican representative James Sensenbrenner with broad support from the anti-immigrant lobby. This draconian bill would criminalise undocumented immigrants by making it a felony to be in the US without documentation. It also stipulated the construction of the first 700 miles of a militarised wall between Mexico and the US and would double the size of the US border patrol. And it would apply criminal sanctions against anyone who provided assistance to undocumented immigrants, including churches, humanitarian groups and social service agencies.

Following its passage by the House, bill HR4437 became stalled in the Senate. Democrat Ted Kennedy and Republican John McCain co-sponsored a "compromise" bill that would have removed the criminalisation clause in HR4437 and provided a limited plan for amnesty for some of the undocumented. It would have allowed those who could prove they have resided in the US for at least five years to apply for residency and later citizenship. Those residing in the US for two to five years would have been required to return home and then apply through US embassies for temporary "guest worker" permits. Those who could not demonstrate that they had been in the US for two years would be deported. Even this "compromise" bill would have resulted in massive deportations and heightened control over all immigrants. Yet it was eventually jettisoned because of Republican opposition, so that by late April the whole legislative process had become stalled. In May, the Senate renewed debate on the matter and seemed to be moving towards consensus based on tougher enforcement and limited legalisation, although at the time of writing (late May 2006) it appeared the legislative process could drag on until after the November 2006 congressional elections.

However, the wave of protest goes well beyond HR4437. It represents the unleashing of pent-up anger and repudiation of what has been deepening exploitation and an escalation of anti-immigrant repression and racism. Immigrants have been subject to every imaginable abuse in recent years. Twice in the state of California they have been denied the right to acquire drivers' licences. This means that they must rely on inadequate or non-existent public transportation or risk driving illegally; more significantly, the drivers' licence is often the only form of legal documentation for such essential transactions as cashing cheques or renting an apartment. The US-Mexico border has been increasingly militarised and thousands of immigrants have died crossing the frontier. Anti-immigrant hate groups are on the rise. The FBI has reported more than 2,500 hate crimes against Latinos in the US since 2000. Blatantly racist public discourse that, only a few years ago, would have been considered extreme has become increasingly mainstreamed and aired in the mass media.

More ominously, the paramilitary organisation Minutemen, a modern day Latino-hating version of the Ku Klux Klan, has spread from its place of origin along the US-Mexican border in Arizona and California to other parts of the country. Minutemen claim they must "secure the border" in the face of inadequate state-sponsored control. Their discourse, beyond racist, is neo-fascist. Some have even been filmed sporting T-shirts with the emblem "Kill a Mexican Today?" and others have organised for-profit "human safaris" in the desert. One video game discovered recently circulating on the internet, "Border Patrol," lets players shoot at Mexican immigrants as they try to cross the border into the US. Players are told to target one of three immigrant groups, all portrayed in a negative, stereotypical way, as the figures rush past a sign that reads "Welcome to the United States." The immigrants are caricatured as bandolier-wearing "Mexican nationalists," tattooed "drug smugglers" and pregnant "breeders" who spring across with their children in tow.

Minutemen clubs have been sponsored by right-wing organisers, wealthy ranchers, businessmen and politicians. But their social base is drawn from those formerly privileged sectors of the white working class that have been "flexibilised" and displaced by economic restructuring, the deregulation of labour and global capital flight. These sectors now scapegoat immigrants—with official encouragement—as the source of their insecurity and downward mobility.

The immigrant mobilisations have seriously threatened ruling groups. In the wake of the recent mobilisations, the Bush administration stepped up raids, deportations and other enforcement measures in a series of highly publicised mass arrests of undocumented immigrants and their employers, intended to intimidate the movement. In April 2006 it was revealed that KBR, a subsidiary of Halliburton—Vice-President Dick Cheney's former company, which has close ties to the Pentagon and is a major contractor in the Iraq war—won a $385 million contract to build large-scale immigrant detention centres in case of an "emergency influx" of immigrants.

Latino immigration to the US is part of a worldwide upsurge in transnational migration generated by the forces of capitalist globalisation. Immigrant labour worldwide is conservatively estimated at over 200 million, according to UN data.[1] Some 30 million are in the US, with at least 20 million of them from Latin America. Of these 20 million, some 11–12 million are undocumented (south and east Asia are also significant contributors to the undocumented population), although it must be stressed that these figures are low-end estimates. The US is by far the largest immigrant-importing country, but the phenomenon is global. Racist attacks, scapegoating and state-sponsored repressive controls over immigrants are rising in many countries around the world, as is the fightback among immigrant workers wherever they are found. Parallel to the US events, for instance, the French government introduced a bill that would apply tough new controls over immigrants and roll back their rights. In response, some 30,000 immigrants and their supporters took to the streets in Paris on 13 May 2006 to demand the bill's repeal.

The Global Circulation of Immigrant Labour

The age of globalisation is also an age of unprecedented transnational migration. The corollary to an integrated global economy is the rise of a truly global—although highly segmented—labour market. It is a global labour market because, despite formal nation state restrictions on the free worldwide movement of labour, surplus labour in any part of the world is now recruited and redeployed through numerous mechanisms to where capital is in need of it and because workers themselves undertake worldwide migration, even in the face of the adverse migratory conditions.

Central to capitalism is securing a politically and economically suitable labour supply, and at the core of all class societies is the control over labour and disposal of the products of labour. But the linkage between the securing of labour and territoriality is changing under globalisation. As labour becomes "free" in every corner of the globe, capital has vast new opportunities for mobilising labour power where and when required. National labour pools are merging into a single global labour pool that services global capitalism. The transnational circulation of capital induces the transnational circulation of labour. This circulation of labour becomes incorporated into the process of restructuring the world economy. It is a mechanism for the provision of labour to transnationalised circuits of accumulation and constitutes a structural feature of the global system.

While the need to mix labour with capital at diverse points along global production chains induces population movements, there are sub-processes that shape the character and direction of such migration. At the structural level, the uprooting of communities by the capitalist break-up of local economies creates surplus populations and is a powerful push factor in outmigration, while labour shortages in more economically advanced areas is a pull factor that attracts displaced peoples. At a behavioural level, migration and wage remittances become a family survival strategy (see below), made *possible* by the demand for labour abroad and made increasingly *viable* by the fluid conditions and integrated infrastructures of globalisation.

In one sense, the South penetrates the North with the dramatic expansion of immigrant labour. But transnational migratory flows are not unidirectional from South to North and the phenomenon is best seen in global capitalist rather than North-South terms. Migrant workers are becoming a general category of super-exploitable labour drawn from globally dispersed labour reserves into similarly globally dispersed nodes of accumulation. To the extent that these nodes experience labour shortages—skilled or unskilled—they become magnets for transnational labour flows, often encouraged or even organised by both sending and receiving countries and regions.

Labour-short Middle Eastern countries, for instance, have programmes for the importation (and careful control) of labour from throughout south and east Asia and north Africa. The Philippine state has become a veritable labour recruitment agency for the global economy, organising the export of its citizens to over a hundred countries in Asia, the Middle East, Europe, North America and elsewhere. Greeks migrate to Germany and the US, while Albanians migrate to Greece. South Africans move to Australia and England, while Malawians, Mozambicans and Zimbabweans work in South African mines and the service industry. Malaysia imports Indonesian labour, while Thailand imports workers from Laos and Myanmar and, in turn, sends labour to Malaysia, Singapore, Japan and elsewhere. In Latin America, Costa Rica is a major importer of Nicaraguan labour, Venezuela has historically imported large amounts of Colombian labour, the Southern Cone draws on several million emigrant Andean workers and an estimated 500,000 to 800,000 Haitians live in the Dominican Republic, where they cut sugar cane, harvest crops and work in the *maquiladoras* under the same labour market segmentation, political disenfranchisement and repression that immigrant workers face in the United States and in most labour-importing countries.

The division of the global working class into "citizen" and "non-citizen" labour is a major new axis of inequality worldwide, further complicating the well-known gendered and racialised hierarchies among labour, and facilitating new forms of repressive and authoritarian social control over working classes. In an *apparent* contradiction, capital and goods move freely across national borders in the new global economy but labour cannot and its movement is subject to heightened state controls. The global labour supply is, in the main, no longer coerced (subject to extra-economic compulsion) due to the ability of the universalised market to exercise strictly economic discipline, but its movement is juridically controlled. This control is a central determinant in the worldwide correlation of forces between global capital and global labour.

The immigrant is a juridical creation inserted into real social relations. States create "immigrant labour" as distinct categories of labour in relation to capital. While the generalisation of the labour market emerging from the consolidation of the global capitalist economy creates the conditions for global migrations as a world-level labour supply system, the maintenance and strengthening of state controls over transnational labour create the conditions for immigrant labour as a distinct category of labour. The creation of these distinct categories ("immigrant labour") becomes central to the global capitalist economy, replacing earlier direct colonial and racial caste controls over labour worldwide.

But why is this juridical category of "immigrant labour" reproduced under globalisation? Labour migration and geographic shifts in production are alternative forms for capitalists to achieve an optimal mix of their capital with labour. State controls are often intended *not to prevent* but to *control* the transnational movement of labour. A *free* flow of labour would exert an equalising influence on wages across borders whereas state controls help reproduce such differentials. Eliminating the wage differential between regions would cancel the advantages that capital accrues from disposing of labour pools worldwide subject to different wage levels and would strengthen labour worldwide in relation to capital. In addition, the use of immigrant labour allows receiving countries to separate reproduction and maintenance of labour, and therefore to "externalise" the costs of social reproduction. In other words, the new transnational migration helps capital to dispose of the need to pay for the reproduction of labour power. The interstate system thus acts as a condition for the structural power of globally mobile transnational capital over labour that is transnational in actual content and character but subjected to different institutional arrangements under the direct control of national states.

The migrant labour phenomenon will continue to expand along with global capitalism. Just as capitalism has no control over its implacable expansion as a system, it cannot do away in its new globalist stage with transnational labour. But if global capital needs the labour power of transnational migrants, this labour power belongs to human beings who must be tightly controlled, given the special oppression and dehumanization involved in extracting their labour power as non-citizen immigrant labour. To return to the situation in the US, the immigrant issue presents a contradiction for political and economic elites: from the vantage points of dominant group interests, the dilemma is how to deal with the new "barbarians" at Rome's door.

Latino immigrants have massively swelled the lower rungs of the US workforce. They provide almost all farm labour and much of the labour for hotels, restaurants, construction, janitorial and house cleaning, child care, gardening and landscaping, delivery, meat and poultry packing, retail, and so on. Yet dominant groups fear a rising tide of Latino immigrants will lead to a loss of cultural and political control, becoming a source of counter-hegemony and instability, as immigrant labour in Paris showed itself to be in the late 2005 uprising there against racism and marginality.

Employers do not want to do away with Latino immigration. To the contrary, they want to sustain a vast exploitable labour pool that exists under precarious conditions, that does not enjoy the civil, political and labour rights of citizens and that is disposable through deportation. It is the *condition of deportability* that they wish to create, or preserve, since that condition assures the ability to super-exploit with impunity and to dispose of this labour without consequences should it become unruly or unnecessary. The Bush administration opposed HR4437 not because it was in favour of immigrant rights but because it had to play a balancing act by finding a formula for a stable supply of cheap labour to employers with, at the same time, greater state control over immigrants.

The Bush White House proposed a "guest worker" programme that would rule out legalisation for undocumented immigrants, force them to return to their home countries and apply for temporary work visas, and implement tough new border security measures. There is a long history of such "guest worker" schemes going back to the *bracero* programme, which brought millions of Mexican workers to the US during the labour shortages of the Second World War, only to deport them once native workers had become available again. Similar "guest worker" programmes are in effect in several European countries and other labour-importing states around the world.

The contradictions of "immigrant policy reform" became apparent in the days leading up to the May 1 action, when major capitalist groups dependent on immigrant labour—especially in the agricultural, food processing, landscaping, construction, and other service sectors—came out in support of legalisation for the undocumented. Such transnational agro-industrial giants as Cargill, Swift and Co, Perdue Farms, Tyson Foods and Goya Foods, for instance, closed down many of their meat-packing and food processing plants and gave workers the day off.

Neoliberalism in Latin America

If capital's need for cheap, malleable and deportable labour in the centres of the global economy is the main "pull factor" inducing Latino immigration to the US, the "push factor" is the devastation left by two decades of neoliberalism in Latin America. Capitalist globalisation—structural adjustment, free trade agreements, privatisations, the contraction of public employment and credits, the break-up of communal lands and so forth, along with the political crises these measures have generated—has imploded thousands of communities in Latin America and unleashed a wave of migration, from rural to urban areas and to other countries, that can only be analogous to the mass uprooting and migration that generally take place in the wake of war.

Just as capital does not stay put in the place it accumulates, neither do wages stay put. The flip side of the intense upsurge in transnational migration is the reverse flow of remittances by migrant workers in the global economy to their country and region of origin. Officially recorded international remittances increased astonishingly, from a mere $57 million in 1970 to $216 billion in 2005, according to World Bank data. This amount was higher than capital market flows and official development assistance combined, and nearly equalled the total amount of world FDI (foreign direct investment) in 2004. Close to one billion people, or one in every six on the planet, may receive some support from the global flow of remittances, according to senior World Bank economist Dilip Ratha.[2] Remittances have become an economic mainstay for an increasing number of countries. Most of the world's regions, including Africa, Asia, Latin America and southern and eastern Europe, report major remittance inflows.

Remittances redistribute income worldwide in a literal or geographic sense but not in the actual sense of *redistribution,* meaning a transfer of some added portion of the surplus from capital to labour, since they constitute not additional earnings but the separation of the site where wages are earned from the site of wage-generated consumption. What is taking place is a historically unprecedented separation of the point of production from the point of social reproduction. The former can take place in one part of the world and generate the value—then remitted—for social reproduction of labour in another part of the world. This is an emergent structural feature of the global system, in which the site of labour power and of its reproduction have been transnationally dispersed.

Transnational Latino migration has led to an enormous increase in remittances from Latino ethnic labour abroad to extended kinship networks in Latin America. Latin American workers abroad sent home some $57 billion in 2005, according to the Inter-American Development Bank.[3] These remittances were the number one source of foreign exchange for the Dominican Republic, El Salvador, Guatemala, Guyana, Haiti, Honduras, Jamaica and Nicaragua, and the second most important source for Belize, Bolivia, Colombia, Ecuador, Paraguay and Surinam, according to the Bank. The $20 billion sent back in 2005 by an estimated 10 million Mexicans in the US was more than the country's tourism receipts and was surpassed only by oil and *maquiladora* exports.

These remittances allow millions of Latin American families to survive by purchasing goods either imported from the world market or produced locally or by transnational capital. They allow for family survival at a time of crisis and adjustment, especially for the poorest sectors—safety nets that replace governments and fixed employment in the provision of economic security. Emigration and remittances also serve the political objective of pacification. The dramatic expansion of Latin American emigration to the US from the 1980s onwards helped to dissipate social tensions and undermine labour and political opposition to prevailing regimes and institutions. Remittances help to offset macroeconomic imbalances, in some cases averting economic collapse, thereby shoring up the political conditions for an environment congenial to transnational capital.

Therefore, bound up with the immigrant debate in the US is the entire political economy of global capitalism in the western hemisphere—the same political economy that is now being sharply contested throughout Latin America with the surge in mass popular struggles and the turn to the Left. The struggle for immigrant rights in the US is thus part and parcel of this resistance to neoliberalism, intimately connected to the larger Latin American—and worldwide—struggle for social justice.

No wonder protests and boycotts took place throughout Latin America on May 1 in solidarity with Latino immigrants in the US. But these actions were linked to local labour rights struggles and social movement demands. In Tijuana, Mexico, for example, *maquiladora* workers in that border city's in-bond industry marched on May 1 to demand higher wages, eight-hour shifts, an end to "abuses and despotism" in the *maquila* plants and an end to sexual harassment, the use of poison chemicals and company unions. The workers also called for solidarity with the "Great American Boycott of 2006 on the other side of the border" and participated in a protest at the US consulate in the city and at the main crossing, which shut down cross- border traffic for most of the day.

The Nature of Immigrant Struggles

Labour market transformations driven by capitalist globalisation unleash what McMichael calls "the politics of global labor circulation"[4] and fuel, in labour-importing countries, new nativisms, waves of xenophobia and racism against immigrants. Shifting political coalitions scapegoat immigrants by promoting ethnic-based solidarities among middle classes, representatives of distinct fractions of capital and formerly privileged sectors among working classes (such as white ethnic workers in the US and Europe) threatened by job loss, declining income and the other insecurities of economic restructuring. The long-term tendency seems to be towards a generalisation of labour market conditions across borders, characterised by segmented structures under a regime of labour deregulation and racial, ethnic and gender hierarchies.

In this regard, a major challenge confronting the movement in the US is relations between the Latino and the Black communities. Historically, African Americans have swelled the lower rungs in the US caste system. But, as African Americans fought for their civil and human rights in the 1960s and 1970s, they became organised, politicised and radicalised. Black workers led trade union militancy. All this made them undesirable labour for capital—"undisciplined" and "noncompliant."

Starting in the 1980s, employers began to push out Black workers and massively recruit Latino immigrants, a move that coincided with deindustrialisation and restructuring. Blacks moved from super-exploited to marginalized—subject to unemployment, cuts in social services, mass incarceration and heightened state repression—while Latino immigrant labour has become the new super-exploited sector. Employers and political elites in New Orleans, for instance, have apparently decided in the wake of Hurricane Katrina to replace that city's historically black working class with Latino immigrant labour. Whereas fifteen years ago no one saw a single Latino face in places such as Iowa or Tennessee, now Mexican, Central American and other Latino workers are visible everywhere. If some African Americans have misdirected their anger over marginality at Latino immigrants, the Black community has a legitimate grievance over the anti-Black racism of many Latinos themselves, who often lack sensitivity to the historic plight and contemporary experience of Blacks with racism, and are reticent to see them as natural allies. (Latinos often bring with them particular sets of racialised relations from their home countries.)[5]

White labour that historically enjoyed caste privileges within racially segmented labour markets has experienced downward mobility and heightened insecurity. These sectors of the working class feel the pinch of capitalist globalisation and the transnationalisation of formerly insulated local labour markets. Studies in the early 1990s, for example, found that, in addition to concentrations in "traditional" areas such as Los Angeles, Miami, Washington DC, Virginia and Houston, Central American immigrants had formed clusters in the formal and informal service sectors in areas where, in the process of downward mobility, they had replaced "white ethnics," such as in suburban Long Island, the small towns of Iowa and North Carolina, in Silicon Valley and in the northern and eastern suburbs of the San Francisco Bay Area.[6]

The loss of caste privileges for white sectors of the working class is problematic for political elites and state managers in the US, since legitimation and domination have historically been constructed through a white racial hegemonic bloc. Can such a bloc be sustained or renewed through a scapegoating of immigrant communities? In attempting to shape public discourse, the anti-immigrant lobby argues that immigrants "are a drain on the US economy." Yet, as the National Immigrant Solidarity Network points out, immigrants contribute $7 billion in Social Security a year. They earn $240 billion, report $90 billion, and are only reimbursed $5 billion in tax returns. They also contribute $25 billion more to the US economy than they receive in health-care and social services. But this is a limited line of argument, since the larger issue is the incalculable trillions of dollars that immigrant labour generates in profits and revenue for capital, only a tiny proportion of which goes back to them in the form of wages.

Moreover, it has been demonstrated that there is no correlation between the unemployment rate among US citizens and the rate of immigration. In fact, the unemployment rate has moved in cycles over the past twenty-five years and exhibits a comparatively lower rate during the most recent (2000–2005) influx of undocumented workers. Similarly, wage stagnation in the United States appeared, starting with the economic crisis of 1973, and has continued its steady march ever since, with no correlation to increases or decreases in the inflow of undocumented workers. Instead, downward mobility for most US workers is positively correlated with the decline in union participation, the decline in labour conditions and the polarisation of income and wealth that began with the restructuring crisis of the 1970s and accelerated the following decade as Reaganomics launched the neo-liberal counterrevolution.

The larger backdrop here is transnational capital's attempt to forge post-Fordist, post-Keynesian capital-labour relations worldwide, based on flexibilisation, deregulation and deunionisation. From the 1970s onwards, capital began to abandon earlier reciprocities with labour, forged in the epoch of national corporate capitalism, precisely because the process of globalisation allowed to it break free of nation state constraints. There has been a vast acceleration of the primitive accumulation of capital worldwide through globalisation, a process in which millions have been wrenched from the means of production, proletarianised and thrown into a global labour market that transnational capital has been able to shape. As capital assumed new power relative to labour with the onset of globalisation, states shifted from reproducing Keynesian social structures of accumulation to servicing the general needs of the new patterns of global accumulation.

At the core of the emerging global social structure of accumulation is a new capital-labour relation based on alternative systems of labour control and diverse contingent categories of devalued labour—sub-contracted, outsourced, casualised, informal, part-time, temp work, home-work, and so on—the essence of which is cheapening and disciplining labour, making it "flexible" and readily available for transnational capital in worldwide labour reserves. Workers in the global economy are themselves, under these flexible arrangements, increasingly treated as a sub-contracted component rather than a fixture internal to employer organisations. These new class relations of global capitalism dissolve the notion of responsibility, however minimal, that governments have for their citizens or that employers have towards their employees.

Immigrant workers become the archetype of these new global class relations. They are a naked commodity, no longer embedded in relations of reciprocity rooted in social and political communities that have, historically, been institutionalised in nation states. Immigrant labour pools that can be super-exploited economically, marginalised and disenfranchised politically, driven into the shadows and deported when necessary are the very epitome of capital's naked domination in the age of global capitalism.

The immigrant rights movement in the US is demanding full rights for all immigrants, including amnesty, worker protections, family reunification measures, a path to citizenship or permanent residency rather than a temporary "guest worker" programme, an end to all attacks against immigrants and to the criminalisation of immigrant communities. While some observers have billed the recent events as the birth of a new civil rights movement, clearly much more is at stake. In the larger picture, this goes beyond immediate demands; it challenges the class relations that are at the very core of global capitalism. The significance of the May 1 immigrant rights mobilisation taking place on international workers' day—which has not been celebrated in the US for nearly a century—was lost on no one.

In the age of globalisation, the only hope of accumulating the social and political forces necessary to confront the global capitalist system is by transnationalising popular, labour and democratic struggles. The immigrant rights movement is all of these—popular, pro-worker and democratic—and it is by definition transnational. In sum, the struggle for immigrant rights is at the cutting edge of the global working-class fight back against capitalist globalisation.

Notes

1. Manuel Oruzco, "Worker remittances in an international scope," *Working Paper* (Washington, DC, Inter-American Dialogue and Multilateral Investment Fund of the Inter-American Development Bank, March 2003), p. 1.

2. For these details, see Richard Boudreaux, "The new foreign aid; the seeds of promise," *Los Angeles Times* (14 April 2006), p. 1A.

3. Inter-American Development Bank, *Remittances 2005: promoting financial democracy* (Washington, DC, IDB, 2006).

4. Philip McMichael, *Development and Social Change: A Global Perspective* (Thousand Oaks, CA, Pine Forge Press, 1986), p. 189.

5. In a commentary observing that mainstream Black political leaders have been notably lukewarm to the immigrant rights movement, Keeanga-Yamahtta Taylor writes: "The displacement of Black workers is a real problem—but not a problem caused by displaced Mexican workers . . . if the state is allowed to criminalize the existence [of] immigrant workers this will only fan the flames of racism eventually consuming Blacks in a back draft of discrimination. How exactly does one tell the difference between a citizen and a non-citizen? Through a massive campaign of racial profiling, that's how. . . . In fact, the entire working class has a stake in the success of the movement." She goes on to recall how California building owners and labour contractors replaced Black janitors with largely undocumented Latino immigrants in the 1980s. But after a successful Service Employees International Union drive in the "Justice for janitors" campaign of the late 1980s and 1990s, wages and benefits went up and the union's largely Latino members sought contractual language guaranteeing African Americans a percentage of work slots. See Taylor, "Life ain't been no crystal stair: Blacks, Latinos and the new civil rights movement," *Counterpunch* (9 May 2006), downloaded 18 May 2006 (http://www.counterpunch.org/taylor0508 2006.html).

6. See the special issue of NACLA *Report on the Americas*, "On the line: Latinos on labor's cutting edge" (Vol. 30, no. 3, November/December 1996).

THINKING ABOUT THE READING

What is the "global circulation of immigrant labor"? What are some of the issues and concerns that face immigrant workers? According to Robinson, recent immigrant labor demonstrations reflect the growing consciousness of a "global working class." Who or what are the "global working class" described in this reading? What are some of the ways they resist global capitalism? What will be some of the implications of these strategies if they are successful?

Credits

Chapter 1

From *The Sociological Imagination* by C. Wright Mills, copyright © 2000 by Oxford University Press, Inc. Reprinted by permission of the publisher.

From *Invitation to Sociology* by Peter Berger, copyright 1963 by Peter L. Berger. Used by permission of Doubleday, a division of Random House, Inc.

From "The My Lai Massacre: A Military Crime of Obedience," by Herbert Kelman and V. Lee Hamilton. In *Crimes of Obedience* (pp. 1–20), edited by Herbert Kelman and V. Lee Hamilton. © 1989 by Yale University Press. Reprinted by permission.

Chapter 2

From Georg Simmel, "The Metropolis and Mental Life," *International Quarterly 10*, 1903.

From Bauman, Z. (1990). Gift and exchange. In Bauman, Z., *Thinking sociology* (pp. 89–106). Cambridge, MA: Basil Blackwell.

From *Culture of Fear* by Barry Glassner. Copyright © January 1, 1999 Glassner, Barry. Reprinted by permission of Basic Books, a member of Perseus Books Group.

Chapter 3

Reprinted by permission of Waveland Press, Inc. from Earl Babbie, *Observing Ourselves: Essays in Social Research.* (Long Grove, IL; Waveland Press, Inc., 1986 [reissued 1998]) All rights reserved.

From Joel Best, *Missing Numbers,* University of California Press, 2004.

Chapter 4

From "Body Ritual among the Nacirema" by Horace Miner. *American Anthropologist* 58:3, June 1956, pp. 503–507.

Excerpts from "The Melting Pot," from *The Spirit Catches You and You Fall Down* by Anne Fadiman. Copyright © 1997 by Anne Fadiman. Reprinted by permission of Farrar, Straus, and Giroux, LLC.

From *Golden Arches East, McDonald's in East Asia,* 2nd edition. Edited by James L. Watson. Copyright © 1997, 2006 by the Board of Trustees of the Leland Stanford Jr. University. All rights reserved. Used with the permission of Stanford University Press, www.sup.org.

Chapter 5

From "Life as the Maid's Daughter: An Exploration of the Everyday Boundaries of Race, Class, and Gender" by Mary Romero, from *Feminisms in the Academy* by Mary Romero, Abigail J. Stewart, and Donna Stanton (eds.) (pp. 157–179). Copyright © 1995. Used by permission of The University of Michigan Press.

From "Introduction: The Making of Culture, Identity, and Ethnicity Among Asian American Youth," by Min Zhou and Jennifer Lee from *Asian American Youth Culture, Identity, and Ethnicity* by Jennifer Lee and Min Zhou (eds.) (pp. 1–30). Copyright 2004. Used by permission of Routledge.

From "Working 'the Code': On Girls, Gender, and Inner-City Violence" by Nikki Jones from *Australian and New Zealand Journal of Criminology,* Vol. 41, No. 1, April 2008, pp. 63–83. Copyright 2008 by Australian Academic Press. Reprinted with permission.

Chapter 6

From "The Presentation of Self in Everyday Life" by Erving Goffman, copyright © 1959 by Erving Goffman. Used by permission of Doubleday, a division of Random House, Inc.

From Karyn Lacy, *Public Identities: Managing Race in Public Spaces,* University of California Press 2007.

From David Grazian, *The Girl Hunt,* University of California Press, 2007.

Chapter 7

From "The Radical Idea of Marrying for Love" from *Marriage, a History* by Stephanie Coontz © 2005 by the S. J. Coontz Company. Used by permission of Viking Penguin, a division of Penguin Group (USA) Inc.

From Stacey, J., (2011). Gay parenthood and the end of paternity as we knew it. In Stacey, J., *Unhitched.* New York, NY: New York University Press. 8 pages of abridged content.

Fee, D. (2011). Covenant marriage: Reflexivity and retrenchment in the politics of intimacy. In Seidman, S., Fischer, N., & Meeks, C. (Eds.), *Introducing the new sexuality studies* (2nd ed.) (pp. 215–220). New York, NY: Routledge.

Chapter 8

From "Watching the Canary," reprinted with permission of the publisher from *The Miner's Canary: Enlisting Race, Resisting Power, Transforming Democracy* by Lani Guinier and Gerald Torres, pp. 254–267, Cambridge, Mass.: Harvard University Press, Copyright © 2002 by Lani Guinier and Gerald Torres. All Rights reserved.

McGann, P.J., (2011). Healing (disorderly) desire: Medical-theraputic regulaltion of sexuality. In Seidman, S., Fischer, N., & Meeks, C., (Eds.), *Introducing the new sexuality studies* (2nd ed.) (427–437). New York, NY: Routledge.

From "Patients, 'Potheads,' and Dying to Get High" by Wendy Chapkis from *Production of Reality* by Jodi O'Brien (ed.) (pp. 484–492). Copyright 2006 Pine Forge Press. Used with permission. For more information visit: http://www.dyingtogethigh.net.

Chapter 9

From "These Dark Satanic Mills," by William Greider. Reprinted with the permission of Simon & Schuster, Inc., from *One World, Ready or Not: The Manic Logic of Global Capitalism* by William Greider. Copyright © 1997 by William Greider. All rights reserved.

From Van Maanen, J. (1991). The Smile Factory: Work at Disneyland. In P. Frost *Reframing Organizational Culture.* Copyright © Sage Publications. Reprinted by permission of Sage Publications, Inc.

From Murry Milner, *Creating Consumers: Freaks, Geeks, and Cool Kids*, Routledge, 2006.

Chapter 10

From "Making Class Invisible" by Gregory Mantsios from *Race, Class, and Gender in the United States, 4/e.* Copyright © 1998 Gregory Mantsios. Reprinted by permission.

From "The Compassion Gap in American Poverty Policy" by Fred Block, Anna C. Korteweg, and Kerry Woodward with Zach Schiller and Imrul Mazid from *Contexts,* Vol. 5, No. 2, Spring 2006, pp. 14–20. Copyright 2006 by University of California Press. Used with permission.

From "Branded with Infamy: Inscriptions of Poverty and Class in America" by Vivyan C. Adair. *Signs.* Volume 27, Issue 2 (2002). © University of Chicago Press.

Chapter 11

From "Racial and Ethnic Formation" by Michael Omi and Howard Winant from *Racial Formation in the US.* Copyright 1994. Reprinted by permission of Routledge via Copyright Clearance Center.

From "Optional Ethnicities" by Mary Waters from Pedraza / Rumbaut. *Origins and Destinies,* 1E. © 1996 Wadsworth, a part of Cengage Learning, Inc. Reproduced by permission. www.cengage.com/permissions.

From Barbara Trepagnier, *Silent Racism: Passivity in Well-Meaning White People*, Paradigm, 2010.

Chapter 12

From "Black Women and a New Definition of Womanhood" by Bart Landry, from *Black Working Wives: Pioneers of the American Family Revolution* by Bart Landry. Berkley: University of California Press. Copyright © 2000 by the Regents of the University California. Reprinted by permission.

Excerpts from *Still a Man's World: Men Who Do "Women's Work"* (pp. 1–5, 81–108) by Christine L. Williams. Berkeley: University of California

Press. Copyright © 1995 by the Regents of the University of California. Reprinted by permission.

From Eve Shapiro, *New Biomedical Technologies, New Scripts, New Genders,* Routledge, 2010.

Chapter 13

From "Age-Segregation in Later Life: An Examination of Personal Networks" by Peter Uhlenberg and Jenny de Jong Gierveld from *Ageing and Society,* Vol. 24, No. 1, 2004, pp. 5–28. Copyright 2004 Cambridge University Press. Used with permission.

From "Love and Gold" by Arlie Russell Hochschild from *Global Woman: Nannies, Maids and Sex Workers in the New Economy* by Barbara Erhrenreich and Arlie Russell Hochschild. Copyright © 2002 by Barbara Ehrenreich and Arlie Russell Hochschild. Reprinted by arrangement with Henry Holt and Company, LLC

From Felicity Schaeffer-Grabiel, *Cyberbrides and Global Imaginaries: Mexican Women's Turn from the National to the Foreign,* SAGE, 2007.

Chapter 14

From "Muslim American Immigrants after 9/11: The Struggle for Civil Rights" by Pierrette Hondagneu-Sotelo, from *God's Heart Has No Borders: How Religious Activists Are Working for Immigrant Rights* by Pierrette Hondagneu-Sotelo. Berkley: University of California Press. Copyright © 2008 by the Regents of the University of California. Reprinted by permission.

From Walter Winslow, *The Seattle Solidarity Network: A New Approach to Working Class Social Movements,* 2011.

From "'Aquí estamos y no nos vamos!' Global Capital and Immigrant Rights" by William I. Robinson from *Race & Class,* Vol. 48, No. 2, 2006.